Lecture Notes in Geoinformation and Cartography

Publications of the International Cartographic Association (ICA)

More information about this series at http://www.springer.com/series/10036

Michael P. Peterson
Editor

Advances in Cartography and GIScience

Selections from the International Cartographic Conference 2017

ICC 2017 WASHINGTON
JULY 02-07, 2017

ICA ACI

Springer

Editor
Michael P. Peterson
Department of Geography and Geology
University of Nebraska Omaha
Omaha, NE
USA

ISSN 1863-2246 ISSN 1863-2351 (electronic)
Lecture Notes in Geoinformation and Cartography
ISSN 2195-1705 ISSN 2195-1713 (electronic)
Publications of the International Cartographic Association (ICA)
ISBN 978-3-319-57335-9 ISBN 978-3-319-57336-6 (eBook)
DOI 10.1007/978-3-319-57336-6

Library of Congress Control Number: 2017937270

Printed on acid-free paper

This Springer imprint is published by Springer Nature
The registered company is Springer International Publishing AG
The registered company address is: Gewerbestrasse 11, 6330 Cham, Switzerland

Preface

What a pleasure—and tremendous amount of work—it was to put together this volume of works submitted to the 2017 International Cartographic Conference. With contributions from every continent and 16 countries, the volume represents a celebration of maps, and map-related research. A common thread that connects all chapters is the fascination with maps as agents of communication. More broadly, the works reflect the dynamic and ever-evolving discipline while demonstrating a commonality in research topics and methods from diverse parts of the world.

156 papers were submitted to the book track for the 2017 International Cartographic Conference. All papers were reviewed by the Scientific Program Committee consisting of 79 pre-eminent scholars associated with the various commissions of the International Cartographic Association (ICA). They ranked the works and provided valuable comments, including whether the submission should be included in the book. I personally examined all 156 submissions to make sure that none had mistakenly fallen through cracks in the reviewing process. Based on this, 49 papers were selected for further consideration. Of these, 36 made it through the rigorous editing process to be included in this book. A final count shows that there are five chapters each from China, Germany, and the United States. France, Hungary, and South Africa are represented by three chapters. Countries with two chapters include Brazil, Czechia, and Poland.

In the editing process, all authors were asked to address the comments of the reviewers and re-submit their entries. An important concern for this volume on cartography was the quality of the illustrations. All illustrations containing line-work had to be submitted in a vector format (e.g., SVG or similar). Raster images needed to be of sufficient resolution for printing.

The diversity of topics represented a challenge for organizing this volume into meaningful sections. In the end, chapters were grouped primarily based on the underlying methodology and the broader purpose of the research. For example, in terms of methodology, a number of contributions employed some type of user testing in the evaluation of maps. An example of a category based on the underlying purpose is the use of maps in the analysis of geographic distributions.

The book starts with two chapters that demonstrate the span of research in cartography. The first deals with an examination and preservation of a historical map for a city in Brazil. By sharp contrast, the second chapter deals with location-spoofing—the ability to fool location-based devices about their actual location. This is followed by a series of chapters on crowdsourcing and data. The next section deals with map design, followed by a series of chapters from many different countries evaluating maps through user testing.

The fifth part of the book examines how maps are used to examine particular geographic distributions. The sixth part uses various forms of numerical methods to generalize or evaluate maps. The last part examines routes, their analysis, display, and encoding in our brains. The book concludes with a chapter evaluating the essence of cartography. Taken together, the book makes for interesting reading and serves as a good reflection on current research in cartography.

On a personal note, I first attended the biennial conference of the ICA in 1978, only the 9th meeting of the organization and the last in the United States. The meeting was held in College Park, Maryland, not far from its meeting place in 2017. At that time, all presentations were given in a large auditorium and translated simultaneously into English or French. Translators sat in a small, windowed room in the back of the auditorium. It was a daunting experience for a young scholar presenting research from a recently completed master's thesis. Addressing an international audience in this setting was a great honor. It is also an honor, 39 years later, to put these works together into a cohesive volume for the 28th International Cartographic Conference.

Omaha, NE, USA Michael P. Peterson

Acknowledgements

Putting a volume like this together is an enormous task. I would like to thank the Scientific Program Committee for reviewing all submissions. As chair of that committee, Cindy Brewer did a tremendous job organizing the online submission process. I would particularly like to thank her for having the foresight to put together a book template file. The book would not have come together without the help of numerous individuals at Conference Managers and Cadmium CD. Finally, I would like to thank the chapter authors who all responded quickly to my requests for revisions to the graphics and the text.

Scientific Program Committee

Cynthia A. Brewer (chair)
Mirela Altic
Masatoshi Arikawa
Yifang Ban
Franz-Josef Behr
Dirk Burghardt
Rex Cammack
Sébastien Caquard
Sidonie Christophe
Keith Clarke
Serena Coetzee
Alejandra Coll Escanilla
Jeremy Crampton
Imre Josef Demhardt
Urška Demšar
Cécile Duchene
Sara Fabrikant
David Fairbairn
Ján Feranec
Kenneth Field
Michael Finn
David Forrest
Ron Furness
Georg Gartner
Mátyás Gede
Amy Griffin
Henrik Hargitai
Francis Harvey
Anja Hopfstock

Haosheng Huang
Bin Jiang
Liu Jiping
Markus Jobst
Peter Jordan
Alexander Kent
Pyry Kettunen
Milan Konecný
Menno-Jan Kraak
Horst Kremers
Jukka Krisp
Miljenko Lapaine
Jonathan Li
Christophe Lienert
Yaolin Liu
Yuefeng Liu
Evangelos Livieratos
Lucia Lovison-Golob
Paulo Menezes
Harold Moellering
Ian Muehlenhaus
Kristien Ooms
Cosimo Palagiano
Tom Patterson
Ammatzia Peled
Stefan Peters
Dušan Petrovic
Irina Petrovna Karachevtseva
Silvana Philippi Camboim
Nathan Piekielek
José Jesús Reyes Nuñez
Waldirene Ribeiro
Anthony Robinson
Robert Roth
Pilar Sánchez-Ortiz Rodríguez
Peter Schmitz
Carla Sena
Monika Sester
René Sieber
Artemis Skarlatidou
Trudy Suchan
Vladimir Tikunov
Lysandros Tsoulos

Lynn Usery
Laurene Vaughan
Alena Vondráková
Vít Voženílek
Xiaobai Angela Yao
Long Yi
László Zentai

Contents

Part I The Span of Cartography

Cartographic Memory Preservation of the Petrópolis City in Brazil: Koeler Map Scanning Using Photographic Survey 3
Manoel do C. Fernandes, Tainá Laeta, Deivison F. dos Santos and Paulo M.L. de Menezes

Location Spoofing in a Location-Based Game: A Case Study of Pokémon Go 21
Bo Zhao and Qinying Chen

Part II Crowdsourcing and Data

Educational Aspects of Crowdsourced Noise Mapping 35
Andrea Pődör and László Zentai

Crowd and Community Sourced Data Quality Assessment 47
Laurence Jolivet and Ana-Maria Olteanu-Raimond

Crowdsourcing Mapping and Participatory Planning Support System: Case Study of Brno, Czechia 61
Jiří Pánek and Vít Pászto

A Framework for Enhancing Real-Time Social Media Data to Improve the Disaster Management Process 75
Syed Attique Shah, Dursun Zafer Şeker and Hande Demirel

Building a Real-Time Geo-Targeted Event Observation (Geo) Viewer for Disaster Management and Situation Awareness 85
Ming-Hsiang Tsou, Chin-Te Jung, Christopher Allen, Jiue-An Yang, Su Yeon Han, Brian H. Spitzberg and Jessica Dozier

The Academic SDI—Towards Understanding Spatial Data Infrastructures for Research and Education 99
Serena Coetzee, Stefan Steiniger, Barend Köbben, Adam Iwaniak, Iwona Kaczmarek, Petr Rapant, Antony Cooper, Franz-Josef Behr, Govert Schoof, Samy Katumba, Rumiana Vatseva, Kisco Sinvula and Harold Moellering

Part III Map Design

Introducing Leader Lines into Scale-Aware Consistent Labeling 117
Hsiang-Yun Wu, Shigeo Takahashi, Sheung-Hung Poon and Masatoshi Arikawa

On the Way to Create Individualized Cartographic Images for Online Maps Using Free and Open Source Tools 131
Csaba Szigeti, Gáspár Albert, Virág Ilyés, Dávid Kis and Dávid Várkonyi

Hebrus Valles—The Mars Exploration Zone Map 145
Mateusz Pitura

XY Domain: A Sound Map Artwork for Communicating Big Data Characteristics 159
Antoni B. Moore and Charlotte Parallel

Reproducible Cartography 173
Timothée Giraud and Nicolas Lambert

Part IV Evaluating Map Quality

Effectiveness and Efficiency of Using Different Types of Rectangular Treemap as Diagrams in Cartography 187
Mengjie Zhou, Yan Cheng, Ning Ye and Jing Tian

The Usability of a GeoVisual Analytics Environment for the Exploration and Analysis of Different Datasets 207
Irma Kveladze, Menno-Jan Kraak and Corné P.J.M. van Elzakker

Characterizing Maps from Visual Properties 223
Catherine Dominguès and Laurence Jolivet

How Hard Is It to Design Maps for Beginners, Intermediates and Experts? 239
Gáspár Albert, Virág Ilyés, Csaba Szigeti, Dávid Kis and Dávid Várkonyi

Interaction Problems Found Through Usability Testing on Interactive Maps 255
Elaine Gomes Vieira de Jesus, Patricia Lustosa Brito and Vivian de Oliveira Fernandes

The Apprehension of Overlaid Information in a Web Map 269
Dzenan Dumpor and Terje Midtbø

Visualization of Environment-related Information in Augmented Reality: Analysis of User Needs 283
Kateřina Chmelařová, Čeněk Šašinka and Zdeněk Stachoň

Part V Geographic Analysis

Analysis and Visualization of the Urban Residents' Income-Related Happiness Index in China 295
Ying Song, Yang Yu, Yanfang Liu, Zixi Liu, Qianyi Li and Guoguang Xu

Displaying Voter Gains and Losses: Local Government Elections in South Africa for 2011 and 2016 309
Peter M.U. Schmitz

Mapping Community Vulnerability to Poaching: A Whole-of-Society Approach 325
Peter M.U. Schmitz, Duarte Gonçalves and Merin Jacobs

Mapping Urban Landscapes Along Streets Using Google Street View 341
Xiaojiang Li, Carlo Ratti and Ian Seiferling

Part VI Numerical Analysis

Cross-Scale Analysis of Sub-pixel Variations in Digital Elevation Models 359
Mehran Ghandehari, Barbara P. Buttenfield and Carson J.Q. Farmer

Extraction of Ridge Lines from Grid DEMs with the Steepest Ascent Method Based on Constrained Direction 375
Wenping Jiang, Daping Xi, Xiaolong Deng, Lina Huang and Shen Ying

Using the A* Algorithm to Find Optimal Sequences for Area Aggregation 389
Dongliang Peng, Alexander Wolff and Jan-Henrik Haunert

Quantitative Expressions of Spatial Similarity in Multi-scale Map Spaces 405
Haowen Yan, Liming Zhang, Zhonghui Wang, Weifang Yang, Tao Liu and Liang Zhou

Balanced Allocation of Multi-criteria Geographic Areas by a Genetic Algorithm 417
Shahin Sharifi Noorian and Christian E. Murphy

Rethinking the Buffering Approach for Assessing OpenStreetMap Positional Accuracy 435
Qi Zhou

Data Classification for Highlighting Polygons with Local Extreme Values in Choropleth Maps 449
Jochen Schiewe

Part VII Routing

A Confidence-Based Approach for the Assessment of Accessibility of Pedestrian Network for Manual Wheelchair Users 463
Amin Gharebaghi, Mir-Abolfazl Mostafavi, Geoffrey Edwards, Patrick Fougeyrollas, Patrick Morales-Coayla, François Routhier, Jean Leblond and Luc Noreau

Accessibility in Pedestrian Routing 479
Megen Brittell, Christine Grummon, Amy Lobben, Masrudy Omri and Nicholas Perdue

Visualization of Traffic Bottlenecks: Combining Traffic Congestion with Complicated Crossings 493
Andreas Keler, Jukka M. Krisp and Linfang Ding

Psychogeography in the Age of the Quantified Self—Mental Map Modelling with Georeferenced Personal Activity Data 507
Sebastian Meier and Katrin Glinka

Part VIII Final Reflections

In Search of the Essence of Cartography 525
Marek Baranowski, Dariusz Gotlib and Robert Olszewski

Author Index 537

Subject Index 539

Part I
The Span of Cartography

Cartographic Memory Preservation of the Petrópolis City in Brazil: Koeler Map Scanning Using Photographic Survey

Manoel do C. Fernandes, Tainá Laeta, Deivison F. dos Santos and Paulo M.L. de Menezes

Abstract Historical cartographic documents have been increasingly essential to several studies, especially when understanding how the current forms of a landscape are structured. In this regard, these documents, along with the assistance of new digital technologies, are preserved and made available to the public. This research focused on investigating the Koeler Plan, a document providing guidance on urban planning for the development of the city of Petrópolis. The plan's milestone is the Petrópolis Map (1846), which became known as Koeler Map, in honor of its creator. The map was made at a scale of 1:5,000 and contains different cartographic elements such as hydrography, public parks, streets, blocks, and land reserved for public and religious buildings. This research aims to digitally reproduce this map, since its restoration and further use are quite difficult due to its current conditions. The methodology comprises photographic survey techniques using a high resolution digital camera and a track system which moves the camera vertically and horizontally. Several tests were made using multiple georeferencing algorithms to check which of them would better adjust into a mosaic the 191 photographs taken and preserve the features for the digital map generation. The process was developed in five pairs of photographs distributed along the map area, using ArcGIS 10.1 software with different transformations: Zero Order Polynomial, 1st Order Polynomial (Affine), 2nd Order Polynomial, 3rd Order Polynomial, Adjust, Projective Transformation, and Spline. After georeferencing these pairs, a test was made based on the distortions produced in area and length, in addition to

M.C. Fernandes (✉) · T. Laeta · D.F. dos Santos · P.M.L. de Menezes
Laboratory of Cartography, Geography Department, Federal University of Rio de Janeiro, Rio de Janeiro, RJ, Brazil
e-mail: manoel.fernandes@ufrj.br

T. Laeta
e-mail: tainalaeta@gmail.com

D.F. dos Santos
e-mail: sirdeivison@gmail.com

P.M.L. de Menezes
e-mail: pmenezes@acd.ufrj.br

M.P. Peterson (ed.), *Advances in Cartography and GIScience*, Lecture Notes in Geoinformation and Cartography, DOI 10.1007/978-3-319-57336-6_1

the RMS for each transformation. From these results, it was found that the 1st Order Polynomial (Affine) transformation had lower distortion values for the photographs, which allowed a highest quality photo mosaicking.

Keywords Koeler map • Photographical survey • Preservation of archives

1 Introduction

The municipality of Petrópolis, located in the mountainous region of the state of Rio de Janeiro, was subjected to an unusual planning when compared to other Brazilian cities: it was established upon imperial decree and defined by a map, the Petrópolis Map (1846)—also known as Koeler Map—which served as basis for the occupation plan of the future city. Today, the municipality consists of five districts, namely: Petrópolis (1st District), Cascatinha (2nd District), Itaipava (3rd District), Pedro do Rio (4th District), and Posse (5th District). The city was founded at and originated in the first district, where the area covered by Koeler Map (see Fig. 1) is located. Currently known for its tourism potential, the city's background is directly related to the plans of the Imperial Court to establish, in the nineteenth century, a summer house for the Imperial Family in the mountains, along a section of Variante do Caminho Novo road. Thus, Petrópolis achieved the status of "city" straightaway,

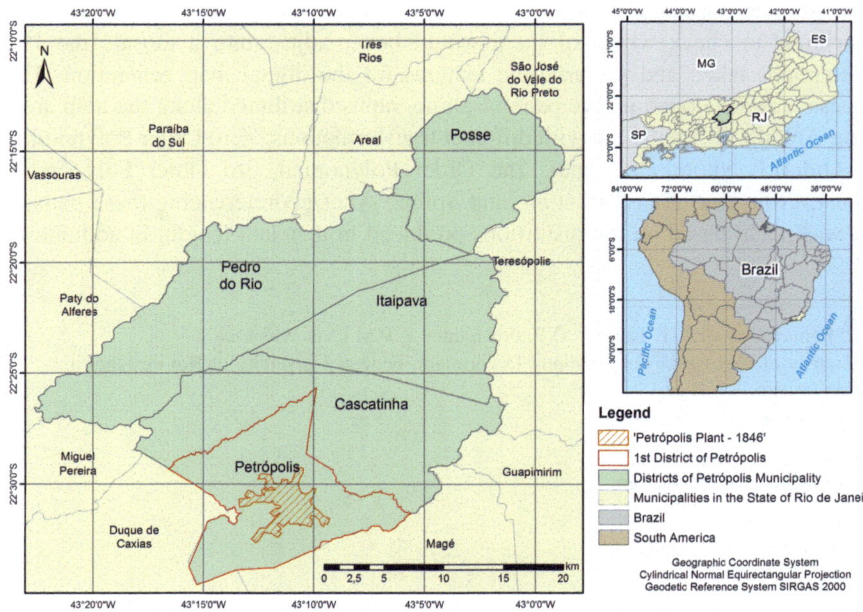

Fig. 1 Location of Petrópolis city and Koeler map

having never been a village. After its founding, the city accrued an important cultural and natural heritage. Because of these assets, it was granted the title of Imperial City, a popular nickname among its residents, scholars and tourists, officially established by decree on March 6th, 1981 by João Figueiredo, then President of Brazil (Souza 2013).

The city was founded on March 16th, 1843, by imperial decree signed by Emperor Pedro II and registered in the Book of Administration under No. 155. The decree provided for the creation of a settlement and the building of an imperial house in the area of the former Córrego Seco farm, acquired by Pedro II's father, Emperor Pedro I, 13 years earlier (Ambrozio 2012). However, although the date of signature of the decree is defined by some historians as the date of the city's foundation, IBGE (1948) states that the title of city was officially granted on September 29th, 1857, upon approval of Law No. 961 on the same date.

The 1843 imperial decree approves the plan suggested by the Imperial House Butler, Paulo Barbosa. The plan provided for leasing the Imperial Farm, formerly known as Córrego Seco Farm and Concórdia Farm, to Major Julio Frederico Koeler, a German immigrant who had joined the military engineering sector of the Brazilian army. This decree resulted in a document signed on July 26th, 1843 that constituted an integral part of the previous original decree. Said document provided for the lease rules of Pedro II's farm, indicating the general rules for the granting of the Imperial House farm. The document is widely reported by Paula Buarque (1940). According to Ambrozio (2012), the territorial power of the Imperial House over its mountainous soil is made evident in this "Deed for Lease …" prepared by Paulo Barbosa and containing 16 clauses. Ambrozio (2012) also states that the lessee, Koeler, was only the farm administrator: a privileged foreman or agent of the Administration with some benefits including, as a tax collector, the right to 10% of the collections sent to the Imperial House.

This decree became known as the "Village-Summer Palace" plan, or Koeler plan. The lessee, Julio Koeler, was required to prepare the future map of Petrópolis, the Emperor's palace and its outbuildings, dividing the imperial lands in lots or spots numbered for granting of privileges (Paula Buarque 1940, 112). The map contains different cartographic elements such as hydrography, streets, public parks, lots of land in which the blocks were divided, and land reserved for public and religious buildings (Sá Earp 1996; Souza 2013). These spots were distributed to Germanic tenant farmers who would erect the main buildings in the city of Petrópolis, such as the Imperial Palace, now Imperial Museum. According to Laeta and Fernandes (2015), the original map graphic scale was in *braças*, an old Portuguese unit. Preliminary georeferencing found that the plan had been made at a 1:5,000 scale. Laeta and Fernandes (2015) also found that the area covered by the map (15.91 km^2) represented 12% of the current first district of Petrópolis, and 2% of the entire current city.

Rabaço (1985) raises important points that define the Koeler Map as a project with concerns regarding responsible environmental use. According to the author, this plan could be seen as the first code of works of Petrópolis, prohibiting buildings on hilltops and the subdivision of lots horizontally, that is, parallel to their front

Fig. 2 Details of the Koeler map with the subdivision into numbered lots and the names of the tenant farmers who occupied them

line. The plan also set forth that the front of all houses should face the rivers, and sewage should be directed to tanks in the back of the lots. The author considers this feature as one of the most original characteristics in the Koeler plan, as it waived the traditional colonial style wherein houses had their backs to rivers, which were used for waste disposal. Finally, one must highlight the planning of green areas that would contain tree alleys and gardens with hydrangeas, magnolias, and silk floss trees (see Fig. 2).

The Koeler Plan can be seen as an urban plan whose spatial reference element is the Koeler Map (1846), which also serves as a guide for the former's management. Rabaço (1985) points out that Koeler took two roles in this project: as Director of Petrópolis Imperial Colony, he would plan and carry out the urban plan, and monitor educational, health, safety, and even religious activities; and as Superintendent of the Imperial Farm of Petrópolis, he acted as a kind of legal entity that controlled the rents of the estates in which the city was originally divided, representing one of the existing emphyteuses in Brazil. This company is currently known as Companhia Imobiliária de Petrópolis, and the object of this paper, the Koeler Map, is kept by it.

Given the historical importance of this map, this paper aims at digitally reproducing the cartographic document. The original, framed in wood with a glass protection, is located at Companhia Imobiliária de Petrópolis. Due to its current conservation conditions, it is difficult to restore and, consequently, make use of it (see Fig. 3). Thus, the principal goal of this paper is to create a memory of this map before its complete degradation, to serve as basis for future studies on the origin of this city. It is noteworthy that the use of historical cartographic documents in several studies resulted in growing concerns about their preservation, considering that such documents are increasingly made available to the public by means of new technologies, as stated in the works of Collier (2013), Menezes et al. (2013) and (2015).

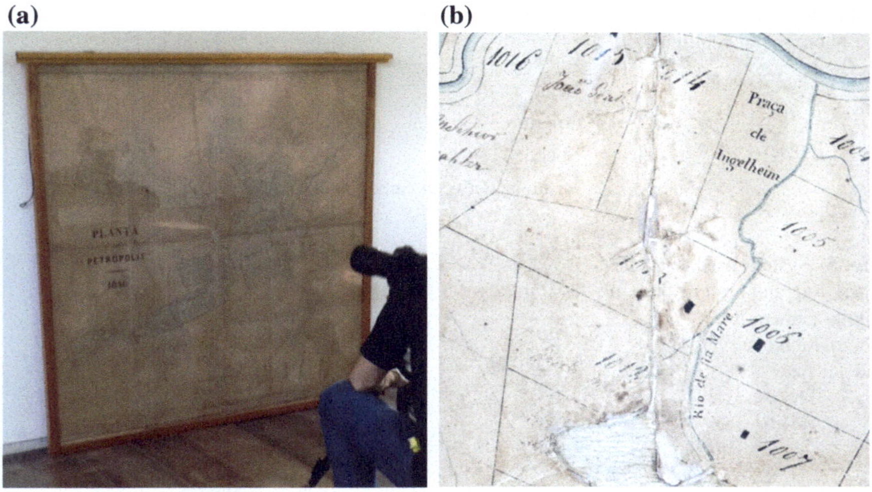

Fig. 3 Current conditions of the Koeler map (**a**) Overview of the map framed in wood and covered with glass (**b**) One of several damaged parts of the map

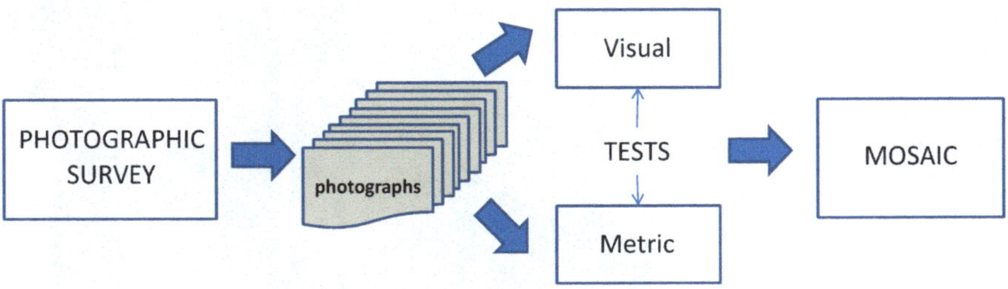

Fig. 4 Methodological flowchart used in this paper

2 Materials and Methods

The methodology was based on capturing high-resolution images from the original map and further applying two types of tests to assess the georeferencing methods among the images, to ensure their best arrangement and, therefore, create a mosaic to reconstitute the map in satisfactory conditions. These steps include photographic survey and visual and metric tests. The methodology flowchart can be seen in Fig. 4.

The photographic survey was made at Companhia Imobiliária de Petrópolis, thus avoiding transport and the risk of damage. For this purpose, high-resolution photographic equipment was used with characteristics described in Table 1. The camera was secured on a rod, which was placed on a track system on the ground for capturing photographs at varying positions (see Fig. 5). From these materials, we defined a focal length that allowed us to capture pictures with a defined spatial

Table 1 Characteristics of the photographic equipment used

Item	Description
Camera	Canon EOS 5D Mark II
Lens	Fixed Canon 100 mm
Aperture	f/25 to f/18
Speed	3.2"–30"
Distance between camera and map	1 m

Fig. 5 System used with the track, rod, and high resolution digital camera for photographic survey on Koeler map

resolution on their own, and to facilitate the identification of relevant details, such as toponyms and other planimetric elements of the map. Digital photographic files were generated in RAW image format and then processed to equalize the colors in the pictures using image editing software. The output images were saved in JPG format without compression.

The result of this survey generated 191 digital photographs. From the fragmented photographs, a photographic index was created to allow the visualization and identification of the photographs by codes.

After the photographic survey, visual and metric tests were made to identify the best georeferencing method for the photographs, using different algorithms. The goal was to understand which georeferencing algorithm provided the best adjustment for the photographs, avoiding major distortions, while preserving their features. It is worth to mention that this process aimed to georeference one photo in relation to another, and no terrestrial positioning system was adopted at this time. The idea was to perform this terrestrial positioning only on the mosaic resulting from this work. Testing on georeferencing methods is necessary because the photographs were not symmetrical. It was difficult to make proper adjustments among them, even taking due care during the photographic survey.

Georeferencing was performed by identifying known points in two photographs, which are referred to as control points. These control the adjustment of coordinates for each pixel (lines, columns—i, j) between two photographs. Before choosing these points, some criteria for location and quantity of control points to be used in georeferencing were analyzed. For georeferencing a pair of photographs, 10 points of terrestrial control were selected distributed as evenly as possible within the overlapping area between the photographs, corresponding to 1/5 of the total area of a photograph. For this purpose, ArcGIS 10.1 was used with different transformations: Zero Order Polynomial, 1st Order Polynomial (Affine), 2nd Order Polynomial, 3rd Order Polynomial, Adjust, Projective Transformation and Spline. Thus, one photograph was used as reference (B) and the other (A) was georeferenced using the ten points mentioned and present in both photographs (see Fig. 6).

The first test, named visual test, aimed to visually determine the impact in the area of one of the photographs after being subjected to georeferencing transformations, that is, to identify which of these transformations better preserved the area of the georeferenced photograph. Thus, after georeferencing, a comparative visual test was made based on the distortion produced by area of the georeferenced image with reference to the original area of the photograph.

The second test, named metric test, is more specific as it aims to make a detailed evaluation of the impact of each change after georeferencing the photographs. Thus, distance and area were measured before and after the georeferencing. The test measured the distance between two known points in the photographs and the area of the whole photograph. In addition to these two measurements, the Root Mean Square (RMS) was further assessed for each of the transformations. These values were tabulated and listed in a graph for assessment and classification of

Fig. 6 Mosaic with a pair of photographs. The total area of the two photographs used is outlined in *red*. The overlapping area where the control points were extracted is outlined in *blue*. *Red* crosses indicate control points

transformations. This test was performed on five pairs of photographs considering the presence of several cartographic elements and on areas distributed along the Koeler Map in order to make its assessment as uniform as possible, using one pair of photographs per area and 10 control points per pair.

The distance between the two known points was measured by marking the same pixels in each transformation. After defining the points, the point-to-line tool was used to generate a straight line connecting the points. Then, the length of this line was measured.

3 Results and Discussion

In the first test, the georeferenced pair generated 6 photographs, one for each transformation used. From the visual analysis of these photographs, it was found that georeferencing using 1st Order Polynomial (Affine) transformations, Projective Transformation, Spline and Zero Order Polynomial (Shift) algorithms presented the least distortions regarding the area and adjustment between the photographs (see Fig. 7a, d, e and f). The other transformations—except for Adjust, which did not generate a georeferenced image—are shown in Fig. 7b and c, where in the position of photograph A in relation to the red box in its original area is highlighted.

It is worth mentioning that this visual assessment generated an initial result that will be analyzed in the following test, the metric test. However, each algorithm of the transformations provided results that must be further analyzed. For example, the result of the visual test of transformation Zero Order Polynomial (Shift) visually presented the least distortion, since this transformation only promotes a translation process (Shift) on the photograph, fully preserving its area. However, the adjustment to the other photograph in the pair was impaired, since the homologous features in the two pictures did not fit together, even in the points that were used for georeferencing (see Fig. 8). It occurs because this transformation has a global characteristic in which georeferencing occurs taking into account the photograph as a whole, aligned with the premise of generating only one translation in the resulting image. Thus, this kind of transformation was not useful for the work proposed.

In the second test, after georeferencing the pairs, the generation of 35 images was expected. Once again, however, the Adjust transformation did not generate any images for any of the five pairs after its application. Furthermore, images resulting from the 3rd Order Polynomial transformation were distorted which made it impossible to identify the pixels to be used in the metric distance test (see Fig. 7c). Thus, it was not possible to survey data from images that were not generated and from those with distortions that would make testing unfeasible. This occurred because the Adjust and 3rd Order Polynomial transformations work globally and locally, requiring a large number (compared to other methods) of control points, well distributed throughout the image to be georeferenced, to ensure a quality process. These characteristics make these transformation methods unsuitable for the purpose of this work, since the methodology is restricted to a small area on the

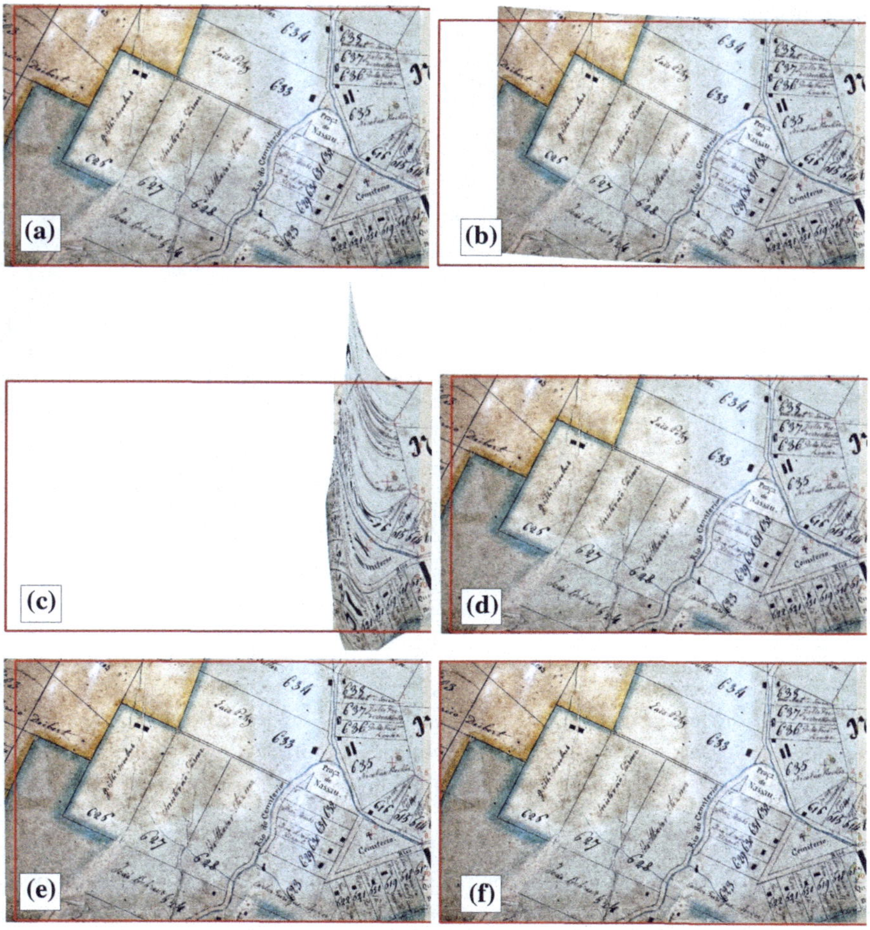

Fig. 7 Examples of the first test after georeferencing. Result (**a**) was obtained using transformation 1st order polynomial (affine), (**b**) using 2nd order polynomial, (**c**) using 3rd order polynomial, (**d**) using projective transformation, (**e**) using spline, and (**f**) using zero order polynomial (shift)

border of a photograph, that is, the overlapping area between a pair of photographs used for collecting control points.

Figure 9 shows examples resulting from georeferencing on a pair of photographs with the inclusion of elements that were used to implement the metric tests. For the area test, the green portion in the image represents the area before processing and the blue portion represents the area at the end of georeferencing. For the distance test, the yellow line presents the original measurement and the red line represents the area after georeferencing.

With the initial assessment of the Root Mean Square (RMS) values for each one of the transformations, from the ten points used (see Fig. 10), one can see that the

Fig. 8 Junction area of two photographs subjected to zero order polynomial (shift) transformation

transformation methods with best results (excluding the Zero Order Polynomial (Shift), Adjust, and 3rd Order Polynomial methods, which were discarded in the previous test (visual test)) were the Spline, 2nd Order Polynomial, Projective Transformation, and 1st Order Polynomial methods. Their average pixel RMS was 0.00; 1.33; 4.85; and 5.36, respectively.

It is worth emphasizing that the Root Mean Square (RMS) value is the difference between the place of origin of a control point and the location of the same point when it is subjected to a geometric transformation. Thus, this amount is the measure of positioning error between the control points, in this case, the ten points used. Thus, this value is not suitable for assessing accuracy, since it uses only the control points, ignoring the errors that may occur along the resulting image. In this case, this value is restricted to the points that are distributed on the overlapping area of the pairs, which is equivalent to a small portion of only 1/5 of the photograph.

Moreover, it is noteworthy that each transformation has characteristics that may affect this value, such as the Spline method—an interpolation method that estimates values using a mathematical function that minimizes the curvature of the surface, resulting in a smooth surface that fits precisely in-between the entry points (Longley et al. 2005). The spline transformation is a method intended for location accuracy, and does not ensure global accuracy. When the control point is entered, the transformation fits the target exactly in the place where the point was marked. However, it does not guarantee any accuracy for pixels far from the control points. Thus, RMS will always be equal to zero, yet it does not mean that the image as a whole maintains the same behavior in terms of accuracy.

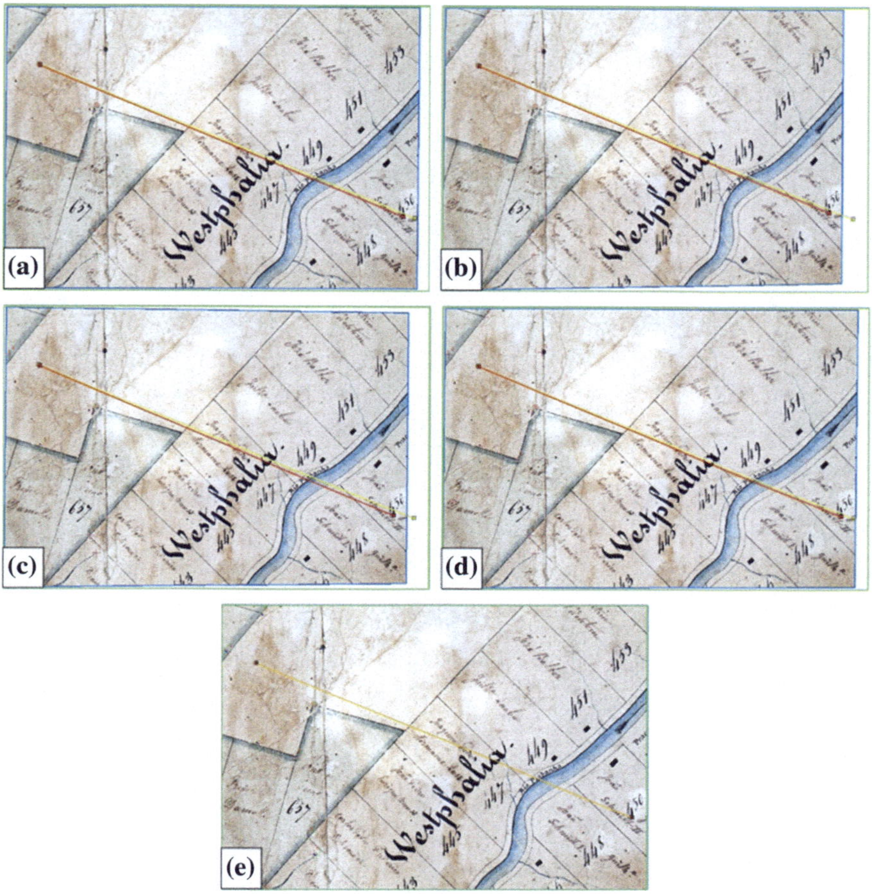

Fig. 9 Examples of photograph 5067 after georeferencing. Result (**a**) was obtained using the 1st order polynomial (affine) transformation, (**b**) using 2nd order polynomial, (**c**) using projective transformation, (**d**) using spline, and (**e**) using zero order polynomial (shift). The *yellow line* is the line generated before georeferencing and the *red line* is the line generated after georeferencing

The metric distance test indicated that the Affine transformation method had the least average percentage of difference in distances before and after georeferencing (2.36%), followed by the Spline method (2.89%) and Projective Transformation (4.48%) (see Fig. 11). The other methods showed average percentage errors with very high values, indicating that they are not adequate to the work proposed, as seen in 2nd Order Polynomial (25.59%), and Adjust and 3rd Order Polynomial, which reached values of 100%. It is worth highlighting that these methods did not generate images, or generated very distorted images, which made this test unfeasible. Therefore, a value of 100% was assigned. In addition, as the Zero Order Polynomial method has no scale geometric transformation or rotation associated to it, only translation (Shift), the error is equal to zero as defined in the area test. However,

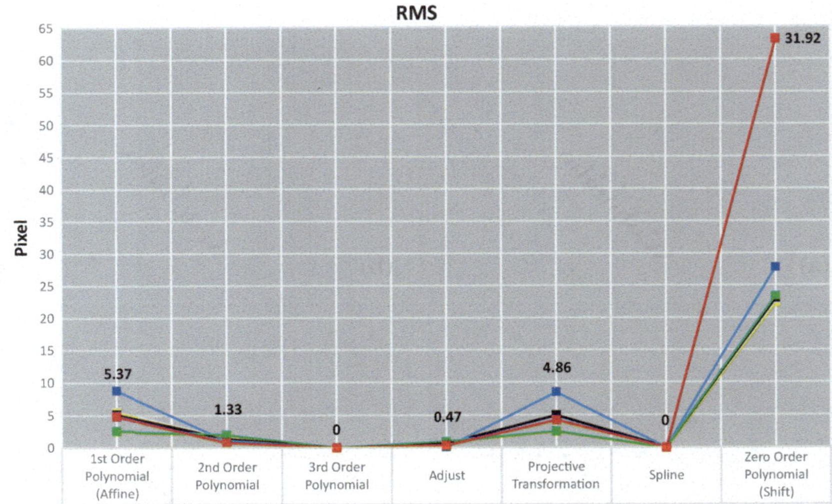

Photograph	1st Order Polynomial (Affine)	2nd Order Polynomial	3rd Order Polynomial	Adjust	Projective Transformation	Spline	Zero Order Polynomial (Shift)
5067	5.50576	1.45379	0	0.425195	4.19396	0	22.4031
5104	5.07599	1.32842	0	0.559624	4.88776	0	22.9812
5134	8.7986	1.07617	0	0.049206	8.47388	0	27.7823
5168	2.59255	2.03687	0	0.965744	2.52551	0	23.3093
5205	4.87167	0.7777	0	0.354557	4.21084	0	63.1384

Fig. 10 Root mean square (RMS) of each method used for the five pairs of photographs and their average values

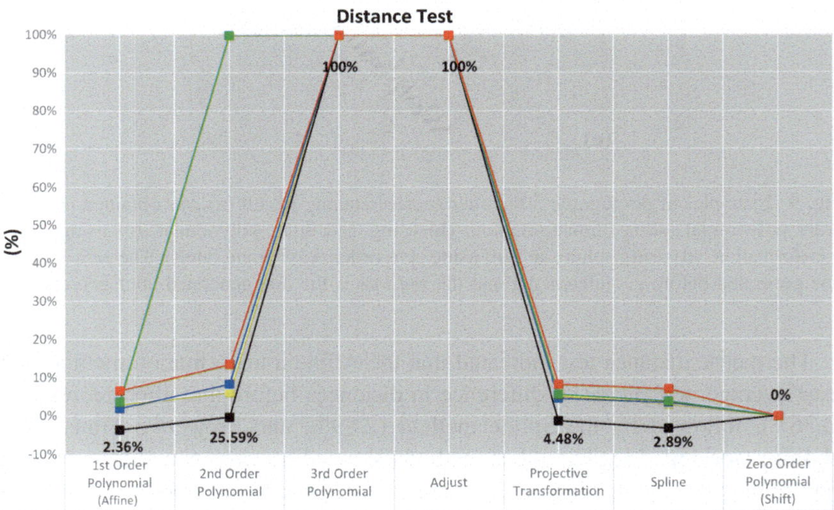

Photograph	1st Order Polynomial (Affine)	2nd Order Polynomial	3rd Order Polynomial	Adjust	Projective Transformation	Spline	Zero Order Polynomial (Shift)
5067	3.03%	6.34%	100.00%	100.00%	5.28%	3.21%	0.00%
5104	-3.68%	-0.36%	100.00%	100.00%	-1.42%	-3.34%	0.00%
5134	1.95%	8.41%	100.00%	100.00%	4.61%	3.51%	0.00%
5168	3.77%	100.00%	100.00%	100.00%	5.68%	3.98%	0.00%
5205	6.74%	13.56%	100.00%	100.00%	8.23%	7.11%	0.00%

Fig. 11 Relative values of distance error for each method used for the five pairs of photographs and their average values

this result is not applicable because there is an accurate adjustment in the junction area of the pairs, as seen before. This transformation is commonly used when the image is already georeferenced and must be adjusted, displacing it for a better alignment of data. Only one point of control is necessary for execution.

The metric area test indicated the same trend shown in the distance test, that is, the Affine transformation method had the least average percentage of difference in areas before and after georeferencing (3.06%), followed by the Spline method (3.40%) and Projective Transformation (5.56%) (see Fig. 12). The other methods showed average percentage errors with very high values, indicating that they are not adequate to the work proposed. That was the case for 2nd Order Polynomial (25.59%), and Adjust and 3rd Order Polynomial, which reached values of 25.66% and 100%, respectively. In this test, once again the Zero Order Polynomial transformation had an area-relative error value equaling zero, due to its characteristics mentioned before.

The results of the tests suggest the use of the 1st Order Polynomial (Affine) transformation, since it had the best behavior in metric tests (area and distance) and was one of the best methods in the visual test. It is still worth emphasizing that these metric tests were the ones that most influenced the choice of method to use in the photographic reconstitution of the Koeler Map.

Using the 1st Order Polynomial (Affine) transformation requires the translation, change of scale and rotation of data. Thus, squares and rectangles in data are usually

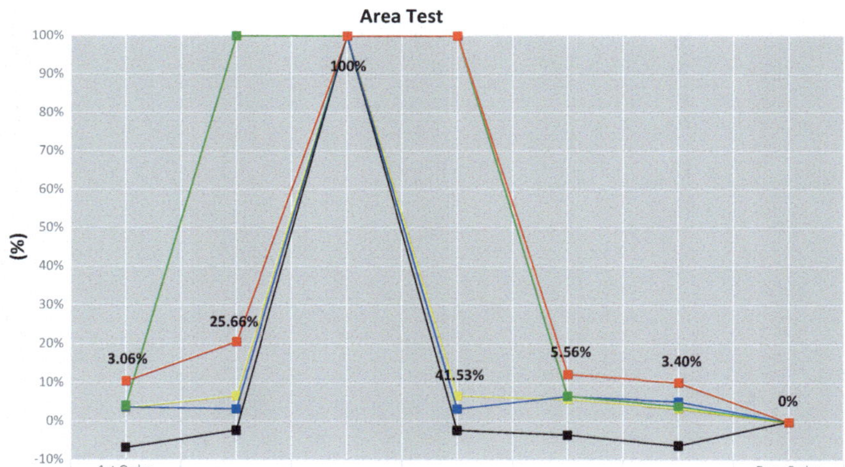

Photograph	1st Order Polynomial (Affine)	2nd Order Polynomial	3rd Order Polynomial	Adjust	Projective Transformation	Spline	Zero Order Polynomial (Shift)
5067	3.58%	6.70%	100.00%	6.70%	5.82%	3.55%	0.00%
5104	-6.85%	-2.36%	100.00%	-2.36%	-3.56%	-6.29%	0.00%
5134	3.78%	3.29%	100.00%	3.29%	6.58%	5.29%	0.00%
5168	4.20%	100.00%	100.00%	100.00%	6.67%	4.11%	0.00%
5205	10.58%	20.67%	100.00%	100.00%	12.28%	10.34%	0.00%

Fig. 12 Relative values of area error for each method used for the five pairs of photographs and their average values

changed by random sizing and angular orientation parallelograms. This transformation requires at least 3 control points for the algorithm to be applied accurately. With an additional control point, residual errors begin to be produced and distributed along all points to adjust them according to their position. Therefore, although the mathematical transformation error increases with the number of control points, the accuracy of the set is more reliable. The higher the order of polynomial transformation, the more complex the distortion to be corrected. Transformations beyond third order are rarely necessary, because they require a greater amount of control points and thus involve more processing time. In general, polynomial transformations of second or third order are used when the image needs to be bent to fit in, which is not the case in this work.

From this definition, our work was started with the goal of facilitating the georeferencing of 191 photographs that were divided into four quadrants. The division into quadrants has characteristics that directly affected their format, as the distortion

Fig. 13 Partial mosaics of the first quadrants of the Koeler map

generated by the transformation during the georeferencing process behaves individually for each photograph, so each picture is deformed in a way to ensure better adjustment. As the set of photographs increases, there is a trend to high deformation on the photographs on the borders when compared to the photographs closest to the reference photograph used as basis for georeferencing (see Fig. 13).

These four mosaics were combined and adjusted to the measurement references of the original map and resulted in the total composition of the map (see Fig. 14). It is important to note that the photographs will go through a color equalization process before the composition of the final map. In addition, the map will also be submitted to georeferencing using coordinates from a current cartographic base and will be completely converted to vectors. This way, it will form, in addition to an image collection, another map layer with its characteristics mapped according to the time when the map was made.

Fig. 14 Mosaic resulting from the four quadrants

4 Conclusions

Choosing the best georeferencing method is subject to several tests and criteria, and in this work it resulted in the choice of the 1st Order Polynomial (Affine) method. This choice was also assessed from surveying the algorithm of each method and its characteristics, with the purpose of deciding which of them is most suitable to the specific objective of this work. This does not mean that the 1st Order Polynomial (Affine) transformation is appropriate for other works with different goals and working conditions.

It is worth highlighting that this whole procedure had to be carried out because the map cannot be removed from its place of guardianship, even due to its fragility. Other scanning processes would have been more appropriate, but there were transport restrictions and lack of resources to carry out other in situ processes.

Finally, it is worth mentioning that the method was satisfactory, yet additional checks are necessary to ensure the final geometry of the map in digital media. However, all that is measured or modelled is subject to errors, and such errors result from the quality of the product under analysis.

Acknowledgements We are grateful to CNPq—Brazil (National Counsel of Technological and Scientific Development) and FAPERJ (Foundation Carlos Chagas Filho Research of the Rio de Janeiro State).

References

Ambrozio, J. C. G. (2012). O Território da Enfteuse e a Cidade de Petrópolis–RJ, Brazil. *Revista Electrónica de Geografía y Ciencias Sociales*, vol. XVI, *418*(39), 1–7.

Collier, P. (2013). The military map of the United Kingdom and its impact on mapping in the twentieth century. *The Cartographic Journal, 50*(4), 324–331. doi:10.1179/1743277412Y.0000000023.

IBGE. (1948). *Sinopse Estatística do Município de Petrópolis—Estado do Rio de Janeiro*. Rio de Janeiro, RJ: Serviço Gráfico do Instituto Brasileiro de Geografia e Estatística.

Laeta, T. & Fernandes, M. C. (2015). Cartografia histórica de Petrópolis (RJ): levantamento dos documentos cartográficos no período de 1846 a 1861. *Atas do VI Simpósio Luso-Brasileiro de Cartografia Histórica* (pp. 107–123). Braga: Universidade do Minho.

Longley, P., Goodchild, M. F., Maguire, D. J., & Rhind, D. W. (Eds.). (2005). *Geographical information systems: Principles, techniques, management and applications*. England: John Wiley & Sons.

Menezes, P. M. L., Santos, C. J. B., Fernandes, M. C., Santos, K. S., Sardella, A. B., Cardoso, R. S. P., Alves, S. L. P. (2013). Linking Rio de Janeiro State. In: *Brazil through historical cartography and toponymy, workshop on historical maps, atlases and toponomy, Leipzig, Germany*.

Menezes, P. M. L., Fernandes, M. C., Santos, K. S., & Sardella, A. B. (2015). Historical-Toponymic Analysis of the 1823 Carta Geographica of the Rio de Janeiro Province. In C. R. Sluter, C. B. M. Cruz, P. M. L. Menezes (Eds.), *Lecture notes in geoinformation and cartography*: *Maps connecting the world* (1st ed.) (pp. 235–247). Springer International Publishing.

Paula Buarque, A. (1940). *História e Historiographos da Cidade de Petrópolis*. Rio de Janeiro: O Livro Vermelho dos Telefones.

Rabaço, H. J. (1985). *História de Petrópolis*. Petrópolis: Instituto Histórico de Petrópolis.

Sá Earp, A. L. (1996). Os quarteirões. *Revista de Petrópolis, 1*(3), September 1996. http://www.ihp.org.br/lib_ihp/docs/alse19941025.htm. Retrieved April 15 from 2014.

Souza, B. C.P. (2013). A Toponímia como uma forma de resgatar a memória do lugar. *Anais do V Simpósio Luso-Brasileiro de Cartografia Histórica*, Petrópolis. http://www.cartografia.org.br/vslbch/trabalhos/77/100/beatriz_simposio-luso-brasileiro-2013_1379385701.pdf. Retrieved December 19 from 2015.

Location Spoofing in a Location-Based Game: A Case Study of Pokémon Go

Bo Zhao and Qinying Chen

Abstract The worldwide Pokémon Go fever has brought location spoofing into the public's spotlight. Location spoofing is an intentional act to masquerade locational information to somewhere other than the actual location where a network communication takes place. In the realm of Cartography and GIScience, compared with well-studied spatial quality issues, our knowledge of location spoofing is still quite limited. By reviewing five frequently-conducted location spoofing techniques for Pokémon Go, this paper critically examines location spoofing as an emerging spatial data quality issue. To unveil the hidden motivation for location spoofing, we discuss the uneven spatial distribution of game rewards through mapping a large volunteered data set of worldwide Pokémons, and gaze at the major actors in the location-based game, including location spoofers, game designers, hackers and bots. Though the research scope of this paper lies in a location-based game per se, location spoofing widely exists in Internet applications and is not fundamentally different from other deceptive phenomena in the real world. We encourage cartographers and GIScientists to face this spoofing phenomena head-on in order to promote more effective and trustworthy uses of geospatial big data.

Keywords Location spoofing • Location-based game • Spatial data quality • Actor-network theory

B. Zhao (✉)
College of Earth, Ocean, and Atmospheric Sciences, Oregon State University, Corvallis, OR 97331, USA
e-mail: bzhao@coas.oregonstate.edu

Q. Chen
School of Media and Communication, Temple University, Philadelphia, PA 19122, USA
e-mail: qinying.chen@temple.edu

M.P. Peterson (ed.), *Advances in Cartography and GIScience*, Lecture Notes in Geoinformation and Cartography, DOI 10.1007/978-3-319-57336-6_2

1 Introduction

Pokémon Go (PG) is a location-based game for mobile phone users developed by the game company—Niantic. Initially released in selected countries in July 2016, PG has quickly become popular. By September 2016, it has been downloaded more than 500 million times (Gilbert 2016). As a location-based game, the location of a game player in the digital game world is identical to her/his location in the real world. With the positioning capability of a mobile phone, a game player can locate, catch, train and battle with Pokémon (a virtual cartoon creature), who appears on the screen as if being in the same location.

Based on our observation, a number of PG fans from countries where the game had not been released yet have falsified their locational information to download and play this game. Some PG players spoofed their locations to join in remote battles or to accelerate the process of training Pokémons. To our surprise, PG players adopted and even invented several kinds of location spoofing techniques (Zhao and Sui 2017), including, but not limited to, drone spoofing (Sofka 2016), VPN spoofing (Jenner 2014), WiFi spoofing (Harvey 2016), GPS spoofing (Schurgot et al. 2015), and bots with location spoofing functions (Kang et al. 2013).

In the era of big data, spatial data quality is a critical concern for cartographers and GIScientists. Compared with various studies on uncertainty (Hunsaker et al. 2013), noise, and outlier (Lovelace et al. 2016), our knowledge on location spoofing is still limited. Using Pokémon Go—a popular location-based game as the research scope—we closely scrutinize this deceptive phenomenon as a severe spatial data quality issue, discuss game players' hidden motivation for falsifying location, and reflect on the behaviors of other stakeholders related to this game.

Section 2 of this paper is a review of location-based game and actor-network theory. Section 3 explores the location spoofing phenomena in PG; Sect. 4 explains location spoofing under the context of spatial data quality; Sect. 5 discusses the motivation for spoofing; this is followed by the concluding remarks.

2 Related Works

This section introduces two strands of works related to this study: a review of location-based gaming and a brief introduction to Actor-Network Theory. The review of location-based gaming provides a general background of the interaction between the digital and real world of PG. Actor-Network Theory helps place the network of actors related to PG, and further unveils the hidden reasons why game players falsify their locations.

2.1 Location-Based Game

The location-based game (Wetzel et al. 2012) is a mobile game using location to link the real world (where the game player lives and plays) and the digital world (where the avatar of the game player exists). The concept of location-based games originated from an outdoor activity named geocaching (O'Hara 2008). The geocache is a waterproof container having a logbook, a pen, and some trade items. It can be hidden anywhere in the world and its location is shared in the form of longitude and latitude on an openly available website. With the help of a GPS-enabled mobile device, a player is able to look for geocaches. Once a cache is found, the player records this exploits both in the logbook and online, but must return the cache to the same coordinates so that other players can find them.

In a location-based game, locational information is acquired by the positioning function of mobile devices (e.g., iPhone, iPad, Android phones). To accurately position a location, most mobile devices need to receive and parse a series of radio frequency readings, such as GPS signal, cell tower signals, crowdsourced WiFi signals, or even other sensors (Sommers and Barford 2012). GPS is the most accurate method. Though its measurement error can be as small as a few meters, its accuracy is vulnerable to several factors: the quality of GPS chipset, the exposure area to the sky, building materials, and the number of accessible satellites. So, when the GPS signal is weak, mobile devices turn to an assisted-GPS algorithm. This algorithm combines the available GPS with the locational information of nearby cell towers in order to estimate a result. The accuracy is approximately within 1000 m. Today, crowdsourced WiFi data are frequently used for positioning. It is faster than GPS and more reliable than cell tower data, especially in urban areas where WiFi hotspots are ubiquitous but GPS signals may be blocked by buildings.

2.2 Actor-Network Theory

To further examine why game players falsify their locational information, we bring in Actor-Network Theory. Location spoofers can be treated as agents who interact with other actors and have their own motivation in the network. Actor-Network Theory was developed in the 1980s in Paris and is associated with scholars such as Callon (1999), Latour (1987) and Law and Hassard (1999). Different from traditional views that humans are in charge of technologies, Actor-Network Theory treats humans and non-humans (e.g., technologies, materials, and media) as equal and interactive agents or actors. Also, the theory assumes that each agent or actor achieves their shape and attributes by interacting with other elements. In other words, non-human technical actors mediate the way in which people react to them, so human and non-human agents or actors coordinate and evolve collectively rather than separately in social processes.

In Actor-Network Theory, actions are not the special properties of humans, but the properties of the associations of human and/or non-human agents. The intentions, meaning and values of actions keep adjusting and altering through the dynamic relationships among agents or actors. As Latour says, "when the two agents join, any number of unintended goals can emerge which are made possible only by their association in the collective." (Latour 1999, pp. 179).

By applying Actor Network Theory into game studies, we can overcome the limitation of only focusing on game players, and expand our view to a holistic picture depicting a mixed net of all human and non-human actors. Games are simple in terms of their rules, but games could become super complex when multiple human and non-human actors are involved, such as game players and their communities, game makers, the game itself, platforms, technologies, hackers, and even outsiders who are not directly related to the game. Many popular digital games have a huge number of users, and are embedded in sophisticated technical infrastructures, which largely increases the complexity of the games.

3 Location Spoofing in Pokémon Go

As a typical location-based game, PG uses location to link the game frame and the real world. In the game, one can easily locate oneself using the location of one's avatar in the game frame. The terrain and environment of the game frame are built by the geospatial data from Google Maps; and major locations, such as PokéStops, gyms, are collected from crowdsourced geographic data. PokéStops recharge players with game items, such as eggs, Pokéballs, and potions; Gyms serve as battle locations for matches between teams. A player may throw a Pokéball to catch a nearby wild Pokémon. Once the Pokémon is caught, the player gains some game rewards—Pokécoins. These game rewards can be used to evolve or level up Pokémons. Since players have limited Pokéballs, if players need more or cannot wait to hatch the eggs by devoting extra time and energy, they can buy Pokécoins—approximately 0.99 dollars for 100 Pokécoins.

PG published strict location-based game rules. For example, a player must walk continuously to level up a Pokémon, and physically travel to a location to enjoy the game resources associated with that location. Since walking a long way is tedious and some places are difficult to access, a number of game players choose to falsify their locations. Below, we list some frequently-used location spoofing techniques on PG.

PokéDrone: PG players use drone mounted smartphones to catch remote Pokémons, visit distant gyms, or hatch eggs. To do that, a smartphone, attached to a drone, is remotely accessed by a computer through the Internet (see Fig. 1a). A player controls the drone with a app named AirDroid. The drone can take the phone to any nearby places with Internet access. Especially in urban areas, the drone can almost visit any places because of the well mobile data coverage. Based on the same idea, a company named TRNDlabs developed another drone-based device—PokéDrone

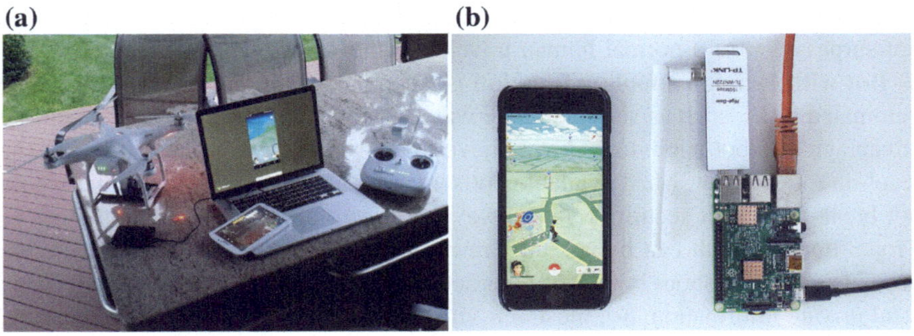

Fig. 1 (**a**) Drone for PG (Muoio 2016), (**b**) SKYLIFT (Harvey 2016)

(Sofka 2016). It has an external camera to navigate to a designated location. TRNDlabs did not treat PokéDrone as a device for location spoofing, arguing that the PokéDrone merely brought GPS chipsets out of the phone, and players could travel to nearby places, which were not approachable by walking.

VPN Spoofing: The initial release of PG was only available in the U.S. version of Google Play Store and Apple App Store, meaning that the PG fans outside of U. S. could not even download the game app. Therefore, in order to play this game, PG fans in other countries spoofed their IP addresses through VPN (virtual private network). While connecting to a VPN, a device will be tunneled to another IP address. Since the IP address is bound to a physical location, most web services will treat the device at the new location. As a result, VPN enables PG players outside of U.S. to download and play PG.

WiFi Spoofing allows the user to spoof to a particular place using a simulated WiFi environment. SKYLIFT (Harvey 2016) is such a low-cost WiFi spoofing device consisting of a Raspberry Pi Computer, a WiFi dongle and an Ethernet cable. It was alleged that a SKYLIFT had enabled a PG player to spoof to a PokéStop at London's Ecuadorian embassy—the home of Julian Assange. To spoof, SKYLIFT scans and collects the records of all surrounding WiFi devices at the target location (i.e., the embassy) with each record containing a MAC address (BSSID) and a WiFi name. Then, all the WiFi records are rebroadcast to simulate the WiFi environment at the target location. Since PG prioritizes GPS signals over cell-tower and crowdsourced Wi-Fi positioning methods, a PG user would need to shield the GPS signals while conducting WiFi spoofing. Consequently, a player can successfully cheat a PG app to believe that they are at the reported location.

GPS Spoofing Apps trick a phone's built-in GPS by transmitting a false set of coordinates. By this means, players can virtually go anywhere in the world. Specifically, one would need a GPS spoofing app to take over the GPS chipset of a mobile phone and to report a designated location instead of the actual location parsed by GPS. PG players are permanently banned if they are caught using a GPS spoofing app. So, to avoid being caught, some players try to spoof their locations at a reasonable speed. For example, the speed between two reported locations should not surpass that of a normal human travelling with an available vehicle.

Bot with location spoofing function is a computer program which can execute automated tasks, such as simulating the way a real human being plays a game. The advantage of a bot is obvious, it can move to anywhere anytime, which makes the level-up much faster. Most available bot services claim themselves safe and would not break any cyber security rules, however, bots are designed in the same idea as Trojan horses—both can bypass the firewall or anti-virus functions, and there are growing concerns about bot users' Internet security. For example, a bot may easily steal private information. So, most anti-virus tools treat bots as malicious software.

There are a variety of location spoofing techniques used by PG players. Basically, location spoofing represents a deliberate locational inconsistency between the reported location and actual geographic location where a specific network communication is made to location-based game or other kinds of Internet applications. Indeed, location spoofing does not only manifest itself as a locational inconsistency between the reported location and the ground truth, but also associates with a spoofing motivation that triggers the inconsistency.

4 Technical Nuisance or Intentional Plot?

To further understand location spoofing, we aim to analyze the inherent locational inconsistency. In Fig. 2, the x axis indicates the value of locational information, and the y axis indicates the probability that a value is reported. When making a measurement of the object, the operator generally assumes that some ground truth exists based on how they define what is being measured. The operator reports the result usually by specifying a range of values, and expects this ground truth to fall within the range. Since the readings are more likely to fall near the mean than further away, the spread of all the measured results would possibly represent a Gaussian

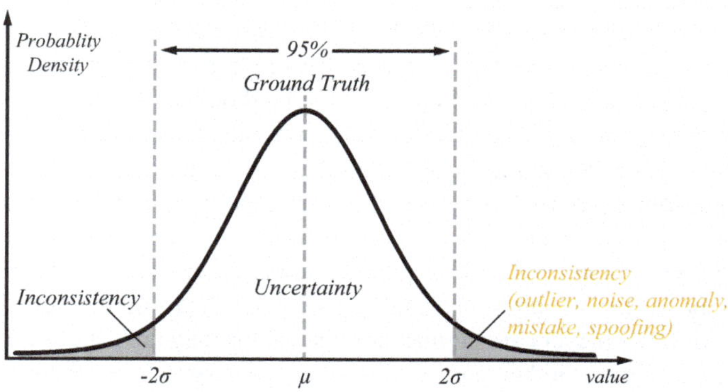

Fig. 2 Comparing uncertainty to spoofing

distribution with the mean being the best estimation of the ground truth. The systematic error is consistent and repeatable, and is a constant offset between the best estimation and the ground truth, whereas the random error arises from random fluctuations. The range of uncertainty is bounded by the two limits of random error, and the inconsistent readings fall out of the range of uncertainty.

Contingent upon different circumstances, the operator sets the limit of random error as multiple times the standard deviation to rule out inconsistent readings. For example, if another reading of the same object has a 95% chance that the measurement is consistent, it will fall within the two limits of double standard deviations to the mean; however, if the reading has 5% chance that the measurement is inconsistent, it will fall outside of these limits.

In most quantitative studies, the inconsistency is often referred to as a data quality issue, such as outlier, anomaly, or noise. However, spoofing is seldom studied in the context of spatial quality, and is often categorized as an outlier or noise. It is worth noting that the terms outlier or noise have a certain degree of ambiguity. They mostly describe inconsistencies caused by some unknown, unclear or inherent factors, such as the environment uncertainties, the measurement uncertainties, or limited knowledge about measurement. Intentional errors/uncertainties are not included. In this regard, location spoofing, as a deliberate locational inconsistency, cannot be regarded as an outlier or noise. To differentiate spoofing from other types of inconsistencies, we suggest investigating the hidden motivations for spoofing, which are discussed in the next section.

5 Generative Mechanisms for Spoofing

After presenting the location spoofing techniques and differentiating location spoofing phenomena from pure noises, the following two issues warrant our further discussion in terms of understanding why game players falsify their locational information.

5.1 Uneven Distribution of Pokémons

We mapped the Pokémons at both global and local level. To do that, we use an API from pokemapper.co to get the Pokémon locations. Pokemapper.co is a volunteered web-map of Pokémons. Once a wild Pokémon is found on the street, a game player can voluntarily upload the information of this Pokémon (e.g., location, timing, and species) to pokemapper.co. In this way, pokemapper.co has acquired a tremendous dataset of Pokémons. On October 21 2016, we visited its data API at https://pokedata.c66.me/sightings (site no longer active), a list of Pokémons' locations in GeoJSON format streamed to our client (data available here: https://github.com/jakobzhao/pokemongo). In the end, we collected 77,445 Pokémons for the whole world. Using QGIS, we plotted this large dataset as maps at different scales.

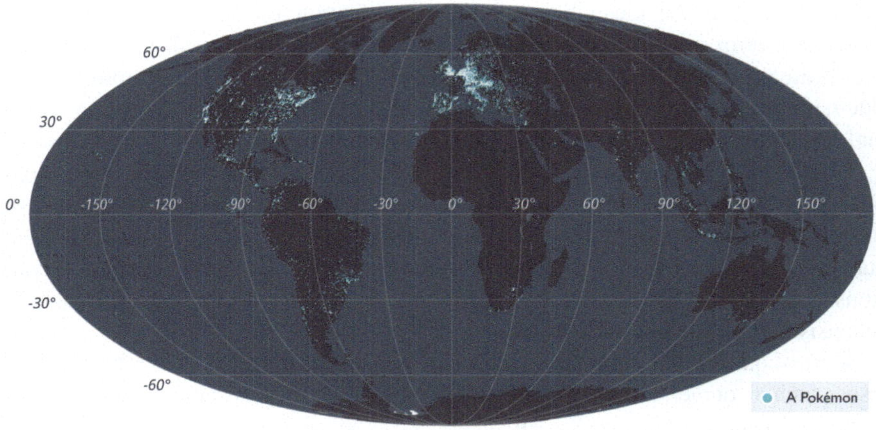

Fig. 3 Global distribution of Pokémons in the Mollweide projection

Figure 3 illustrates the the world-wide Pokémon resources. As observed, most of the Pokémons are located across Europe and North America, and are also clustered in a few cities in Central and South America, Japan, India, Southeast Asia, Australia, New Zealand and South Africa. We also observed that a few Pokémons scattered in oceans, and especially a large amount of Pokémons were around Galápagos Islands of Ecuador.

The map clearly illustrates an uneven global distribution of Pokémon resources among different countries. This uneven spatial pattern suggests why PG fans in the countries with fewer Pokémons want to spoof their locations. Only by spoofing a location to a country where Pokémon Go is released, can a fan download and play this game. Valid locational information becomes the *passport* to access the game. Moreover, in a country with fewer Pokémon resources, it becomes very competitive to catch a Pokémon; this situation urges PG players to find places which possess more Pokémons. Obviously, location spoofing is a much more convenient way to discover the digital landscape of PG, even at the risk of being blocked.

The Pokémon resources also indicate unique spatial characteristics in urban areas. In New York City, for example, (see Fig. 4), most Pokémons are clustered around places like World Trade Center, Time Square, Central Park, while less Pokémons are scattered around the suburban areas. Though PG does encourage people to play outside in the parks and crowded streets, and meet with people there, PG players who live in suburban areas have very limited access to Pokémons. This uneven resource distribution at the local level would possibly result in long travel times or location spoofing to places with more Pokémon resources.

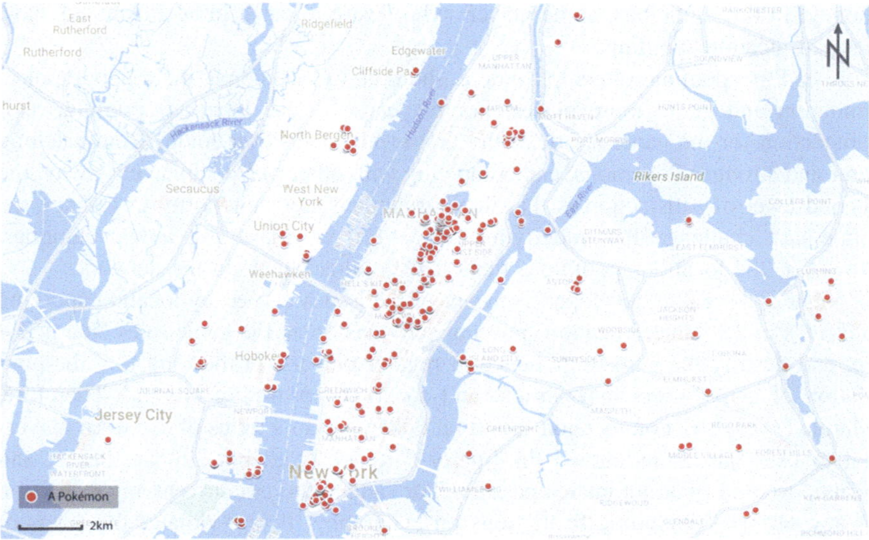

Fig. 4 Pokémons in New York City. The base map uses the Google map in the light political theme

5.2 Individual Motivations

To further comprehend the motivations for spoofing, we apply Actor-Network Theory to analyze PG, focusing on location spoofing and anti-location spoofing behaviors. The experience of autonomy (the extent to which the game prefers flexibility over strategies, choice over tasks and goals, and rewards that provide feedback and not control) and competence (the extent to which tasks offer ongoing challenges and opportunities) influence the level of motivation and satisfaction a player can gain from the game (Przybylski et al. 2010). In PG, players collect Pokémons and battle with each other to obtain rarer and stronger Pokémons. If PG was fully virtual, like many other video games, players might just walk around the virtual map by their mouse-clicks or keyboard-touches. However, PG requires players to move in the real world, which brings many uncertainties to users, such as the places where they live, their daily routines, weather, and the safety of the surrounding areas.

What is more crucial to players is the unequal distribution of Pokémons, which is illustrated in Sect. 5.1. The uneven distribution on one hand motivates players to go to further places, and on the other hand frustrates some players, especially the ones who live in the places having less Pokémons and are not capable to overcome their geographical limitations. In fact, PG offers players one alternative way to overcome the limitations, that is buying incense or lure module with PokéCoins. While using incense or lure module, players have higher chances to encounter wild Pokémons, instead of moving to places further away. However, PokéCoins cost real money. In this situations, some players who desire to catch more Pokémons have devised location spoofing.

Location spoofing offers players a shortcut to overcome their geographical limitation and go to more places. For players who use location spoofing, their satisfaction for the game are dual: the first-level satisfaction comes from catching rarer and stronger Pokémons, and competing with other players in an easy way; the second-level of satisfaction is that they strengthen their autonomy to reshape the game and challenge the underlying logic of it by mastering new techniques. However, the location-spoofing actions irritate other agents in the network.

Niantic, the game developer, is aware of the issues caused by location spoofing activities in the game. Location spoofing activities break the game rules Niantic set up, and destroy the game experience of regular players. In the long run, the population of active users may shrink, and so do the profits of Niantic. Therefore, Niantic is very strict on location spoofing. Pokémon Go's terms of use warn players not to use "any unauthorized third-party software (e.g., bots, mods, hacks, and scripts) to modify or automate operation," or "attempt to circumvent any restriction in any service," including restrictions on geography. If one violates the terms of use, Niantic can "suspend or terminate your access to some or all" functions of the game. At the same time, Niantic works on strategies to detect location spoofing behavior more effectively, and improving the program to make it harder to spoof.

Although Niantic does not condone cheating, Niantic does not ban online volunteer-based maps, such as Pokemapper.co and a few similar ones like Go Radar, Poké Radar, and Pokécrew. These maps, being embedded in websites or apps, track all Pokémons around the world in a real-time fashion by using crowdsourced data from users, and can show users the exact locations of nearby Pokémons. One reason for not banning the volunteered maps may be that these volunteer activities can help promote friendly player communities, and keep players active and engaged in the game.

Hackers are a special group of location spoofers in the network. Various kinds of bots were developed by hackers to simulate humans playing PG. Most hackers are anonymous in online communities, but some hackers offer simple contact information to help players contact them for bots or services. Many hackers do not profit from their activites and, only some sell bots to get profits, or charge plug-in services. Hackers are rule breakers and get a feeling of fulfillment from acquiring the root privilege.

The dynamic network which is built upon PG makes us revisit the main concept of location. In PG, location spoofers and other users are competing for Pokémons, which are distributed unevenly at global and local level. Pokémons, as a type of resource, are attached to locations. They are locational resources in the digital space, and also connected to the locational resources in the real world. Each agent has different capabilities to turn the locational resources into the capital they want (Zaheer and Nachum 2011). Location spoofers use advanced techniques to break the connections between locational resources in the digital spaces and in the real world, and occupy more locational resources in the digital space without moving to the real locations. Location spoofing in PG warns us to pay attention to the disconnectedness between "the digital" and "the physical" in the geospatial data and other real-world situations.

6 Concluding Remarks

Location spoofing is defined here as an intentional act of falsifying one's true location to somewhere else. After reviewing several techniques for location spoofing with Pokémon Go, this paper regards location spoofing as an emerging spatial data quality issue that is closely connected with the operator's motivation. To holistically comprehend location spoofing, we further discuss its generative mechanism: a fundamental mechanism which triggers players' motivation to spoof is the uneven distribution of game rewards. At the individual level, we analyze a variety of motivations for location spoofing from players, game designer, bots, and hackers.

Though the research scope of this paper lies in a location-based game, location spoofing widely exists in Internet applications and is not fundamentally different from other forms of deception. Contingent upon specific circumstances, location spoofing could have significant impacts on our daily lives, social developments, and even national security. We hope this paper can raise awareness of the multiple dimensions of location spoofing when geospatial data are used in research or policy-making. We encourage cartographers and GIScientists to face this spoofing phenomena head-on whenever necessary in order to promote more effective and trustworthy uses of geospatial data.

References

Callon, M. (1999). Actor-network theory—The market test. *The Sociological Review, 47*(S1), 181–195.

Cypher, M., & Richardson, I. (2006). An actor-network approach to games and virtual environments. In *Proceedings of the 2006 International Conference on Game research and Development. ACM International Conference Proceeding Series, 223*. Retrieved January, 30, 2017 from http://dl.acm.org/citation.cfm?id=1234382.

Gilbert, B. (2016). *Pokémon Go has been downloaded over 500 million times*. Retrieved Janurary, 30, 2017 from http://www.businessinsider.com/pokemon-go-500-million-downloads-2016-9.

Harvey, A. (2016). *SKYLIFT-WiFi SSID geolocation emulator*. Retrieved Janurary 30, 2017, from https://ahprojects.com/projects/skylift/.

Hunsaker, C. T., Goodchild, M. F., Friedl, M. A., & Case, T. J. (2013). *Spatial uncertainty in ecology: Implications for remote sensing and GIS applications*. Springer Science & Business Media.

Jenner, M. (2014). Is this TVIV? On Netflix, TVIII and binge-watching. *New Media & Society* doi:10.1177/1461444814541523.

Kang, A. R., Woo, J., Park, J., & Kim, H. K. (2013). Online game bot detection based on party-play log analysis. *Computers & Mathematics with Applications, 65*(9), 1384–1395.

Latour, B. (1987). *Science in action: How to follow scientists and engineers through society.* Harvard university press.

Latour, B. (1999). *Pandora's hope: Essays on the reality of science studies.* Harvard University Press.

Law, J., & Hassard, J. (1999). *Actor network theory and after.*

Lovelace, R., Birkin, M., Cross, P., & Clarke, M. (2016). From big noise to big data: Toward the verification of large data sets for understanding regional retail flows. *Geographical Analysis, 48*(1), 59–81.

Muoio, D. (2016). *One 'Pokémon GO' player had a genius idea to try and catch 'em all.* Retrieved Janurary, 30, 2017 from http://www.businessinsider.com/man-uses-drone-to-catch-pokemon-in-pokemon-go-2016-7.

O'Hara, K. (2008). Understanding geocaching practices and motivations. *Paper presented at the Proceedings of the SIGCHI Conference on Human Factors in Computing Systems.*

Przybylski, A. K., Rigby, C. S., & Ryan, R. M. (2010). A motivational model of video game engagement. *Review of General Psychology, 14*(2), 154.

Schurgot, M. R., Shinberg, D. A., & Greenwald, L. G. (2015). Experiments with security and privacy in IoT networks. *Paper presented at the World of Wireless, Mobile and Multimedia Networks (WoWMoM) 2015 IEEE 16th International Symposium on a.* 1–6.

Sofka, S. (2016). TRNDLabs *Pokédrone will solve some of your Pokémon go problems.* Retrieved Janurary, 30, 2017 from http://nerdist.com/trndlabs-pokedrone-will-solve-some-of-your-pokemon-go-problems/.

Sommers, J., & Barford, P. (2012). Cell vs. WiFi: on the performance of metro area mobile connections. *Paper presented at the Proceedings of the 2012 ACM Conference on Internet Measurement Conference.*

Wetzel, R., Blum, L., & Oppermann, L. (2012). Tidy city: A location-based game supported by in-situ and web-based authoring tools to enable user-created content. *Paper presented at the Proceedings of the International Conference on the Foundations of Digital Games.*

Zaheer, S., & Nachum, L. (2011). Sense of place: From location resources to MNE locational capital. *Global Strategy Journal, 1*(1–2), 96–108.

Zhao, B., & Sui, D. Z. (2017). True lies in geospatial big data: Detecting location spoofing in social media. *Annals of GIS.* 1–14 doi:10.1080/19475683.2017.1280536.

Part II
Crowdsourcing and Data

Educational Aspects of Crowdsourced Noise Mapping

Andrea Pődör and László Zentai

Abstract Examined here is a project-based method for integrating of crowd-sourced noise mapping in the teaching of geoinformatics. Practical examples of a GIS course are designed to use crowdsourced noise measurements in the curricula. The approach showed students how GIS integrates data acquisition, processing, analyses and visualization. Implementing noise measurement tasks in the curricula also helped students gain skills in detecting and resolving errors. The teacher acts as a mentor during the classes in order to help students acquire knowledge in a way that best suits them. A project like this is helpful because it makes the problem easily understandable and offers several cartographic visualization methods that the students can test and compare in practice. In addition, we found that the quality and the accuracy of students' performance improved during the semester. The students share their knowledge and their experiences. They also learn to cooperate because the success of their project is dependent on the collective work of each student. The most recent methods are presented to the students during the whole working process and lets them discover good and bad solutions.

Keywords Noise mapping · Project-based learning · GIS

A. Pődör
Institute of Geoinformatics, University of Óbuda, Alba Regia Technical Faculty, Székesfehérvár, Hungary
e-mail: podor.andrea@amk.uni-obuda.hu

L. Zentai (✉)
Eötvös Loránd University, Department of Cartography and Geoinformatics, Budapest, Hungary
e-mail: lzentai@caesar.elte.hu

M.P. Peterson (ed.), *Advances in Cartography and GIScience*, Lecture Notes in Geoinformation and Cartography, DOI 10.1007/978-3-319-57336-6_3

1 Introduction

Noise pollution is one of the main and growing environmental problems in urban areas. Noise affects everyday life; well-being and noise can cause severe psychological issues (Pődör and Révész 2014). According to the Environmental Noise Directive of the European Union 2002/49/EG (END), each agglomeration pursuant to article 7—with a population more than 250,000—should create a noise map. The Hungarian Government Decree 280/2004 (X. 20.) in 2004 directed all settlements with a population more than 100,000 to create a noise map and an action plan to reduce noise pollution exceeding the values set in the EU Directive. All these maps should have been prepared by 2012. An order prescribes that the maps should be renewed every five years from 2010 onwards (Pődör and Révész 2014). However, due to the lack of financial resources, local authorities have been unable to fund the renewal of their maps.

A possible method to regularly update noise maps is crowdsourced data collection. Thanks to the growing availability of location-enabled smartphones with a range of capabilities including sound recording, this data acquisition technique looks promising.

Teachers already have good experiences in project-based learning (Pődör 2011). Therefore, it seemed to be a practical idea to involve students in this activity who had received a scholarship from the gSmart project (http://em-gsmart.zgis.net/gsmart). The project offered mobility scholarships for Central Asian (CA) students until 2017. With the help of the project, students could participate in B.Sc. and M.Sc. courses in spatial sciences in the EU countries. Within the framework of the project, eight master students studied the basics of GIS at Óbuda University and at the University of West Hungary in Sopron.

The project-based learning (PBL) method, which is an effective alternative to subject-based teaching, has been widely used in secondary schools and in higher education. The method helps students learn by participating and completing projects (Demirci et al. 2011). As Enemark (2009) defined, the importance of this method is that it teaches students how to learn. The aim of the PBL process is to give to the students a broad understanding of interrelationships and to make them able to deal with new ideas and cope with unknown solutions. Barge (2010) defined the essence of the project as a complex effort in which problem analyses, planning, management and binding cooperation is necessary for a team on successful completion of the project. He also noted that a project is a sequence of different tasks resulting in a final product.

As Pődör (2011) described, a possible sequence of tasks in a GIS project could be as follows: Geodatabase design, Planning the field work, Data acquisition; Processing of the collected data, Control of the accuracy of the data, Analysis of the data, Querying of the data, and Publication of the data. Following this sequence, students are able to see how theories connect to empirical and practical knowledge. Projects in PBL should focus on problems, involve students in a constructive investigation, be student-centred and be realistic by focusing on real-life activities Thomas (2000).

The goal of involving Central Asian students in noise measurement activities was twofold. They would learn aspects of GIS including data acquisition, analysis and mapmaking. On the other hand, they would became familiar with the settlement that they were studying.

1.1 The Study Area

Sopron is a middle-sized Hungarian settlement with a medieval town centre and a population of approximately 60 thousand. Sopron and its surroundings include four small physical geographical regions: the Sopron Mountains, the Sopron Basin, the Fertő Hills and Lake Fertő.

Sopron lies near to two main roads of the Hungarian national network: road 84, leading to Lake Balaton, and road 85 that connects Sopron with Győr town. In addition, a railway line crosses the town. City administrators are aware of the environmental noise pollution caused by roads and the need to make traffic safer and faster. To address this, Sopron plans to construct a road bypass between Pereszteg and Sopron (Fig. 1). It is no surprise that the heavy transport has a significant effect on air and noise pollution (Sopron Municipality Sustainability Plan 2010).

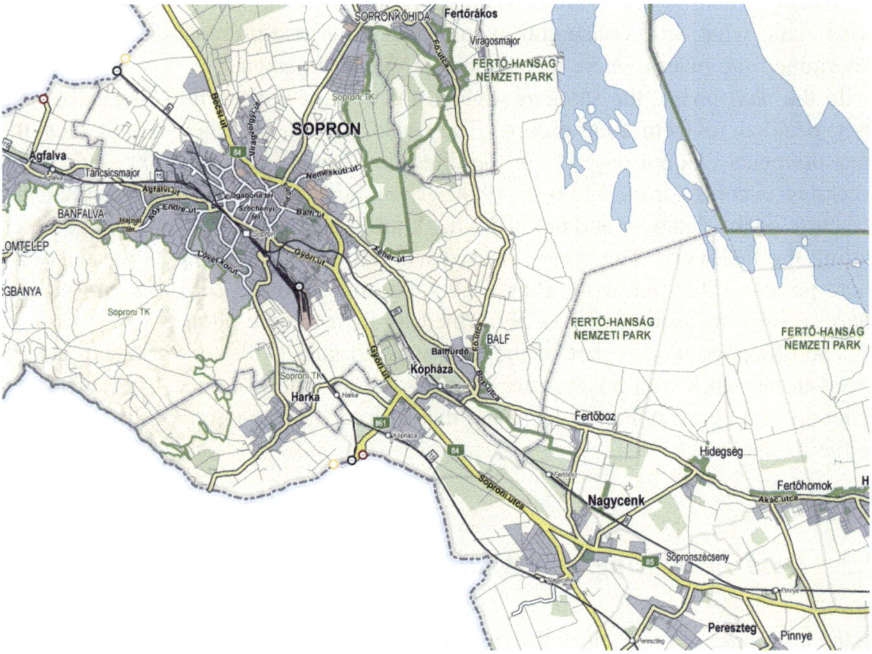

Fig. 1 Sopron and its surroundings (http://terkep.varosterkep.hu/map/)

2 Method

Students at the masters level worked with the same commercial GIS program during the semester. They learned about data modelling, database building, data manipulation, and processing and analyses of spatial data.

The noise measurement project was done over two years in a spatial data modelling course. The course consists of both theoretical and practical lessons. Four CA students started taking measurements in the spring semester of 2015. In the 2016 spring semester, five students repeated this work and made additional measurements.

Prior to starting the project, students learned about the theory and practice of official noise measurements, as well as about crowdsourcing. They also became familiar with database modelling tasks. During the project, the teachers acted as mentors to help the students fulfil the project goal in a way that best suited them. Students shared their knowledge and their experiences with each other and how to cooperate in a project.

2.1 Project Planning, Database Modelling and Fieldwork

The first, preparatory step of the project was a discussion how the students would organise the noise measurement of Sopron, specifically how they should perform the measurement based on the predefined existing points, how they should collect noise data, what time constraints would affect the measurements, etc. Then, we determined the places where measurements would be taken.

In the next phase, the students divided the area of Sopron into four parts, and they decided to form four groups. They also agreed that they would take the measurements twice a day: (1) in the afternoon from 13:00 to 15:00, and (2) on weekday evenings from 17:00 to 19:00. Then, still in the lab, they designed the structure of the database and defined what kind of attribute data would be gathered during the fieldwork. The project started with a detailed planning of the measurement points. In the fieldwork, they used a Trimble Juno SC GPS to identify these places. Noise measurement was performed by their mobile phone.

The objective of this part of the tasks was to gain skills to work with a GPS receiver and work with noise meter software on the mobile device. They measured three times on each point for 15 s, and they recorded the average of dB (A) values in the database.

2.2 Data Processing and Analyses

In the last phase of the project, the students returned to the lab. First, they processed their raw data in MS Excel. They calculated the means, correlations, and detected data outliers. After data processing, they started to create their GIS database. First, they used OpenStreetMap (OSM) to acquire the most recent customised, up-to-date layers of Sopron in a raw .osm format. They converted this data into an SQLite database. In the end, they created a layer (or multiple layers) that included only the feature types and tags. These layers could be used as shapefiles in QGIS.

Afterwards, the students continued with visualisation experiments (see Sect. 2.3). The process was followed by creating an interpolated map in ArcGIS using the IDW interpolation method and by implementing equal intervals in order to use a predefined colour scheme. This had to be as similar as possible to the symbology of professional noise maps, for which the symbolisation is defined in the END directive.

Subsequently, they continued with the analysis of maps and data. They used visual interpretation and some statistical analyses in R program (a free software environment for statistical computing and graphics.) The objective of this part was to gain skills in data processing, create a noise measurement map of Sopron town, and draw conclusions.

2.3 *Visualization*

This part of the project aimed at helping the students to understand the importance of correct visualization for perfect interpretation. They realized that if someone raises the question of where the noisiest areas are, the most convincing answer is to show these areas on a map.

Different cartographic thematic mapping methods (Fig. 2) were used to show the measurements on the map. As the students already had a course in cartography, they had now the opportunity to discuss the advantages and disadvantages of each mapping technique.

Experiment with various classifications of the acquired data were done using the built-in classification methods in the software. Students needed to understand each method or risk choosing an inappropriate method that would have produced thematic maps with misleading results. In interpreting the differences between the various classification methods, histograms were found to be useful. The x-axis of histograms shows the range of values in the fields, the y-axis is a count of features. ArcGIS has six classification methods. We analysed the impact of the different classification methods with the help of noise measurement data.

By this time, the students had learnt that the visualization of a professional or a strategic noise map is strictly regulated. Although the students did not prepare an “official” noise map, they were asked to follow pre-defined rules for noise map construction. The classification of noise data requires an equal interval representation as each five-decibel change results in a new colour code in END.

IDW interpolation was used for the interpretation of noise measurement data. This form of interpolation leads to a better understanding of Tobler’s first law of Geography. Students tested different interpolation methods and understood that,

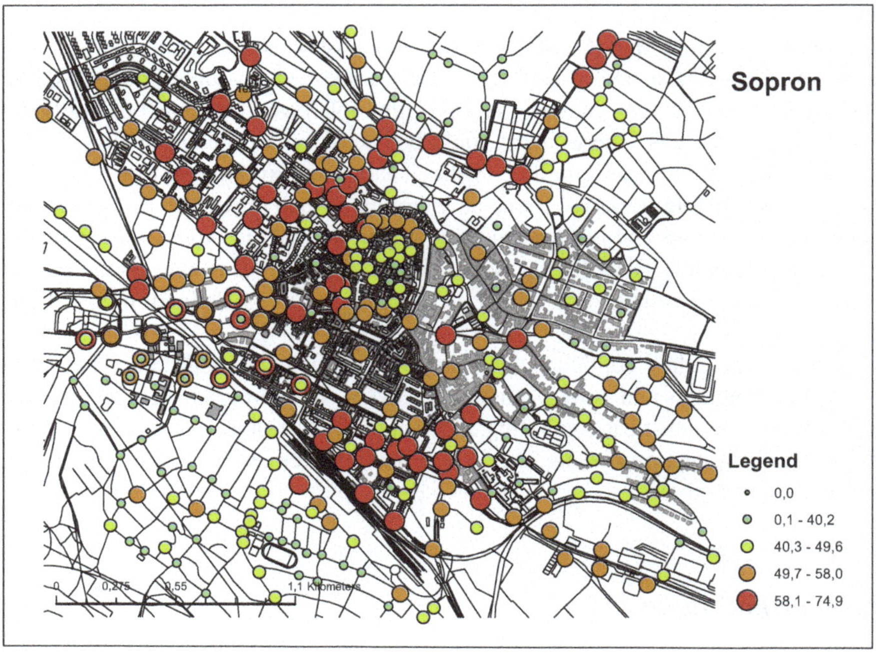

Fig. 2 Noise measurements visualized with graduated symbols

although, cartographers can use several cartographic methods, displaying the spatial context was most important. 3D maps were also made using the TIN model from the measurements and added other 3D objects such as houses, trees, rivers and lakes to their map (Fig. 3).

The project demonstrated to students that several mapped representations could be appropriate. There are different visualization methods for decision-makers, for noise measurement experts, and for the local citizens. Students could discuss the various visualization methods used by other students, either in the same class or from previous years. The teacher could give the students guidance on the visualization methods by presenting their advantages and disadvantages.

A discussion of good and bad practices can be the best solution from a didactive perspective. Noise mapping can also be a good example of testing the use of different colours and their effect on the visualization, and making students understand all aspects of visualization. Teachers can also emphasize for the students the difference between eye-catching and effective representational methods. Both approaches have relevance depending on the kind of users for which the maps are intended.

Experiencing different thematic cartographic methods for presenting a certain theme proved to be a good tool in analysing how users interpret the maps depending on their experience and previous knowledge (Pődör et al. 2015).

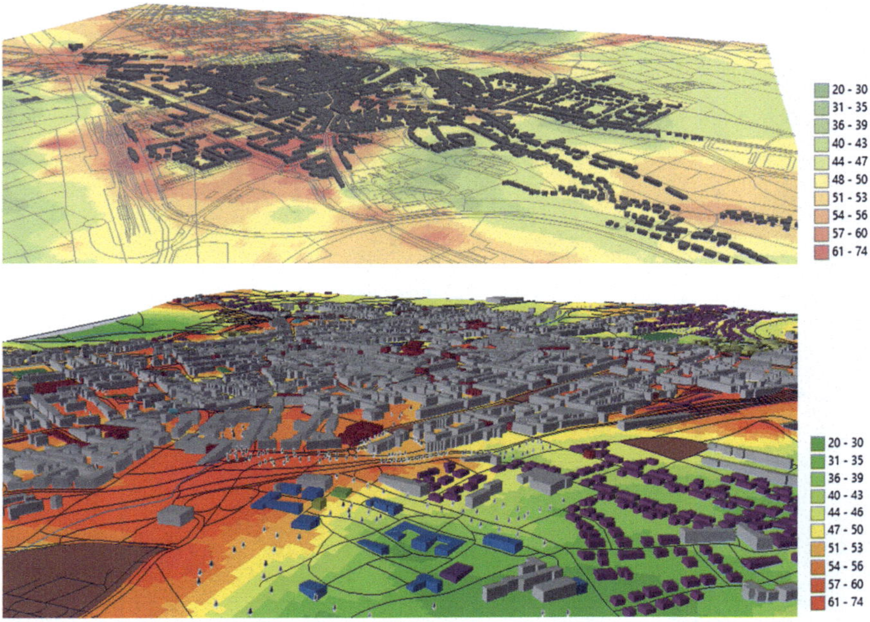

Fig. 3 Two solutions for the 3D visualization of buildings and measured noise

3 Results

Long ago, John Dewey promoted the idea of learning by doing in *In My Pedagogical Creed* (1897). Project-based learning supports this type of learning process, but in our case, we had other results as well. This type of project-based learning produced significant results in two areas: exploring the noisy and quiet areas in the city, and deepening the students' GIS knowledge by improving their practical skills.

In this process, the students produced an extensive noise measurement campaign, which resulted in a complex noise measurement map (Figs. 4 and 5). The students measured 343 points in 2015. In the following year, 370 measurements were taken at the same points. Two maps were prepared in each year—one for the rush hour and one for a quieter tie in the early afternoon. With mentorship of the teachers, students analysed the problem areas. It was revealed that the noisiest areas were associated with the main arterial roads (road 84 and the railway). Also, the main and busiest roads had the highest decibel values.

Comparing the results of 2015 to that of 2016, each measurement showed that the highest measured values could be found at the intersections of the roads. Unfortunately, the measurements could not be fully comparable because measurements in 2015 were in the morning and in the afternoon, while afternoon and evening were used in 2016. Comparing the results, we found that the quietest period is the morning, the noisiest is the evening (Fig. 6).

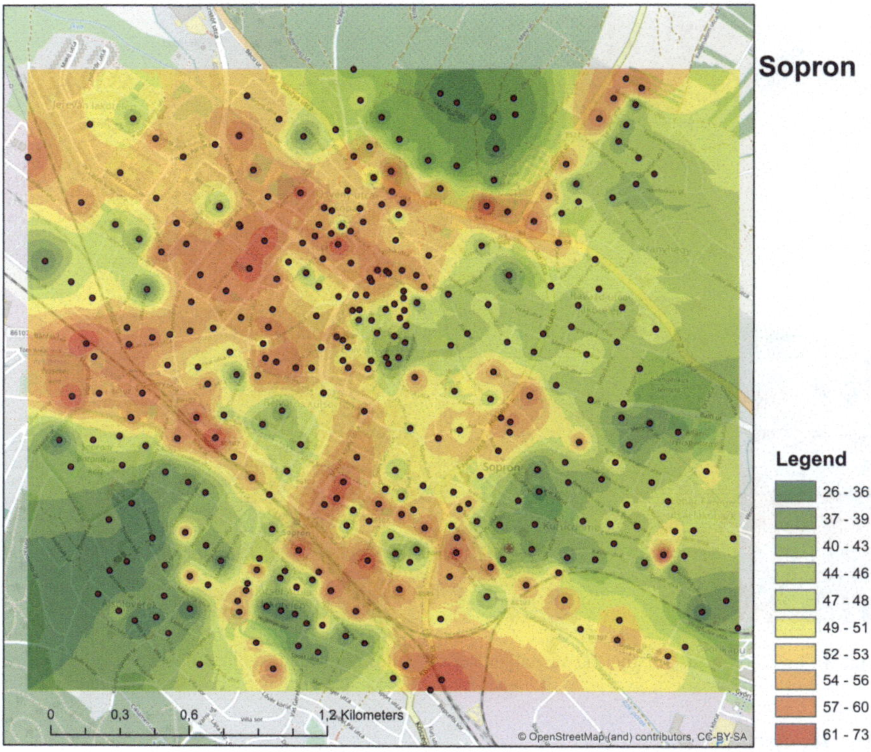

Fig. 4 Noise measurement map of the city of Sopron in 2015

This project-based learning method improved the skills of the students in many fields. First of all, this approach enhanced the collaborative activity of the students and it augmented the quality of their work. In addition, students deepened their knowledge in using open source and commercial software. At the same time, they learned how to access OpenStreetMap data as the base map for their study area. They acquired certain specific skills in noise measurement as well. They are now able to make a distinction between noise maps and noise measurement maps as well as identifying accuracy problems and discrepancies in measurements.

In general, they gained greater understanding of concepts and a broader knowledge base. With the integrated collaborative work, they improved communication and social skills, with some enhancing their leadership skills as well. They needed to find creative solutions to solve problems. These activities helped the student to engage in investigations in a student-centred way. During the analysis of their results, they applied their knowledge in an interdisciplinary way. They not only used GIS software, but also became familiar with statistical programs such as R or ggplot2. They created boxplots (Fig. 6), scatter plots and histograms (Fig. 7) for an in-depth understanding of the data they gathered, so they understood and drew their own conclusion through creative experimentation.

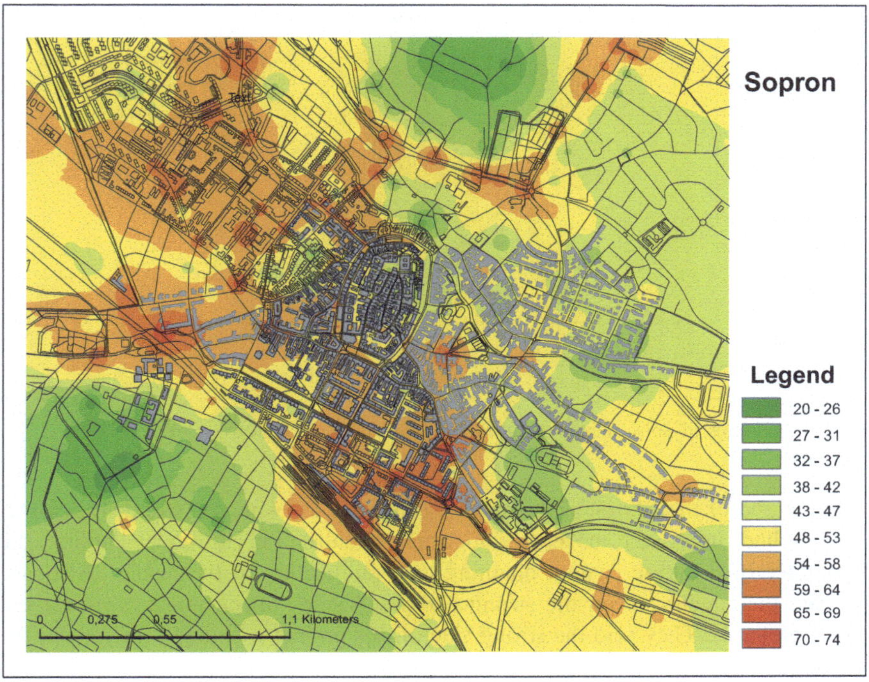

Fig. 5 Noise measurement map of the city of Sopron in 2016

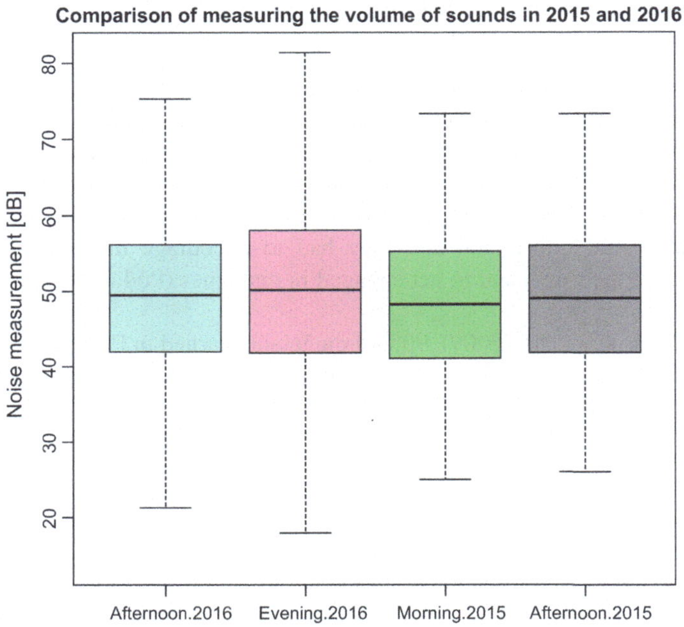

Fig. 6 Comparison of noise measurement in 2015 and 2016 in Sopron

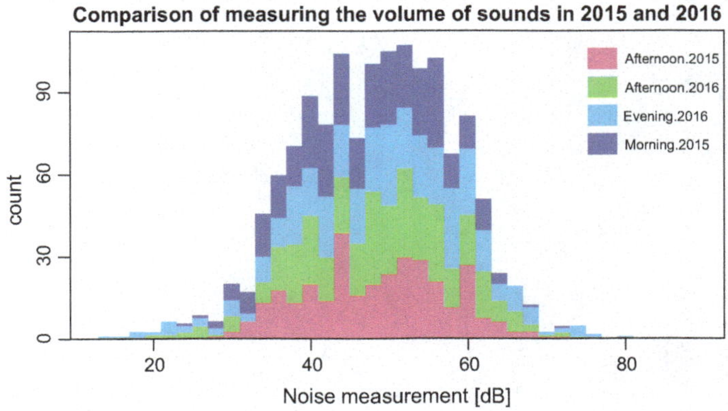

Fig. 7 Histograms of noise measurement in 2015 and 2016 in Sopron

At the end of the semester, we made a formative assessment. We identified that students had improved in team-work and visualisation, in understanding the importance of database design, planning, and publication of the results. Also, we analysed their feedback on the strong and weak sides of the noise measurement project methods. They said the strong side of this project was that it gave them the chance to:

- work in teams;
- following a project from the beginning to end;
- test different interpolation methods;
- visit unknown and beautiful places.

Students reported that the main disadvantage of the project was that different mobile devices produced different noise measurement results. Therefore, the accuracy could not be judged. In this process, the students became able to select information, form hypothesis and make decisions. On the other hand, the teacher had a different role. The teachers had to transform lesson resources into a form that the students could understand and they had to encourage the students to gain experiences on their own and to get engaged in dialogues (Pődör 2011; Pődör et al. 2015).

According to Enemark (2009), 90% of the lesson learned in PBL may be retained with problem simulations. This is because of the fact that knowledge accomplished by ourselves will be easily recalled and learnt.

4 Conclusion

The findings demonstrate that implementing crowdsourced noise measurement in a GIS course has several advantages. The pilot project revealed that during one semester we could reach concrete results detecting where the noisiest areas were in Sopron city. These findings help the local authorities implement the results in actions and to reveal environmental problems for the citizens. The results confirm the hypothesis that crowdsourced noise measurements can identify problem areas, although they are not fully substituting official noise measurements. We conclude that this type of noise measurement campaign can help other municipalities as well.

Besides these results, we can confirm that this type of project-based approach as underlined in other studies (Blumenfeld et al. 1991; Thomas 2000) is appropriate to engage students in investigation: they should apply what they had learnt during the lessons and they are able to acquire skills that cannot be learnt from textbooks.

This project-based approach is also very helpful to understand the whole process of map-making, especially to practice the cartographic visualization in a GIS course. A collection and presentation of good and bad practices can be one of the best tools from the didactical aspect. The re-evaluation and comparison of the maps created by other student groups and/or by other providers can enhance the visualization skills of the students. Practicing cartographic visualization in different software environment can also enhance the students' skills and makes them understand the essential elements and aspects of thematic cartography.

Acknowledgements This work was supported by COST Action TD1202 Mapping and the Citizen Sensor and gSmart project (545696-EM-1-2013-1-AT-ERA MUNDUS-EMA21). The authors also thank the Central Asian M.Sc. students for their participation in the project.

References

Barge, S. (2010). *Principles of problem and project based learning.* Retrieved January 14, 2017, from https://www.aau.dk/digitalAssets/62/62747_pbl_aalborg_modellen.pdf.

Blumenfeld, P. C., Soloway, E., Marx, R. W., Krajcik, J. S., Guzdial, M., & Palincsar, A. (1991). Motivating project-based learning: Sustaining the doing, supporting the Learning. *Educational Psychologist, 26*(1), 369–398.

Demirci, A., Karaburun, A., Ünlü, M., & Özey, R. (2011). Using GIS-based projects in learning: Students help disabled pedestrians in their school district. *European Journal of Geography, 2* (2), 48–61.

Enemark, S. (2009). Surveying education: Facing the challenges of the future. In *Navigating the future of surveying education.* Retrieved January 14, 2017, from http://vbn.aau.dk/en/publications/surveying-education(be36e5a0–1dd2-11de-bddb-000ea68e967b).html.

Environmental Noise Directive of the European Union 2002/49/EG (2002). Retrieved January 02, 2015, from http://ec.europa.eu/environment/noise/directive.htm.

Pődör A. (2011). The methodological advantages of using web server in teaching GIS. In A. Ruas (Ed.), *Advances in cartography and GIScience. Volume 2: Selection from ICC 2011*, Paris, July

03–08, 2011 (pp. 73–82). Lecture Notes in Geoinformation and Cartography, Vol. 6. Berlin, Heidelberg: Springer Science+Business Media.
Pődör, A., & Révész, A. (2014). Noise map: Professional versus crowdsourced data. In Huerta, Guerta, Schade, Granell (Eds.), *Connecting a digital Europe through location and place. Proceedings of the AGILE'2014 International Conference on Geographic Information Science*, June 3–6, 2014. Castellón. ISBN 978-90-816960-4-3, 06, 2014.
Pődör, A., Révész, A., Ócsai, A., & Ladomerszki Z. (2015). Testing some aspects of usability of crowdsourced smartphone generated noise maps. In: A. Car, G. Griesebner, J. Strobl, T. Jekel (Eds.), GI_forum 2015: Geospatial minds for society. Salzburg: Universität Salzburg p. 354. (ISBN 978-3-87907-558-4).
Pődör, A., Zentai, L., Révész, A., & Dobos, M. (2015). Thematic maps in analysing the prejudice and preconception in the fear of crime of citizens of a typical hungarian small town. In International Cartographic Association (Ed.), *Proceedings of the 27th International Cartographic Conference, ICC 2015*. Rio de Janeiro: International Cartographic Association, Paper T25-788_1426191672. (ISBN 978-85-88783-11-9).
Sopron Municipality Sustainability Plan (2010). Retrieved September 17, 2016, from http://www.sopron.hu/upload/content/43/_4308/Fenntarthat%C3%B3s%C3%A1gi%20terv%20Sopron.pdf.
Thomas, J. W. (2000). *A review of research on project-based learning*. http://www.newtechnetwork.org.590elmp01.blackmesh.com/sites/default/files/dr/pblresearch2.pdf.

Crowd and Community Sourced Data Quality Assessment

Laurence Jolivet and Ana-Maria Olteanu-Raimond

Abstract Data quality assessment of different volunteered initiatives and platforms presents several challenges for data validation given the high amount of data collected. This paper focuses on two goals. The first consists in defining both a generic workflow and data quality indicators for validation of reports coming from crowd and community sourcing platforms. In the proposed workflow, a qualified report can be even described by each indicator separately or by a combination of them. Here, we focus mostly on analyzing the results obtained for each indicator separately. The second goal is to learn more information about contributors who has engaged in a platform proposed by a public body (i.e., the French National Mapping Agency): Who are they? How are they contributing? What are their motivations? More is known about contributors to OpenStreetMap than of any other VGI platform. Indeed, knowing the contributors is a crucial task for both motivation and data quality, especially now that public institutions are engaging with VGI.

Keywords Data quality · VGI · Community and crowdsourcing platform

1 Introduction

In the field of GIS, Volunteered Geographic Information (VGI) describes the process defining the collection of geographic information by volunteers (Goodchild 2007). In this paper, the term of VGI is used to define volunteered geographic information coming from community and crowdsourcing activities.

Local governments are increasingly interested in using VGI for different purposes (Haklay et al. 2014). In many cases, their own VGI platforms are created to engage with citizens rather than to simply gain or share information (Johnson and

L. Jolivet (✉) · A.-M. Olteanu-Raimond
University of Paris-Est, LASTIG COGIT, IGN, ENSG, 94160 Saint-Mandé, France
e-mail: laurence.jolivet@ign.fr

A.-M. Olteanu-Raimond
e-mail: ana-maria.raimond@ign.fr

M.P. Peterson (ed.), *Advances in Cartography and GIScience*, Lecture Notes in Geoinformation and Cartography, DOI 10.1007/978-3-319-57336-6_4

Sieber 2013). Several national mapping agencies (NMA) also have experience in organizing their own VGI collection processes by developing tools to collect data from citizens or public partners (Olteanu-Raimond et al. 2016). The necessity in defining protocols for VGI data collection are also pointed out by many authors (Johnson and Sieber 2013; Mooney et al. 2016). Data collection protocols are relevant in both upstream and downstream data collection processes. Protocols can improve the quality of data by defining guidelines and rules about how data should be collected. They are also useful for quality assessment by proposing new fields (e.g., attributes, comments, or metadata) characterizing the contributions and to define new data quality criteria.

The goal of this paper is to propose a methodology for crowd and community sourced data quality assessment by defining indicators measuring the reliability of data, to identify new elements that can be introduced in data collection and to improve the knowledge regarding contributors in a community crowdsourcing system. Our approach is tested with the *Espace Collaboratif* platform developed by IGN France.

The paper is organized as follows: Sects. 2 and 3 describe the current state of the art regarding VGI data quality assessment. Our proposed quality assessment methodology is described in Sect. 4. This is following by a description of first results in Sect. 5. Conclusions and future work are outlined in Sect. 6.

2 State of the Art

VGI data quality assessment has been intensively studied in the last decade. Most of the works are focused in studying OpenStreetMap (OSM) data quality. Different elements of ISO data quality indicators are studied such as positional accuracy (Hackley et al. 2010; Girres and Touya 2010), thematic accuracy (Brando and Bucher 2010), temporality and evolution (Antoniou and Skopeliti 2015) or logical consistency (Touya and Brando 2013; Peyman and Rahim 2015). For a complete review of VGI data quality assessment methods, see (Antoniou and Skopeliti 2015; Senaratne et al. 2015).

In recent years, many VGI platforms have been proposed through different initiatives managed by an NGO, citizens or public and private bodies (Olteanu-Raimond et al. 2016). In this context, data validation is crucial in the success of such initiatives. Recent work have proposed general workflows for assessing data quality that are generally based on different filters or indicators that allow data validation (Leibovici et al. 2015; Fonte et al. 2015; Ballatore and Zipf 2015). For example, Bonter and Cooper (2012) proposed filters to automatically check the validity of bird species by taking into account historical data. In the eBird Project, the observations are validated by using their spatial location and date of observation (Sullivan 2014). Other criteria are proposed such as consistency of observations with their known spatial distribution (Boakes et al. 2010), specifications compliance or integrity constraints (Brando 2013). Goodchild and Li (2012) proposed three categories of indicators based on crowdsourcing revision, contributors' assessment of and spatial context. Although the mentioned workflows can be too conceptual or they are only suitable for some specific types of data and projects, we believe that they are a good starting point and that they can be adapted or combined to define a data validation workflow to fit some specific purposes, as in our work.

3 Report Platform Description

Starting to 2008, IGN France has developed different applications to make reports to highlight errors and changes in authoritative databases and to offer to public bodies the opportunity to collect data suited to their needs. Nowadays, IGN France propose a unique citizen and community sourcing portal *Espace collaboratif* (https://espacecollaboratif.ign.fr/).

A data collection protocol is defined to describe how to localize a report, to fill in the list of attributes describing the report, and how to submit the report. Each report should be characterized by: a point location (latitude, longitude) and a set of attributes; the name of the town to which the reports belongs to; the theme to which the report belongs to; and a comment. Note that the comment is an open field leaving free expression to contributors to describe the reports. However, the name of the town and type of theme are normalized by being selected from a list. The theme can be from an authoritative database or a specific theme from professional databases (e.g., fire pumps for fire services). Changes in data collection protocol have occurred over time, mainly concerning the definition of themes corresponding to the reports and the number of themes: from multiple choices, to two choices and finally to one choice in the current version.

Registration is mandatory for contribution, a valid email and password being required. The contributor has the possibility to be independent or to join one or many groups. In addition to this information, reports are also described by metadata with respect to the process of data collection such as contributor's identifier, date, group membership, the type of support, version of the report protocol (e.g., to relate with the evolution in defined themes), number of attached sketches and files.

Concerning data validation, reports are first checked and validated by the surveyors with respect to data specifications, quality expectations by visual checking, comparison with different data sources (e.g., construction permits given by municipalities) which make this process time consuming.

4 Reports Reliability Assessment Methodology

4.1 Data Quality Assessment Workflow

As described in Sect. 3, reports are used as triggers to update authoritative databases. Our goal is to propose quality indicators to assess the reliability of reports. We propose to measure the reliability of reports by evaluating the consistency of different characteristics of the reports among themselves or compared to the spatial context (Goodchild and Li 2012).

The reports reliability assessment workflow is illustrated in Fig. 1. For each report, qualitative reliability indicators are computed using spatial data (e.g., administrative boundaries, topographic databases), ontologies representing topographic elements or report attributes. For each indicator, three values are possible: yes (i.e., is consistent), maybe (i.e., possibly consistent) and no (i.e., not consistent).

A qualified report (QReport) can be even described by each indicator separately or by a combination of them. For each report, the combination consists in computing the number of qualitative values obtained for each criterion. Thus, depending on the number of Yes, Maybe and No responses, a reliability assessment is set. Three final levels are proposed: consistent, inconsistent and moderately consistent. The first two values automatically validate the report (as approved or as declined) whereas the last one will generate a message to manually check. A combination of

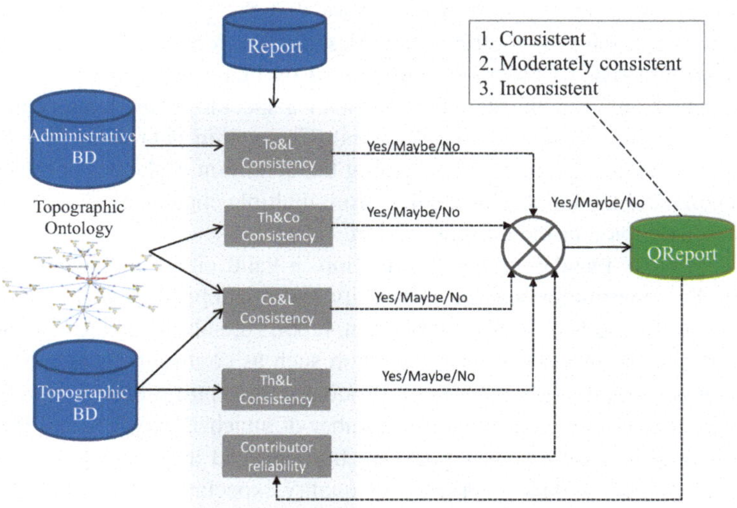

Fig. 1 Reports data reliability workflow (To = Township, L = Location, Th = Theme and Co = Comment)

rules must be adapted to data and tuned by the final user to be more or less permissive depending if the final user prefers over- or under-estimation validation.

The proposed workflow is iterative. First, the reports are qualified by using the defined criteria, second the qualified reports are used to define the contributor reliability indicator. Finally, all indicators can be combined again as explained previously. In this paper, only the first iteration is implemented.

4.2 Data Quality Indicators

Five qualitative indicators are proposed based on the report's location, attributes and their spatial context defined by different topographic elements.

Township and location consistency: This indicator deals with the assessment of the consistency between the computed name of the town (*CName*), in which the report is spatially included and the name is a town from a township attribute (*AName*). Thus, the comparison is based on a string similarity between *CName* and *AName,* when possible. Indeed, it is possible that the report location is not included in any township boundaries, especially in coastal areas. In this case, no value exists for *CName*. The indicator values are:

- If *CName* is not null: 1/**Yes**, is similarity higher than a threshold; 2/**No**, if similarity is below a threshold
- If *CName* is null, the shortest distance between the report location and the *AName* boundaries is computed. In this case, the result is 3/**Maybe**, if the distance is less than a threshold T; 4/**No**, if the distance is more than a threshold T.

Theme and location consistency. This indicator verifies whether the report location is consistent with its theme attribute by taking into account the spatial context of the report. Spatial context is described by an authoritative database and defined by both topological operators and metrics. First, intersections between reports and topographic elements are investigated. To take into account the imprecision of reports, a buffer is created for each report. Second, the similarity is calculated between the theme attribute (*ATheme*) and the themes of features intersecting the reports $\{ITheme\}_i$, where i represents the number of features intersecting the report. The indicator values are:

- If intersection is not null: 1/**Yes**, if all similarities between *ATheme* and *ITheme* (i) is below a threshold; 2/**No**, if similarity between *ATheme* and one of the *ITheme* (i) is higher than a threshold
- If intersection is null: **Maybe**.

Comment and location consistency. It determines whether the report location is consistent with its comment attribute by using the spatial context of the report. The comment attribute is split to extract a list of meaningful words $\{AComment\}_j$, $j = 1 \ldots$number of words. Due to the fact that the comment attribute is an open field, the variety of terms can be very wide. Thus, a dictionary containing related topographic elements extracted from an existing ontology is associated to each *ITheme*. For example, for ITheme = buildings, the dictionary may contains {house, town hall, etc.}. The computation of this indicator follows the approach of the *Theme and location consistency* indicator except that similarities are calculated between *AComment* (j) and each dictionary element associated to *ITheme* (i). In addition to that, if the intersection is not null and the comment is not filled in, then the value of the indicator is **Maybe**.

Theme and comment consistency: This indicator compares the contents of the theme and comment attributes of each report. Like in the previous indicator, the comment attribute is split into a list of meaningful words $\{AComment\}_j$ and a dictionary of topographic elements coming from an ontology is associated to *ATheme*. The similarities between *ATheme* and each element of the dictionary associated to $\{AComment\}_j$ are computed. The indicator values are:

- **Yes**: if at least one similarity is higher than a threshold
- **No**: if all similarities are below a threshold
- **Maybe**: if no comment is filled in.

Contributor reliability. This indicator is calculated from the previous four indicators (e.g., the more the reports of a contributor are reliable, the more the contributor is reliable), the characteristics of each contributor described in Sect. 5.2 such as inclusion in a group, number of accepted reports and length of time since registration, and the local knowledge and experience of contributors (van Exel et al. 2010). The indicator is going to be implemented in nearly further work by investigated also other approaches (Foody and Boyd 2012; Comber et al. 2013).

5 Reports Description and Results

5.1 Reports Data Description

This section deals with geographic information of reports mainly implying the following attributes: declared township, theme and comments.

Spatial distribution. The 59.680 reports collected between 2009 and 2015 are declared in 3.766 townships, representing about 10% of the French metropolitan and overseas territories. Reports are included in 94% of the departments which are larger administrative areas. Spatial distribution of the reports is uneven between the townships as well as between the departments. For example, 20% of the departments have more than 100 reports and only 5% have more than 1000 reports. Moreover, 75% of reports belong to one department (i.e., the Pas-de-Calais department in the North of France) and 99% of reports concern only 33% administrative departments. This spatial distribution aspect is directly related to the groups of contributors and to the interest in the collaborative system.

Theme analysis. The theme attribute representing the type of reports contains 32 distinct values of labels. From the total number of values, 76% of reports are dedicated only to specific needs of the contributors (e.g., "DECI" for firefighting or fire pumps). The left labels concern themes from authoritative databases or references to the databases themselves. Figure 2 illustrates occurrences of themes except those dedicated to specific professional needs. The most frequent themes are: Addresses or Places, Roads, Buildings, and POI. It is important to note that some reports have no specific associated themes (e.g., other or null).

Comments analysis. Only 26% of the reports have a comment. Comments can be studied from a word cloud (see Fig. 3). Values are quite diverse, as expected. Words with high occurrence concern administrative and features ('town'; 'street'; 'lane'; 'town hall'), spatial relation ('direction', 'in front of'), toponyms ('name'),

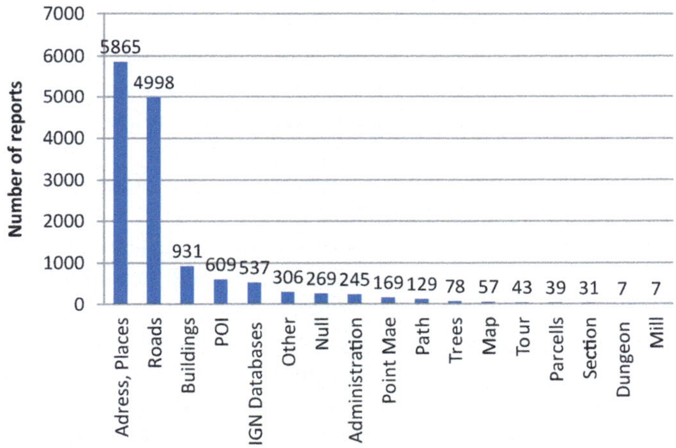

Fig. 2 Type of themes characterizing reports

Fig. 3 Word cloud from comments

and data collection ('source'; 'new'; 'modification'). The joint analysis of themes and comments has highlighted the fact that the two fields should be respectively taken into account in the context of topographical databases update.

Reports with themes that do not correspond directly to a targeted database can yet be associated with a useful comment about that database.

5.2 Contributors Description

Contributors carry out reports during professional activities or on their spare time. Depending on their background, they are more likely to report topographic elements related to their concerns (e.g., contributors belonging to fire departments tends to report elements useful for firefighting purpose). Two fields are exploited for taking into account background of contributors. First, the unique anonymized identifier is useful for considering reports within a human aspect at individual level. Second, contributors can claim to belong to a group that represents interesting indications about their background. Those are mostly professional groups from public organizations (state and local departments, firefighters). The other groups correspond to targeted actions: initiative between several organizations and campaigns toward citizens for a short period of time.

In *Espace collaboratif* platform, 97% of reports are generated by contributors belonging to public bodies. Reports are considered both according to groups of contributors and to contributors as individuals. Among 520 contributors, 46% of contributors are part of a group. Among the 43 existing groups of contributors, 40 of them belong to public bodies, and 3 of them concern a test group, a time-limited gaming group and citizens group. The number of contributors per group is quite heterogeneous varying from 1 to 277 contributors per groups (mean = 13, standard deviation = 45).

Reports can be both counted by groups or by distinct contributors. Concerning the contributors and their engagement, the distribution presented in Fig. 4 shows that 20% of the contributors have made ~90% of the reports with more than 1000 reports by contributor. This result shows that, even it can be comparable, the contributor's participation does not fit the known behavior encountered in crowdsourcing projects such as Wikipedia, the famous '90-9-1' low (i.e., 90% of contributors are not actives, just testing, 9% have a temporary interest and 1% are the active contributors). In OSM, these figures are about '81-14-5' (Neis and Zipf 2012). In our case, it seems to be '65-15-20'.

Regarding the groups activity it was noted that one group is very active and has collected 77% of the reports. Inside the group, more or less the same low is following with less inactive percentage of contributors. Relation between number of the contributors and the number of reports by group was investigated. For all groups, there is a positive correlation between both variables ($r^2 = 0.13$, p-value < 0.05). Though, two groups overwhelm the dataset of reports and if left aside, there is no longer correlation. Duration of contributions by contributor

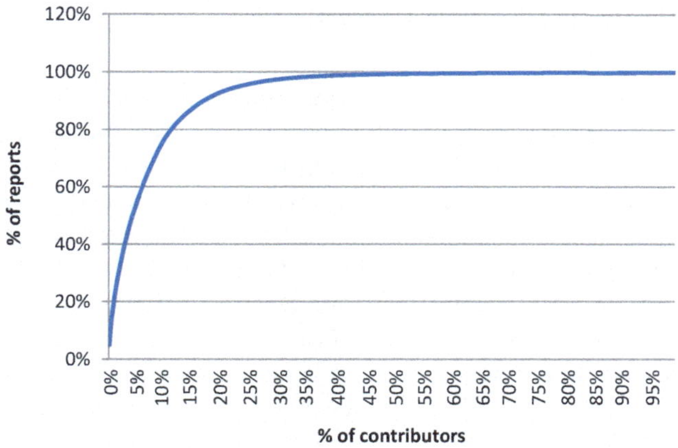

Fig. 4 Contributor's behavior with respect with their activity

corresponds to the time interval between the first day of contribution and the last day. Durations are between 1 and 2500 days. There is no strong relation between the number of contributions and the duration but there is a low significant positive correlation (Pearson value = 0.37, 99.9% confidence interval).

At group level, five groups contribute only for 1 day, 23 groups for more than 1 year. Most of the groups (6) have started to contribute since the beginning of 2013. Among all groups, 31 of them are still contributing during the last year of the study. In department 62, the contributors are mostly from the firefighting local structure (SDIS62), so all the results will be interpreted in that context. This group contains 125 contributors (23%) and the reports have started in early 2013.

5.3 Topographic Database

The database BDTopo from IGN is used both as basemap in the collaborative platform and as reference to define the spatial context of reports. BDTopo contains data about infrastructures-building, roads, railways- and various topographic elements-forest, hydrography, and relief- in France with metric accuracy (<1 m). The version of BDTopo is from 2015, which means that no reports were taken into account after that date. As there is continuous update of the database, it means that we followed the proposed workflow on reports that have been already taken into account. It implies some adaptations of the workflow, in the case of spatial intersections between reports and topographic features that have logically been updated.

Another database exploited to assess reports is GEOFLA that describes the administrative boundaries.

5.4 Reliability Assessment Results

Results are presented by indicators of whole reports or a selection of them (i.e., reports located in the Pas-de-Calais department (62) in the North of France. For indicators based on theme and comment attributes, a focus is made on buildings.

Township and location consistency (To&L). Results on whole reports give a majority of consistency, though some mismatches are detected (see Fig. 5). Distances between reports and the boundaries of the indicated towns (*AName*) are less than 16 km. In 29 cases, inconsistency are detected because they correspond to landscape elements located outside any French municipal administrative boundaries, for instance: new bridge above an estuary, impassable path in an island outside the coast not in the GEOFLA database, beaches, water line in a foreign enclave. In department 62, there are 4 reports outside the coast in the English Channel corresponding to interesting points for launching. Partial conclusion is that indicator is useful to detect new features or changes in a given town and that, when in the field, the name of the town concerning the report might be difficult to give (reports near boundaries of a town).

Theme and location consistency (Th&L). Buffers (3 m) are assigned to the location of each report. For this reports dataset, the consistency value had to be adapted since the database was likely up to date. *ATheme* and at least a theme of intersected $\{ITheme\}_i$ (i = 1 and *ITheme* = 'Building') should be similar. Spatial intersections with buildings include 10% of the reports. Buildings can be residential, administrative and industrial or entities like bridges and locks. The theme from attribute *ATheme* is analyzed for that selection of reports. Thus, 98.5% of the values are dedicated to firefighting groups rather than the types of features. 1.5% corresponds to themes in those DB including 0.5% of a theme directly meant for buildings. Thus, this corresponds to a small proportion of reports corresponding to building that can be directly qualified as consistent. Figure 6 illustrates consistent

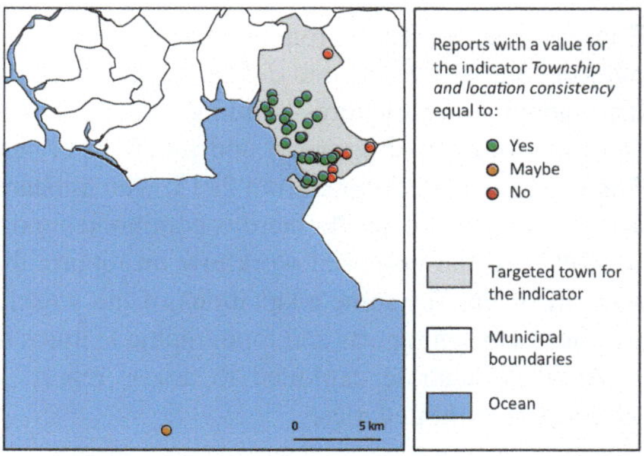

Fig. 5 To&L consistency indicator applied to a town (in grey). There are 41 consistent reports from this indicator, 6 inconsistent and 1 maybe. Inconsistent reports are mainly close to the boundaries of the town filled in the attribute *AName*

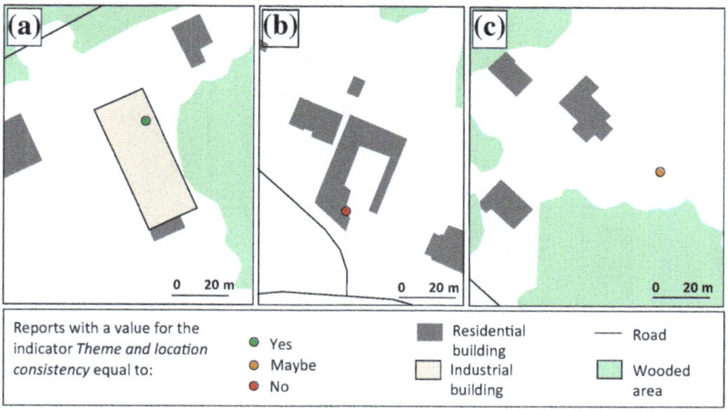

Fig. 6 Th&L consistency for buildings. (**a**) **Yes** as the report intersects a building and its *ATheme* is 'Building'; (**b**) **No** as the report intersects a building but its *ATheme* is a theme for firefighting; (**c**) **Maybe** as report does not intersect any building even if it *ATheme* indicates a building

and inconsistent reports by using Th&L indicator. There is a possibility that reports in buildings are not exactly inside the building but correspond to an element attached to that building (e.g., water tap). Besides, reports for buildings might be associated to non-explicitly related themes. In that case, the comments might be crucial to fill in the qualifying assessment workflow.

Comment and location consistency (CO&L) exploits an external ontology to analyze comment attribute $\{AComment\}_j$. Reports are selected according to their positive or non-intersection with a building. A dictionary of words related to the *ITheme*(i) 'Buildings' was determined {house, church, town hall, farm, center, hospital, etc.}. The spatially intersected reports are investigated from the similarity between their $\{AComment\}_j$ and the element of the dictionary. In department 62, 50% of the comments contain meaningful words, setting half of the reports with an indicator value at **Maybe**. Most comments concerning reports intersecting buildings contains indication about firefighting features or a change in an attribute of the building. Very few are in fact new buildings.

Theme and comment consistency (Th&Co) uses the pretreatments from previous indicators and the same external sources of assessment. We looked specifically of the consistency between the *ATheme* value 'Building' and the correspond $\{AComment\}_j$. The indicator value is equal to **Yes** for 78% of the reports, to **No** for 9% and to **Maybe** for 13%.

Table 1 illustrates the pattern of results obtained after the combination of the indicators. If all values of indicators are either **Yes** or **No,** then the decision is respectively consistent or inconsistent. If both values **Yes** and **No** are in the combination, then the decision is moderately consistent. This last balanced decision has to be investigated and refined in future work.

Table 1 Pattern of results obtained after the combination of the indicators

To&L	Th&L	CO&L	Th&Co	Decision
Yes	Yes	Yes	Yes	Consistent
Yes	Yes	No	No	Moderately consistent
Yes	No	Yes	No	Moderately consistent
No	Yes	Yes	Yes	Moderately consistent
No	Yes	Yes	Yes	Moderately consistent
No	Yes	No	Yes	Moderately consistent
No	Yes	No	Yes	Moderately consistent
No	No	Yes	Yes	Moderately consistent
No	No	No	No	Inconsistent

6 Conclusions

In this paper, we have proposed a global workflow based on quality indicators measuring the consistency of reports within their own characteristics or with their spatial context to validate reports coming from a collaborative platform. Both data and contributors have been analyzed for three global purposes: to know the contributors better, to characterize the VGI data for a more adapted data quality validation and to improve the data collection protocol.

The methodology is tested on reports coming from *Espace Collaboratif* platform developed by IGN France. Five indicators are defined with respect to the datasets characteristics which fit to data collection protocol and databases specifications. According to this, the authors believe that the methodology can be applied on data coming from other community or crowdsourced systems. Indeed, the proposed criteria can be adapted to better fit to other types of VGI data on the one hand. Other criteria can also be defined if additional information is available given the data collection protocol.

From the first results, the authors believe that the indicators based on spatial context can be very useful and discriminatory. The initial work in the future is to improve the definition of the spatial context by using not only the proximity in terms of buffers but also the distance between each observation and its spatial context defined by different types of topographic features (e.g., building, roads, and rivers).

As it was mentioned in Sect. 4.2, an effort will be made to define the contributor reliability based on data history and other metadata automatically generated by the platform (e.g., type of support used to contribute).

Finally, the comments provided by contributors are very rich and useful. In this paper, only words are used to study the consistency of comments, but we have noticed that relative location (e.g., 'on the left', 'the report concerns the building near to the bridge'), toponyms and other valuable information for validation are given. Thus, in the future, some language processing methods will be developed to better exploit the meaning of this text.

In the perspective of a citizen observatory for land use land cover monitoring (Landsense H2020 project) a series of campaigns will run to collect observations from citizens. In this context, an important issue concerns data collection protocol for both contributors' motivation and data quality assessment. From the study presented in this paper some elements considered as good practices and some others that should be improved in the next protocol have been identified. Registration to the platform, engaging with citizens, public bodies and giving the possibility to a contributor to be a part of a group are considered such as good practices. The attribute that defines the type of an observation should be mandatory and the value should be chosen from a predefined list. The balance between an exhaustive list that can be heavy and demotivating and a shorter list that can be considered not sufficient should be provided. Even if completely transparent for contributors, a link between this list and the existing themes in authoritative data should be defined in order to be compliant and interpretable with authoritative data.

In our case, the field "comment" has no constraints except that it should not be empty. Though, it is always possible to have a free text field where contributors can describe his/her contribution. Nevertheless, we believe that it should not be mandatory. We have noticed that quite a lot of comments are not always valuable (e.g., contributors just put numbers and letters to respect the constraints) and brings no added value to quality assessment.

Acknowledgements We are grateful to Maryame Rhezali for her valuable inputs for this work. Our work is funded under grant #689812 (H2020, LandSense project) by European Commision.

References

Antoniou, V., & Skopeliti, A. (2015). Measures and indicators of VGI quality: An overview. *ISPRS annals of the photogrammetry, remote sensing and spatial information sciences, ISPRS geospatial week 2015* (pp. 345–351). France: La Grande Motte.

Ballatore, A., & Zipf, A. (2015). A conceptual quality framework for Volunteered Geographic Information. In S. I. Fabrikant, M. Raubal, M. Bertolotto, C. Davies, S. Freundschuh, S. Bell (Eds.), *Spatial information theory* (pp. 89–107). Cham: Springer International Publishing.

Boakes, E. H., McGowan, P. J. K., Fuller, R. A., Chang-qing, D., Clark, N. E., O'Connor, K., et al. (2010). Distorted views of biodiversity: spatial and temporal bias in species occurrence data. *PLoS Biology, 8,* e1000385.

Bonter, D. N., & Cooper, C. B. (2012). Data validation in citizen science: a case study from project FeederWatch. *Frontiers in Ecology and the Environment, 10,* 305–307.

Brando, C., & Bucher, B. (2010). Quality in user generated spatial content: A matter of specifications. In *13th AGILE International Conference on GIS.*

Brando, C. (2013). Un modèle d'opérations réconciliables pour l'acquisition distribuée de données géographiques. Thèse de doctorat de l'université Paris Est.

Comber, A., See, L., Fritz, S., Van der Velde, M., Perger, C., & Foody, G. (2013). Using control data to determine the reliability of volunteered geographic information about land cover. *International Journal of Applied Earth Observation and Geoinformation, 23,* 37–48.

Fonte, C., Bastin, C., Foody, G., Kellenberger, T., Kerle, N., Mooney, P., et al. (2015). VGI quality control. *ISPRS annals of photogrammetry, remote sensing and spatial information sciences II-3/W5*, 317–324.

Foody, G., & Boyd, D. S. (2012). Using volunteered data in land cover map validation: Mapping tropical forests across West Africa. In *Geoscience and remote sensing symposium (IGARSS)*, pp. 6207–6208.

Girres, J.-F., & Touya, G. (2010). Quality assessment of the French OpenStreetMap dataset. *Transactions in GIS, 14,* 435–459.

Goodchild, M. F. (2007). Citizens as sensors: the world of volunteered geography. *GeoJournal, 69,* 211–221. doi:10.1007/s10708-007-9111-y.

Goodchild, M. F., & Li, L. (2012). Assuring the quality of volunteered geographic information. *Spatial Statistics, 1,* 110–120. doi:10.1016/j.spasta.2012.03.002.

Haklay, M. (2010). How good is volunteered geographical information? A comparative study of OpenStreetMap and Ordnance Survey datasets. *Environment and Planning B: Planning and Design, 37,* 682–703. doi:10.1068/b35097.

Haklay, M. E, Antoniou, V, Basiouka, S, Soden, R, & Mooney, P. (2014). Crowdsourced geographic information use in government. In *Global facility for disaster reduction & recovery (GFDRR).* London, UK: World Bank.

Johnson, P., & Sieber, R. (2013). Situating the adoption of VGI by government. In D. Sui, S. Elwood, & M. Goodchild (Eds.), *Crowdsourcing Geographic Knowledge* (pp. 65–81). Netherlands: Springer.

Leibovici, D. G., Evans, B., Hodges, C., Wiemann, S., Meek, S., Rosser, J., et al. (2015). On data quality assurance and conflation entanglement in crowdsourcing for environmental studies. In *ISPRS annals of photogrammetry, remote sensing and spatial information sciences II-3/W5.*

Mooney, P., Minghini, M., Laakso, M., Antoniou, V., Olteanu-Raimond, A.-M., & Skopeliti, A. (2016). Towards a Protocol for the Collection of VGI Vector Data. *ISPRS Int. J. Geo-Inf., 5* (11), 217. doi:10.3390/ijgi5110217.

Neis, P., & Zipf, A. (2012). Analyzing the contributor activity of a volunteered geographic information project—the case of OpenStreetMap. *ISPRS International Journal of Geo-Information, 1*(2), 146–165.

Olteanu-Raimond, A.-M., Hart, G., Touya, G., Kellenberger, T., Foody, G., & Demetriou, D. (2016). The scale of VGI in map production: A perspective of European National Mapping Agencies. *Transactions in GIS*. doi:10.1111/tgis.12189.

Peyman, H., Rahim, A. A. (2015). Assessment of logical consistency in OpenStreetMap based on the spatial similarity concept. In J. Jokar Arsanjani, A. Zipf, P. Mooney, & M. Helbich (Eds.), *OpenStreetMap in GIScience* (pp 19–36). *Lecture notes in geoinformation and cartography.* Cham: Springer International Publishing. doi:10.1007/978-3-319-14280-7_2.

Senaratne, H., Mobasheri, A., Loai Ali, A., Capineri, C., & Haklay, M. (2015). A review of volunteered geographic information quality assessment methods. *IJGIS*, 1–29.

Sullivan, B. L. (2014). The eBird enterprise: An integrated approach to development and application of citizen science. *Biological Conservation, 169,* 31–40.

Touya, G., & Brando C. (2013). Detecting level-of-detail inconsistencies in volunteered geographic information data sets. *Cartographica: The International Journal for Geographic Information and Geovisualization*, *48*(2), 134–143. doi:10.3138/carto.48.2.1836.

van Exel, M., Dias, E., & Fruijtier, S. (2010). The impact of crowdsourcing on spatial data quality indicators. In *Proceedings of GIScience 2010*, Zurich, Switzerland.

Crowdsourcing Mapping and Participatory Planning Support System: Case Study of Brno, Czechia

Jiří Pánek and Vít Pászto

Abstract Many cities have started to engage their citizens in collaborative planning and in re-creating the urban spaces where they live. The case study presented in this paper crowdsourced subjective/emotional responses concerning the environment of the city of Brno in the Czech Republic, from 2,087 respondents, who marked 24,065 points on six different topics and added over 3,000 further comments. The paper presents not only the case study, but also the authors' own web-application called Emotional Maps—built for crowdsourcing spatial data, mainly about city development and subjective perceptions of urban spaces. The collected data showed three main hot-spots in the city, with a significant number of marked points reported by citizens. The paper comments on the reasons these spots are important for the citizens of Brno and their suggestions for future activities there. Furthermore, the authors argue that ad hoc flexible web-mapping platforms such as the one presented in the paper can be an asset to city administrations as well as urban planners.

Keywords Emotional mapping · Urban planning · Perceptions · Neocartography

1 Introduction

Urban planning is often elitist and non-participatory (Brooks 1988; Winter and Brooke 1993; Corburn 2003; Galdós 2010). But, the neglect of citizens as participants in the planning process is slowly changing as local political representatives

J. Pánek (✉)
Department of Development Studies, Palacký University Olomouc, Olomouc, Czech Republic
e-mail: jiri.panek@upol.cz

V. Pászto
Department of Geoinformatics, Palacký University Olomouc, Olomouc, Czech Republic
e-mail: vit.paszto@upol.cz

M.P. Peterson (ed.), *Advances in Cartography and GIScience*, Lecture Notes in Geoinformation and Cartography, DOI 10.1007/978-3-319-57336-6_5

start to understand that citizens have a relevant role in the processes of town planning and administration (Čermák and Vobecká 2011). Furthermore, current trends in urban planning support innovative instruments such as participatory urban planning (Kahila and Kyttä 2009; Kahila-Tani et al. 2015), emotional maps (Pánek et al. 2017) or open data (Jäppinen et al. 2013).

The idea of citizen participation in e-planning via geospatial technologies is grounded in the term GeoParticipation, described as a collection of methods through which communities can be involved in the creation of maps that help them be advocates for their environment and living space (Pánek et al. 2014). GeoParticipation is often linked with the process of community mapping as *"local mapping, produced collaboratively by local people and often incorporating alternative local knowledge"* (Perkins 2007, p. 127). Besides GeoParticipation, the concepts of Public Participation GIS (Obermeyer 1998) as well as Volunteered Geographical Information (Goodchild 2007) and Crowdsourcing (Howe 2006) play an important role. For detailed reviews about the history of Public Participation GIS please see (Schuurman 2000; Sieber 2006; Dunn 2007; Pánek 2016).

There is legislative support across Europe for public participation in planning and local decision-making processes. The first such ideas were published within the Aarhus Convention (1998) and subsequently, the Leipzig Charter on Sustainable European Cities (European Commision 2007). In the Czech Republic, participation is granted by the Constitution of the Czech Republic as well as the Act of Parliament 128/2000—Act concerning Municipalities (Czech Republic 2000).

Although research has questioned the function of participatory mapping (Reyes-García et al. 2012), public participation in urban planning can smooth the process and reduce the number of possible conflicts (Mapedza et al. 2003; Sandström et al. 2003; Cronkleton et al. 2010). It can also strengthen the understanding and acceptance of different values for the various stakeholders (Healey 1997).

The aim of this paper is to present the methods and the data gathered during the crowdsourcing participatory mapping event in Brno. The main research question is: Are there any significant perceived hot-spots in the city? Are there areas that draw controversy among the citizens of Brno?

2 Methods, Case Study Location and Data

The authors worked with the method and tool called *Emotional Maps*. The method is a version of participatory mapping that involves Gould-style (Gould and White 1974) sketch-able mental maps (Jankowski et al. 2015) used mainly in a web interface, combined with the answers to the questions that are placed directly onto the map. Opposite to comparative (Lynch 1960) preferential (Gouldian) mental maps are used to evaluate the quality of selected locations. The idea of using mental maps in geographical research is focused on recording a cognitive representation on paper, although GIS is often used to analyse the data (Waterman and Gordon 1984; Janoška et al. 2014).

There is an ongoing discussion as to whether emotional maps capture map emotions, perceptions, ideas and experiences. Nevertheless, the authors support the term emotional mapping, partly as a legacy of the book *Emotional Cartography* by Christian Nold (2009), and partly as their own research preference. Some authors prefer to use terms such as sentiment mapping (Caragea et al. 2014), hedonic mapping (Ennis and Ennis 2013), ephemeral mapping (Art and Cartography Commission of ICA 2015), perceptual mapping (Doran and Burgess 2011) and many others. The theoretical discussions and uncertainty about the terminology show that emotional mapping research and practice is not clearly defined or anchored in its respective disciplines. Nevertheless, Perkins (2009, p. 130) states that "*emotional maps chart human feelings onto a cartographical landscape … and allow users to devise and customise their own emotional landscape, to choose what kinds of thoughts and experiences, feelings and passions to map*".

The *Emotional Maps tool* (see Fig. 1; available at http://www.pocitovemapy.cz/index-en.html) is a web application built as a computer-assisted web interview (CAWI) with a map-based background. Similar to other web-based tools for crowdsourced mapping (Ushahidi, ArcGIS Online, or Maptionnaire), *Emotional Maps* allows users to collect spatial data on a multi-scale pannable map background, but it does not require the registration or installation of any specific software, plug-in or virtual server. The application is created as a single-page web application using two open-source JavaScript libraries. The crowdsourced data are saved in a MySQL database, available on most webhosting sites. Therefore, there is no need to have specialised hosting or one's own server with the geodatabase installed. Each entry in the database contains a unique user-ID (randomly generated), question identifier, number of points/lines/polygons marked and geometry.

Fig. 1 The web application *Emotional Maps* used for collection of data. (*Source* http://www.pocitovemapy.cz/brno-2016-original/nahled)

These entries are later merged by a *GeoJSON PHP library script* (Mikola 2015) that allows GeoJSON to maintain data from multiple users.

The case study presented in this paper was implemented in May and June 2016 in Brno, the second largest city in the Czech Republic. The city has about 380,000 inhabitants (Czech Statistical Office 2015) and its greater metropolitan area is home to more than 600,000 people (SPF Group and Hope Group 2015). In total, 2,087 respondents took part in the survey. First, the authors met with municipal representatives to negotiate the themes that were to be researched through the emotional mapping activity, both online (through *Emotional Maps)* as well as offline (paper-based map survey). Hence the data were collected in two ways (1) via online survey (n = 1,093) and (2) during four days of the "*Week of the City*" event in one of the main squares. A total of 994 respondents (n = 994) created a series of paper-based emotional maps.

The link to the map was distributed via the city webpage and city profile on social networks, through mailing lists of local action groups, local newspapers, etc. This effort represented resulted in a snowball sample.

In May 2016, the city of Brno, via its department of Strategic Development, took part in the event called "Week of the City" where, over four days, citizens had the opportunity to create an emotional map of the city (the topics were the same as in the online (geo)questionnaire), with the results being merged afterwards. The questions asked via the online form, as well as via the emotional map, were intended to find the general public's opinions of the city. The respondents were asked six spatial questions as specified below and additional questions intended to identify the demographics of the respondents. With each spatial question, it was also possible to add comments where necessary in order to further explain an answer. Each answer was represented by a coloured pin (either digital or analogue) on the map. Up to now, the predominant method for spatially explicit preference mapping was marking points for locations or sketching polygons annotated with expressions (Jankowski et al. 2015). Brown and Pullar (2012) suggested that points instead of polygons be used in future PPGIS applications and it is also possible to use a fuzzy multi-point (Huck et al. 2014) feature to collect the information.

The spatial questions asked:

- Which places in Brno do you like? (4,327 points)
- Which places in Brno do you dislike? (4,079 points)
- Where do you spend your free time in Brno? (4,062 points)
- Where in Brno would you like to live? (3,829 points)
- Which places in Brno are you proud of? (3,868 points)
- What would you change in Brno? (3,900 points).

Although the idea was that both questionnaires would be identical and fully comparable, there were minor differences between the digital and analogue versions. When completing the analogue map on the streets of Brno, respondents were only able to mark one place per question, as opposed to the digital version, where the number of points was limited to three per question. This limit was set by the

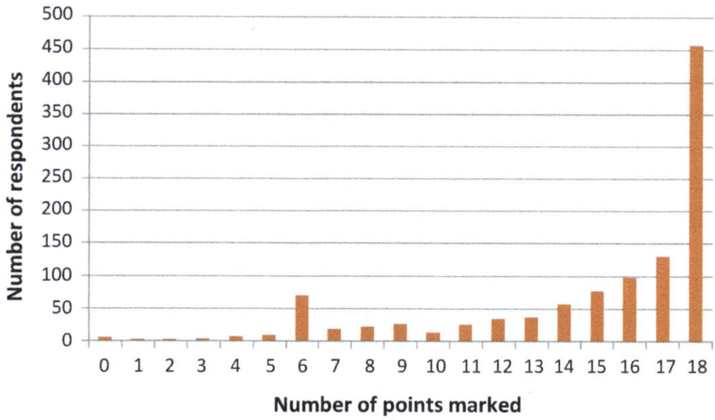

Fig. 2 Number of points used by respondents in the online mapping platform

Brno representatives and as can be seen in Fig. 2, the majority of respondents marked three points per question (i.e., 18 points in total). The digital version allowed the direct linking of respondents' profiles with the points they marked, but with the analogue version the points from demographic profiles had to be separated as the profiles were conducted separately.

The main disadvantage of the emotional mapping action on the streets of Brno was the need to manually digitise all the data via the online web application used for the digital survey. Nevertheless, the authors decided to use the street collection of data to ensure a broader range of answers as well as to gather information from a variety of target groups. For example, the street data collection attracted a higher percentage from the *older* generation, while the online form was mainly preferred by respondents under 30 years of age.

3 Results

The results from the case study can be divided into three sections: (1) demographic analysis of the respondents; (2) spatial distribution of marked answers; and (3) analysis of comments attached to marked points.

3.1 Demographics of the Respondents

Due to the sampling and the mixed methods of data collection, the results cannot be considered as a fully representative sample. Nevertheless, the sample is sizable enough to draw some conclusions. The variety of respondents also allowed the extraction of feedback that will be used in the forthcoming *Strategy of Brno* document, as prepared by the Brno administration. The demographic structure of respondents from the analogue map is specific, with slightly more answers from females (59.4%) and a higher participation of respondents over 46 years of age. The youngest respondent was 7 years old and the oldest was 90. Almost half of the participants had a university degree. Most of the respondents (86%) lived in Brno and 2/3 had lived in Brno for more than 10 years. Of those respondents who answered that they live in Brno, the majority (74.6%) also plan to live in Brno in the future.

The online survey had slightly different structure of respondents as it had more diverse forms of promotion—the city website, social networks, etc. With the online mapping aspect, there were slightly more answers from males (55.7%) and, as was expected, the majority of answers came from respondents under 30 years old (54.3%). There was a much lower engagement of respondents over 45 years of age. Respondents who used the online version of the survey were most likely to be university educated (71%) and younger than 46 years old (89%). There were no other significant differences between the respondents between the paper and the online version of the questionnaire.

3.2 Spatial Distribution of Marked Points

In total, 2087 respondents marked 24,065 points on six different spatial questions. The respondents mainly marked places in the city centre, but there were also locations marked on the outskirts of the city, such as the Brno reservoir and the dam on the Svratka river just outside the city—a favourite spot for leisure activities. In the hotspot analysis, several areas were identified. Due to a lack of space, only three (see Fig. 3 for localisation) that contained the most points will be analysed in more detail: (A) the area around the main train station; (B) Špilberk castle; and (C) the excluded locality between Bratislavská and Cejl streets.

The area of the main train station in Brno (A) has been the centre of discussion for several years. Due to the frozen situation on the Brno municipal council regarding either the construction of a new train station or the renovation of the current train station, there is a lack of investment in this area. This situation is also confirmed by the 2093 points marked relating to the train station—the highest number acquired by any location. Of those, 1107 points were marked in the question *What would you change in Brno?* 28% of all answers to this question (1st in the spatial category), while another 986 points were marked under the question *Which places in Brno do you dislike?* (24% of all answers on the topic—1st in the category).

Špilberk is the castle on the hill in Brno (B) and its construction began as early as the first half of the 13th century. It was gradually turned into a huge baroque fortress, considered to be the harshest prison in the Austro-Hungarian empire, and then into barracks. In 1960, Špilberk became the seat of the Brno City Museum. The place was marked highly in three categories—*Which places in Brno are you*

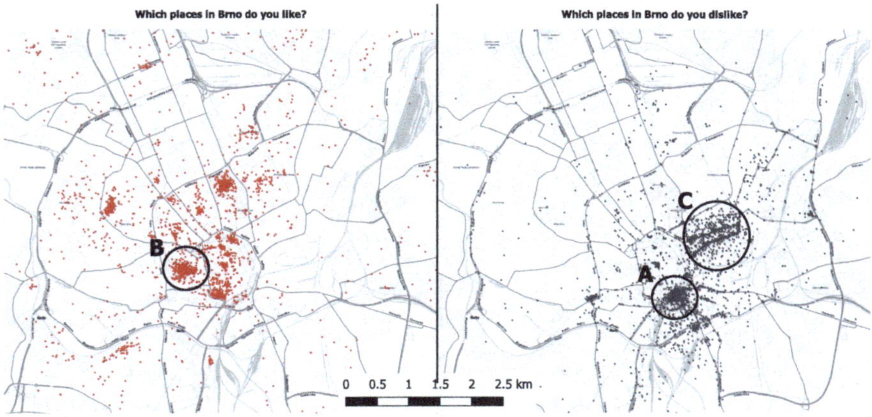

Fig. 3 Maps of answers to the questions *Which places in Brno do you like?* (*left*) and *Which places in Brno do you dislike?* (*right*). The letters (*A*, *B*, *C*) refer to locations described in the text (*Source* Background data—Stamen Toner Lite/OpenStreetMap)

proud of? (843 points marked—22% of all answers on the topic—1st in the category), *Which places in Brno do you like?* (492 points—11% of all answers on the topic—1st in the category), and *Where do you spend your free time in Brno?* (361 points—9% of all answers on the topic—3rd in the category).

The third area (C) that was heavily marked by points in same categories as the main train station is the excluded locality between Bratislavská and Cejl streets. Although close to the city centre, the area is widely considered as a place that needs change (664 points—17% of all answers on the topic—2nd in the category) and also a place that is disliked (967 points—24% of all answers on the topic—2nd in the category). With beginnings in the 16th century, the area used to be an independent village but now belongs to Brno. The locality is famous for the large community of Roma people (gypsies) as well as number of pawn shops, bars and gambling clubs (Fig. 4).

3.3 Why Were Certain Locations Marked?

Participants in the survey had the opportunity to attach a comment to each point they marked, both on the analogue map and on the digital mapping platform. In total, respondents added comments to 2,550 locations in the web map interface and about 600 further comments were recorded with the analogue map on the streets of Brno.

Some positive comments were:

- Multicultural city—*"like Brno more every day. Especially the development of cultural life and renovation of the environment (coffee places + gardens)."*

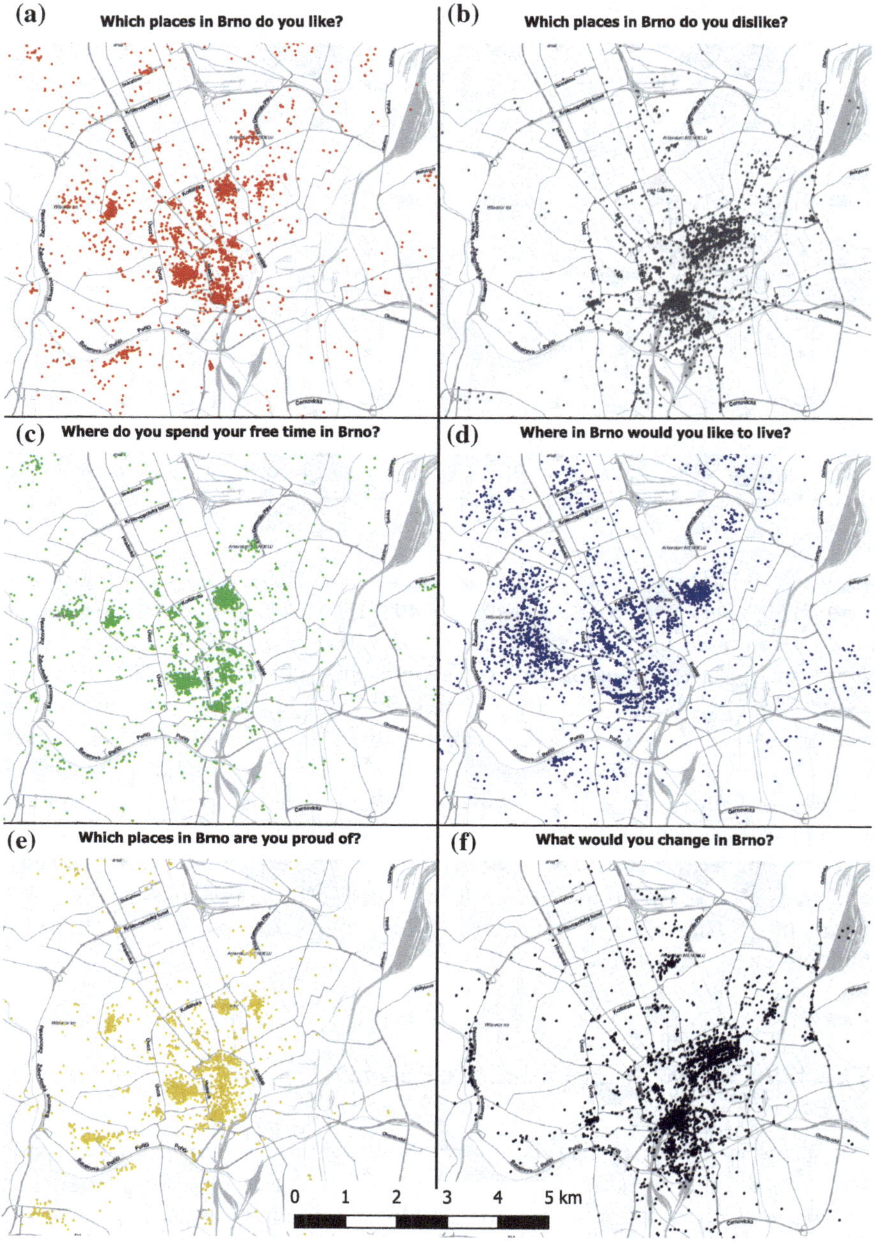

Fig. 4 Spatial distribution of answers for the centre of Brno. (**a**) Which places in Brno do you like? (**b**) Which places in Brno do you dislike? (**c**) Where do you spend your free time in Brno? (**d**) Where in Brno would you like to live? (**e**) Which places in Brno are you proud of? (**f**) What would you change in Brno?

- A city that is growing dynamically—*"I like it in Brno. The city works well. I think it is time for the citizens to move the city further, especially in the private sector in order to set up high-quality companies, shops, businesses. We have to create wealth and at the same time feel free. The city can only promote a good atmosphere, infrastructure and ideas. It is important to preserve the creative character of the town. The only specific thing needs to be addressed is the parking in the city."*
- Love and relationship to the city—*"Brno is a nice city and for the past 25 years has come a long way forward. I love Brno very much. The people are very friendly and seem to have very high standards and morals. Everyone has a high degree of respectability. When I am in Brno I feel lighter and relaxed."*

Some negative comments were:

- Insufficient cleanliness of streets and squares (garbage in the streets, chewing gum on the pavement, full trash cans, etc.)—*"Brno is a very dirty city (compared to other countries). Homeless on the streets/in a dumpster. Remove trash."*
- The poor state of the area around the main railway station—*"Brno has always been for me a matter of heart. I wanted to study here, furthermore, I found a great job, so I'm really happy here. But what I regret the most is the situation at the main railway station. The first thing people see when stepping off a train and come to Brno, the benches are full of the homeless and strange people. Is it really what we want to show tourists at the first sight? That's probably not a very good impression of an otherwise beautiful city full of great people."*
- Poor traffic conditions—*"Brno urgently needs to solve the traffic. Not only trains, but also tracks that are blocking the development of the city. Public transport around the train station needs improvement. Cycling—not just isolated pieces of cycling. Road—city circuit. Parking. The existing train station is a shame. It would be great at least to redecorate and clean it before the new one will be built." "More cycling safe zones! More lanes for cyclists on the roads. More bike racks (controlled camera system)."*

Comments specifically linked to the three most marked locations described above were:

- Main train station—*"I am ashamed of the main train station building" "Shame of the city", "Dirty, lots of homeless people, I never feel well here".*
- Špilberk—*"Špilberk is a symbol of successful defence against Swedish occupation (during Thirty Years' War)", "Impregnable fortress, nicely renovated", "Nice castle, view of the city, nicely renovated".*
- Bratislavská and Cejl street—*"Dirty, I would be afraid to go there alone", "Very busy and dirty part of the city", "Bronx in Brno, but it has a future", "Resolve Roma ghetto around Cejl and Bratislavská".*

4 Conclusions

The purpose of this research was for the residents of Brno to map a subjective image of their city. In collaboration with the city administration and in concurrence with the preparation of the new *Strategy of the City*, six map-based questions were asked and over 2,000 respondents marked more than 24,000 spots in the city and surroundings. The results of the case study will be further used by the city administration to move towards a more participatory and sustainable city planning.

Furthermore, the case study can serve as an example of crowdsourcing emotional participatory mapping used in the process of city planning and in negotiations over the future shape of urban spaces. Emotions and perceptions are a crucial part of modern cartography and a relevant source of information for city planners and decision-making stakeholders. Participatory crowdsourcing mapping applications, such as the one presented in this case study, can create a link between administrations and the public. As almost every place can evoke an emotional response, it is only logical that maps are used to collect these emotional reactions. Maps also help to communicate the perceptions and preferences in a space and with the tools of 'neocartography' and web-mapping, they could become a new platform for participatory planning.

We argue that there is enormous potential in participatory mapping and that crowdsourcing map-applications created ad hoc for specific mapping activities can bridge the technocratic gap between planners and the public. We are aware of mobile mapping apps such as EmoMap, SmellScaper and CitySourced, nevertheless, we understand that urban planning is a relatively complicated process, where citizens need time for reflection and to organise their thoughts prior to engaging in the mapping activity. Therefore, we do not see potential in mobile apps that are designed to collect immediate responses. Personal meetings (with the collaborative creation of community maps) and computer-assisted web interviews allow respondents to express the full depth of their thoughts and ideas.

There are three possible lines of future research with the dataset collected: (1) to explore mutual spatial correlation and demographic specifics within the data, (2) to compare our data with official census indicators, and (3) to analyse if place of living influences perceptions of the city.

Acknowledgements We are grateful to the city administration of Brno for the cooperation with data collection and the demographic analysis of the data. This paper is supported by the project Spationomy (2016-1-CZ01-KA203-024040) and GoGoGozo (2014-1-UK01-KA203-001642) funded by the European Union within the Erasmus + programme.

References

Aarhus Convention. (1998). The UNECE convention on access to information, Public participation in decision-making and access to justice in environmental matters. http://www.unece.org/fileadmin/DAM/env/pp/documents/cep43czech.pdf.

Art & Cartography Commission of ICA. (2015). Mapping Ephemeralities| Art & Cartography on WordPress.com. https://artcarto.wordpress.com/2014/11/10/mapping-ephemeralities/. Retrieved April 28 2015.

Brooks, M. (1988). Four critical junctures in the history of the urban planning profession: An exercise in hindsight. *Journal of the American Planning Association, 54*(2), 241–248. doi:10.1080/01944368808976481.

Brown, G., & Pullar, D. V. (2012). An evaluation of the use of points versus polygons in public participation geographic information systems using quasi-experimental design and Monte Carlo simulation. *International Journal of Geographical Information Science, 26*(2), 231–246. doi:10.1080/13658816.2011.585139.

Caragea, C., Squicciarini, A., Stehle, S., Neppalli, K., & Tapia, A. (2014, May). Mapping moods: Geo-mapped sentiment analysis during hurricane sandy. In *ISCRAM 2014 Conference Proceedings—11th International Conference on Information Systems for Crisis Response and Management*, (pp. 642–651).

Corburn, J. (2003). Bringing local knowledge into environmental decision making: Improving urban planning for communities at risk. *Journal of Planning Education and Research, 22*(4), 420–433. doi:10.1177/0739456X03022004008.

Cronkleton, P., Albornoz, M. A., Barnes, G., Evans, K., & de Jong, W. (2010). Social geomatics: Participatory forest mapping to mediate resource conflict in the bolivian amazon. *Human Ecology, 38*(1), 65–76. doi:10.1007/s10745-009-9296-4.

Czech Republic. (2000). Act concerning Municipalities.

Czech Statistical Office. (2015). Population of municipalities of the Czech republic, 1 January 2015. https://www.czso.cz/documents/10180/20556287/1300721503.pdf/33e4d70e-e75f-4596-930c-63406c9068d0?version=1.1. Retrieved September 25 2015.

Čermák, D., & Vobecká, J. (2011). *Spolupráce, partnerství a participace v místní veřejné správě : význam, praxe, příslib*. Sociologické nakladatelství (SLON).

Doran, B. J., & Burgess, M. B. (2011). *Putting fear of crime on the map: investigating perceptions of crime using geographic information systems*. Springer Science & Business Media.

Dunn, C. E. (2007). Participatory GIS a people's GIS? *Progress in Human Geography, 31*(5), 616–637. doi:10.1177/0309132507081493.

Ennis, D. M., & Ennis, J. M. (2013). Mapping hedonic data: A process perspective. *Journal of Sensory Studies, 28*(4), 324–334. doi:10.1111/joss.12048.

European Commision. (2007). Leipzig charter on sustainable european cities. Leipzig. http://ec.europa.eu/regional_policy/archive/themes/urban/leipzig_charter.pdf. Retrieved July 22 2016.

Galdós, M. I. (2010). The Czech Republic: Local government in the years after the reform. In J. Loughlin, F. Hendriks, & A. Lidström (Eds.), *The Oxford handbook of local and regional democracy in Europe* (pp. 505–527). Oxford University Press.

Goodchild, M. F. (2007). Citizens as sensors: the world of volunteered geography. *GeoJournal, 1–15.*

Gould, P., & White, R. (1974). *Mental Maps*. Harmondsworth: Pelican Books.

Healey, P. (1997). *Collaborative planning: Shaping places in fragmented societies*. UBc Press.

Howe, J. (2006). The rise of crowdsourcing. *Wired magazine, 14*(6), 1–4.

Huck, J., Whyatt, D., & Coulton, P. (2014). Spraycan: A PPGIS for capturing imprecise notions of place. *Applied Geography, 55,* 229–237. doi:10.1016/j.apgeog.2014.09.007.

Jankowski, P., Czepkiewicz, M., Młodkowski, M., & Zwoliński, Z. (2015). Geo-questionnaire: A method and tool for public preference elicitation in land use planning. *Transactions in GIS, online first.*. doi:10.1111/tgis.12191.

Janoška, Z., Voženílek, V., & Tuček, P. (2014). Interval delimitation of mean line of geographical phenomena from a set of polygons: Case study of mental maps for haná region. *Geografie–Sborník CGS, 119*(1), 91–104. http://geography.cz/sbornik/wp-content/uploads/downloads/2014/03/g14-1-s91-104_janoska.pdf. Retrieved April 28 2015.

Jäppinen, S., Toivonen, T., & Salonen, M. (2013). Modelling the potential effect of shared bicycles on public transport travel times in Greater Helsinki: An open data approach. *Applied Geography, 43,* 13–24. doi:10.1016/j.apgeog.2013.05.010.

Kahila-Tani, M., Broberg, A., Kyttä, M., & Tyger, T. (2015). Let the Citizens map—Public participation GIS as a planning support system in the Helsinki master plan process. *Planning Practice & Research,* 1–20. doi:10.1080/02697459.2015.1104203.

Kahila, M., & Kyttä, M. (2009). SoftGIS as a bridge-builder in collaborative urban planning. In S. Geertman, & J. Stillwell (Eds.), *Planning Support Systems Best Practice and New Methods* (Vol. 95, pp. 389–411). Dordrecht: Springer Netherlands. doi:10.1007/978-1-4020-8952-7.

Lynch, K. (1960). *The image of the city* (Vol. 1). MIT press.

Mapedza, E., Wright, J., & Fawcett, R. (2003). An investigation of land cover change in Mafungautsi Forest, Zimbabwe, using GIS and participatory mapping. *Applied Geography, 23* (1), 1–21. doi:10.1016/S0143-6228(02)00070-X.

Mikola, J. (2015). GeoJSON implementation for PHP. https://github.com/jmikola/geojson.

Nold, C. (Ed.). (2009). Emotional cartography: Technologies of the self.

Obermeyer, N. (1998). PPGIS: the evolution of public participation GIS. *Cartography and Geographic Information Science, 25,* 65–66. http://dusk.geo.orst.edu/ucgis/web/oregon/ppgis.pdf. Retrieved February 14 2013.

Pánek, J. (2016). From mental maps to GeoParticipation. *The Cartographic Journal, 53*(4), 1–8. doi:10.1080/00087041.2016.1243862.

Pánek, J., Hrubeš, M., Kubásek, M., Valůch, J., & Zahumenská, V. (2014). *GeoParticipace - jak používat prostorové nástroje v rozhodování o lokalitách ve kterých žijeme?*. Olomouc: Univerzita Palackého v Olomouci.

Pánek, J., Pászto, V., & Marek, L. (2017). Mapping emotions: Spatial distribution of safety perception in the city of Olomouc. In I. Ivan, A. Singleton, J. Horák, & T. Inspektor (Eds.), *Lecture notes in geoinformation and cartography: The rise of big spatial data* (pp. 211–224). Ostrava, Czech Republic: Springer International Publishing. doi:10.1007/978-3-319-45123-7.

Perkins, C. (2007). Community mapping. *Cartographic Journal, The, 44*(2), 127–137. doi:10.1179/000870407X213440.

Perkins, C. (2009). Performative and embodied mapping. *International Encyclopedia of Human Geography, Oxford: Elsevier,* 126–132.

Reyes-García, V., Orta-Martínez, M., Gueze, M., Luz, A. C., Paneque-Gálvez, J., Macía, M. J., et al. (2012). Does participatory mapping increase conflicts? A randomized evaluation in the Bolivian Amazon. *Applied Geography, 34,* 650–658. doi:10.1016/j.apgeog.2012.04.007.

Sandström, P., Granqvist Pahlén, T., Edenius, L., Tømmervik, H., Hagner, O., Hemberg, L., et al. (2003). Conflict resolution by participatory management: remote sensing and gis as tools for communicating land-use needs for reindeer herding in Northern Sweden. *AMBIO: A Journal of the Human Environment, 32*(8), 557–567. doi:10.1579/0044-7447-32.8.557.

Schuurman, N. (2000). Trouble in the heartland: GIS and its critics in the 1990s. *Progress in Human Geography, 24*(4), 569–590. doi:10.1191/030913200100189111.

Sieber, R. (2006). Public participation geographic information systems: A literature review and framework. *Annals of the Association of American Geographers, 96*(3), 491–507. http://www.tandfonline.com/doi/abs/10.1111/j.1467-8306.2006.00702.x. Retrieved September 2 2012.

SPF Group, & Hope Group. (2015). *Integrated strategy of Brno Metropolitan development.* http://www.brno.cz/fileadmin/user_upload/sprava_mesta/Strategie_pro_Brno/dokumenty/ISR_BMO_ITI.pdf. Retrieved September 30 2016.

Waterman, S., & Gordon, D. (1984). A quantitative-comparative approach to analysis of distortion in mental maps. *The Professional Geographer, 36*(3), 326–337. doi:10.1111/j.0033-0124.1984.00326.x.

Winter, I., & Brooke, T. (1993). Urban planning and the entrepreneurial state: The view from Victoria, Australia. *Environment and Planning C: Government and Policy, 11*(3), 263–278. doi:10.1068/c110263.

A Framework for Enhancing Real-Time Social Media Data to Improve the Disaster Management Process

Syed Attique Shah, Dursun Zafer Şeker and Hande Demirel

Abstract Social Media datasets are playing a vital role providing information that can support decision-making in nearly all domains. This is due to the fact that social media is a quick and economical approach for collecting data. It has already been proved that in case of disaster (natural or man-made) the information extracted from Social Media sites is very critical to Disaster Management Systems for response and reconstruction. This study comprises of two parts: The first proposes a framework that provides updated and filtered real time input data for the disaster management system through social media, and the second consists of a designed web user API for a structured and defined real time data input process. The aim of this study is to propose a framework that can filter and organize data from the unstructured social media sources through recognized methods and bring this retrieved data to the same level as that acquired through structured and predefined mechanisms, such as a web API. Both components are designed such that they can potentially collaborate and produce updated information for a disaster management system to carry out accurate and effective decision-making.

Keywords Crowd-sourcing · GIS · Disaster management · Design science · Social media · Real-time data · API

S.A. Shah (✉) · D.Z. Şeker · H. Demirel
Faculty of Civil Engineering, Department of Geomatics, Istanbul Technical University, Maslak Sarıyer, Istanbul, Turkey
e-mail: shah@itu.edu.tr

D.Z. Şeker
e-mail: seker@itu.edu.tr

H. Demirel
e-mail: hande.demirel@itu.edu.tr

M.P. Peterson (ed.), *Advances in Cartography and GIScience*, Lecture Notes in Geoinformation and Cartography, DOI 10.1007/978-3-319-57336-6_6

1 Introduction

During disasters (e.g., floods, earthquake, storms, large fire, etc.), people tend to report and share their observations, findings and suggestions on various social media sites. However, it is still a challenge for researchers as how to automatically filter out useful information and make that information searchable and accessible for emergency services.

Social Media applications are considered very useful to collect information in case of any disaster because it is the fastest and the cheapest source to provide effective, updated and relevant information for decision-making. The practical use of such applications is resulting in a large number of academic studies to research many aspects of social media in Disaster Management (Veil et al. 2011). Social Media provides the opportunity to contribute and disseminate valuable information be it in the shape of text, pictures, audio and video that is necessary for disaster management processes and communications (Turel and Serenko 2012).

Current research states that the communication services such as Short Message Services (SMS) or social media (Facebook, Twitter) have the ability to improve the regular and updated transmission of valuable information and provide the effective resources in all of the disaster management life-cycle phases and aid in developing a disaster resilient community (Chan 2013; Ahmed 2011). In case of any disaster, the emergency service authorities should be able to access the social media networks and blogs to identify the source and scale of the disaster and develop the recovery plans according to the requirements of affected community. In addition, authorities should be able to observe online communities to detect mounting trends and possible hotspots that can substantiate a disaster (Chan 2013). When it comes to managing disasters efficiently, the main thing for government and emergency agencies is to be provided with accurate, updated and complete information. There may be serious consequences if the information is incorrect and or not delivered in a timely manner (Erskine and Gregg 2012; Ostermann and Spinsanti 2011).

Real-time Geospatial Information Systems use crowd-sourcing to map social feeds using geotag metadata with longitude/latitude coordinates. With hardware-based sensors, GIS can be combined with social media to create a compressive source of information to understand the disaster situation accurately, as general public can be added in the process to improve the overall awareness (Adam et al. 2012; Yin et al. 2012). However, there are numerous challenges associated with social media data collection methods, such as variable quality of the data, intelligently managing the big volume of social media feeds, checking and verification of the data, accurately geo-parsing map information and the need to find the right balance between verifying and responding to new information.

Prior research shows the practicality of this research using frameworks such as Twitcident (Abel et al. 2012) and SensePalce2 (MacEachren et al. 2011). Moreover, crowd-souring is emerging as a corresponding research area that works as a virtual sensor to collect data from every possible potential source, and to support activities from basic mapping (e.g., Open-Street Map) to providing data for a disaster management system (e.g., CrisisMappers.org, Ushahidi).

A very good example of web API for collecting social media data in case of an earthquake and then model earthquake activity according to that accessed information is a website named as 'Did You Feel It?' by U.S Geological Survey (USGS) (http://earthquake.usgs.gov/earthquakes/dyfi/).

2 Social Media in Relation with Disaster Management

Social Media are applications that are totally depended on user-generated content or applications in which the user-generated contents and activities play an important role in increasing the overall value of that application or service (Kaplan and Haenlein 2010).

On the other hand, a disaster is a sudden event that seriously affects the normal routine of a community or society. It could be natural calamities such as earthquakes, tornadoes or hurricanes, but also man-made destructive activities such as terrorist attacks or industrial accidents. It has not only an economic and environmental impact, but also an important humanitarian component.

Disaster management can be modeled into four phases, namely mitigation, preparedness, response, and reconstruction. Having a good strategy for each of the phases is essential for efficient disaster management. In order to accomplish this, managers need proper information about the different activities within each of the four disaster management phases (Jayaraman et al. 1997). With social media, information is now accessible in real-time, so those activities can be planned more accurately. With disaster management models and its phases a lot of research work needs to be done to map where and how social media information can be used to improve the decision-making (Cozzolino 2012). Research has identified that the use of social media after a disastrous incident increase and even surpasses the use of other conventional communication methods such as fixed phones (Velve and Zlateva 2012). Social Media sites like Facebook, Twitter and YouTube can be very useful for real time data collection when tsunamis, earthquakes, floods and other natural disasters strike.

According to Crystal Washington (2016) "Social media is the application that;

- Provides valuable information to those in a disaster area pre- and post-disaster (via the Internet, if available, or SMS updates).
- Drives awareness to those outside the affected areas, generating volunteers and/or donors.
- Connects displaced family and friends.
- Provides information about unclaimed property and in worst case, scenarios, bodies.
- Offers information about aid centers and other resources available to those affected."

Five discrete uses for social media in disaster management are identified by (MacEachren et al. 2011) as,

(a) “to disseminate information to the public (e.g., for alerts)”
(b) “to gather information from the public (e.g., crowd-sourcing)”
(c) “to coordinate with crisis management professionals”
(d) “to monitor activities of crisis management professionals”
(e) “as input to situational assessment for crisis management”.

Instead of categorizing the existing research into disaster management phases directly, we added the social media application because they represent a more fine-grained perspective of social media within a disaster. Moreover, applying the social media applications to traditional disaster management phases allows us to integrate pervious research about potential social media activities into the disaster management model. This will act as a theoretical lens to classify the existing research into disaster management phases.

3 Proposed Research Framework

This research work contributes to the discipline of design science for the information systems domain. Based on reviewed literature and detailed perception of related technologies, we propose a model that can provide updated and essential input data for the disaster management system through two different potential crowd-sourced data platforms; the social media component and a designed web user API component.

The proposed model is divided in two sections based on the method for data input. As seen in Fig. 1, the first section is the web user API, named as ‘Disaster Analytic API’ which is a designed web template to take the data according to some structured parameters already defined and to somewhat enforce users to follow the designed format for feeding data. This second section is the ‘social media system component’ which is taking in the unstructured social media data and applying selective recognized methods to filter the data and present it for further processing to the disaster management system component. This model emphasizes the quality of real time data gained through crowd-sourcing, which is normally considered noisy and unfit to use for accurate decision-making processes. This model can be a good example to compare and measure data quality gained through a designed API for structured data and real time data gained through social media. There is no such comparison currently found in the literature and this area needs to be focused to make the coward-sourcing data gathering mechanism more effective and open to both structured (web data entry API) and unstructured (social media) for new actionable insights.

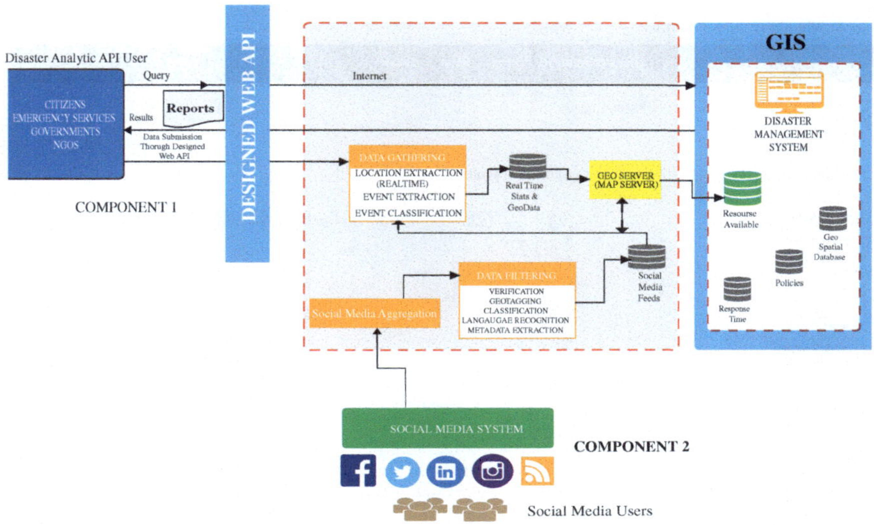

Fig. 1 Proposed research model

3.1 Designed Web User API Component

This "Disaster Analytic" component is a proposed web API for receiving information for any disaster in a structured data entry template. It includes, all potentially required sections such as location, time, scale, description, attachments (photo or video) etc. Additionally, the Disaster Analytic API can also manage queries from the user requesting reports regarding the disaster situation. A rough outline for the interface is proposed in Fig. 2. The potential users of this web API can be citizens/volunteers, governments' officers, emergency services and NGOs who can provide data with a defined format and query. Reports can be searched and acquired through the disaster management system.

The main aim for the design and implementation of this web API is making sure that accurate and machine-readable information is received. It also should be user friendly, and can be operated without much setup or training. Every user needs to be identified and an authorized login should be created. There are many different types of format constraints to be followed in order to get the structured and machine readable information. The notable constraints can be user's log in ID (who), type of incident (what), photo of the identified incident (what), scale (what), date and time (when), location of the incident (where), classification and description (how). To increase the user-friendliness, constraints such as date and time need to be provided automatically so that user should be able to select the options rather than make manual entries in the form. The user can request the forms directly from the disaster management system whenever they want and are able to see the recent verified social media feedbacks and trends being shared regarding the disaster.

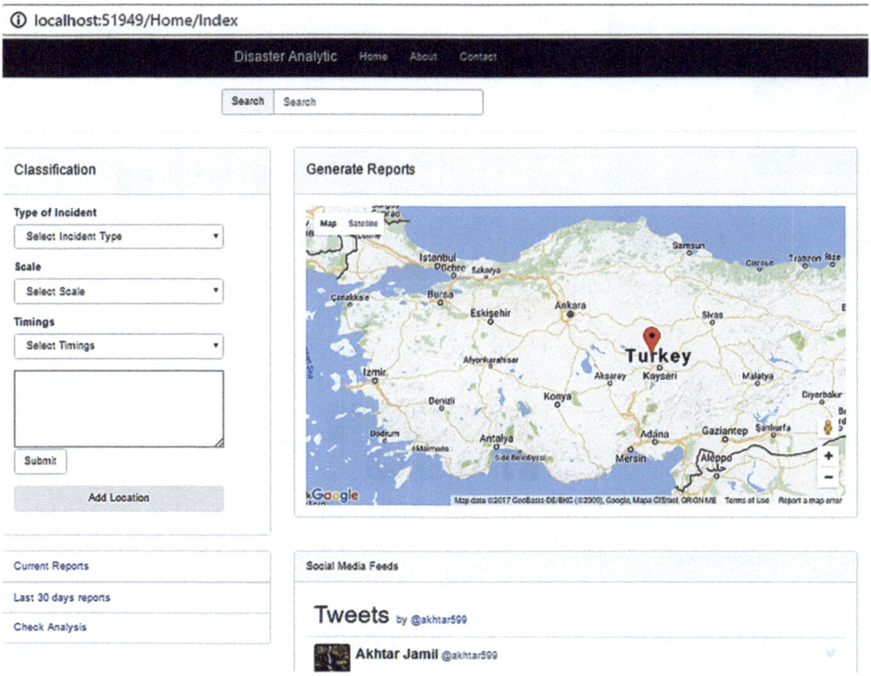

Fig. 2 Designed web API (Disaster analytic)

3.2 Social Media System Component

We are living in an information age where people tend to report, discuss and share the ground facts, observations, and their experiences on different social media forums. The aim of this social media component is to filter and extract the required data gained through crowd-sourcing, and make it beneficial for disaster management (Fig. 3).

The data filtering process is following the semantic methods as it supports filtering and extraction of social media feeds to recognize only the suitable feeds relevant to the incident and provides means to organize information about the incidents for performing real-time analysis. After intensive review of the literature some notable filtering methods are selected and their details are presented in sequence in the following sections.

3.2.1 Data Capture

Whenever an incident is reported the system is triggered to get real time data through social media. Through this framework, the data is captured from different social media sources and translated into an incident profile. The main aim of this

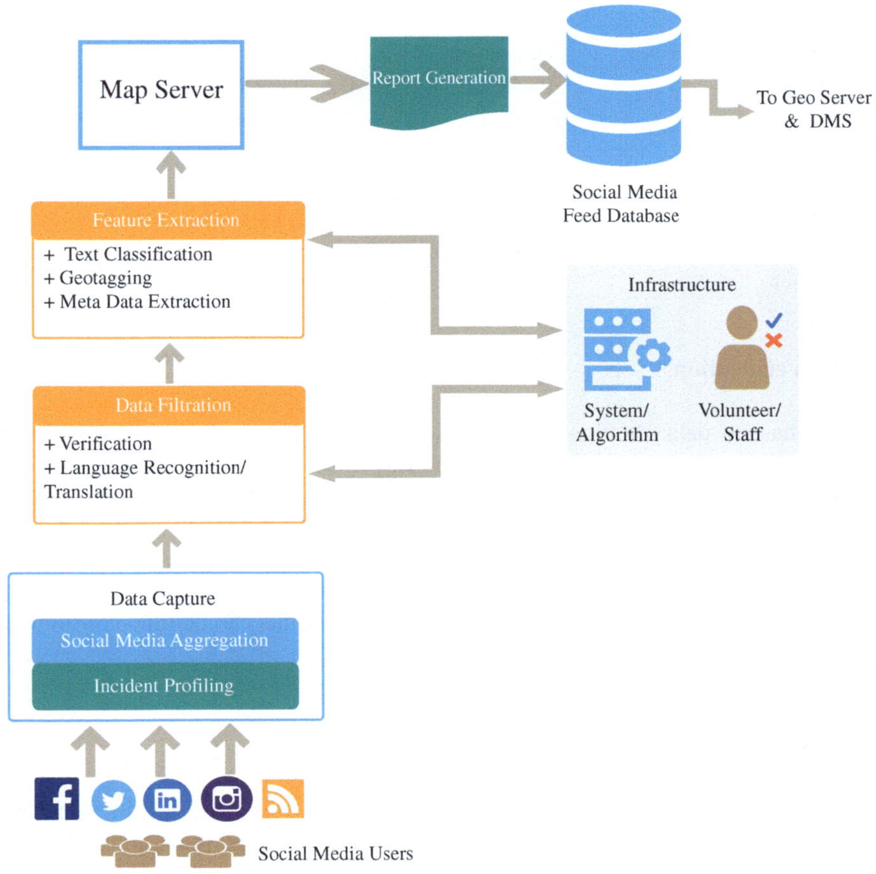

Fig. 3 Proposed social media component

incident profiling is to produce a profile that can provide the valid raw data for data filtering and extraction. "An incident profile is a set of weighted attribute-value pairs that describe the characteristics of the incident" (Abel et al. 2012). These attributes contain data regarding the incidents and might have specific weights to highlight the importance of each attribute according to the type of incident. Incident profiles are kept open to change as it needs to be updated according to the modifications that may occur during the incident. Keeping in view the incident profile designed for the disaster, social media aggregation is performed to capture any type of feed that fulfills the criteria. These feeds are then reported and processed for extracting the useful information for the system. Some common attributes for an incident profiling are mentioned in Table 1.

Table 1 Incident profiling attributes

Incident profiling attributes	Description
Classification	Earthquake, fire, flood, power failure, accident
Address	State, city, street, zip code, coordinates (if extractable)
Descriptive location	Any location name used
Time of occurrence	Time stamp
Scaling	Size or value reported
Image reporting	Any image or video reported
Observations	Comments regarding the incident

3.2.2 Verification

In this phase of data filtration the authenticity and reliability of the crowd-sourced data are checked because of the anonymous way through which it is collected. This remains an issue with social media data and a lot of algorithms are designed to cope with this concern. Privacy concerns are also an issue and need to be addressed with proper procedures. A lot of social media sites have already addressed their users in satisfaction regarding location and feeds in their user agreements. Verification can also be performed manually by volunteers or stuff, but this can be time-consuming and laborious.

3.2.3 Language Recognition

Regions having multiple languages can get social media feeds in different languages and hence affect the source data by adding noise. This issue highlights the need for improved filtering techniques to translate feeds in a common language (i.e., English). Neuro-Linguistic Programming (NLP) techniques can be used to manage multilingual situations.

3.2.4 Metadata Extraction

Metadata Extraction is a vital part of these social feeds as it provides additional information on each feed. The metadata table can contain information about the originator of the feed. To verify the source it may contain the profile picture, number of followers and feeds, location and timestamp when sharing the feed. Such type of Metadata can strengthen the reliability and accuracy of the data provided for processing and decision-making (Stronkman 2011).

Table 2 Text classification attributes

Classification type	Description
Impact area	Map highlighting the potential affected areas
Status	Current status of the incident
Threats	Possible future threats
Related news	Ongoing news about the situation
Casualties and injured	Number of casualties and injuries reported
Image or video reporting	Any images of the incident received
Respond time	Possible time to respond to rescue
Instructions	Any precautions needed

3.2.5 Geotagging

Through a designed API, we can integrate map source data that provides street and building level locations using coordinates. Social media feeds generally describe locations at regional levels if GPS is not activated. As we know, the precise location of the incident plays an important role in disaster management. Therefore, to enable spatial exploration of social media feeds, a geotagging model should be used to display the narrative of the feeds at its exact location on the map. The latitude/longitude coordinates of the user profile can be used to get the desired location.

3.2.6 Text Classification

This phase is the core part of feature extraction as it indicates the extracted contents and provides the required information for reporting. It may contain reports regarding casualties and possible threats and damages about the incident. The classification can further be categorized according to the feeds if the publisher was witness, hearing the news, observing or smelling something. Handcrafted rules are used for the classification that can work in both the attribute-value pairs and the plain words stated in the social media feeds (Abel et al. 2012).

Some common attributes, keeping in mind the filtered social media feeds for classification, are described in Table 2.

4 Conclusion

This study presents a design model for the development of an integrated system consisting of social media crowd-sourced component and a designed web API component through which organized and reliable data can be provided for real-time disaster management. This design-science research demonstrates that the concept of social media crowd-sourcing can effectively be used for real-time disaster management and tries to aid the theory of making crowd-sourced data as trustworthy as other data sources. The basic theme of this design is to make the unstructured crowd-sourced data processable so that it can be compared and merged with a structured data sources such as a web API. The effectiveness of real-time crowd-sourced disaster management systems has been proven but there are many gaps and challenges in this research domain. The design science to model integrating frameworks plays a key role for providing the basis for interdisciplinary research to be carried out.

References

Abel, F., Hauff, C., Houben, G.J., Stronkman, R. & Tao, K. (2012, April). Twitcident: fighting fire with information from social web streams. In *Proceedings of the 21st International Conference on World Wide Web* (pp. 305–308). ACM.

Adam, N. R., Shafiq, B., & Staffin, R. (2012). Spatial computing and social media in the context of disaster management. *IEEE Intelligent Systems, 27*(6), 90–96.

Ahmed, A. (2011). Use of social media in disaster management. *ICIS 2011 Proceedings*. Paper 16.

Chan, J. C. (2013). 1 The role of social media in crisis preparedness, Response and Recovery.

Cozzolino, A. (2012). *Humanitarian logistics: Cross-sector cooperation in disaster relief management*. Springer Science & Business Media.

Crystal Washington, N. P. (2016). The role of social media during natural disasters. http://crystalwashington.com/the-role-of-social-media-during-natural-disasters/. Retrieved October 13, 2016.

Erskine, M., & Gregg, D. (2012). Utilizing volunteered geographic information to develop a real-time disaster mapping tool: A prototype and research framework. In *CONFIRM 2012 Proceedings*.

Jayaraman, V., Chandrasekhar, M. G., & Rao, U. R. (1997). Managing the natural disasters from space technology inputs. *Acta Astronautica, 40*(2), 291–325.

Kaplan, A. M., & Haenlein, M. (2010). Users of the world, unite! The challenges and opportunities of Social Media. *Business Horizons, 53*(1), 59–68.

MacEachren, A. M., Jaiswal, A., Robinson, A. C., Pezanowski, S., Savelyev, A., Mitra, P., et al. (2011, October). Senseplace2: Geotwitter analytics support for situational awareness. In *2011 IEEE Conference on Visual Analytics Science and Technology (VAST),* (pp. 181–190). IEEE.

Ostermann, F. O., & Spinsanti, L. (2011, April). A conceptual workflow for automatically assessing the quality of volunteered geographic information for crisis management. In *Proceedings of AGILE* (pp. 10–14).

Stronkman, R. J. P. (2011). Exploiting Twitter to fulfill information needs during incidents. (Doctoral dissertation, TU Delft, Delft University of Technology).

Turel, O., & Serenko, A. (2012). The benefits and dangers of enjoyment with social networking websites. *European Journal of Information Systems, 21*(5), 512–528.

Veil, S. R., Buehner, T., & Palenchar, M. J. (2011). A work-in-process literature review: Incorporating social media in risk and crisis communication. *Journal of Contingencies and Crisis Management, 19*(2), 110–122.

Velve, D., & Zlateva, P. (2012). Use of social media in natural disaster management. *International Proceedings of Economic Development and Research, 39*, 41–45.

Yin, J., Lampert, A., Cameron, M., Robinson, B., & Power, R. (2012). Using social media to enhance emergency situation awareness. *IEEE Intelligent Systems, 27*(6), 52–59.

Building a Real-Time Geo-Targeted Event Observation (Geo) Viewer for Disaster Management and Situation Awareness

Ming-Hsiang Tsou, Chin-Te Jung, Christopher Allen, Jiue-An Yang, Su Yeon Han, Brian H. Spitzberg and Jessica Dozier

Abstract Situation awareness plays an important role in disaster response and emergency management. Displaying real-time location-based social media messages along with videos, pictures, and hashtags during a disaster event could help first responders improve their situation awareness. A geo-targeted event observation (Geo) Viewer was developed for monitoring real-time social media messages in target areas with four major functions: (1) real-time display of geo-tagged tweets within the target area; (2) interactive mapping functions; (3) spatial, text, and temporal search functions using keywords, spatial boundaries, or dates; and (4) manual labeling and text-tagging of messages. Different from traditional web GIS maps, the user interface design of GeoViewer provides the interactive display of multimedia content and maps. The front-end user interface to visualize and query tweets is built with open source programming libraries using server-side MongoDB.

M.-H. Tsou (✉) · C. Allen · J.-A. Yang · S.Y. Han · J. Dozier
Department of Geography, The Center for Human Dynamics in the Mobile Age, San Diego State University, San Diego, CA, USA
e-mail: mtsou@mail.sdsu.edu

C. Allen
e-mail: ccchris.allen@gmail.com

J.-A. Yang
e-mail: jyang@sdsu.edu

S.Y. Han
e-mail: suhanmappingideas@gmail.com

J. Dozier
e-mail: jessdozr@me.com

C.-T. Jung
Esri (Beijing) R & D Center, Beijing, China
e-mail: chinte.jung@gmail.com

B.H. Spitzberg
School of Communication, San Diego State University, San Diego, CA, USA
e-mail: spitz@mail.sdsu.edu

M.P. Peterson (ed.), *Advances in Cartography and GIScience*, Lecture Notes in Geoinformation and Cartography, DOI 10.1007/978-3-319-57336-6_7

GeoViewer is built for assisting emergency responses and disaster management tasks by tracking disaster event impacts, recovery activities, and residents' needs in the target region.

Keywords Social media • Web maps • Disaster management • Situation awareness • Geovisualization • Open source

1 Introduction

Situation awareness involves short-term, immediately responsive human behaviors by acquiring geospatial information in the field to facilitate operational and tactical decision-making. Situation awareness can help decision makers identify key environmental elements and their meanings within a volume of time and space (Endsley 1995). Situation awareness plays an important role in disaster response, recovery and emergency management for first responders and emergency response providers (Tsou and Sun 2007; Wang et al. 2016). For example, a firefighter can view a map showing nearby wildfire perimeters and decide to create a backfire plan in order to control the spread of wildfires (immediate operational plan). A police chief can identify the locations of evacuation shelters after a major earthquake and dispatch police officers to evacuation shelters (short-term tactical plan).

There are a number of good examples of using social media for emergency management, including the 2010 Haiti Earthquake (Veil et al. 2011), 2012 Waldo Canyon Fire (Sutton et al. 2014), 2013 Typhoon Haiyn in the Philippines (Humanity Road and Statistics Without Borders 2014), Hurricane Sandy (Guan and Chen 2014), and the 2014 Ebola outbreaks in West Africa (Moore and Verity 2014). Disaster-relevant social media has been used to analyze how individuals or decision-makers enhance their situational awareness (Vieweg et al. 2010; de Albuquerque et al. 2015; Huang and Xiao 2015).

Previous studies have indicated the need of developing an integrated, user-friendly web application to collect and display relevant location-based social media messages in real-time. Many relief efforts during disastrous events will need to provide aid promptly to target areas. However, there are usually a significant delay of communication between the affected communities and their formal emergency response agencies (Tapia et al. 2013; OCHA 2014; VSMWG 2014; Sutton et al. 2014). Location-based social media messages can facilitate the prompt response and decision making for disaster response and recovery because social media is free and highly accessible. The messages provided by people in affected areas can be collected and interpreted by emergency responders immediately while also opening direct lines of communication between responders and affected communities (Goolsby 2010; Sakiki et al. 2013; Tomaszewski 2015).

One important metadata for disaster response stored in social media messages is the location of users. Twitter posts with geographic locations are referred to as "geo-tagged tweets". However, only between 1–3% of tweets include location (Morstatter et al. 2013) as users must activate the geolocation function in their mobile devices. Utilizing geo-tagged tweets with content analysis can help improve situational awareness and further disaster management efforts.

One early software example of using geo-tagged social media for disaster relief is Ushahidi (http://www.ushahidi.com), which was an open source crowdsourcing crisis map platform that utilizes online volunteers for information validation (Gao et al. 2011). Created in 2007 in Kenya to provide situational information from the online volunteers (Okolloh 2009), many disaster relief agencies used Ushahidi to help coordinate emergency responses in such places as Haiti, Chile, Russia, Kenya, and India. However, there are several major challenges of using geo-tagged social media for disaster relief, including the accuracy of geo-tagged locations, verification of reports, summarizing all reports automatically, social behavior prediction, and the scalability and safety (Gao et al. 2011). Sahana is another well-known open source Disaster Management System for emergency management and disaster preparedness. It provides limited mapping functions to add information on disaster-relevant event locations or requests for help. Users can add notes or upload photos with specific locations on a web map, so that different agencies can collaboratively work on the disaster response and relief activities (Careem et al. 2006).

Since both Ushahidi and Sahana platforms cannot provide a real-time social media messages display functions, the main design goal of GeoViewer is to deliver real-time map display functions for disaster events ground truth observation. A product of the Center for Human Dynamics in the Mobile Age (HDMA) at San Diego State University, GeoViewer is a web-based real-time social media mapping application that links GPS locations, texts, photos, and time stamps in social media (Twitter) messages. GeoViewer helps users visualize historical and real-time social media content, provides easy-to-use geospatial query functions, and displays multimedia sources (see Fig. 1).

The first beta version of GeoViewer was developed on July 30, 2014 after the San Diego Wildfire (http://vision.sdsu.edu/hdma/wildfire/). The official version of GeoViewer was released by the HDMA Center on December 6, 2014 (http://vision.sdsu.edu/hdma/geoviewer/sanDiego) with the real-time mapping function. The key design principle of GeoViewer follows the Knowledge Discovery in Cyberspace

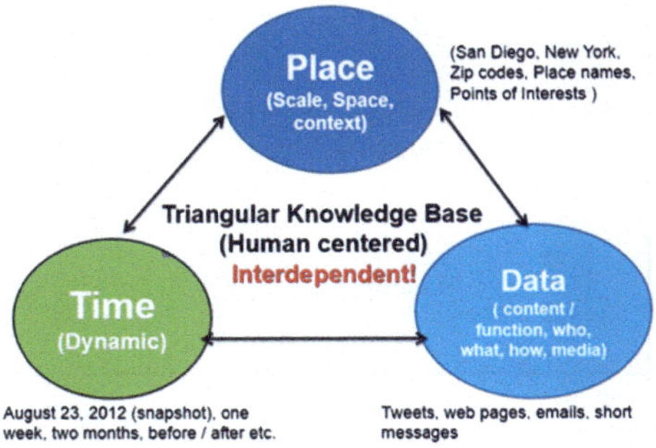

Fig. 1 The design of GeoViewer is based on the analytic framework of Knowledge Discovery in Cyberspace (KDC). (Modified from Tsou and Leitner 2013, 56)

(KDC) conceptual framework (Fig. 1). In order to acquire geospatial knowledge and situation awareness in social media and cyberspace, GeoViewer needs to help users discover new patterns, new trends, and new spatial knowledge from aggregated social media records (Tsou and Leitner 2013). One unique characteristic of KDC is a triangular knowledge base consisting of place, time, and messages. The three components of KDC are interdependent and human-centered, which means they are originally defined or created by human beings (messengers).

2 The Design of GeoViewer System Architecture

The system design of GeoViewer follows the client-server architecture as shown in Fig. 2. On the server side, NodeJS was selected to develop the GeoViewer server functions because of (1) its multiple plugin modules; (2) the same programming language with the client side (JavaScript); and (3) its lightweight and efficient features. The GeoViewer server includes the following components.

- *Twitter Streaming Engine*: The engine maintains a streaming connection with Twitter Streaming Application Programming Interface (API) to collect real-time geo-tagged tweets. The engine specifies a bounding box in the POST statuses/filter endpoint of the public Twitter streams to collect geo-tagged tweets if they are located within the targeted region.
- *Database*: MongoDB was selected as the database for GeoViewer to store the collected tweets because it: (1) is an open source and document-based database with ease of development and scaling; (2) supports spatial query; and (3) provides full-text search functions.

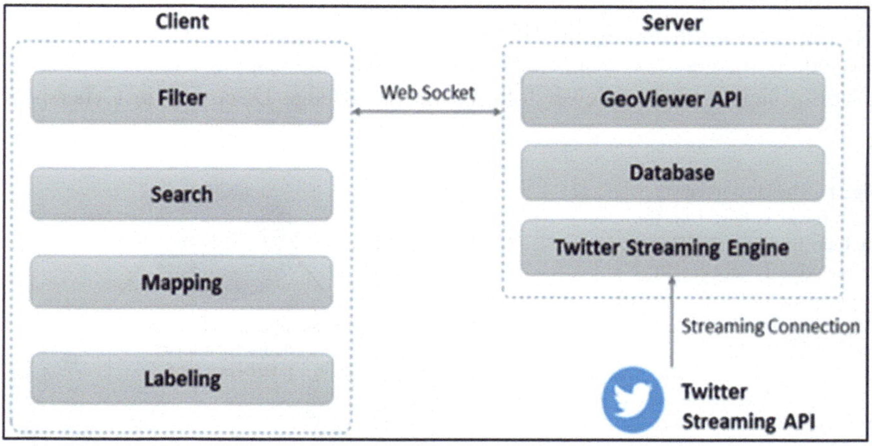

Fig. 2 The client/server system architecture of GeoViewer

Table 1 The JavaScript (JS) modules used in the client side of GeoViewer

JS modules	Purpose
jQuery	jQuery makes things like HTML document traversal and manipulation, event handling, animation, and Ajax much simpler (https://jquery.com/)
Bootstrap	Bootstrap is a popular HTML, CSS, and JS framework for developing responsive projects on the web (http://getbootstrap.com/)
Leaflet	Open source JavaScript map library (http://leafletjs.com)
Moment.js	Format date and time functions (http://momentjs.com/)
Masonry	Show tweets in a grid layout (http://masonry.desandro.com/)
introJS	Show step by step tutorial functions in GeoViewer (https://usablica.github.io/intro.js/)

- *GeoViewer API*: The GeoViewer server provides several JavaScript APIs (such as search historical tweets, spatial search, and sending new tweets) to communicate messages between the client and server sides. The messages are delivered by a WebSocket (i.e., a persistent connection between clients and servers to increase interactions). Once GeoViewer is launched in a Web browser, a WebSocket is initiated to communicate messages in real-time between the client and the server.

On the client side, GeoViewer was created based on HTML5, CSS, JavaScript, and JavaScript modules (Table 1). Besides the key mapping functions (such as searching, mapping, and labelling), the client side provides a filter function, which filters out unmatched tweets based on the search criteria (keywords, spatial boundary, or dates). If a real-time tweet does not match the criteria, the tweet will be invisible until a user resets the search criteria.

GeoViewer is an easy-configurable package. Developers can easily customize GeoViewer functions to target different areas by customizing the configuration file. GeoViewer is planned to become an open source project, so that more researchers and institutes can develop or utilize GeoViewer for their own research project.

3 User Interface Design and Key Functions

Different from traditional web maps, GeoViewer provides interactive display of multimedia content in addition to map display. The web-based user interface (front-end) is built by open source programming libraries, jQuery and Leaflet APIs, to visualize and query tweets from a server-side database (back-end). The user interface design of GeoViewer follows the concept of user-centered design (UCD) in cartography (Tsou 2011), which emphasizes providing the most important (and simplified) information for map readers with a user-friendly interface and data analytic functions. Four major components of the user interface design are highlighted: (1) real-time display of new geo-tagged tweets within the target area; (2) interactive mapping functions for geovisualization; (3) spatial, text, and temporal search functions using keywords, spatial boundary, or dates; and (4) labeling and text-tagging for individual messages and for further sentiment analysis.

3.1 Real-Time Display of Geo-Tagged Tweets

When users launch GeoViewer, the web app automatically collects geo-tagged tweets within the selected region with the public Twitter Streaming API. Although the public Streaming API has a 1–3% limitation of total tweets within a region, GeoViewer can collect 40–50% of all geo-tagged tweets within the defined region based on our previous testing results. Some researchers also indicated that over 90% coverage of total geo-tagged tweets can be collected via public Streaming APIs (Morstatter et al. 2013). When a new geo-tagged tweet is collected, GeoViewer automatically displays the tweet at the top (newest first) or at the end (oldest first) of the Tweet window (left-side). Users can select the refresh rate of new tweets by time or by the number of tweets. Users can also pause the real-time feed if needed (Fig. 3).

The dual-view interface is adjustable to increase or decrease map view window (right-side), which also adjusts the number of columns in the tweet window. Each tweet in the Tweet window has an orange *Map* button to identify its exact location in the Map Window. Disaster responders and resource coordinators can query and analyze these real-time tweets and identify the location and time of urgent requests. Each affected area can have one GeoViewer instance with a pre-defined area for

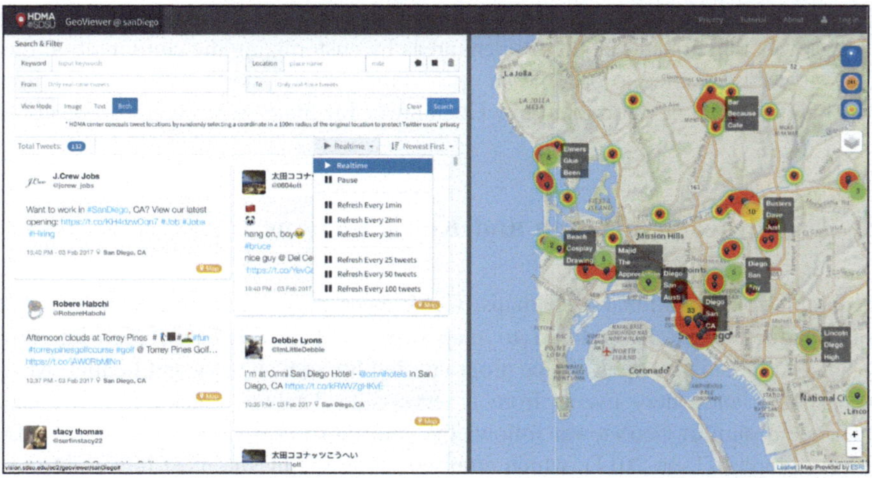

Fig. 3 Displaying and arranging real-time tweets in GeoViewer (San Diego region) (http://vision.sdsu.edu/ec2/geoviewer/sanDiego#)

tweet collection. The bounding box is defined in the web server configuration file for each GeoViewer and can be visualized in the front-end as well by clicking the *Twitter Search Boundary* in the Map Layer option.

3.2 Interactive Mapping Functions for Geovisualization

GeoViewer utilizes *Leaflet*, an open source JavaScript mapping library for the development of interactive mapping function. In addition to the standard zoom-in (+), zoom-out (−), and pan function, four key mapping functions for better geo-visualzation are provided: map marker selection, cluster mapping, hot spot mapping, and map layers overlay.

- ***Map marker selection*** allows users to pick the color and size of map markers (Fig. 4, left). The markers depict the proximate location of each geo-tagged tweet. In order to protect Twitter users' locational privacy, GeoViewer applies geo-masking to conceal the actual tweet locations by randomly selecting a coordinate within the 100 m radius of the original location.
- ***Cluster mapping*** provides a generalized view for the clustered distribution of tweets within a region to simplify the display of point data. In addition, a *popular text display* function is associated with the cluster mapping to present the ***top three keywords*** from the message clusters (Fig. 4, right). When users zoom-in or zoom-out, tweet clusters are re-arranged and top keywords are dynamically generated based on the new cluster groups. One motivation for this function is to provide a quick overview of popular issues or topics within the clustered groups. For example, during a new wildfire event, the popular keywords used near fire perimeters can be used to identify the name of wildfire for the future reference and public announcement. Also, if there are multiple urgent requests from a collapsed building or a landmark, the cluster labels can help decision-makers identify the key issues within the target area.

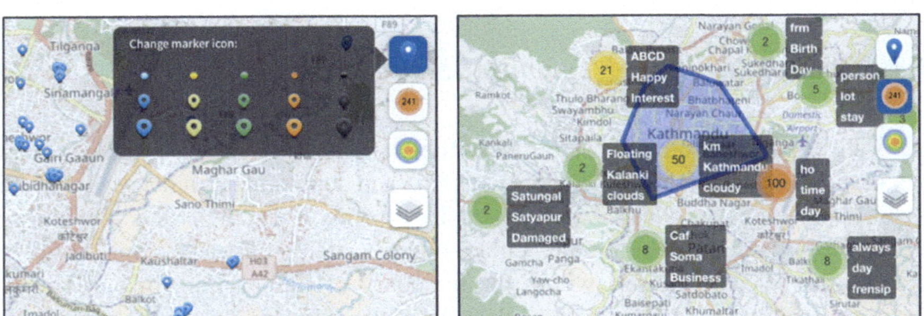

Fig. 4 The *map marker selection* function (*left*) and the *cluster mapping function* with *top three keywords* within each cluster (*right*)

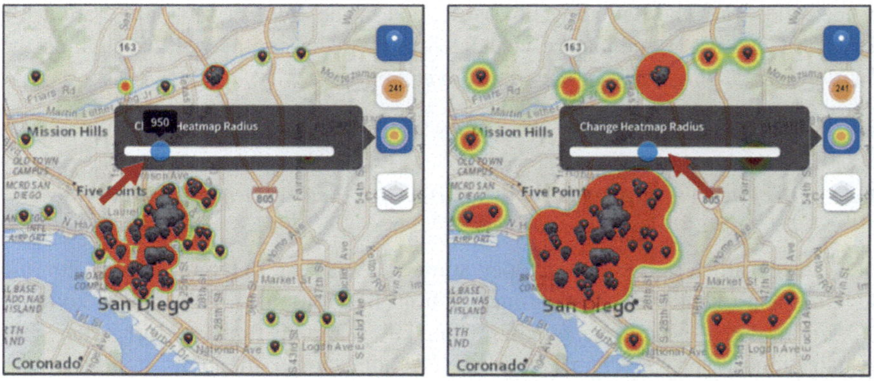

Fig. 5 Hot spot mapping functions with adjustable radius by using kernel density estimation (KDE)

- ***Hot spot mapping*** (heat maps) highlights the clustered areas in red. Users can turn on/off the hot spots in the map and modify the radius of hot spots by using kernel density estimation (KDE). Figure 5 illustrates two hot spot maps with different radius settings for the KDE function. The scalable hot spot function is designed for disaster relief work since it will help decision-makers evaluate the significance of damages or requests. The modifiable radius function assists users to identify urgent requests in small areas and provides an overview of spatial patterns in larger regions.
- ***Map layers overlay*** function combines multiple types of external web services, including WMS, WFS, and ArcGIS Online Mapping Services. Currently, GeoViewer has added the following external maps: NOAA Red Flag Warning Live Update, NOAA NWS Weather Alert CAP (polygons), and several San Diego County OES's ArcGIS online mapping services. This function allows decision-makers to more efficiently integrate multiple data resources, operation plans, and collaboration tasks.

3.3 Spatial, Text, and Temporal Search

GeoViewer provides three types of query functions for collected geo-tagged social media: spatial search, text (keyword) search, and temporal search based on the KDC framework. Users can search tweets by drawing a rectangle or a polygon or typing place names within a radius circle. Figure 6 (left) illustrates a spatial query example by selecting tweets within the boundary of Balboa Park in San Diego. Users can also type keywords and select dates to query the historical tweets collected by GeoViewer. Figure 6 (right) illustrated a combined query result by using a keyword ("beach") and a temporal window (from 2016-01-01 to 2016-02-01)

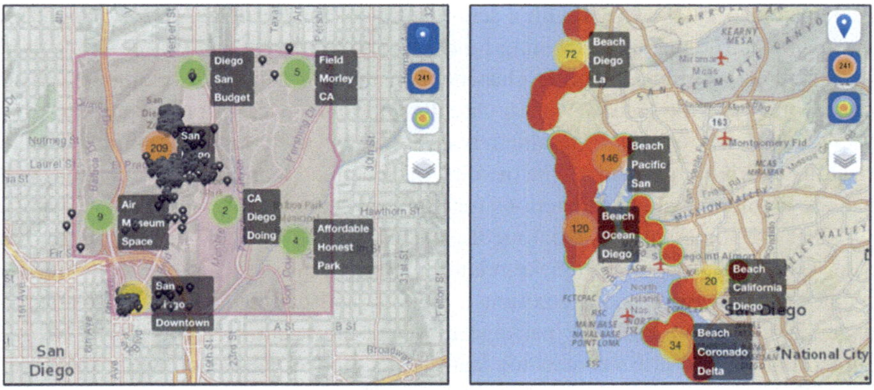

Fig. 6 The spatial query function (drawing a polygon) (*left*) and the text search function (keyword: beach) (*right*) in GeoViewer

Fig. 7 Labeling tweet (tag tweet) function and sentiment labels in GeoViewer

together. The search results indicated that most tweets containing the text of "beach" were located close to beach areas. One interesting finding is that the top three popular keywords reflected the actual name of beach in the clustered map. Researchers may be able to use this function to retrieve local place names associated with different land use types.

3.4 Labeling and Text-Tagging Function

One unique feature of GeoViewer is labeling and text-tagging functions for authorized individual users. This function helps users identify important messages related to disaster management tasks, such as warnings or medical help. Ideally, users can create their unique labeling system for different types of disasters or events. The current version of GeoViewer provides five pre-defined labels (Fig. 7): *Safe* (green), *Info* (blue), *Medical Help* (cyan), *Warning* (orange), and *Danger* (red).

Users can also label a tweet with a sentiment label for *Positive*, *Neutral*, *Negative*, or *Unknown*. The manually-labeling functions and sentiment analysis could be used for the future machine learning classification method (under development). Volunteers may use the GeoViewer to classify important tweets during and after disaster events. The manual classified tweets can be used for the training sets in machine learning methods. The results of machine learning classification will provide automatic labeling functions for future tweets related to the disaster events.

There are some similar social media text mining projects for disaster relief. Qu et al. (2011) classified earthquake-related messages into four categories (situation update, opinion expression, emotional support, and calling for action); Cameron et al. (2012) built a software package to detect emergent incident-related tweets; Imran et al. (2013) utilized machine learning methods to classify Twitter messages into different classes (caution and advice, casualties and damage, donations, people, information sources and others). However, most of these applications did not utilize real-time feeds of tweets and the data mining methods and take one or two days to generate results.

We hope to develop near-real time mechanism for automatically analyzing and classifying disaster-relevant tweets. There are two key technical challenges for near-real time text mining and classification in GeoViewer: (1) combining multiple volunteers' labeling results (message ranking, classification, and error checking) into a single system; and (2) developing different types of classification schemes for different types of disasters. We hope to solve these technical issues by developing a new text analytic framework for GeoViewer in future releases.

4 Nepal Earthquake Case Study

In order to assess the usability of GeoViewer for disaster events, we designed and implemented an instance of GeoViewer for the 2015 Nepal earthquake as one case study. On April 25, 2015, an earthquake with a magnitude of 7.8 devastated Nepal causing 8,999 fatalities and 19,009 injuries. Despite the geographic distance, people were able to help internationally by volunteering online to produce crisis maps. Crisis maps can be used to aid in emergency relief efforts by providing first responders with detailed online maps that use information gathered from social media, satellite imagery, and drones which all contribute to increasing situation awareness (Gao et al. 2011). Organizations such as the International Network of Crisis Mappers, the Digital Humanitarian Network, and Humanitarian OpenStreetMap Team have formed to help by utilizing crowdsourcing and humanitarian efforts (Meier 2014).

The HDMA Center instantiated a GeoViewer instance for Nepal Earthquake on April 26 immediately after the event (Fig. 8) (http://humandynamics.sdsu.edu/NepalEarthquake.html). Information collected was shared with officers in the United Nations (UN) and affiliated international organizations. GeoViewer was used in two major ways: (1) analysis of historic tweets collected during and after the

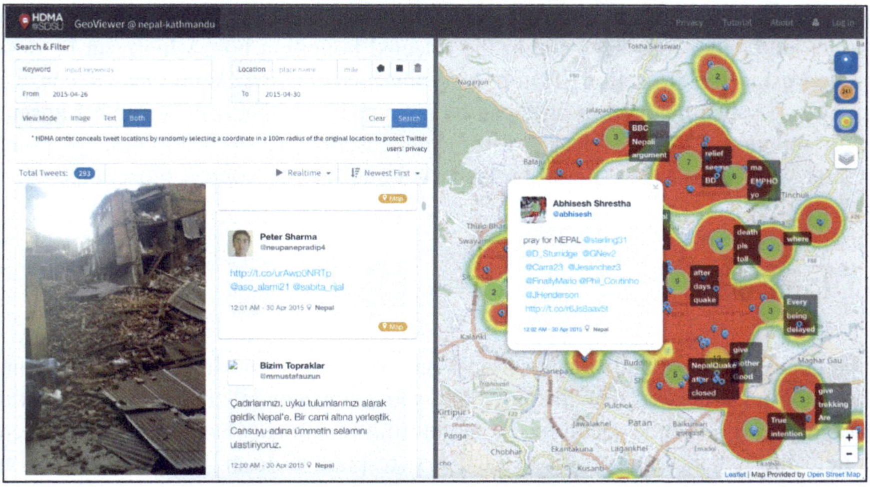

Fig. 8 The new GeoViewer for 2015 Nepal Earthquake. (http://vision.sdsu.edu/ec2/geoviewer/nepal-kathmandu#)

disaster by keywords, dates, and locations; and (2) real-time monitoring of tweets during the disaster. During analysis, some keywords needed to be translated since the official and most commonly used language in the area is Nepali, with English used by those with more education.

The Nepal Earthquake GeoViewer found that the majority of communication via Twitter is in English with only 20% of tweets in Nepali. However, translation was difficult since Google Translate does not provide the translation option between Nepali and English.

Furthermore, we noticed many messages included trending hashtags for the Nepal earthquake, such as "*#NepalEarthquake*", "*#Nepal*", and "*#NepalQuake*". Applying these popular hashtags could help users broadcasting messages more effectively. There are many important tweets indicating people requesting assistance or providing local help information for those who may be seeking assistance. These tweets can be manually labeled using the GeoViewer labeling functions. We can also identify the exact geographic coordinates of these tweets. Some examples of Twitter messages are:

- *plz help us we are stuck in Nepal Kathmandu. Vinod lila 9851035###. 10:15 pm, 25 April 2015 (geo-tagged location).*
- *This shelter need help* http://t.co/####, *03:16 am, 26 April 2015 (geo-tagged location).*

The cluster mapping function in GeoViewer displays the top three trending keywords from the nearby tweets. During the disaster event, these trending keywords (Fig. 8) might indicate where the main damage has occurred, shelters or supplies are needed, or the locations where aftershocks were felt. Another method to collect disaster related tweets is to use the download function in GeoViewer to extract all Twitter messages within the selected bounding box, which can be designated using the "Location" feature or typing place names. With GeoViewer, we hope these social media messages can be reviewed by volunteers before being sent to, or monitored by, the local governmental response teams who could pinpoint exactly where assistance should be dispersed and how to contact those affected.

5 Conclusion

Emergency response and rescue services are critical tasks in disaster management. By combining cluster map display, hot spots, manual labeling, spatiotemporal querying, and real-time social media monitoring functions, GeoViewer is designed to help firefighters, police officers, and first responders to enhance their situation awareness and make better decisions for disaster relief. GeoViewer provides a more integrated geovisualization solution by analyzing disaster-relevant messages with scalable hot spots and cluster mapping/labels functions. These visualizations help users to validate major damaged areas. For example, a hot spot with over 10 reports from different users of a collapsed building becomes more credible than from just one message.

After the case study of 2015 Nepal Earthquake, we identified a few research challenges and opportunities for the next version of GeoViewer. The first challenge is how to validate information (reports) provided by social media users. Previous research applications (such as Ushahidi) only emphasized on crowd-sourced manual classification to filter noises and errors. GeoViewer will need to combine both crowd-sourced manual classification and real-time machine learning methods to validate and filter disaster-relevant messages in near real-time. How to effectively recruit online volunteers and generate training datasets for machine learning within a short period of time is an additional challenge. There were also significant noises and errors in geo-tagged tweets from job advertisement, weather reports, traffic updates, etc. These need to be filtered before conducting the hybrid manual and machine learning classification.

Another challenge is how to increase the coverage of total geo-tagged tweets by using Firehose or PowerTrack APIs from Twitter directly, which are not affordable for most local disaster response agencies or NGOs. The sampling process of the public Twitter Streaming APIs are a black box and the underlying algorithm has been changed frequently. We need to understand more about these algorithms to establish a stable and full coverage stream. On the other hand, to increase the percentage of geo-tagged tweets, disaster relief agencies, such as Red Cross or local government agencies, should ask people in affected areas to enable GPS on their devices when submitting relevant reports via location-based platforms.

The language translation and interpretation is another key challenge for future development (Tsou 2015). For example, if we want to use GeoViewer to guide disaster relief in San Diego, we need to provide both English and Spanish versions because there is a significant Hispanic population. Some language translation tasks can be facilitated by using Google Translator. But, in the Nepal case study, we found that there is no appropriate translation software for Nepali. Besides technological challenges, we also need to consider human and ethical issues. Sometime, people at the most damaged areas may not have electricity or wireless communications to report their urgent needs. Removing digital divides and building accessible wireless communication networks for everyone may be the most important challenge before we can effectively utilize social media for disaster relief and management tasks in the future.

Acknowledgements This material is based upon work supported by the U.S. National Science Foundation under Grant No. 1416509 and Grant No. 163464. Any opinions, findings, and conclusions or recommendations expressed in this material are those of the author and do not necessarily reflect the views of the National Science Foundation.

References

de Albuquerque, J. P., Herfort, B., Brenning, A., & Zipf, A. (2015). A geographic approach for combining social media and authoritative data towards identifying useful information for disaster management. *International Journal of Geographical Information Science, 29,* 667–689. doi:10.1080/13658816.2014.996567.

Cameron, M. A., Power, R., Robinson, B., & Yin, J. (2012). Emergency situation awareness from twitter for crisis management. In *Proceedings of the 21st International Conference on World Wide Web* (pp. 695–698). NY, USA: ACM.

Careem, M., De Silva, C., De Silva, R., Raschid, L., & Weerawarana, S. (2006). Sahana: Overview of a disaster management system. In *2006 International Conference on Information and Automation* (pp. 361–366). IEEE.

Endsley, M. R. (1995). Toward a theory of situation awareness in dynamic systems. *Human Factors Journal of Human Factors and Ergonomics Society, 37*(1), 32–64.

Gao, H., Barbier, G., & Goolsby, R. (2011). Harnessing the crowdsourcing power of social media for disaster relief. *IEEE Intelligent Systems, 26*(3), 10–14.

Goolsby, R. (2010). Social media as crisis platform: The future of community maps/crisis maps. ACM Transactions on Intelligent Systems and Technology (TIST), 1, 1, pp. 1–11, ACM Digital Library. DOI: 10.1145/1858948.1858955.

Guan, X., & Chen, C. (2014). Using social media data to understand and assess disasters. *Natural Hazards, 74*(2), 837–850.

Huang, Q., & Xiao, Y. (2015). Geographic situational awareness: mining tweets for disaster preparedness, emergency response, impact, and recovery. *ISPRS Int J Geo-Inf, 4*(3), 1549–1568.

Humanity Road and Statistics Without Borders. (2014). A Guide to Social Media Emergency Management Analytics. Understanding the Place of Analytics through Typhoon Haiyan Social Media Analysis. http://humanityroad.org/smemanalyticsguide/.

Imran, M., Elbassuoni, S., Castillo, C., Diaz, F., & Meier, P. (2013). Practical extraction of disaster-relevant information from social media. In *Proceedings of the 22nd International Conference on World Wide Web companion* (pp 1021–1024). International World Wide Web Conferences Steering Committee.

Meier, P. (2014). Digital humanitarians: How big data is changing the face of humanitarian response. CRC Press.

Moore, R. & Verity, A. (2014). Hashtag standards for emergencies. United Nations Office for the Coordination of Humanitarian Affairs (OCHA) Policy and Studies Series 012. http://reliefweb.int/report/world/hashtag-standards-emergencies. Retrieved January 10, 2015.

Morstatter, F., Pfeffer, J., Liu, H., & Carley, K. (2013). Is the sample good enough? Comparing data from Twitter's Streaming API with Twitter's Firehose. In *Proceedings of ICWSM* (p. 9).

OCHA (2014). Hashtag standards for emergencies (pp. 1–18). United Nations Office for the Coordination of Humanitarian Affairs. http://reliefweb.int/sites/reliefweb.int/files/resources/Hashtag%20Standards%20For%20Emergencies.pdf.

Okolloh, O. (2009). Ushahidi, or 'testimony': Web 2.0 tools for crowdsourcing crisis information. *Participatory Learning and Action, 59*(1), 65–70.

Qu, Y., Huang, C., Zhang, P., & Zhang, J. (2011). Microblogging after a major disaster in China: A case study of the 2010 Yushu earthquake. In *Proceedings of the ACM 2011 Conference on Computer Supported Cooperative Work*, (pp. 25–34). ACM.

Sakaki, T., Okazaki, M., & Matsuo, Y. (2013). Earthquake shakes twitter users: Real-time event detection by social sensors. In *IW3C2*, (pp. 851–860). DOI: 10.1145/1772690.1772777.

Sutton, J., Spiro, E. S., Johnson, B., Fitzhugh, S., Gibson, B., & Butts, C. T. (2014). Warning tweets: serial transmission of messages during the warning phase of a disaster event. *Information, Communication and Society, 17*(6), 765–787.

Tapia, A., Moore, K., Johnson, N. (2013). Beyond the trustworthy tweet: A deeper understanding of microblogged data use by disaster response and humanitarian relief organizations. In *ISCRAM* (pp. 770–779).

Tomaszewki, B. (2015). *Geographic information systems (GIS) for disaster management*. LLC, Boca Raton, FL: CRC Press Taylor and Francis Group.

Tsou, M. H. (2011). Revisiting web cartography in the United States: The rise of user-centered design. *Cartography and Geographic Information Science, 38*(3), 249–256.

Tsou, M. H. (2015). Research challenges and opportunities in mapping social media and Big Data. *Cartography and Geographic Information Science, 42*(sup1), 70–74. doi:10.1080/15230406.2015.1059251.

Tsou, M. H., & Leitner, M. (2013). Editorial: Visualization of social media: seeing a mirage or a message? In special content issue: Mapping cyberspace and social media. *Cartography and Geographic Information Science, 40*(2), 55–60.

Tsou, M. H. & Sun, C. H. (2007). Mobile GIServices applications in disaster management, book chapter in dynamic and mobile GIS: Investigating change in space and time. In J. Drummond, R. Billen, D. Forrest & Joao (Eds.), *Innovations in GIS book series* (pp. 213–236) London: CRC Press Taylor and Francis.

Veil, S. R., Buehner, T., & Palenchar, M. J. (2011). A work-in-process literature review: Incorporating social media in risk and crisis communication. *Journal of Contingencies and Crisis Management, 19,* 110–122. doi:10.1111/j.1468-5973.2011.00639.x.

Vieweg, S., Amanda, L. H., Starbird, K., & Palen, L. (2010). Microblogging during two natural hazards events: What Twitter may contribute to situational awareness. In *Proceedings of the SIGCHI conference on human factors in computing systems* (pp. 1079–1088). ACM.

Virtual Social Media Working Group (VSMWG). (2014) Using social media for enhanced situational awareness and decision support. Homeland Security Science and Technology.

Wang, Z., Ye, X., & Tsou, M. H. (2016). Spatial, temporal, and content analysis of Twitter for wildfire hazards. *Natural Hazards, 83*(1), 523–540. doi:10.1007/s11069-016-2329-6.

The Academic SDI—Towards Understanding Spatial Data Infrastructures for Research and Education

Serena Coetzee, Stefan Steiniger, Barend Köbben, Adam Iwaniak, Iwona Kaczmarek, Petr Rapant, Antony Cooper, Franz-Josef Behr, Govert Schoof, Samy Katumba, Rumiana Vatseva, Kisco Sinvula and Harold Moellering

Abstract The demand for geospatial data across different disciplines and organisations has led to the development and implementation of spatial data infrastructures (SDI) and the theory and concepts behind them. An SDI is an evolving concept about facilitating and coordinating the exchange of geospatial data and

S. Coetzee (✉) · A. Cooper · K. Sinvula
Department of Geography, Geoinformatics and Meteorology,
Centre for Geoinformation Science, University of Pretoria, Pretoria, South Africa
e-mail: serena.coetzee@up.ac.za

A. Cooper
CSIR Built Environment, Pretoria, South Africa
e-mail: acooper@csir.co.za

K. Sinvula
e-mail: ksinvula@gmail.com

S. Steiniger
CEDEUS & Department of Transportation & Logistics,
Pontifical Catholic University of Chile, Santiago de Chile, Chile
e-mail: ssteiniger@uc.cl

B. Köbben
ITC—University of Twente, Enschede, Netherlands
e-mail: b.j.kobben@utwente.nl

A. Iwaniak · I. Kaczmarek
Wroclaw University of Environmental and Life Sciences, Wroclaw, Poland
e-mail: adam.iwaniak@up.wroc.pl

I. Kaczmarek
e-mail: kaczmarek.iw@gmail.com

P. Rapant
Institute of Geoinformatics, Faculty of Mining and Geology,
VSB—Technical University of Ostrava, Ostrava, Czechia
e-mail: petr.rapant@vsb.cz

M.P. Peterson (ed.), *Advances in Cartography and GIScience*, Lecture Notes in Geoinformation and Cartography, DOI 10.1007/978-3-319-57336-6_8

services between stakeholders from different levels in the spatial data community. Universities and other research organisations typically have well-established libraries and digital catalogues for scientific literature, but catalogues for geospatial data are rare. Geospatial data is widely used in research, but geospatial data produced by researchers is seldom available, accessible and usable, e.g., for purposes of teaching or further research after completion of the project. This chapter describes the experiences of a number of SDI implementations at universities and research institutes. Based on this, the Academic SDI, an SDI for research and education, is defined and its stakeholders are described. The purpose, scope and stakeholders of the Academic SDI are described based on the formal model of an SDI developed by the International Cartographic Association (ICA) Commission on SDIs and Standards (formerly the Commission on Geoinformation Infrastructures and Standards). The results contribute to understanding the state-of-the-art in SDI implementations at universities and research institutes; how the Academic SDI differs from a 'regular' SDI; and which role players need to be involved in a successful SDI implementation for research and education.

Keywords Spatial data infrastructure · SDI · University · Research · Education · Stakeholder

F.-J. Behr
Stuttgart University of Applied Sciences, Stuttgart, Germany
e-mail: franz-josef.behr@hft-stuttgart.de

G. Schoof
GIS Centrum Voor Informatie Technologie, Rijksuniversiteit Groningen, Groningen, Netherlands
e-mail: g.schoof@rug.nl

S. Katumba
Gauteng City-Region Observatory, A Partnership of University of Johannesburg, University of The Witwatersrand (Johannesburg), Gauteng Provincial Government and Organised Local Government in Gauteng, Johannesburg, South Africa
e-mail: samy.katumba@gcro.ac.za

R. Vatseva
National Institute of Geophysics, Geodesy and Geography, Bulgarian Academy of Sciences, Sofia, Bulgaria
e-mail: rvatseva@bas.bg

H. Moellering
Harold Moellering, Department of Geography, Ohio State University, Columbus, OH, USA
e-mail: moellering.1@osu.edu

1 Introduction

A national spatial data infrastructure (SDI) facilitates the sharing of public geospatial data within a country. Similarly, SDIs maintained by regional and local governments permit data sharing at different administrative levels. The situation is different in the field of education and research. Like books and journals, geospatial data is also widely used in research, but geospatial data produced by researchers is rarely distributed to others or archived properly, and therefore seldomly registered in specialized data catalogues.

Implementations of SDIs and associated catalogues of geospatial data at universities and research institutes would improve the availability, accessibility and (re-)usability of geospatial data by scientists and others (Bernard et al. 2014). This would not only make it possible to use the same data in subsequent research, but—with an eye on publications and research integrity—would also enable reproducible research, and perhaps, most importantly, make research results accessible to funders and to the public.

This chapter presents a brief overview of SDI implementations at research institutes in Chile and South Africa and at universities in Czechia, the Netherlands, Poland and South Africa. The implementations are described in more detail in Coetzee et al. (2017). For each Academic SDI, the purpose, stakeholders and lessons learned are presented based on the formal model of an SDI developed by the International Cartographic Association (ICA) Commission on SDIs and Standards (formerly the Commission on Geoinformation Infrastructures and Standards) (Hjelmager et al. 2008).

2 Demand for SDIs at Universities and Research Institutes

Research in many disciplines requires geospatial data that describe urban, rural or natural landscapes. For instance, geography ranges from cultural geography over physical geography to economic geography and requires a multitude of data, including demographic, economic, urban infrastructure, land cover and land use data (Haggett 2001). As knowledge evolves, newer methods and approaches are developed and influenced by research(ers) from different disciplines, resulting in inter-disciplinary methods and approaches. For instance, for transportation planning nowadays one needs to take geographical, demographical/social, psychological and environmental aspects into account (Sagaris 2014). Academic SDIs can help to fulfill these wider data needs in different ways.

Apart from demand for SDIs in research, there are two distinct roles for SDIs in education. Firstly, as a tool to make data searchable and accessible for exercises, projects and thesis work. This is relevant for students across a broad range of disciplines where geospatial data is used. Secondly, in programmes where geographic information technology itself is the subject of study, the teaching methods are founded in engineering science. Here, the SDI can serve as a best practice example by exposing students to SDI technology as a subject of study, and by creating a realistic working environment.

Implementing an SDI presents the opportunity to make data and research results available internally and externally. Publicly accessible results increase transparency towards the public, who often funds research (indirectly), and also allows other (external) researchers to reproduce experiments with the original data. The frequency of non-reproducible research results has raised critical discussions in the science community (e.g., Rey 2014), led to retractions of published papers and prompted publishers (e.g., Nature 2016) into setting new policies on the availability of data, methods and materials presented in research publications.

Data providers also benefit from the availability of, and access to, core geospatial data (such as, topographical data, administrative boundaries, geology, hydrology) in academic research and teaching. There is the opportunity for feedback on data quality and fitness for use; for developing innovative applications and business opportunities; and for students to become 'real' geospatial data users, who are more likely to use such data again when they enter the labour market (Medycky-Scott et al. 2011).

An example of making geospatial data available for universities is the Virtual Map Forum 2.0 at the Saxon State and University Library (SLUB) (Saxon State and University Library 2016) in Dresden (Germany) which provides a map–based graphical user interface allowing spatial-temporal browsing through a collection of historical maps. University library portals with links for accessing geospatial data include MIT Libraries (2016), Michigan State University Libraries (2016), Stanford University Libraries (2016), and the Harvard Geospatial Library (2016). Another example is the Geographic and Statistical Information Centre (Centre GéoStat) of Laval University's Library hosts a collection of raster and vector datasets useful for many different fields. A powerful semantic search and discovery engine, *Géoindex +* , combines traditional text-based search and spatial filters (Ouellet and Biondo 2012). Dodsworth and Nicholson (2015) describe how Google Earth can be used as a data retrieval tool in libraries. Scanned maps are georeferenced and presented together with digital geospatial data (in KML) on a virtual globe.

Despite the many benefits, most universities do not have an SDI for research and education. For example, at the Stuttgart University of Applied Sciences (Germany), where GIS and geospatial science is taught in several Bachelors and Masters degrees, the geospatial data resources are listed in the university's central e-learning platform without search or download capabilities.

3 The ICA's SDI Model

The ICA Commission on SDI and Standards has developed a conceptual model of an SDI (Hjelmager et al. 2008; Cooper et al. 2011, 2012). The model describes an SDI from different viewpoints specified in the Reference Model for Open

Table 1 SDI stakeholders (*Source* Hjelmager et al. 2008)

Stakeholder	Description
PolicyMaker	Sets the policy pursued by an SDI and all its stakeholders
Producer	Produces SDI data or services
Provider	Provides data or services to users throughout SDI
Broker	Brings users and providers together and assists in the negotiation of contracts between them
Value-added reseller (VAR)	Adds some new feature to an existing product or group of products, and then makes it available as a new product
EndUser	Uses the SDI for its intended purpose

Distributed Processing (RM-ODP) (ISO 1998). Each viewpoint provides a different abstraction of the SDI. The model describes six SDI stakeholders (see Table 1). It has been applied to understand several SDIs, e.g., Sinvula et al. (2017); here we use it to describe stakeholders in an Academic SDI.

4 SDI Implementations at Universities and Research Institutes

4.1 University of Twente, The Netherlands

At ITC (now part of the University Twente), an initiative was started, already in 1999, to set up what was then called a Common Database Project (Hootsmans et al. 1999). Although not yet named an SDI, it was one for all intents and purposes, as can be seen from the setup in Fig. 1.

The purpose was to gather experience in-house with the integrated approach that we nowadays call SDI. A secondary purpose was to demonstrate the benefits of such an SDI in comparison to the (at that time) conventional ways of structuring, storing and distributing geospatial data. Stakeholders are described in Table 2; lessons learned are listed in Table 3.

4.2 University of Groningen, the Netherlands

Geodienst successfully implemented a spatial expertise group with the goal of *'Enabling better research through the use of spatial information'*. Since this has proven to be difficult for many institutions it is an interesting case. The focus is not on the SDI or tools, but rather on the applications of customers. Tasks are to promote the application of spatial information actively; to provide expertise; and to manage data, infrastructure and software.

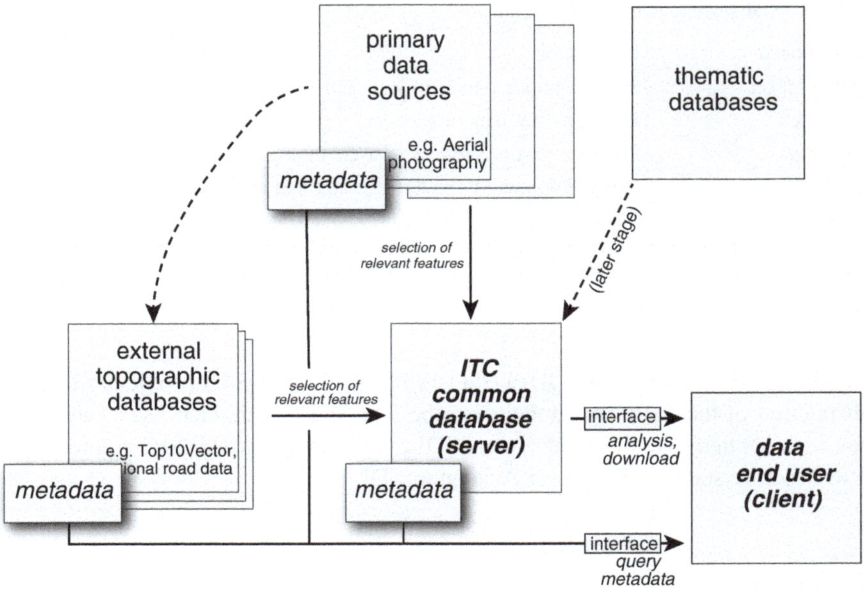

Fig. 1 Setup of the Common Database Project (CDP) (from Hootsmans et al. 1999)

Table 2 SDI stakeholders: University of Twente

PolicyMaker	Four-person steering group
Producer	(National) Mapping and Cadastral agencies ITC staff
Provider	ITC
Broker	n/a
VAR	Students and researchers
EndUser	Students and staff (researchers and educators)

Table 3 Lessons learned: University of Twente

PolicyMakers were mostly interested in the technological side and did not manage to identify, let alone activate, the main stakeholders, especially the EndUsers
SDI technology was still in its infancy and standards did not exist, e.g., the WMS 1.0.0 standard was only published a year later (OGC 2010). Probably, the CDP came a little too early to be successful
CDP as a learning tool was a success because ITC staff gained early and hands-on exposure to the technology and organisational aspects of SDI
CDP has led to the integration of the SDI^{light} concept (Köbben et al. 2010) in all teaching at ITC: the principles of SDI are applied so that the SDI serves as subject of study, as well as a working environment

Table 4 SDI stakeholders: University of Groningen

PolicyMaker	Geodienst management, Faculties, CIT Board
Producer	Data is harvested from different sources Own productions or alterations
Provider	Geodienst
Broker	n/a
VAR	n/a
EndUser	Researchers, teachers, students and administrative staff

Table 5 Lessons learned: University of Groningen

An SDI involves not only the hard- and software but also GIS-expertise and communication
Focusing on the SDI itself does not lead to success; one has to focus on the barriers and remove them, whatever they may be
Promotion is important; one has to take responsibility for selling the idea of an SDI, don't wait for others to do it
Students are capable of handling responsibilities in an SDI implementation
Results, together with political goodwill, lead to permanent funding

Geodienst is organised like a start-up, with very little organizational hierarchy and a lot of freedom and responsibility. This leads to quick and pragmatic responses to support calls and projects, fast internal learning, but a bit of chaos. More information is available at http://rug.nl/geo. Stakeholders are described in Table 4; lessons learned are listed in Table 5.

4.3 VSB—Technical University of Ostrava, Czechia

The Institute of Geoinformatics, VSB—Technical University of Ostrava (VSB-TUO), Czechia, participated in the development and deployment of the meta-information system, MIDAS (MetaInformation Database System) (Pauknerova et al. 2002), which was established, amongst others, as a pilot meta-information system of public administration.

Many teams at VSB-TUO work intensively with geospatial software and data from both internal (e.g., collected through research) and external (e.g., licensed to the whole VSB-TUO) sources. The Institute addressed other teams with the offer to establish a meta-information system for VSB-TUO based on MIDAS with the aim of becoming the foundation of the university's infrastructure for geospatial data and services.

Unfortunately, the *EndUsers* could not be convinced of the usefulness of this system, and eventually it was decided to terminate the project. Stakeholders are described in Table 6; lessons learned are listed in Table 7.

4.4 CSIR, South Africa

The CSIR, a public science council, operates through a number of largely independent business units. The CSIR contributes in various ways to base geospatial data sets of several South African SDI custodians. A basic SDI makes base geospatial data available internally. Sharing of geospatial data across business units outside of projects is limited, because of commercial sensitivities, protecting personal data, and military and other secrets. Because costs are incurred at the lowest level in the organisation, it is difficult to establish large and expensive organisation-wide systems, such as a comprehensive SDI. Stakeholders are described in Table 8.

4.5 Research Centre for Sustainable Urban Development (CEDEUS), Chile

The CEDEUS research centre unites university researchers that work on different aspects of urban development, e.g., use of natural resources, transport and mobility,

Table 6 SDI stakeholders: VSB—Technical University of Ostrava

PolicyMaker	A group of interested people with no institutional support
Producer	External data producers data collected by EndUsers during research or teaching
Provider	MIDAS (MetaInformation Database System)
Broker	n/a
VAR	n/a
EndUser	Research and educational groups, researchers and students

Table 7 Lessons learned: VSB—Technical University of Ostrava

To ensure wider acceptance, institutional support from the VSB-TUO would be required
No awareness of the importance of data (not only geospatial data) at the university level and consequent absence of university policies in this area
Concerns of faculties, researchers and educators that they would have to fund the SDI
Reluctance to share data
Fear of losing control over datasets when they are stored centrally

Table 8 SDI stakeholders: CSIR

PolicyMaker	CSIR Executive
Producer	Custodians
Provider	CSIR Computing Services
Broker	n/a
VAR	n/a
EndUser	Naive to expert, some source and maintain themselves all their data

Table 9 SDI stakeholders: CEDEUS

PolicyMaker	GIS/SDI team and CEDEUS board of directors (i.e., principal investigators)
Producer	Data producers: GIS/SDI team, researchers, also 'external sources', such as governmental authorities (i.e., ministries) and other research centres or universities Service producer: GIS/SDI team
Provider	GIS/SDI (original intentions were that all (40) CEDEUS researchers and technical staff would be Providers)
Broker	GIS/SDI team, and the data platform of the hosting university
VAR	CEDEUS (sustainability indicators for cities based on SDI data)
EndUser	GIS/SDI team, researchers, and the public: mainly (thesis) students, postdocs, a small portion of centre and external researchers and educators, but most importantly, also the public

Table 10 Lessons learned: CEDEUS

CEDEUS researchers turned out to be fairly resistant to learn and use the SDI. The number of student EndUsers is growing.
Unexpectedly, some documents (i.e., municipal development plans from some municipalities) experience demand from the public, most likely because some municipalities have not made these documents available online
Success of data demand depends on (a) attractive data (i.e., focused on certain users and their demand); and (b) public relation strategies to make the SDI known to others (e.g., social networks, introductory GIS workshops for students, indexing of the SDI content by Google's Search engine)
The SDI has not reached its objective of researchers sharing their data. The country's (research) culture with a strong thinking of 'data ownership = research advantage' plays a strong role

built environment and social segregation. The centre's SDI, called Observatory was funded to support the centre's proclaimed goals—amongst others, to encourage inter-disciplinary research; to monitor urban development; and to propose public policies—in the form of a service unit. That is, the SDI team provides a 'drop box' and catalogue of geospatial data and offers training and support to students and researchers not familiar with GIS. Stakeholders are described in Table 9; lessons learned are listed in Table 10.

4.6 University of the Witwatersrand (Wits), Johannesburg, South Africa

The GIS and remote sensing group is a multi-disciplinary group with members across Wits who develop and use GIS and remote sensing for teaching and research.

Table 11 SDI stakeholders: Wits

PolicyMaker	School of geography, archeology and environmental studies (GAES) approves policies GIS and remote sensing group champions the initiative and also plays the role of secretariat for the initiative
Producer	Gauteng city-region observatory (GCRO) and other private and public sector producers
Provider	Data: GAES (on ftp server) Services: members of the GIS and remote sensing group
Broker	n/a
VAR	Data providers may act as VARs
EndUser	Students, academic staff, group members and the general public

Table 12 Lessons learned: Wits

Even though there is much enthusiasm among the members of the GIS and remote sensing group, who continue to increase in numbers, there is currently no buy-in GAES to support the efforts with funding

The fields of expertise include geography, archaeology, environmental studies, geosciences, statistics, computational sciences, engineering, health and social sciences. The group of approximately 20 members (lecturers, researchers and IT professionals) produces various freely available services, tools and datasets for spatio-temporal analysis. Data from external sources and value-added data, obtained as output from research work, are shared through an ftp server. In 2013, the group initiated a project of constructing a spatial geodatabase and also expressed the need for a university-based server that can meet the needs of the wider group of users across campus. Stakeholders are described in Table 11; lessons learned are listed in Table 12.

4.7 Academic Geo Hub Platform, Wroclaw University of Environmental and Life Sciences (Poland)

The main objective of the Academic Geo Hub Platform was to create a common platform for the exchange and sharing of geospatial data for scientists in Poland. It was established in 2015 as a part of the PLGrid NG project ('New generation domain-specific services in the PL-Grid Infrastructure for Polish Science', 2014–2015) carried out by the PL-Grid Consortium.

Today, the Academic Geo Hub is a repository, which acts as a science laboratory combined with a social networking service and semantic web solutions. Users can transfer, download, integrate and publish data, as well as manage access to them. It

Table 13 SDI stakeholders: Wroclaw University of Environmental and Life Sciences

PolicyMaker	PLGrid NG project steering group
Producer	Public administrators Academic community (both researchers and students)
Provider	Wrocław Centre for Networking and Supercomputing, Wrocław University of Technology
Broker	n/a
VAR	n/a
EndUser	Academic community (both researchers and students)

Table 14 Lessons learned: Wroclaw University of Environmental and Life Sciences

Due to limited awareness of open data at the time, only few public entities agreed to share their data for the project. This has changed and it may be good timing to re-establish cooperation with them now
The main difficulty in the functioning of the platform is the lack of a business model and policies for SDIs in education and research. As a result, there is no permanent funding and therefore maintenance of the platform in the long run is difficult

supports collaboration among research teams, especially for young researchers who have obtained partial results, not yet published in journals. Stakeholders are described in Table 13; lessons learned are listed in Table 14.

5 The Academic SDI

Following the descriptions in the previous section, the purpose of the Academic SDI is to make geospatial data produced for and by research and education discoverable, accessible and usable, primarily within universities and research institutes, but sometimes also for a wider audience, such as the general public. An Academic SDI may be implemented for users from a single organisation (i.e., a single university or institute), or for users from multiple organisations (i.e., several universities or institutes).

Figure 2 provides a first impression of Academic SDI stakeholders, based on the ICA's SDI model and the SDI implementations presented in this chapter. Classes with a transparent background are original classes from the ICA's SDI model; classes with a gray background are specialisations for research and education. In future work, these stakeholders need to be elaborated in more detail, e.g., post-doctoral students and technical support staff are not yet represented.

Data provided by an Academic SDI originates from many different sources and does not only include data created within the organisation; data may also be produced (and maintained) by external entities, such as government entities or other Academic SDIs. Ideally, the Academic SDI integrates geospatial datasets from

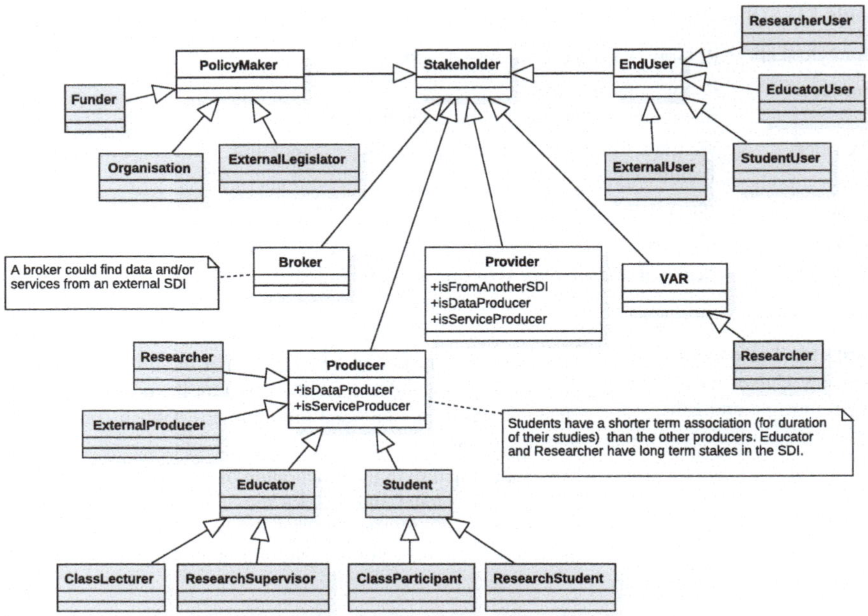

Fig. 2 Stakeholders in the Academic SDI

external SDIs through harvesting metadata. The metadata for datasets produced by researchers has to describe the process or scientific method for developing the dataset.

In many cases, the Academic SDI facilitates access to already existing data produced by other researchers. In these cases, the researcher is considered to be a *VAR*, i.e., the researcher adds value to already published data, though this will not necessarily be for resale.

An Academic SDI is different from a 'normal' SDI in various ways. The SDI implementations described in this chapter show that the relationship between the *Producer*, *Provider*, *PolicyMaker*, and *EndUser* is much closer and more collaborative in nature than in an SDI by a public administration with a top-down data distribution model. Therefore, an Academic SDI often resembles a bottom-up or demand-driven approach and *EndUsers* should therefore be represented among the *PolicyMakers*.

Furthermore, the Academic SDI is not built on the basis of legislation (e.g., administered by a Minister of Education), but in response to project and teaching requirements at universities and research institutes. Due to the silo nature common in academic and research institutions and limited sharing of data sets because of project sensitivities, data hoarding, etc., an Academic SDI tends to consist of a federated set of data islands rather than an integrated network of connected data nodes. Also, there is no central access point to data compiled by researchers. Without a (university) policy that imposes data sharing, establishing an

Academic SDI will always be a challenge because each individual has to be convinced to participate in the SDI.

Due to the close relationship between *Producers*, *Providers* and *EndUsers* in an Academic SDI, the SDI very often contributes to education and capacity building. *Producers* and *Providers* (e.g., researchers, educators and postgraduate students) who maintain the SDI improve their own skills and can train *EndUsers* (e.g., other researchers and undergraduate students) on geospatial information and its applications. Considering the educational environment, the Academic SDI serves not only as data source for teaching, but can also be used as a training platform for students who have to learn how to implement and manage an SDI.

6 Discussion and Conclusion

This chapter presented a first description of an SDI for research and education, the Academic SDI. The description draws on SDI implementations at selected universities and research institutes in the Czechia, Chile, the Netherlands, Poland and South Africa. The Academic SDI is described according to the formal model of an SDI developed by the ICA Commission on SDIs and Standards. The purpose, scope, important stakeholders and first lessons learned from SDI implementations at universities and research institutes are identified. This description of the Academic SDI is a specialization of the general SDI model developed by the Commission.

An Academic SDI can serve as a common data exchange platform in the sense of a geospatial data 'drop box', which researchers use to upload, search, and download data (Bernard et al. 2014). Apart from benefits relating to the availability of data from different sources, there are time and cost savings by avoiding duplication of data collection and acquisition across different universities or research institutes (Rajabifard and Williamson 2001; Crompvoets et al. 2004). With an Academic SDI comes furthermore the opportunity for creating a team of experts who manage and maintain the Academic SDI, train people from different fields in the use of GIS, develop and promote best (mapping) practices and generally raise awareness of the value of geographic information.

A significant barrier to SDI implementations described in this chapter was the absence of sustainable multi-year funding with respect to equipment and personnel. A second significant barrier was the lack of awareness among decision makers in the organisation and/or funding agencies of the value of the information itself, and of sharing information. Consequently, policies to manage data produced by research within the institutions are lacking.

The value of an SDI can be measured in two ways: firstly, by outlining the utility of geospatial information for research and education, and secondly, by estimating the time and money saved through sharing. For an Academic SDI, the value of the SDI as a training platform can be added. Similar to 'regular' SDIs, one has to be aware that the cost-benefit of an SDI is difficult to determine and benefits take years to realize, amongst others, due to resistance or reluctance by researchers to accept such 'new' technology; the reluctance to share data because of the fear to lose control over a publication or further research advantage; and lacking awareness of the SDI among potential users outside the organisation.

The information about stakeholders and their roles, problems and success factors with SDI implementations contributes to the elaboration of the state-of-the-art in the implementation of Academic SDIs. It gives first hints on the stakeholders needed and how they may be involved in an SDI implementation at a university or research institute. Future work should describe the Academic SDI in more detail, e.g., how to implement it.

Acknowledgements Stefan Steiniger is grateful for funding by Conicyt (Conicyt/Fondap/15110020 and Conicyt/Fondecyt/1150239). Samy Katumba acknowledges the assistance and cooperation of Dr. Stefania Merlo, Chair of the GIS and remote sensing group at the University of the Witwatersrand, South Africa.

References

Bernard, L., Mäs, S., Müller, M., Henzen, C., & Brauner, J. (2014). Scientific geodata infrastructures: Challenges, approaches and directions. *International Journal of Digital Earth, 7*(7), 613–633.

Coetzee, S., Steiniger, S., Köbben, B., Iwaniak, A., Kaczmarek, I., Rapant, P., et al. (2017). *SDI implementations at universities and research institutes*. Retrieved January 21, 2017, from http://sdistandards.icaci.org/resources/.

Cooper, A. K., Moellering, H., Hjelmager, J., Rapant, P., Delgado, T., Laurent, D., et al. (2012). A spatial data infrastructure model from the computational viewpoint. *International Journal of Geographical Information Science, 27*(6), 1133–1151.

Cooper, A. K., Rapant, P., Hjelmager, J., Laurent, D., Iwaniak, A., Coetzee, S., et al. (2011). Extending the formal model of a spatial data infrastructure to include volunteered geographical information. *In 25th International Cartographic Conference* (*ICC*). Paris, France.

Crompvoets, J., Bregt, A., Rajabifard, A., & Williamson, I. (2004). Assessing the worldwide developments of national spatial data clearinghouses. *International Journal of Geographical Information Science, 18,* 665–690.

Dodsworth, E. H., &, Nicholson A. (2015). *Using google earth in libraries: A practical guide for librarians*. rowman and littlefield.

Haggett P. (2001). *Geography: A global synthesis*. Longman.

Harvard Geospatial Library (2016). *Welcome to the Harvard Geospatial Library*. Retrieved January 21, 2017, from http://calvert.hul.harvard.edu:8080/opengeoportal/.

Hjelmager, J., Moellering, H., Cooper, A., et al. (2008). An initial formal model for spatial data infrastructures. *International Journal of Geographical Information Science, 22*(11–12), 1295–1309.

Hootsmans, R., Huurneman, G., Lemmens, R., & Paresi, R (1999). *Provisional plan for common database project (CDP)*. Internal report, ITC. Retrieved February 10, 2017, from http://kartoweb.itc.nl/kobben/publications/ProvisionalPlanforCommonDatabaseProject.pdf.

Köbben, B., de By, R., Foerster, T., Huisman, O., Lemmens, R., & Morales, J. (2010). Using the SDIlight approach in teaching a Geoinformatics master. *Transactions in GIS, 14*(s1), 25–37.

Medyckyj-Scott, D., Sutton, E., Higgins, C., & Heywood I. (2011). *The European National Mapping and Cadastral Agencies and Higher Education A Market for Geospatial Information —Myth or Reality.*

Michigan State University Libraries (2016). *GIS data: Worldwide data.* Retrieved January 21, 2017, from http://libguides.lib.msu.edu/gisdata.

MIT Libraries (2016). *Geographic information systems (GIS): General data for the world.* Retrieved January 21, 2017, from http://libguides.mit.edu/c.php?g=176295&p=1161383.

Nature (2016). *Policies: Availability of data, materials and methods.* http://www.nature.com/authors/policies/availability.html. Retrieved21 Jan 2017.

OGC (2010). *OpenGIS web map server interface implementation specification revision 1.0.0.* Technical report 00–028, Open Geospatial Consortium.

Ouellet, M. and Biondo, S. (2012). An open and powerful GIS data discovery engine. In *Proceedings OGRS 2012*. Retrieved October 24, 2016, from http://www.ogrs-community.org/proceedings/2012_proceedings/ogrs2012_symposium_proceedings.pdf.

Pauknerova, E., Horakova, B., & Ruzicka J. (2002). metainformation database system MIDAS as an On-Line Catalogue of Geodata of Public Administration in the transformation process. In *8th EC-GI and GIS WORKSHOP, ESDI—A Work in Progress*, Dublin, Ireland July 3–5, 2002. Retrieved February 10, 2017, from http://www.ec-gis.org/Workshops/8ec-gis/cd/presentations/3_sp_ep.pdf.

Rajabifard, A., & Williamson, I. P. (2001). Spatial data infrastructures: Concept, SDI hierarchy and future directions. In: *Proceedings of Geomatics '80*. Tehran, Iran.

Rey, S. J. (2014). Open regional science. *The Annals of Regional Science, 52*(3), 825–837.

Sagaris, L. (2014). Citizen participation for sustainable transport: The case of 'Living City' in Santiago, Chile (1997–2012). *Journal of Transport Geography, 41,* 74–83.

Saxon State and University Library (2016). *The map collection*. Retrieved October 20, 2016, from https://www.slub-dresden.de/en/collections/maps/the-map-collection.

Sinvula, K. M., Coetzee, S., Cooper, A. K., Owusu-Banahene, W., Nangolo, et al. (2017). A comparative analysis of stakeholder roles in the spatial data infrastructures of South Africa, Namibia and Ghana. Submitted paper, *International Journal of Spatial Data Infrastructures Research.*

Stanford University Libraries. (2016). *Data*. Retrieved January 21, 2017, from https://library.stanford.edu/research/stanford-geospatial-center/data.

Part III
Map Design

Introducing Leader Lines into Scale-Aware Consistent Labeling

Hsiang-Yun Wu, Shigeo Takahashi, Sheung-Hung Poon and Masatoshi Arikawa

Abstract Consistently placing annotation labels across map scales often poses a problem due to the restriction of the screen space. This problem becomes further exacerbated when we navigate by arbitrarily zooming in and out of digital maps on mobile devices. In this paper, we introduce leader lines to conventional techniques for scale-aware consistent labeling to accommodate more annotation labels on the map domain while retaining their plausible arrangement. The overall visibility of annotation labels is optimized using genetic algorithms while avoiding their unwanted popping effects and sudden leaps regardless of the change in the map scale. The feasibility of the proposed approach is demonstrated by experimental results including comparison with relevant techniques.

Keywords Active range optimization • Dynamic labeling • Leader lines • Genetic algorithms

1 Introduction

The development of mobile devices and digital map services allows us to read maps not only on paper but also on electronic display devices. This technology advancement enables us to zoom in and out of the map content for an overview and

---

H.-Y. Wu (✉)
Keio University, Yokohama, Japan
e-mail: hsiang.yun.wu@acm.org

S. Takahashi
University of Aizu, Aizu-Wakamatsu, Japan
e-mail: takahashis@acm.org

S.-H. Poon
Universiti Teknologi Brunei, Gadong, Brunei Darussalam
e-mail: sheung.hung.poon@gmail.com

M. Arikawa
The University of Tokyo, Tokyo, Japan
e-mail: arikawa@csis.u-tokyo.ac.jp

M.P. Peterson (ed.), *Advances in Cartography and GIScience*, Lecture Notes in Geoinformation and Cartography, DOI 10.1007/978-3-319-57336-6_9

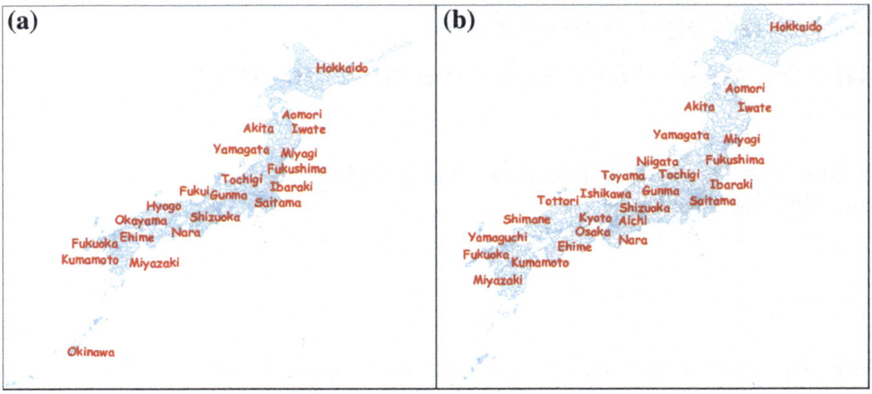

Fig. 1 Naïve inconsistent placement of labels as the scale changes from (**a**) to (**b**). All the labels that appear in (**a**) are expected to stay in (**b**) while some labels disappear (e.g., "Shimane" and "Yamaguchi") and other labels (e.g., "Kumamoto" and "Miyazaki") unexpectedly jump to different positions between the different map scales

details of areas of interest and is especially important to dynamically explore the map content across multiple scales within limited screen space. In addition, annotation labels help precisely locate specific geographic positions on maps and clearly discriminate them from the names of other places. Thus, laying out such labels consistently across the map scales will provide better readability of the map content on digital devices. However, consistency in label placement leads to technically challenging problems. Figure 1 shows specific types of such unwanted behaviors in which we can see popping effects and sudden leaps by the annotation labels as the scale changes. Avoiding visual distractions in label placement across multiple scales usually requires high computational costs because we have to investigate all possible combinations of labels and their spatial placement.

One promising solution to this problem is to retain consistency in the label positions on a map irrespective of the change in the map scale. This consistent map labeling problem was formulated as *active range optimization* (ARO) (Been et al. 2006, 2010). The active range refers to the range of the map scale in which the corresponding label sticks to the unique position on the map domain. Thus, the consistent labeling problem can be solved by maximizing the total sum of active ranges in such a way that as many annotation labels as possible remain and as long as possible within the space spanned by the map domain and scale. To achieve this, we extend the conventional ARO approaches by introducing *leader lines* that connect between the point features and the annotation labels. This permits us to spare a little space between a point feature and its corresponding label, which eventually relaxes the degree of freedom in the label placement and thus is more likely to accommodate more labels within the map domain.

For maximizing the total active range, we employ a genetic-based approach to seek better placement of the annotation labels across the multiple map scales. Specifically, we optimize the order of the labels to be placed and find the best order, where we try to place the labels in a greedy fashion according to the corresponding order. This allows us to finalize the best order of the labels that maximizes the visibility of consistent label placement across the entire range of the map scale.

Our technical contribution lies in the sophistication of previous techniques for consistently placing labels across multiple scales, by introducing leader lines that connect point features and labels to take advantage of the space around the point features. This labeling design is particularly useful for visualizing areas that have heterogeneous distributions of point features in such a way that they form a set of local clusters. This is because we can place annotation labels even when they are slightly away from the corresponding point features with the help of leader lines and thus can increase the total number of labels to be placed in the map. Figure 2 gives an example of such consistent placement of labels together with leader lines. After providing several experimental results, we compare the conventional label placement without leader lines and the proposed label placement with leader lines to demonstrate the feasibility of our labeling approach. While the leader line approach lends itself to relatively narrow land areas surrounded by water, as is the case with Japan, there are many other areas where it would be of use.

The remainder of this paper is structured as follows. In Sect. 2, we briefly summarize the most relevant literature to this study. In Sect. 3, we present an overview of our visualization framework, including the design of the leader drawing styles for consistent dynamic label placement. In Sect. 4, we then detail an algorithm for the extended ARO, which is effectively solved by the genetic-based heuristic and greedy label placement scheme. In Sect. 5, we provide several results generated by our system to demonstrate the feasibility of our approach. Finally, in Sect. 6, we conclude the paper and propose potential future extensions.

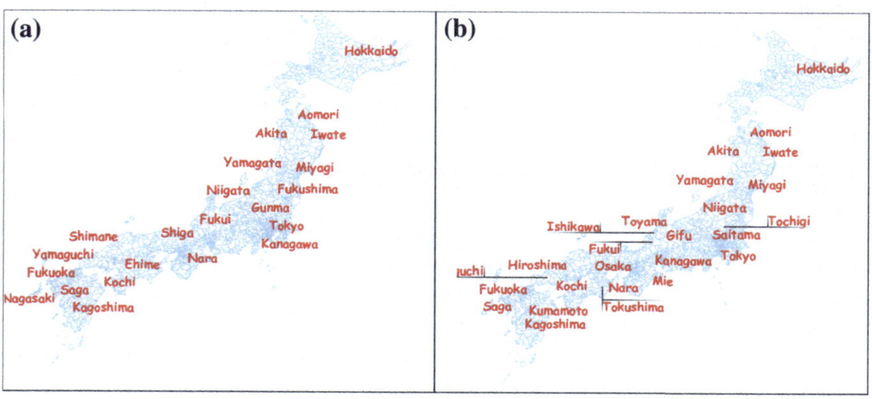

Fig. 2 Label placements (**a**) without leader lines (#{labels} = 22) and (**b**) optionally with leader lines (#{labels} = 26). We can place more labels by selectively employing leader lines to take advantage of available space within the map domain

2 Related Work

2.1 *Static Map Labeling*

Cartographers have investigated the conception of text discrimination, which is evaluated for noticing the difference between features in cartography and has been shown to be a key factor for map design (MacEachren 1995; Chiang et al. 2014). To improve text discrimination, labeling principles have been studied and summarized in several pioneering works (Yoeli 1972; Imhof 1975). Wolff (2009) conducted an extensive survey of techniques for automatic map labeling and thus provided specific instructions for cartographic name placement. For example, gradient-descent optimization (Hirsch 1982) and the integer programming model (Zoraster 1986) are classical static map labeling techniques that follow an empirical investigation to demonstrate the feasibility of each algorithm (Christensen et al. 1995). Afterward, a genetic algorithm was introduced to accelerate the computation time (Raidl 1998). Other map features, such as lines and areas, have also been considered to improve the readability of labels on maps (Edmondson et al. 1997; Luboschik et al. 2008), and domain-assisted label placement called boundary labeling (Bekos et al. 2007, 2015) was developed for this purpose. Appropriately drawing different leader line styles for label placement enhances map readability effectively (Lin 2010; Meng et al. 2015).

2.2 *Dynamic Map Labeling*

In practice, mobile devices are used for viewing maps, and thus, effective navigation, such as annotating labels within a lens, has become a popular solution (Fekete and Plaisant 1999). Fink et al. (2012) enhanced this layout by optimizing multiple design rules, such as minimizing the total leader length and distance between labels, and thus provided beautiful label placement in static and dynamic environments. A two-step labeling approach was introduced by Petzold et al. (2003), where the system computes conflicted labels in a preprocessing process and then applies the results to the user navigation. Gemsa et al. (2013) also conducted trajectory-oriented map labeling, where consistent label placement is computed that refers to moving history. These techniques were developed in a fixed scale and less consider the visual consistency of label placement during user interaction.

Schwartges et al. (2014) presented a slider model to move labels along the associated site points during scale changes, while a waiting function is employed to avoid flickering effects. As pioneering work on consistent label placement, Poon and Shin (2005) conducted another study on consistent label placement; however, in this approach, the orientation of a label is fixed. A significant extension of this approach was then proposed by Been et al. (2006, 2010) in which they relaxed the restrictions on the algorithm to place labels of various sizes. The method optimizes the active ranges of all labels by plotting rectangular pyramids in the three-dimensional (3D) space to compute the conflicts of labels at individual scales. Recently, this work was extended by Zhang et al. (2015) in which the active ranges are designed to be selectable, and the total weight of the associated ranges is maximized.

In this study, we relax the limitation of the approach proposed by Been et al. (2006, 2010) so that labels with leader lines can be employed to improve map readability. A genetic-based approach is introduced to solve the ARO problem (Raidl 1998; Wu et al. 2011), which provides an effective means for the map labeling problem, especially for optimizing fitness functions without being trapped by local extrema.

3 Design Principles for Label Placement

In this section, we provide an overview of the proposed approach with an explanation of the design principles for label placement. The primary technical challenge here is to develop scale-adaptive label placement while introducing leader lines in the ARO model. The key idea is to appropriately select a subset of visible labels at each scale to make full use of the available space around the corresponding point features. Leader lines effectively make it possible to increase such available space by connecting the labels and point features even when they are slightly away from each other.

In our approach, several design rules for placing annotation labels are introduced for this purpose. First, for each point feature, we allow an additional number of possible label positions that are a bit apart from the point feature as shown in Fig. 3b. The conventional label placement does not have leader lines as shown in Fig. 3a, which are all in touch with the point feature and commonly employed in geographical information systems (GISs). However, in this case, we often cannot find available labeling space especially in densely annotated regions where the point features are locally concentrated as a cluster because the four possible positions are more likely to be occupied first by other labels in the vicinity. In our approach, in addition to the conventional four positions (Fig. 3b top), we introduce 12 surrounding label positions (Fig. 3b, bottom) and connect the label and the feature point with a leader line if it can be placed in one of these 12 positions, where the leader line is assumed to be *straight*. Thus, leader lines will help us take advantage of more available space around the respective point features and still maintain the visual correspondence between the labels and the point features.

Please also note that each possible label position is indexed by the ID ranging from 1 to 16 as shown in Fig. 3b, where we employ horizontal leader lines for the positions indexed by 5, 8, 9, and 12, vertical leader lines for the positions indexed

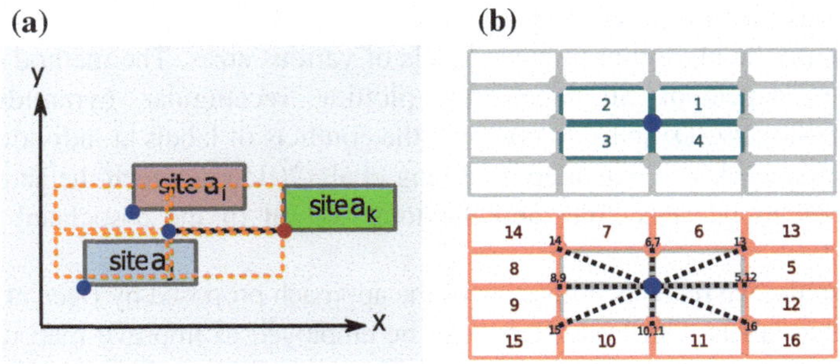

Fig. 3 Possible label positions to a given site point. (**a**) The conventional four-position model (in *orange*) usually suffers from the lack of available space around the point feature. (**b**) Our new 16-position model (*bottom*) consists of not only the conventional four positions (*top*) but also 12 additional positions. For each, we use the leader lines to connect the label and the point feature

by 6, 7, 10, and 11, and diagonal leader lines for the positions indexed by 13, 14, 15, and 16. Employing horizontal and vertical leader lines for the eight new label positions out of 12 effectively helps avoid unnecessary conflicts with existing annotation labels when composing *scale-aware* consistent label placement. Furthermore, these IDs indicate the order of preference for the 16 label positions at each point feature, which means that the smaller ID corresponds to a better label position. Based on this design principle, our algorithm will search for the position of each label in this preference order to arrange the label as close as possible while maximally keeping it to the top right of the point feature. Of course, this search process is conducted under the assumption that all the labels are placed without mutual overlap within the map domain at the respective scales.

Nonetheless, avoiding overlapping labels at multiple map scales inevitably results in inconsistent placement of labels as shown in Fig. 1. To solve this problem, we avoid such overlapping by considering the 3D space spanned by the 2D map domain and scale, by following the conventional ARO formulation (Been et al. 2006, 2010). In cartography, the map scale s often refers to the degree of magnification of the specific map region, which also indicates that the larger *cartographic scale* provides a more detailed representation. However, in this paper, we employ the *zooming scale* r (Been et al. 2006), which is defined as the inverse of the cartographic scale s (i.e., $r = 1/s$), as the map scale for later convenience. This implies that the labels tend to easily collide with each other once the zooming scale becomes larger as illustrated in Fig. 4a, and thus, the leader lines provide a powerful means of alleviating such conflicts among the labels. The idea permits us to utilize available space around the individual point features and encourages us to improve the associated map readability by maximizing the number of labels placed on the map.

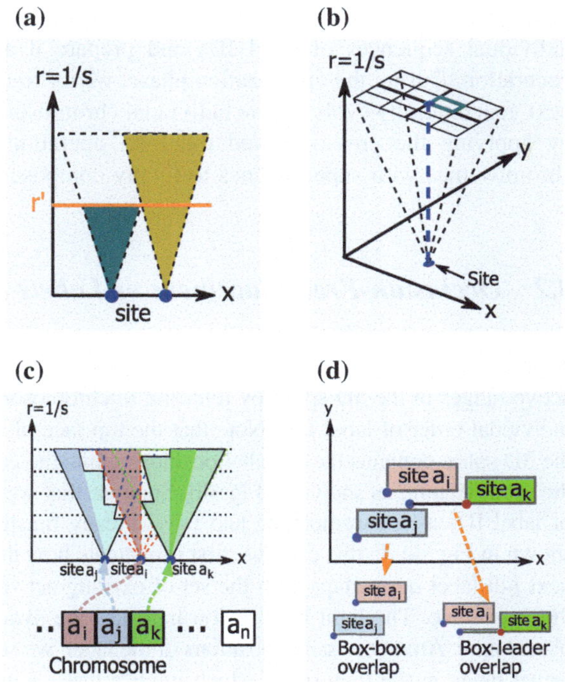

Fig. 4 Active ranges in the 3D space spanned by the map domain (x, y) and the zooming scale r. (**a**) The cross section of the pyramidal active range volume. (**b**) The *top* face of the active range consists of the set of possible label positions and its center corresponds to the feature point. (**c**) Greedy placement of active ranges by referring to a specific chromosome. (**d**) Possible conflicts among labels and leader lines in the active range optimization

4 Consistent Label Placement with Leader Lines

In this section, we describe how we extended Been et al.'s conventional ARO approach for implementing consistent placement of map labels with their leader lines across the scale. We explain our approach in three subsections: genetic-based optimization of active ranges, occlusion-free placement of labels and leader lines, and fitness evaluation of their spatial layout.

4.1 Genetic-Based Optimization of Active Ranges

As described previously, we employ the ARO technique (Been et al. 2006, 2010) to maintain the consistency in the label placement across multiple scales. For this purpose, we introduce genetic algorithms to optimize the placement of pyramidal active range volumes in the 3D space while avoiding unwanted overlap among labels at individual scales. In our formulation, we first encode a specific sequence of label IDs as a chromosome, then try to greedily place the active range volumes in this order (see Sect. 4.2), and finally evaluate the fitness of the chromosome by referring to the resultant label placement (see Sect. 4.3).

In this optimization, we initialize a set of chromosomes by randomizing the individual sequences of label IDs and prepare it as the population of the first generation. During the optimization phase, we try to update the population with the next generation by evolving the individual chromosomes. This is usually conducted by applying the crossover and mutation operations and replacing the existing chromosomes with superior ones to finally compose a set of elite chromosomes.

4.2 Occlusion-Free Placement of Labels and Leader Lines

Now we are ready to consider how we can find the actual placement of pyramidal active ranges in the 3D space by referring to chromosomes, each of which encodes an individual order of label IDs. Note that the top face of each pyramidal active range in the 3D space contains the 16 label positions while the center of the face corresponds to the point feature, as shown in Fig. 4b. Suppose that we currently have a specific order of label IDs as a chromosome and have already finished placing the *i*-th label a_i, as shown in Fig. 4c. In this case, we first investigate how the pyramidal active range of the next *j*-th label a_j overlaps with the set of existing active ranges, by properly sampling the map scale. The light blue region indicates the available space for the label to be placed next. Among possible positions of the label, we select the one that has the largest active range in the map scale, which implies that we employ the active range in blue from the available 3D space. Furthermore, we also have to record the 3D space that is already occupied because this is no longer available for the labels to be placed later.

We also have to avoid unexpected overlap between the labels and leader lines when we select one of the newly introduced 12 label positions as shown in Fig. 4d. For this purpose, we approximate the shape of each leader line as a bounding box that encloses the line. For example, we employ a thin rectangle as the bounding box if the leader line is exactly horizontal or vertical. If the leader line is slanted, we insert a normal rectangle as the leader line's bounding box in such a way that the diagonal of the rectangle coincides with the leader line. This inevitably increases the number of bounding boxes we have to maintain at each sampled map scale while effectively facilitates implementation of consistent label placement with leader lines by taking advantage of existing data structures.

4.3 Fitness Evaluation of the Label Placement

The final task is to formulate the fitness function for each chromosome, so that we can systematically improve the set of chromosomes during the genetic-based evolution. In our approach, we employ the following function

$$\{\text{Weight}\} \times (M - \{\text{Label position ID}\}) \times \{\text{Active range}\}, \qquad (1)$$

to evaluate the fitness of an individual label. Here, {Weight} is the weight assigned to the label, {Label position ID} is the ID among the 16 label positions, and {Active range} is the size of the corresponding active range on the map scale. Moreover, we set $M = 20$ in Eq. (1) so that we can assign larger fitness values for the smaller label position IDs. We sum up the fitness values for all the labels to evaluate the overall label placement of each chromosome. During the chromosome evolution in the optimization, we maximize this fitness function to find a satisfactory solution for the scale-aware label placement.

5 Results

We employed a dataset that has 3, 887 point features over Japan as a demonstration example. Figure 5 shows several screenshots of label placements around Tokyo in Japan. The results were generated using our genetic-based optimization approach, where from top to bottom we zoomed in on the map by progressively reducing the zooming scale. Note that the images on the left were generated using a naïve approach, which optimized the label placement individually at the respective scales and thus could increase the number of visual labels at the cost of inconsistency in the layout of labels and leader lines across multiple scales. However, the image on the right reveals scale-aware consistent placement of labels synthesized using our approach, that successfully avoids the unexpected pop-up effects of labels and leader lines and retains their positions and drawing styles. For example, the county name "Chosei" (in green on the bottom right of the map) changes its position in the naïve inconsistent label placement (on the left of Fig. 5), while it naturally appears as the zooming scale decreases in the consistent label placement (on the right of Fig. 5). The consistent placement is much more reasonable because labels with low priority naturally disappear as we zoom out on the map. Note that here we draw the leader lines together with the corner of the corresponding label boxes by following the styles suggested in Meng et al. (2015).

Figure 6 presents the optimal label placement around Osaka in Japan, where the left and right columns show the label placement using the conventional four label positions and the proposed 16 label positions, respectively. Here, both placements avoid unwanted popping effects and sudden leaps of the labels and successfully maintain consistency across successive scales. Nonetheless, the results in the right column clearly demonstrate that our approach maximally utilizes the space outside the boundary of Japan for accommodating more labels to be placed especially in densely annotated regions. Furthermore, our new approach allows us to avoid distracting conflicts between the labels and the leader lines while keeping the overall layout as tight as possible.

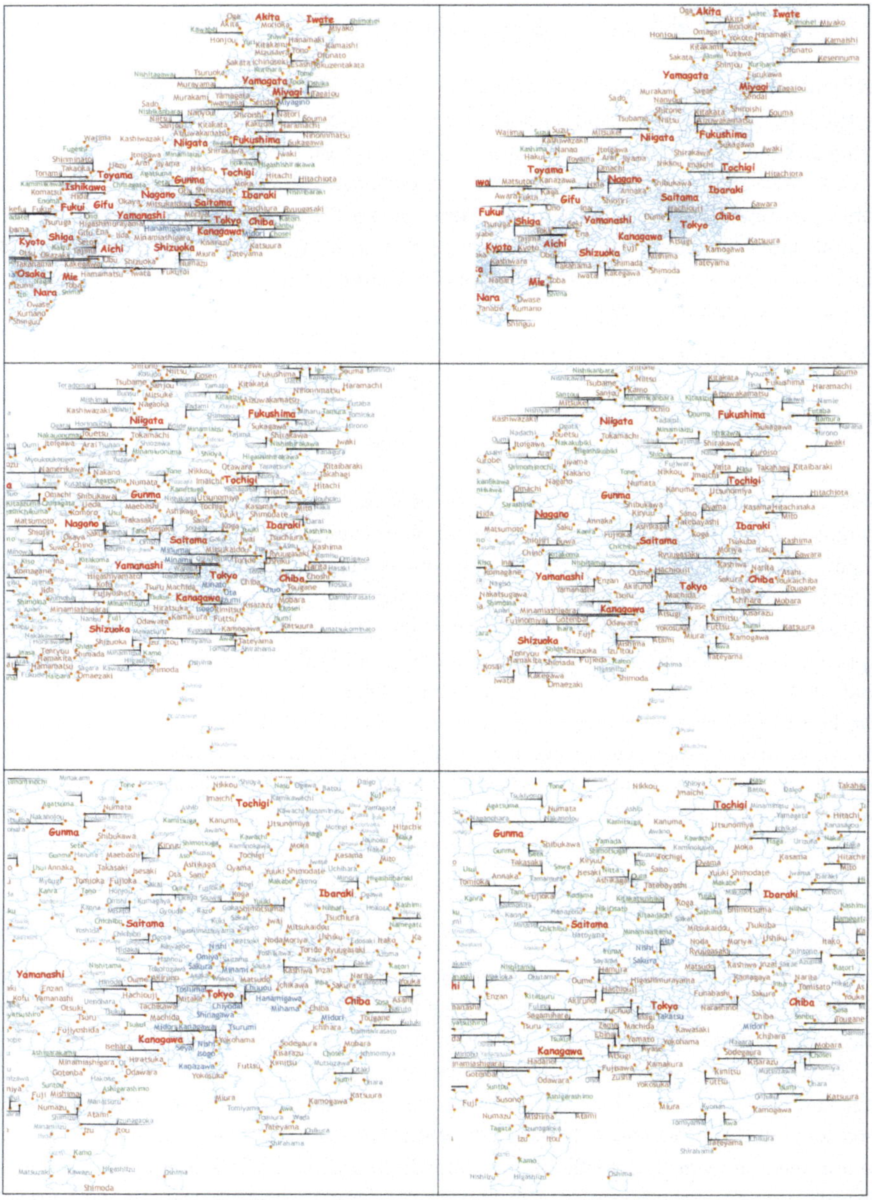

Fig. 5 Label placement around Tokyo, Japan. *Left* Naïve inconsistent placement. *Right* Scale-aware consistent placement. Different colors are assigned to the names according to their administrative levels; the prefecture, city, county, and town names are in *red*, *orange*, *green*, and *gray*, respectively

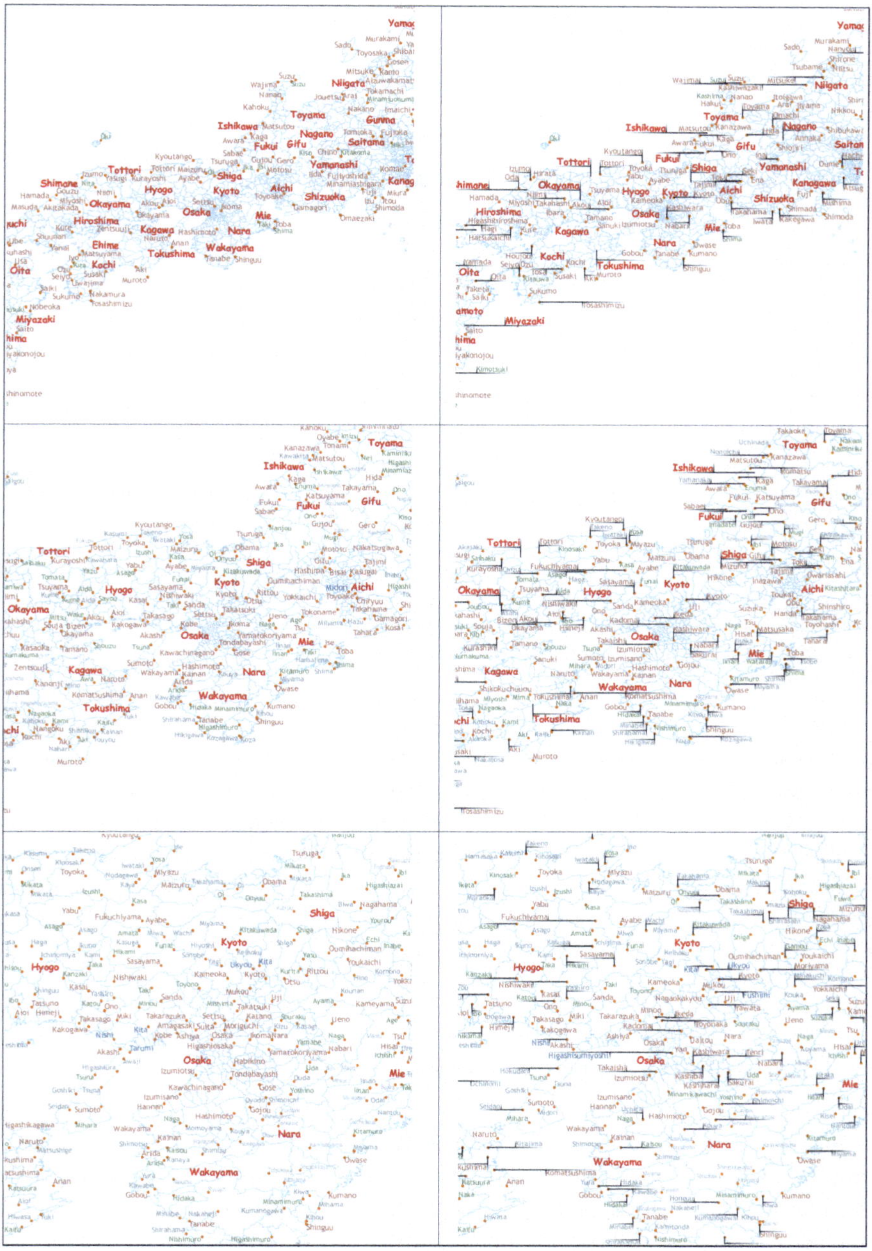

Fig. 6 Label placement around Osaka, Japan. *Left* Consistent placement with four label positions. *Right* Consistent placement with 16 label positions. Different colors are assigned to the names according to their administrative levels; the prefecture, city, county, and town names are in *red*, *orange*, *green*, and *gray*, respectively

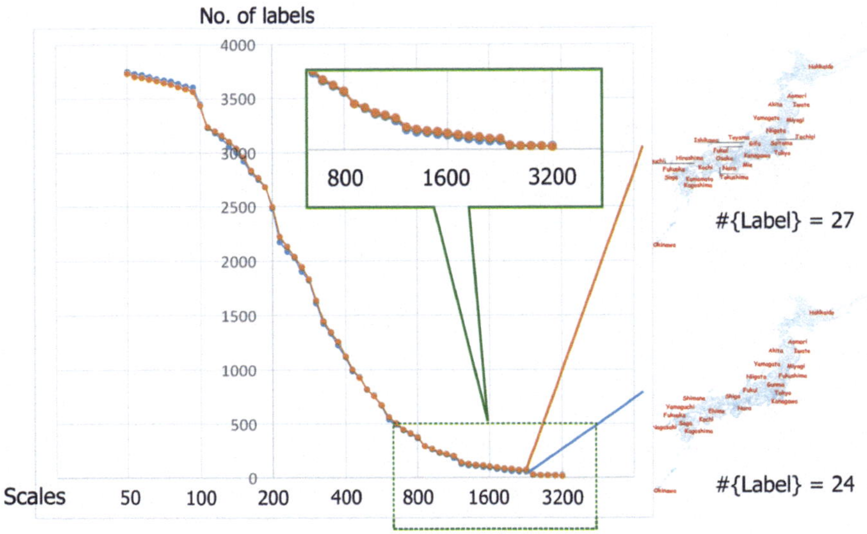

Fig. 7 Comparison between the label placement with the conventional four label positions and the new 16 label positions. The chart summarizes the number of labels that can be embedded in the entire map of Japan (the *vertical axis*) for the respective scales (i.e., distances of the sides of the map domain (km; the *horizontal axis*))

Figure 7 shows the comparison between the results using the conventional four label positions and the proposed 16 label positions, where we summarize the numbers of labels that can be embedded in the entire map of Japan at the respective scales. Note that the horizontal axis of this chart represents the approximate distance of each side of the map square domain, which is proportional to the power of the corresponding scale. The comparison shows that we can increase the number of labels placed on the map especially when the zooming scale is large, to optimize the use of available map space when we zoom out on the map.

6 Conclusion and Future Work

In this paper, we presented an approach for consistently placing labels across multiple scales by introducing leader lines for the conventional active range optimization problem. The approach allows us to incorporate more annotation labels into the map domain without unwanted artifacts, such as overlapping among labels and distracting popping effects across multiple scales. The visibility of the annotation labels is maximized using genetic algorithms by optimizing the order of the labels in which they are embedded in the map domain in a greedy fashion.

Our future directions include accelerating our genetic-based computation by grouping labels into several clusters of reasonable size, as well as combining them together to form the entire label placement. Other potential extensions include investigating additional aesthetic constraints for the label placement in terms of country borders, administrative districts, ponds and rivers, as well as the layout of transportation systems.

Acknowledgements This work was partially supported by JSPS KAKENHI Grant Numbers 16H02825, 26730061, 17K12691, and 15K12032.

References

Been, K., Daiches, E., & Yap, C. (2006). Dynamic map labeling. *IEEE Transactions on Visualization and Computer Graphics, 12*(5), 773–780.

Been, K., Nöllenburg, M., Poon, S. H., & Wolff, A. (2010). Optimizing active ranges for consistent dynamic map labeling. *Computational Geometry: Theory and Applications, 43*(3), 312–328.

Bekos, M. A., Cornelsen, S., Fink, M., Hong, S.-H., Kaufmann, M., Nöllenburg, M., et al. (2015). Many-to-one boundary labeling with backbones. *Journal of Graph Algorithms and Applications, 19*(3).

Bekos, M. A., Kaufmann, M., Symvonis, A., & Wolff, A. (2007). Boundary labeling: Models and efficient algorithms for rectangular maps. *Computational Geometry: Theory and Applications, 36*(3), 215–236.

Chiang, Y.-Y., Leyk, S., & Knoblock, C. A. (2014). A survey of digital map processing techniques. *ACM Computing Surveys, 47*(1), 1:1–1:44.

Christensen, J., Marks, J., & Shieber, S. (1995). An empirical study of algorithms for point-feature label placement. *ACM Transactions on Graphics, 14*(3), 203–232.

Edmondson, S., Christensen, J., Marks, J., & Shieber, S. (1997). A general cartographic labeling algorithm. *Cartographica, 33*(4), 13–23.

Fekete, J.-D., & Plaisant, C. (1999). Excentric labeling: Dynamic neighborhood labeling for data visualization. In *Proceedings of the SIGCHI Conference on Human Factors in Computing Systems* (pp. 512–519).

Fink, M., Haunert, J., Schulz, A., Spoerhase, J., & Wolff, A. (2012). Algorithms for labeling focus regions. *IEEE Transactions on Visualization and Computer Graphics, 18*(12), 2583–2592.

Gemsa, A., Niedermann, B., & Nöllenburg, M. (2013). Trajectory-based dynamic map labeling. In *Proceedings of the 24th International Symposium on Algorithms and Computation* (*ISAAC 2013*) (pp. 413–423).

Hirsch, S. A. (1982). An algorithm for automatic name placement around point data. *The American Cartographer, 9*(1), 5–17.

Imhof, E. (1975). Positioning names on maps. *The American Cartographer,* 2(2), 128–144.

Lin, C.-C. (2010). Crossing-free many-to-one boundary labeling with hyper-leaders. In *Proceedings of 2nd IEEE Pacific Visualization Symposium* (*PacificVis 2010*) (pp. 185–192).

Luboschik, M., Schumann, H., & Cords, H. (2008). Particle-based labeling: Fast point-feature labeling without obscuring other visual features. *IEEE Transactions on Visualization and Computer Graphics, 14*(6), 1237–1244.

MacEachren, A. M. (1995). *How maps work.* The Guilford Press.

Meng, Y., Zhang, H., Liu, M., & Liu, S. (2015). Clutter-aware label layout. In *Proceedings of the 7th IEEE Pacific Visualization Symposium* (*PacificVis 2015*) (pp. 207–214).

Petzold, I., Gröger, G., & Plümer, L. (2003). Fast screen map labeling—Data-structures and algorithms. In *Proceedings of the 23rd International Cartographic Conference* (*ICC 2003*) (pp. 288–298).

Poon, S.-H., & Shin, C.-S. (2005). Adaptive zooming in point set labeling. In *Proceedings of the 15th Fundamentals of Computation Theory* (*FCT 2005*) (pp. 233–244).

Raidl, G. R. (1998). A genetic algorithm for labeling point features. In *Proceedings of the International Conference on Imaging Science, Systems and Technology* (pp. 189–196).

Schwartges, N., Haunert, J., Wolff, A., & Zwiebler, D. (2014). Point labeling with sliding labels in interactive maps. In *Proceedings of the 17th Agile Conference on Geographic Information Science* (pp. 295–310).

Wolff, A. (2009). The map-labeling bibliography. http://i11www.iti.uni-karlsruhe.de/map-labeling/bibliography/.

Wu, H.-Y., Takahashi, S., Lin, C.-C., & Yen, H.-C. (2011). A zone-based approach for placing annotation labels on metro maps. In *Proceedings of the 11th International Conference on Smart Graphics* (*SG 2011*) (Vol. 6815, pp. 91–102).

Yoeli, P. (1972). The logic of automated map lettering. *The Cartographic Journal, 9*(2), 99–108.

Zhang, X., Poon, S.-H., Li, M., & Lee, V. (2015). On maxmin active range problem for weighted consistent dynamic map labeling. In *Proceedings of the 7th International Conference on Advanced Geographic Information Systems, Applications, and Services* (pp. 32–37).

Zoraster, S. (1986). Integer programming applied to the map label placement problem. *Cartographica, 23*(3), 16–27.

On the Way to Create Individualized Cartographic Images for Online Maps Using Free and Open Source Tools

Csaba Szigeti, Gáspár Albert, Virág Ilyés, Dávid Kis and Dávid Várkonyi

Abstract The goal of the research was to create online maps with dynamically changing cartographic images according to the users' map reading skills, by only using open source tools and resources. The maps were designed for three map reader groups (beginners, intermediates and experts). We used OpenStreetMap and Shuttle Radar Topography Mission (SRTM) data to create the maps in QGIS. To display the maps online, we used QGIS Server on the server side, and OpenLayers on the client side. Based on earlier studies on map reading skills and cognitive load, the map design and generalization aimed to maximize the map reading efficiency of the target group. In the grouping process, an online test was used to measure the map reading competences of the map readers, such as the interpretation of hypsography, orientation skills, distance and travel time estimation, interpretation of map symbols, interpretation of geographic names and interpretation of topographic objects. The design and generalization of the map features were based on the map data categories (such as hypsography, coverage, roads, etc.) and they are different on the three maps.

C. Szigeti (✉)
Doctorate School of Earth Sciences, Eötvös Loránd University, Budapest, Hungary
e-mail: szgtcsaba@map.elte.hu

G. Albert
Department of Cartography and Geoinformatics, Eötvös Loránd University, Budapest, Hungary
e-mail: albert@ludens.elte.hu

V. Ilyés
Department of Survey Statistics, Eötvös Loránd University, Budapest, Hungary
e-mail: ilyesvirag@gmail.com

D. Kis · D. Várkonyi
Faculty of Informatics, Eötvös Loránd University, Budapest, Hungary
e-mail: kidraai@inf.elte.hu

D. Várkonyi
e-mail: varkonyidavid91@gmail.com

M.P. Peterson (ed.), *Advances in Cartography and GIScience*, Lecture Notes in Geoinformation and Cartography, DOI 10.1007/978-3-319-57336-6_10

Keywords Open source · Individualized maps · QGIS · Experimental cartography

1 Introduction

Numerous map services can be found online, many of them capable of varying their legends or content to a certain extent (e.g., Google Maps, OpenStreetMap, Yandex Maps, ArcGIS Online, MapQuest, etc.). Most of these services have a fixed map key, and additional information (e.g., traffic, hotels, etc.) can be displayed optionally. OpenStreetMap has map layouts that can be changed based on the required map theme. Swisstopo uses ESRI ArcGIS, enterprise GIS software, to create and display the Swiss topographic maps online. They have rules to create proper cartographic representation. They also use multi-level generalization (model and cartographic generalization) to create a cartographic model for a certain scale band (Neuffer et al. 2006).

Studies found that experience can affect map reading (e.g., Guzmán et al. 2008; Ooms et al. 2013). These studies show that unexperienced map users require more time and effort to understand maps (Ooms et al. 2012). Also, experienced map users have multiple advantages at map reading tasks, for instance in interpreting contour lines (Gilhooly et al. 1988), in symbol knowledge (Guzmán et al. 2008) and other competences such as mental rotation and distance estimation (Albert et al. 2016). During navigational tasks, map reading is also affected by the map reader's gender (Lawton 1994), cultural background (Ito and Sano 2011), cognitive skills and memory (Petchenik 1977; Montello 2002; Guzmán et al. 2008; Murakoshi and Higashi 2016). However, according to the Cognitive Load Theory (CLT) (Sweller 1994; Paas and Merrienboer 2003), the cognitive load capacity of the memory is limited; experienced map readers store information about maps they read earlier in their long-term memories, which can be later retrieved.

It is necessary for cartographers to create maps that not only meet the needs of map readers, but also fit their level of knowledge and skills (Kolácný 1969). By creating individualized maps according to the map readers' skills, we can reduce the map readers' cognitive load. To achieve this, it is important to generalize the number of topographic elements in the maps for users with less experience, and to create a proper map design without visual clutter or ambiguity (Bunch and Lloyd 2006). Creating different maps for each individual person is not yet possible at the present technical level, but by defining map reader groups with similar competencies (Clarke 2003; Albert et al. 2016) it is possible to enhance map reading efficiency with specialized maps. Map readers in user studies are differentiated into categories based on their map reading experience. Making two groups (e.g., Deeb et al. 2012; Ooms et al. 2015), novices and experts is usual if the number of participants is below 100. Three groups (e.g., Gerber 1981; Clarke 2003; Albert et al. 2016), beginners, intermediates and experts, are recommended if the study achieved statistically sufficient data.

The Research Group for Experimental Cartography (RGEC) at Eötvös Loránd University, Budapest, prepared maps to fit to the users' map reading skills and tests the effectiveness of information retrieval from the maps by the subjects. In earlier RGEC studies (Szigeti and Albert 2015; Albert et al. 2016), the map reading skills were measured, and patterns were found amongst different map reader groups regarding four map data categories: hypsography, geographic names, point-like symbols and topography. The topography category can be further divided into natural or artificial linear and areal feature types (Albert 2014; Szigeti and Albert 2015). These categories are also in accordance with the data layer structure, e.g., in digital topographic maps (McEwen and Jacknow 1979; Usery et al. 2009). It was also shown that if the cognitive skills of the map readers are tested with very simple maps displaying only one or two map data categories, the statistical analysis of the results yields a numeric value representing the difficulty level of that category for the actual map reader (Albert et al. 2016).

Based on the results of these experiments, the present paper focuses on the technical and theoretical details of creating online maps with dynamically changing cartographic images, where the factor of change is the skill of the map reader. The maps were applied for user testing research concerning their usability. The details and results can be found later in this volume (see Albert et al. 2017).

2 Data Sources

The maps were created from OpenStreetMap (OSM) and SRTM data. The selected test area is located near Pécs, a city in Hungary, because it contains various types of map objects: multiple road types, different types of coverage (dense city, villages, forests, fields, vineyards), and it also has a diverse relief with both mountains and plains (Fig. 1). The map data was downloaded in .osm format using QGIS 2.14.0 Essen (open source GIS software) because, in this way, the exact size of the required area could be specified in advance. The raw .osm file (map database) was separated into four geometry types: lines, multilines, points and polygons. The contour lines and relief shading were generated from SRTM data with 3 arc seconds resolution.

The cartographic images of the maps change dynamically with the zoom level. At larger zoom (or larger scale) the maps represented city maps in built-up areas and tourist maps on rural areas, while at smaller zoom level (smaller scale) they were similar to road maps. The zoom levels are based on the OSM standards (OpenStreetMap Wiki 2016b), although they were modified to have more details (Table 1).

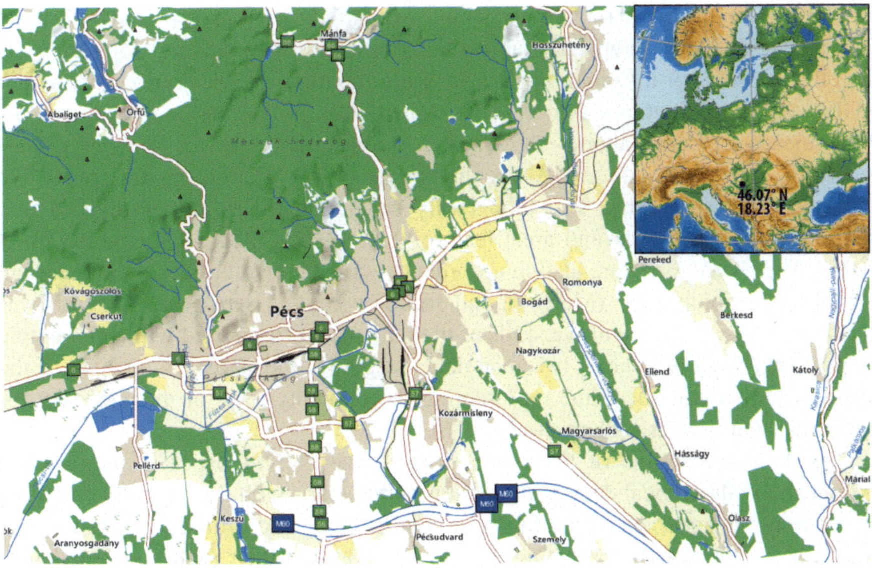

Fig. 1 The test area near Pécs, Hungary, containing various map objects

Table 1 The nine original OSM zoom levels (left), and the more detailed 12-grade zoom levels of the created dynamic maps (right)

	Zoom levels of OSM	Zoom levels of the dynamic maps
1	1: 250,000	1: 250,000
2a	1: 150,000	1: 150,000
2b		1: 75,000
3a	1: 70,000	1: 70,000
3b		1: 50,000
4	1: 35,000	1: 35,000
5a	1: 15,000	1: 15,000
5b		1: 10,000
6a	1: 8,000	1: 8,000
6b		1: 5,000
7	1: 4,000	1: 4,000
8	1: 2,000	1: 1,000
9	1: 1,000	

3 Processing the OpenStreetMap Data

The raw map data was styled and generalized in three different ways to adequately fit them to the map reading skills of the map readers. This can be achieved only if the data is separated into categories (Table 2) and distinguished as separate layers according to the four different geometry types of the OSM data (points, lines,

Table 2 The relationship between the map data categories and the data layers created from the OSM database

Map data category	Description	Layers
Linear features	Road network, transportation, and tourist trails	(1) Paved roads
		(2) Dirt roads
Hydrography	Streams, rivers, lakes, springs	(3) Lines
Land cover	Forest, built-in area	(4) Coverage multipolygon
Hypsography (elevations)	Contours, shading, scarps, peaks	(5) Contours
		(6) Relief shading
Point-like objects (symbols)	All point-like objects in the map	(7) Points
Geographic names	Name attributes of all objects	Labels

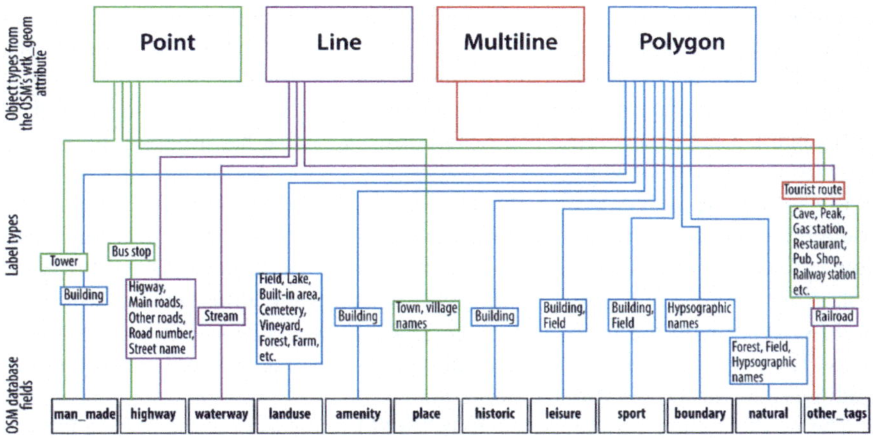

Fig. 2 The labels of object types created in QGIS from the OSM database

multilines, polygons) (OpenStreetMap Wiki 2016a) using the *wtk_geom* attribute (Fig. 2). The point type objects are categorized into 25 QGIS style labels by using the *other_tags*, *highway* and *man_made* attributes (keys) in the OSM's XML database. Each label represents an individual object. The *man_made* attribute contains artificial structures (OpenStreetMap Wiki 2016d). The *highway* attribute is mainly used to identify roads, streets and paths, but it also contains point objects such as bus stops, roundabouts or stop signs (OpenStreetMap Wiki 2016c).

Every attribute that is not strictly identified by OSM is appended to the *other_tags* field (GDAL n.d.). These attributes contained multiple information about one single object, though most of the information was irrelevant from the cartographic aspect, it had to be processed with multiple filtering functions to extract relevant data concerning the defined categories (Fig. 2). The names of peaks and railway stations as well as the names of cities, city quarters and villages were point objects. The created maps inherited this original property even if the object was rather a area (e.g., village) and not a point. The major part of the geographic name content of the maps came from the labels of the point objects.

The multiline type objects (polylines, with graphic symbols on them) contained information about the tourist trails (29 labels). As in the case of points, relevant information was found in the *other_tags* attribute, which had to be data-mined. The linear objects were distinguished by the *highway* attribute for roads, the *waterway* attribute for streams and the *other_tag* attribute for railways. The simple line type objects were separated into two layers to create a proper hierarchy with the multiline objects during rendering the cartographic image: paved roads and railways (14 labels) were below the multilines, while dirt roads (2 labels) were above them. In this way, the symbols of the trails did not cover any road symbols.

The symbols for tourist trails were imported into QGIS in SVG format. Because the trails were multilines, we had to display the symbols as label backgrounds (an option in QGIS to provide a graphic object as a background for the label). It was also necessary to set the option of merging the connected lines to avoid duplicate labels. The map also contained street names, and the road numbers of highways, primary and secondary roads. Since the numbering in the .osm file did not follow the Hungarian road numbering hierarchy consistently, we had to apply regex patterns for the correct labelling.

Regarding the polygon type objects, the QGIS OSM download plugin had trouble downloading too complex multipolygons (polygons with polygon inclusions). This was fixed using Overpass Turbo, which is an online OSM download tool. After solving this problem, the map objects were differentiated using the *landuse*, along with the *amenity*, *barrier, sport, leisure, historic* and *natural* attributes. The geographic names of hills and mountains are also stored as attributes of multipolygons in the OSM database.

The relief shading was generated from SRTM 3 arc second data. Since the test area is relatively small, its resampling method was set to cubic from the nearest neighbour, and the blending mode was set to multiply in order to get smoother visualization. It had 50% transparency in the cartographic image. Contour lines with 5 m intervals were also generated from SRTM.

All the .osm files were converted into shapefiles because this is the supported format in QGIS, and the shaded relief was saved as a GeoTIFF file.

4 Creating Different Cartographic Images

We used the categorized map database to provide base maps with different graphic styles in the study involving several hundred participants. An online test preceded the visualization of the cartographic images presented here and divided the subjects into three groups: experts, intermediates and beginners. After the test, the categorized map data was displayed through a web server. The symbol design was based on the results of earlier studies on the differences of map reading competences: interpretation of hypsography, orientation skills, distance and travel time estimation, interpretation of map symbols, interpretation of geographic names and interpretation of topographic objects (Muir 1985; Clarke 2003; Albert et al. 2016).

We also created 12 zoom intervals where the map legends change dynamically (Table 1). These zoom intervals (or scale levels) define the visibility, graphical style and generalization of each map object. On the test area, the scale varied between 1:250,000 and 1:1,000.

Ultimately, the categorized map data was modified according to graphic variables, defined originally by Bertin (1967). These are colour, pattern, value, size (thickness), position, shape and direction. Modification of a symbol was justified only if significant difference in the interpretation of the given map data type was documented amongst the groups of map readers. In the case of such differences, one or two variables were modified. It was expected that the emphasized symbols would help the map readers to solve map-related tasks.

Albert et al. (2016) found that the interpretation of the linear objects was a difficult task for all map reader groups. For this reason, we increased the legibility of the linear objects, especially the roads, with distinctive colours and line widths for each road type. It was also shown that the beginners make five times as many mistakes on road-related map reading tasks as the experts. To decrease the cognitive load on less skilled map readers, the roads were generalized using "refinement" (McMaster and Shea 1992) on small scale maps. Maps for beginners were generalized the most while maps for experts were generalized the least (Table 3). Practically, it meant that lines shorter than a certain length were not displayed, but this value varied according to the zoom level and the object type. Other automatic generalization methods, for instance, simplification or smoothing (Li 2007), are difficult to use in QGIS and, in some cases, the topology is lost (Stackexchange 2016a, b) or gives a visually unappealing result.

A quite similar approach was used during the styling and generalization of the polygons but here the pattern was modified along with the size. Below a certain area they were not displayed, and their value varied according to the scale level and map object (Table 3). Allen et al. (2006) showed that previous experience of symbols influences comprehension, thus helps the working memory in map reading tasks. Because of this, the patterns were designed to represent familiar shapes like a tree, a grape, or a grape vine (Fig. 3).

On larger scales (>1:75,000), the tree symbols in the forest polygons were detailed in the maps for beginners but more schematic in the map for intermediates. However, as in the case of Hungarian tourist and topographic maps, there were no tree-patterned polygons in the map for experts (Fig. 3) because experts are accustomed to this symbology.

The vineyards in the maps of intermediates and experts were represented by the conventional grape vine symbol. However, we used a schematic grape symbol in the beginners' map (Fig. 3). These figures were imported into QGIS as SVG files.

Intermediate and beginner map readers had difficulties in interpreting hypsography (Eley 1992; Wakabayashi 2013). Thus, to emphasize the plasticity of the

Table 3 An example of the different extensity of generalization on the three maps. The beginners' map was the most, and the experts' map was the least, generalized

	Beginner (number of displayed objects)				Intermediate (number of displayed objects)				Expert (number of displayed objects)			
Scale level	1: 75,000–1: 35,000	1: 35,000–1: 15,000	1: 15,000–1: 5,000	1: 5,000–1: 1,000	1: 75,000–1: 15,000	1: 15,000–1: 1,000	1: 35,000–1: 15,000	1: 5,000–1: 1,000	1: 75,000–1: 15,000	1: 15,000–1: 1,000	1: 35,000–1: 15,000	1: 5,000–1: 1,000
Fields	372		476		393		476		476		476	
Forest	243		277		250		277		277		277	
Buildings	–	165	1893	3570	–	165	4439	18,945	–	165	14,432	18,945
Bridleway, track	502		3345		1091		3345		1981		3345	
Path	188		1182		351		1182		582		1182	

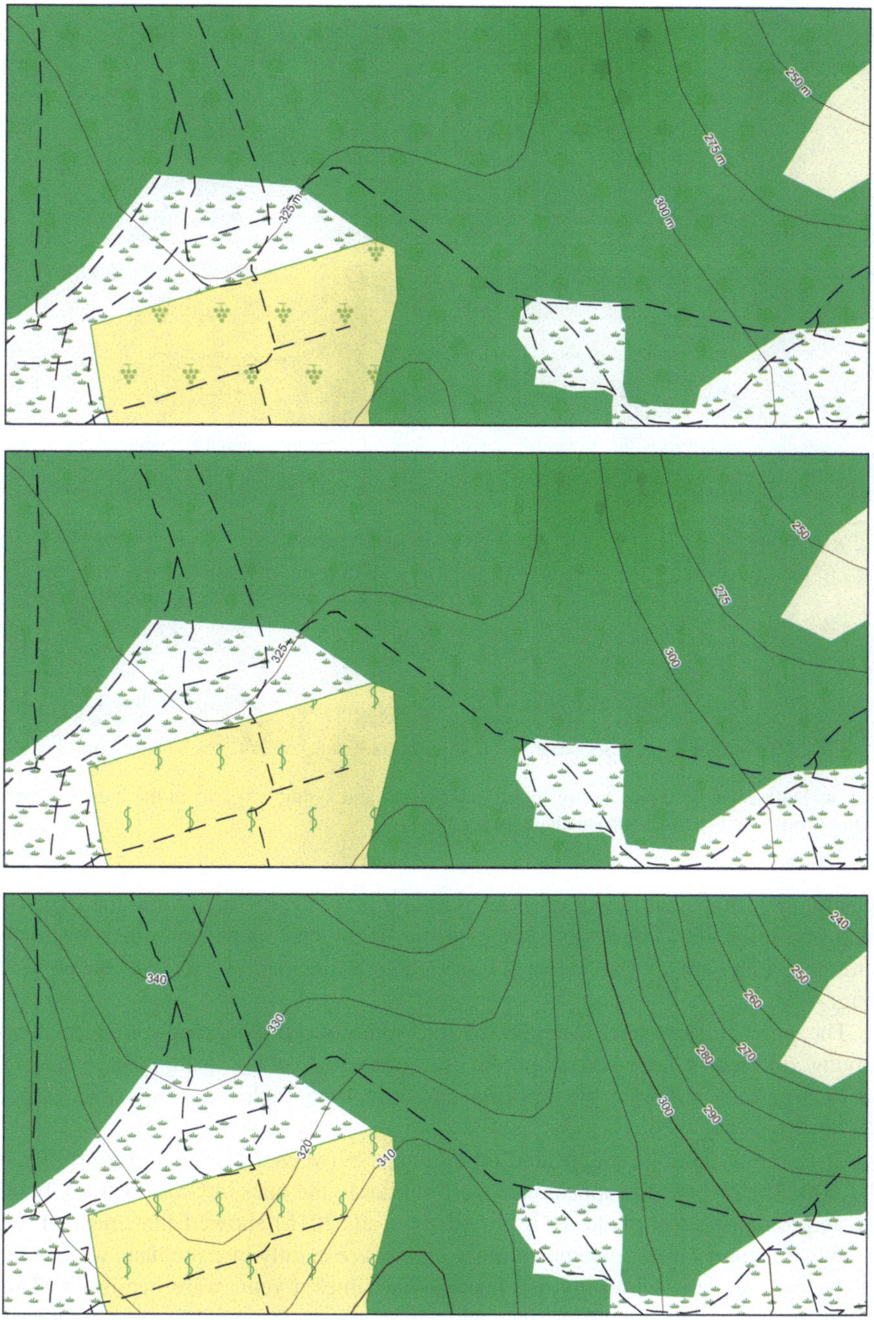

Fig. 3 The difference in the design of polygons in the individualized maps (from *top*: beginner, intermediate, expert). Differences can be seen in the forest and vineyard symbols

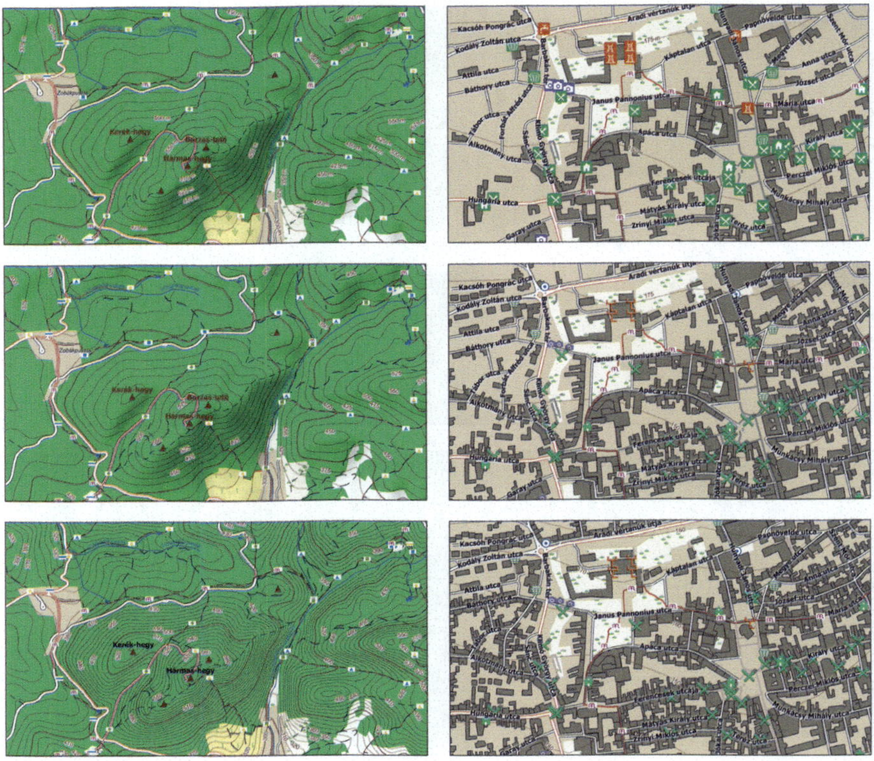

Fig. 4 Differences in the design of hypsography (*left*) and symbols (*right*) in the individualized maps (from *top*: beginner, intermediate, expert)

terrain, their maps had fewer contour lines and less hill shading, while the experts' map had only dense contours. Also, on the beginners' map, we added the measurement unit (m) to the values of the contour lines for better comprehension (Fig. 4).

The point objects were categorized into five conceptual groups, four of them having different colour-codes: public transportation (blue), landmarks (orange), services (green), cultural sights (purple) and other topographic elements, and further 28 label types.

The point type objects appeared on the maps between scale levels 1:70,000 and 1:1,000. The rules of the appearance were similar to the ones in OSM (Tables 1 and 3). A previous study of the RGEC (Albert et al. 2016) showed that most of the weekly map users were experts, monthly users were mainly intermediate, while 45% of the map readers, who only use maps a few times a year, were beginners. This means that experts and intermediates have more experience with conventional map symbols. It was also shown by Easterby and Hakiel (1981) that a symbol is comprehended more easily if it has descriptive details. Thus, the symbols on the experts' and intermediates' maps were simple pictograms commonly used in tourist maps, while for beginners they were more detailed and descriptive for easier comprehension, and had a coloured background (the colour followed the conceptual groups' colour-code). The symbols in this map were also bigger in size to apprehend them more easily (Fig. 4). The symbols were imported into QGIS as separate SVG files.

Sans-serif was preferred by all map reader groups according to Deeb et al. (2012). It was also shown that novice users were more tolerant to differences in shape and texture. The names of cities, villages, districts had the same fonts, but different typefaces (e.g., bold, semi-bold) and sizes, and they also appeared at different zoom levels. The hypsographic names' style varied in the three maps: while the names in the beginners' and intermediates' maps had a brown colour, the names in the experts' map were black to increase their legibility due to the larger number of contour lines.

5 Setting Up Maps Online

The maps were displayed by using QGIS Server on the server side and a JavaScript library called OpenLayers on the client side (web page). The QGIS server is capable of using maps created with the QGIS desktop application. The server contains the map data and can respond to HTTP queries by creating rasterized maps on the fly. The HTTP query is created on the client side in QGIS's own syntax and can specify location, zoom level, legend, etc. To provide such queries, interactive elements on the webpage were supported by a JavaScript library called OpenLayers on the client-side.

OpenLayers is a tool for displaying easy-to-use interactive maps, which is more or less capable of handling all user inputs and queries on its own. Many widely-used features, such as scroll-to-zoom are supported by the JS library. OpenLayers processes these inputs and sends corresponding HTTP queries via asynchronous communication to the server, which in turn sends the corresponding rasterized maps back to the client. The map then appears on the client computer as an HTML element. To decide whether a user is beginner/intermediate/expert, a numeric value was also applied as input data for the server.

The basic elements used on the website are readily implemented in OpenLayers, leaving only a few specific features and/or extensions to be developed by us in JavaScript. By using QGIS Server, it was possible to transfer the QGIS project files from the desktop version without any conversion. In this way, the different styles, and imported SVG files, were displayed similarly on the web page and the desktop.

6 Discussion

The overall task to create dynamically changing cartographic images via web servers was achieved using only free and open source tools and data. However, tasks like this have their own limitations. Downloading, and working with the .osm format in QGIS is not always evident, although there are alternative ways to downloading OSM data, for instance, geofabrik.de, that provides OSM data of countries in both .osm and .shp format. If the problems of data download and conversion are tackled, it is important to process the raw data's attribute table, and create the proper queries in QGIS to distinguish the features required for the map.

The data in the .osm file's attribute table may not be consistent with the national data (e.g.,, road numbering). To fix this, it may be necessary to edit the data, or use regular expressions to properly filter information. In the present study, we aimed to work out automatic methods for this (e.g., filtering and categorizing osm data), but working with a certain pilot area requires more modest hardware and software capacity than working with global scale data, which is the logical ultimate aim.

In terms of creating different designs suited for the skills of map readers, QGIS seemed to be a proper decision. Because custom created SVGs can be imported into QGIS, a large variety of symbols and polygon fills can be applied to the maps. Lines could also be styled well due to the multiple symbol layers (multilines) that can be created.

On the other hand, it was difficult to generalize the maps in QGIS. The simplest way of generalizing the maps for the three map reader groups was applying refinement as a generalization method. In this case, we only selected objects the length/area of which (depending on the object type) were above a certain value.

The testing method to allocate map variations for the user groups is suitable for detecting further differences among the subjects (i.e., gender, age, education). For more details about our testing method and the results of our user study based on the maps of this research, see Albert et al. (2017) (in this book). By creating map style variations according to these variables and relying only on open source tools, it is possible to provide the readers with true individualized maps of the same area having different map legends based on the map readers' skills.

Acknowledgements The study was financed from the grant of the Council for Talent Management at Eötvös Loránd University for multidisciplinary research.

References

Albert, G. (2014). *What does an archive map tell the contemporary map readers?* Paper presented at the 9th International Workshop on Digital Approaches to Cartographic Heritage, Budapest.

Albert, G., Ilyés, V., Kis, D., Szigeti, C., & Várkonyi, D. (2016). Testing the map reading skills of university students. In T. Bandrova & M. Konecny (Eds.), *6th International Conference on Cartography and GIS*. (pp. 188–199.) Albena: Bulgarian Cartographic Association.

Albert, G., Ilyés, V., Kis, D., Szigeti, C., & Várkonyi, D. (2017). How hard is it to design maps for beginners, intermediates and experts? In M. P. Peterson (Ed.), *Advances in Cartography and GIScience—Selections from the International Cartographic Conference 2017*. New York: Springer.

Allen, G. L., Miller Cowan, C. R., & Power, H. (2006). Acquiring information from simple weather maps: Influences of domain-specific knowledge and general visual-spatial abilities. *Learning and Individual Differences, 16*(4), 337–349.

Bertin, J. (1967). *Sémiologie graphique: les diagrammes–les réseaux–les cartes*. Paris: Mouton.

Bunch, R. L., & Lloyd, R. (2006). The cognitive load of geographic information. *The Professional Geographer, 58*(2004), 209–220.

Clarke, D. (2003). Are you functionally map literate? In *Proceedings of the 21st International Cartographic Conference* (pp. 713–719). Durban.

Deeb, R., Ooms, K., & De Maeyer, P. (2012). Typography in the eyes of Bertin, gender and expertise variation. *The Cartographic Journal, 49*(2), 176–185.

Easterby, R. S., & Hakiel, S. R. (1981). Field testing of consumer safety signs: The comprehension of pictorially presented messages. *Applied ergonomics*, 143–152.

Eley, M. G. (1992). Component processing skills in the interpretation of topographic maps. *Cartographica, 29*(1), 35–51.

GDAL. (n.d.). OSM—OpenStreetMap XML and PBF. Retrieved October 24, 2016, from http://www.gdal.org/drv_osm.html.

Gerber, R. V. (1981). Competence and performance in cartographic language. *The Cartographic Journal, 18*(2), 104–111.

Gilhooly, K. J., Wood, M., Kinnear, P. R., & Green, C. (1988). Skill in map reading and memory for maps. *The Quarterly Journal of Experimental Psychology A: Human Experimental Psychology, 40,* 87–107.

Guzmán, J. F., Pablos, A. M., & Pablos, C. (2008). Perceptive-cognitive skills and performance in orienteering. *Perceptual and Motor Skills, 207,* 159–164.

Ito, K., & Sano, Y. (2011). Cultural differences in the use of spatial information in wayfinding behavior. In *Proceedings of the 25th International Cartographic Conference*.

Kolácný, A. (1969). Cartographic information—a fundamental concept and term in modern cartography. *The Cartographic Journal, 6*(1), 47–49.

Lawton, C. A. (1994). Gender differences in way-finding strategies: Relationship to spatial ability and spatial anxiety. *Sex Roles, 30*(11–12), 765–779.

Li, Z. (2007). Digital map generalization at the age of enlightenment: A review of the first forty years. *The Cartographic Journal, 44*(1), 80–93.

McEwen, R. B., & Jacknow, H. R. (1979). *USGS digital cartographic database*. Reston: Paper presented at the AutoCarto TV.

McMaster, R. B., & Shea, K. S. (1992). *Generalization in digital cartography*. Washington: Association of American Cartographers.

Montello, D. R. (2002). Cognitive map-design research in the twentieth century: Theoretical and empirical approaches. *Cartography and Geographic Information Science, 29*(3), 283–304.

Muir, S. P. (1985). Understanding and improving students' map reading skills. *Elementary School Journal, 86*(2), 206–216.

Murakoshi, S., & Higashi, H. (2016). Cognitive characteristics of navigational map use by mountaineers. *International Journal of Cartography, 1*(2), 210–231.

Neuffer, D., Schneider, B., Hardy, P., & Kressmann, T. (2006). Database driven cartography—the "swisstopo" example. In *GICON*. Vienna.

Ooms, K., De Maeyer, P., Fack, V., Van Assche, E., & Witlox, F. (2012). Interpreting maps through the eyes of expert and novice users. *International Journal of Geographical Information Science, 26*(10), 1773–1788.

Ooms, K., De Maeyer, P., & Fack, V. (2013). Study of the attentive behavior of novice and expert map users using eye tracking. *Cartography and Geographic Information Science, 41*(1), 37–54.

OpenStreetMap Wiki. (2016a). Map features. Retrieved October 24, 2016, from http://wiki.openstreetmap.org/wiki/Map_Features.

OpenStreetMap Wiki. (2016b). Zoom levels. Retrieved October 24, 2016, from http://wiki.openstreetmap.org/wiki/Zoom_levels.

OpenStreetMap Wiki. (2016c). Key:highway. Retrieved October 24, 2016, from http://wiki.openstreetmap.org/wiki/Key:highway.

OpenStreetMap Wiki. (2016d). Key:man_made. Retrieved October 24, 2016, from http://wiki.openstreetmap.org/wiki/Key:man_made.

Paas, F., & Merrienboer, J. J. G. (2003). Cognitive load measurement as a means to advance cognitive load theory. *Educational Psychologist, 38*(1), 63–71.

Petchenik, B. (1977). Cognition in cartography. *Cartographica, 14*(1), 117–128.

Sweller, J. (1994). Cognitive load theory, learning difficulty, and instructional design. *Learning and Instruction, 4*(4), 295–312.

Stackexchange. (2016a). Simplify/generalize Polygons and keep topology of different layers PostGIS/QGIS. Retrieved October 24, 2016, from http://gis.stackexchange.com/questions/175930/simplify-generalize-polygons-and-keep-topology-of-different-layers-postgis-qgis.

Stackexchange. (2016b). Generalizing polygon file while maintaining topology in QGIS? Retrieved October 24, 2016, from http://gis.stackexchange.com/questions/20799/generalizing-polygon-file-while-maintaining-topology-in-qgis.

Szigeti, C., & Albert, G. (2015). Measuring the adequacy of maps for field use. In G. Gartner & H. Haosheng, (Eds.), *Proceedings of the 1st ICA European Symposium on Cartography* (pp. 341–343). Vienna: International Cartographic Association.

Usery, E. L., Finn, M. P., & Starbuck, M. (2009). Data layer integration for the national map of the United States. *Cartographic Perspectives*, *62*, 28–41.

Wakabayashi, Y. (2013). Role of geographic knowledge and spatial abilities in map reading process: Impilcations for geospatial thinking. *Geographical reports of Tokyo Metropolitan University, 48,* 37–48.

Hebrus Valles—The Mars Exploration Zone Map

Mateusz Pitura

Abstract Hebrus Valles is one of the 47 candidate Exploration Zones for the first human landing mission of NASA on Mars. The area meets specified criteria for a scientific expedition—among others, it has potential for the water and other resources necessary to perform such a mission. The aim of the project described in this paper was to attempt to create a specialized map for Mars explorers working in the field. The map of Hebrus Valles proposes locations for the landing zone, habitation zone, regions of interests, and traverses. The proposed map also contains descriptions of the geomorphological features of the landscape and warnings about surface hazards. When combined, these elements produce a surface operation map with special requirements that are unlike any map produced on Earth, perhaps the closest analog is hiking maps. This paper presents a detailed characterization of the map, as well as projecting methodology. The process of panchromatic sharpening is the most important part of mapping, mainly due to the effective fusion between a high resolution digital elevation model and a low-resolution color image. A description of tasks carried out using the cartographic tools JMARS, ArcMap and Photoshop is also included. In conclusion, this paper introduces possibilities of implementation of an interactive variant of the map and provides ideas that may be applicable for designing a cartographic system for future operations on the surface of another planet.

Keywords Mars · Exploration zone · Hebrus Valles · Planetary map · Planetary cartography

M. Pitura (✉)
Institute of Geological Sciences, University of Wroclaw,
Wroclaw, Dolnoslaskie, Poland
e-mail: mateusz.pitura@student.uj.edu.pl

M.P. Peterson (ed.), *Advances in Cartography and GIScience*, Lecture Notes in Geoinformation and Cartography, DOI 10.1007/978-3-319-57336-6_11

Fig. 1 Approximate boundaries for Mars regional feature names with selected Hebrus Valles region *circled* in *black*. The map is based on the MOLA digital elevation model with an adjusted color scale. (*Source* Smith et al. 2001)

1 Introduction—Basic Information About Hebrus Valles

Hebrus Valles is a region of Mars on the southern border of Utopia Planitia (see Fig. 1), which is the largest impact plain on the planet (Thomson and Head 2001). The name Hebrus Valles comes from Earth's Balkan River. The trough and valley system extends from south-east to north-west direction for about 250 km. The channels formed by fluid erosion, which emerged from the surface or underground rocks (Christiansen and Hoppin 1987), are characterized by several subchannels and streamlined islands.

Hebrus Valles is located near 19.88°N latitude and 126.74°E longitude (Gazetteer of Planetary Nomenclature 2016), so it is positioned closer to the equator than to the pole of Mars. The area is situated approximately 3000 m below the zero elevation of the Mars surface and the depression extends in the north–east direction.

2 Motivation and Goals

The basic goal of the Hebrus Valles Map is to visualize the Exploration Zone (EZ) surface in a way that could be helpful for future astronauts. There are 47 candidates for EZs (Bussey and Hoffman 2016) and this map is the best medium to show the most valuable characteristics of the site. Another purpose of the map is to allow insightful evaluation of the EZ and assist scientists in their comparison EZ candidates.

Furthermore, the aim of this paper is to introduce cartographic methods or processes which facilitated the development and projection of the Hebrus Valles Map. Additional goals are to explain and describe all of the map elements and to detail possible uses in the field. Finally, this paper suggests application of the map for new technologies (touch screens) and proposed interactive options for the map's use.

3 Exploration Zone Criteria

The EZ is an aggregation of regions of interests (ROIs) placed within 100 km of the landing zone (LZ). The main factors for evaluating EZ candidates, in brief, are as follows:

- the presence of resources that would sustain life of future astronauts in Martian conditions—this includes water in any form (hydrated minerals, ice deposits) and other materials needed for habitation
- the possibility of finding remnants of life in any form in past or present state
- achievable new scientific data which would impact on our knowledge about the planet
- a location between 50°N and 50°S latitude
- terrain located not higher than 2000 m above zero reference surface elevation
- 25 km^2 of surface which is not inclined more than 10°
- area with relatively small number of hazards (e.g., mountains, craters concentration)
- zone shouldn't have thick dust deposits (Bussey and Hoffman 2016)

4 Exploration Zone Map Symbology

The EZ map is not a usual map. The objects, places, buildings, and symbols that the EZ map contains are adjusted for the needs of the first planetary explorers. These necessary objects are defined in the EZ parameters (Bussey and Hoffmann 2016).

The locations and positions of building structures displayed on the Hebrus Valles Map may be arranged differently in order to adjust to the mission. The position of the Science Regions of Interests (SROIs), Resource Regions of Interests (RROIs), Special Regions of Interests (SRs), Landing Sites (LSs), and paths are based on the original EZ proposal (Davila et al. 2015) and other studies (Rodriguez et al. 2012) in which the authors identify the points of significant regions. However, many points were positioned by the author of this map and paper after a thorough analysis of relevant information and maps.

The most important considerations for future astronauts working in the field are surface morphology, potential hazards, building structures of the mission base, and waypoints (places) that may potentially be visited. The Hebrus Valles Map has octahedral symbols that provides information about each specific location.

Table 1 Symbols of the Hebrus Valles Exploration Zone Map

No.	Symbol	Description
1		Landing Sites (symbol occurs also as LS) and Habitation Zone: These areas are places appropriate to spacecraft landing and base establishment. The Habitation Zone is a field station including greenhouse for farming, a power zone, and rover parking
2		Science Region of Interest: This area is actively forming or changing today, erosion is actively exposing old materials. Additionally, the formation of the surface material involved water (sediments of rivers, lakes, or glaciers), and materials may be characterized by fossilized lifeforms
3		Resource Region of Interest: This area contain water in all forms, surface or shallow subsurface ice, slope lineae, hydrated minerals, glaciers, hydrated dunes, and atmospheric H_2O. Geology in these areas may be characterized by cobble sized rocks, regolith (soil), and material that contains iron, aluminum, and silicon ready to be mined by automated systems
4		Special Region: This is the area where terrestrial microorganisms are capable of replicating. These regions should be avoided by humans but can be studied by sterilized robots
5		Paths and Roads: Some of the tracks may be paved for faster transportation
6		Greenhouses for Plant Cultivation: This is the place where plants are grown either hydroponically under artificial light or in local regolith using natural lighting
7		Habitation Zone: This is the area where explorers will have their living quarters
8		Power Zone: Power will be produced in this area and sent to other structures
9		In Situ Resource Utilization: Area where waste will be recycled or destroyed. Encountered materials will also be stored and processed to meet the needs of the mission (Sacksteder and Sanders 2007)

The shape provides a modern method of identifying the locations as well as easy access to details about each place. The labels used on the symbols are projected in Century Gothic font. In order to clearly identify the locations, a unique color of symbology is used for each place (for example, green for Greenhouses, black for Landing Site). The symbols used in the map are explained below in Table 1.

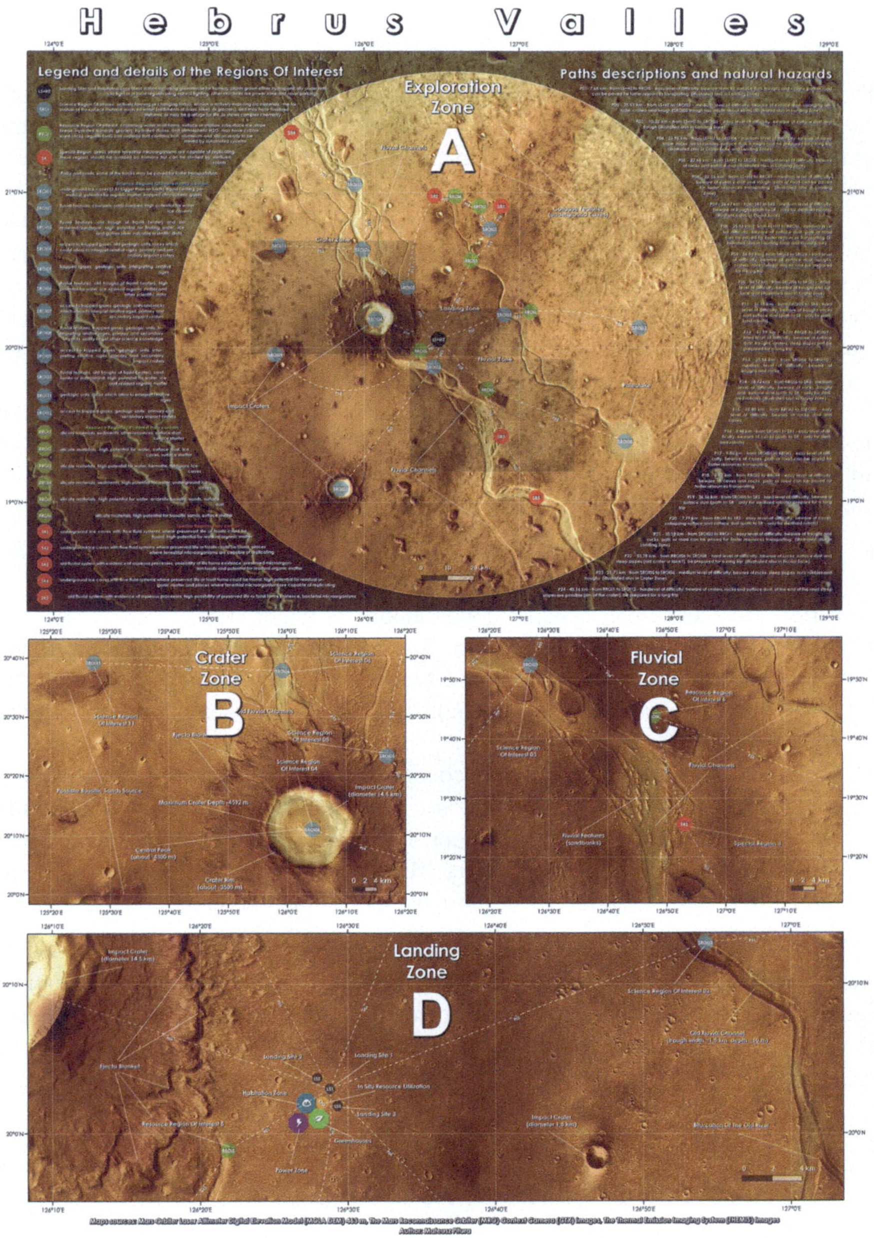

Fig. 2 Full Hebrus Valles Exploration Zone Map layout with signatures

5 Paths (Traverses)

The tracks illustrated on the Hebrus Valles Map are traverses mostly for rovers from the LZ to the SROIs, RROIs, and SRs. Routes were chosen by appointing the shortest possible path across the Mars surface to the destinations. The general method of track selection involved a detailed study of the terrain of Mars. In order to determine and review the tracks, maps from JMARS software were used, including a dust cover index map (Ruff and Christensen 2002), MOLA slope map (Smith et al. 1999), thermal inertia map (Christensen and Moore 1992), rock abundance map (Nowicki and Christensen 2007) and many others. The paths were created primarily on lines where the surface of Mars was clear of obstacles. The length of each path was measured in kilometers using ArcMap.

The paths were divided into easy, medium, and hard levels depending on the hazards that astronauts could encounter and on the length of the track. The levels of difficulty are provided on the map to aid explorers and vehicles (rovers) acting as transporters. The details of the path difficulty levels are as follows:

- Easy—tracks under 15 km. The road does not require special preparations. Obstacles are of a minimum level of difficulty and there is a possibility of surface dust and small craters. Slopes should be under 10°.
- Medium—tracks from 15 to 30 km. The road necessitates preparation of the rover (checking the amount of fuel). Obstacles including rocks, rock rubbles, small craters, surface dust, and troughs will require more attention. Slopes are greater than 10°.
- Hard—tracks above 30 km. The condition of the road necessitates special preparations (rover tanking) and the use of scientific equipment. Some hazards may be present, including steep slopes, rocks greater than 30 cm in size which could present an obstacle for rovers (Bussey and Hoffman 2016), large rock formations, craters, surface dust, troughs, ice, and depressions.

The longest route is 61.99 km (P12) and is marked as hard due to the fact that it would be a long and tiring journey. Some warnings related to geomorphological features of the landscape are also included in the traverse characterizations. Descriptions of the paths depicted on the map are listed on the right side of the main EZ illustration (see Fig. 2).

6 Hebrus Valles Exploration Zone Map Detailed Description

Hebrus Valles was chosen for EZ because of its great potential for the presence of water and remnants of life (Davila et al. 2015). The fluvial-like structures in the center of the zone suggest that fluid once flowed on the surface of the terrain (Schulze-Makuch et al. 2016). What is more, the Hebrus Valles region is rich in impact craters, which will allow astronauts to reach older geological structures of Mars and expand their knowledge about its history. An important concern in planetary cartography is that the map visualize at least three themes: base image, grid, and nomenclature (Hargitai 2005), which give the astronaut the ability to orient himself in the field. The Hebrus Valles Map illustrates everything future Mars explorers will need to plan their journeys and investigate the EZ: terrain as a base image, grids, elements (symbols), paths, and nomenclature referring to the surface. Furthermore, the map has magnified the most important regions for scientists.

The Hebrus Valles Map is divided into four sections (see Fig. 2). The layout contains: (A) the main EZ map in the upper section; (B) a magnified map of the Crater Zone, in the middle on the left; (C) a magnified map of the Fluvial Zone, in the middle on the right; (D) a magnified map of the Landing Zone in the bottom part. The sections are described below:

- A—the main map, area of EZ, 100 km radius from the center point (126.38°E, 20.05°N). The map provides special symbols which are connected to RROIs, SROIs, SRs, LZs, paths, and provides text referencing geomorphological structures. What is more, there are transparent squares covering the magnified areas of the maps below. A legend with significant descriptions is located on the left and right side of the map. A simple, linear scale is situated on the bottom of the map.
- B—the Crater Zone, which is an interesting area for scientists. The map shows symbols (SROIs, paths), text referencing geomorphological structures and other information. A simple, linear scale is situated in the bottom right corner of the map.
- C—the Fluvial Zone, which is an interesting area for scientists. The map shows symbols (SROIs, paths), text referencing geomorphological structures and other information. A simple, linear scale is situated in the bottom right corner of the map.
- D—the Landing Zone, which is the main area of the mission. The map shows symbols (RROIs, SROIs, SRs, paths, and also Habitation Zone, Power Zone, Greenhouses, and In Situ Resource Utilization), text referencing geomorphological structures and other information. A simple, linear scale is situated in the bottom right corner of the map.

All of the maps mentioned above have grids based on a coordinate system named the Interplanetary Mars 2000 Sphere, projection Plate Carree.

The color scale of the EZ Map was selected according to Mars studied hues (see Fig. 3)—yellow (sand) and bronze (Maki et al. 1999). Bronzes characterize the highest elevations on the map (from −3600 to −3400 m) and the yellow-sand colors illustrate the lowlands (from −3700 to −4500 m). By analyzing the map terrain, it can be concluded that the Hebrus Valles region is declining in the north–west and north–east directions. Moreover, from north (north–east) to south, a characteristic wedge-shaped (triangular) ridge is present.

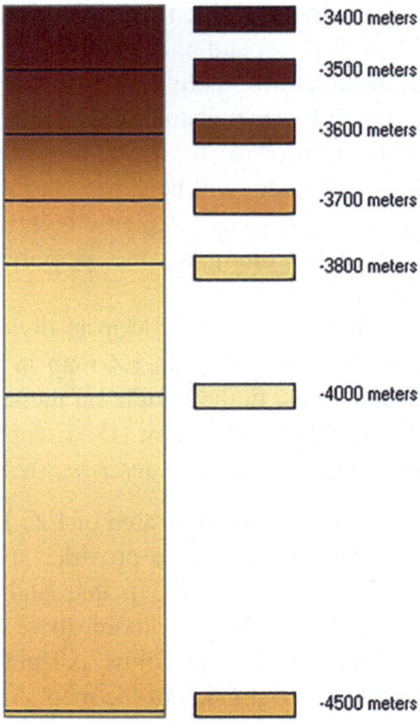

Fig. 3 Color scale used for the Hebrus Valles Exploration Zone Map

7 Creation of the Map, Methodology

The Hebrus Valles Map was designed in three stages and using three different applications. First of all, for finding, viewing, and reviewing EZs, JMARS version 3.5.3 (Christensen et al. 2009) was used. The mapping process was completed using ArcGIS for Desktop (in version 10.2.2). The third part of the work involved performing post-mapping adjustments, modifying the legend, providing descriptions, and adding details in Photoshop CS6.

First, the images of the Hebrus Valles region from The Thermal Emission Imaging System (THEMIS) (Hill et al. 2014) and The Context Camera (CTX) (Malin et al. 2007) instruments were exported in high resolution to raster files from JMARS software. Another common visualization of Mars, the Mars Orbiter Laser Altimeter (MOLA) (Smith et al. 2001) digital elevation model, was obtained from the official source and used to create the main EZ Map.

The first steps in ArcMap were related to rectifying all of the maps exported from JMARS to the Interplanetary Mars 2000 Sphere coordinate system (see Fig. 4). Before that, a WMS (Web Map Service) map was added to layers as a reference. Adjustments were made using the ArcMap Georeferencing Tool.

The next step involved importing the MOLA digital elevation model to ArcMap and adjusting the coordinate system. Clipping to Hebrus Valles zone was necessary to minimize excessive use of computer memory. Then, color symbology was used on a digital elevation model and adjusted for optimum visualization (see Fig. 4).

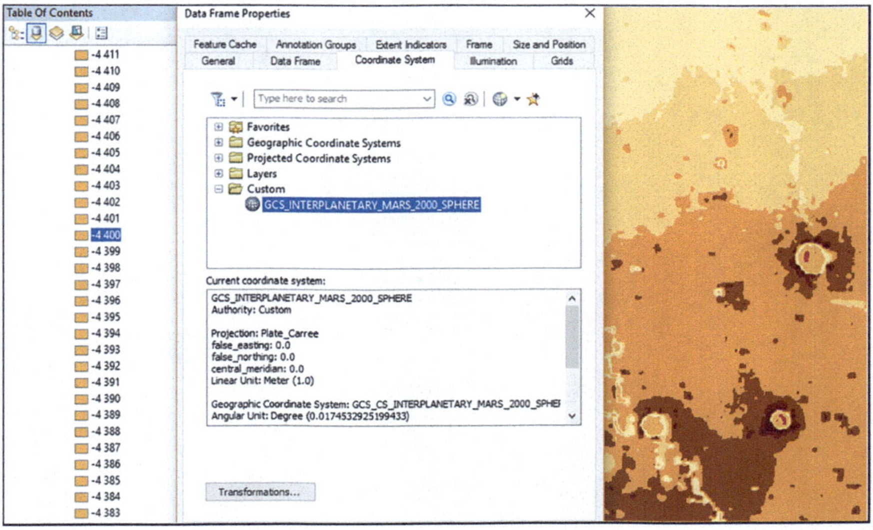

Fig. 4 Coordinate system of the map with color symbology in ArcMap

An important part of creating the map was using the pansharpening tool in ArcMap. Pansharpening (or panchromatic sharpening) is a fusion between two pictures which allows the user to obtain a higher resolution image (Xu et al. 2014). In this case, the No Alteration of Grayscale or Intensity (NAGI) method of pansharpening was used (Nagi 2012). This method uses a regular color distribution and an adjusted gamma value—such technique allows the user to achieve more effective image fusion. After use of the pansharpening tool, images were modified for best performance by changing contrast, color intensity and sharpness (see Fig. 5).

When the main base map was prepared, symbols and others features were created as shape files: LZ+HZ, RROI, SROI, SR and paths. The positions of the points and lines were based on proposed regions of interest (Davila et al. 2015) or, in most cases, on geomorphology, geology, relief of terrain, and many JMARS statistic maps.

A 200 km circle diameter was achieved by inserting the point of Hebrus Valles (126.38°E, 20.05°N) provided by the proposers of the EZ (Davila et al. 2015) and using a buffer tool in ArcMap. The result was a circle with a 100 km radius extending from the center point.

The last stage of the ArcMap process involved preparing the map for printing: creating frame, adjusting scale to 1:800 000 (scale to include all of the EZ), creating grid coordinates and legend with figures to use in later mapping in Photoshop.

Almost all of the above-referenced steps were also applied to the other magnified maps: the Crater Zone, the Fluvial Zone, and the Landing Zone. In contrast, however, the Crater Zone and the Fluvial Zone frames were scaled to 1:350 000. The Landing Zone, the most magnified section of the map, was scaled to 1:140 000.

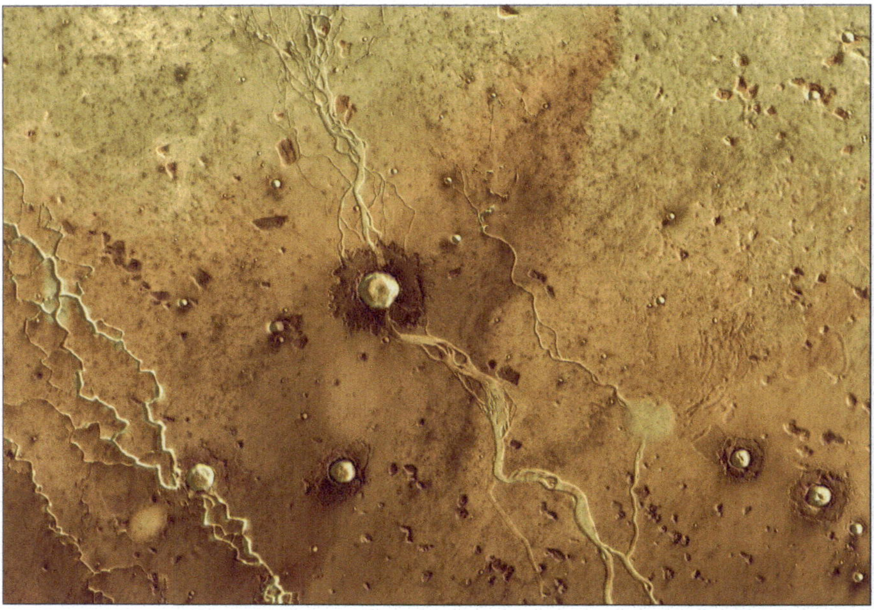

Fig. 5 Final result of the NAGI pansharpening method. Depicted is MOLA DEM in fusion with adjusted color symbology

There was also a need to create new shape files for the Landing Site, Habitation Zone, Greenhouses, Power Zone, and In Situ Resource Utilization areas.

Post-mapping processing was completed using Photoshop, where the transparency of the legend was created. Geomorphological annotations with arrows were also created in this mapping stage. Further actions focused on developing the scales and the arrangement of the map sections. Lastly, color correction was completed using a curve filter.

8 Map Format and Adapting to New Technologies

The Hebrus Valles Map is projected to be printed on paper in portrait mode, so that all four of the sections may be viewed. A remaining question is whether the astronauts should have paper maps with them in the field. Today, digital maps are widely used around the world (including maps of planetary bodies) and their significance has increased in recent years. Maps on screens have many advantages, for example they have the ability to illustrate multiple map sheets, they are interactive (ability to zooming, locating, and tracking), and they can be updated quickly and easily. However, despite the many benefits of digital maps, paper maps also have their advantages. Most of all, easy access to maps is fundamental in the field and electronic maps are not always available, such as when a power source is not present or the device projecting the map is not working. A paper map does not depend on electricity, which could be a crucial factor in a hostile, alien environment.

Additionally, paper is lightweight, it can be folded into a small size and expanded for viewing, and it enables users to visualize large areas in high resolution at once (large screens are generally not available for field work, however, future technologies may develop virtual glasses that project maps to the entire field of view, or in an augmented reality environment). Hence, on a paper map more details may be included for a larger area. Furthermore, the continuous availability of a paper map in emergency situations can be mission critical. Therefore, paper maps should be essential components of an astronaut's field equipment—in fact, the success of the mission could depend on the availability of a paper map.

Above all, the Hebrus Valles Map has implications and potential for new technologies (mainly touch screens). For example, using touch screens, astronauts could zoom into specific regions of the map on their electronic computer pads by selecting the area to be magnified. This feature would allow astronauts to see more details of a specified region. Examples of this solution are available on the EZ section of the Hebrus Valles Map where transparent black rectangles are present (see Fig. 6).

Additionally, the feature of selecting and combining different layers is a common component of interactive maps. Until now, we have had numerous maps of the Mars surface that would be necessary for future astronauts. Furthermore, every computer containing the map ought to include a real-time information system that

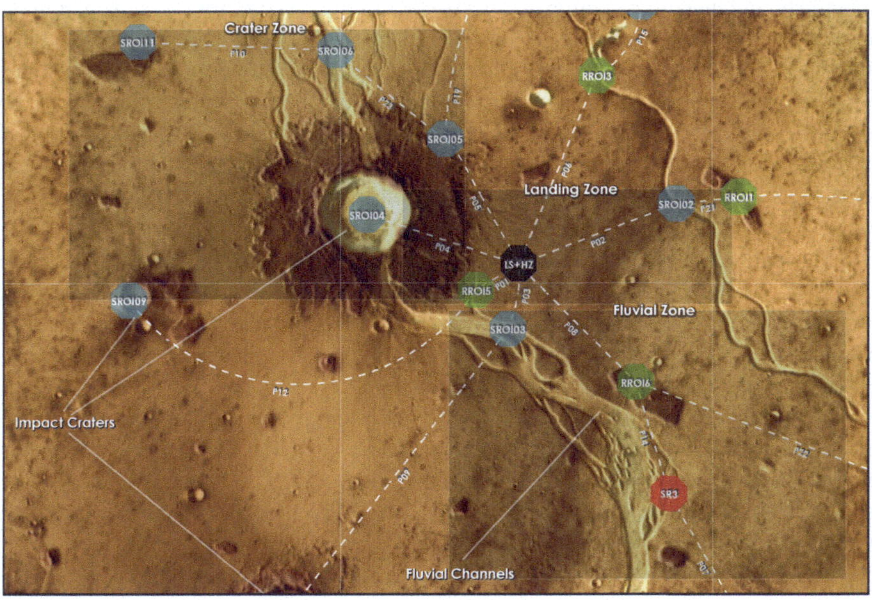

Fig. 6 Transparent *black rectangles* on the Hebrus Valles Map selecting the magnified areas

provides details about current surface conditions—this could be accomplished through the map image and through an alarm received by the astronaut.

Another idea is to enable astronauts to create profile lines to display terrain morphology along specified tracks and identify what hazards are expected on the road. Further analysis of terrain lying on the line would require knowledge about geology and rock abundance—implementing such an option would result in better route planning and would improve the safety of explorers.

What is more, an essential feature would be the ability to choose a point of interest, view the legend and significant information near it on the map. Rather than limiting this only to SROIs, ROIs, SRs, and other symbols, all of the map terrain could be interactive.

Finally, the most important and useful option is positioning an astronaut in real time—using local transmitters for triangulation or orbiting spacecrafts—and orienting the area based on the direction the astronaut is looking. The location of the astronauts should also be available and visible on maps for others in the Habitation Zone.

9 Summary

The Hebrus Valles Map displays one of the 47 Exploration Zones that could be the area of the first human steps on Mars. Future explorers will need access to all of the latest mapping technologies in order to gain invaluable knowledge about our universe. The detailed Exploration Zone visualization is crucial for mission planning, as well as for safety on the surface of Mars. Methods and software used for mapping (JMARS, ArcMap, and Photoshop) apply the techniques of modern cartography and ensure that maps will be sufficiently detailed. Moreover, the use of the map on computers or other modern equipment would enhance the map's usefulness to astronauts. Finally, the addition of interactive and support elements to the map should also be considered in order to provide astronauts with the best mapping tools.

This map could be successfully used in the field, where astronauts will make efforts to travel and survive in unfavorable environment. In order to plan traverses and journeys for revealing secrets of Mars, astronauts will need a map that includes details such as distances, points of interest, dangerous places, and safe paths—everything that the Hebrus Valles Map illustrates.

The success of the mission is the priority. Cartographers can help astronauts by providing them as many significant details as can be found and by presenting this important information in an easily accessible way.

The Hebrus Valles Mars Exploration Zone Map is available online at http://www.hebrusvalles.com.

Acknowledgements I am grateful to the International Cartographic Association—Commission on Planetary Cartography for organizing the Mars Exploration Zone Map Contest, the jury of the Mars Exploration Zone Map Contest for the chance to write this paper, Henrik Hargitai for helping in writing this paper, International Cartographic Association and University of Wroclaw for sponsorship and Artur Sobczyk, my supervisor, for support.

References

Bussey, B., & Hoffman, S. J. (2016). Human Mars landing site and impacts on Mars surface operations. In *IEEE Aerospace Conference Proceedings*. Big Sky, MT.

Christensen, P. R., & Moore, H. J. (1992). The Martian surface layer. In H. H. Kieffer, et al. (Eds.), *Mars* (pp. 686–729). Tucson: University of Arizona Press.

Christensen, P. R., Engle, E., Anwar, S., Dickenshied, S., Noss, D., Gorelick, N., et al. (2009). JMARS—A planetary GIS. AGU Fall Meeting Abstracts.

Christiansen, E. H., & Hoppin, R. A. (1987). *The origin of channels and associated deposits in the Elysium region of Mars*. Washington, DC: National Aeronautics and Space Administration, National Technical Information Service.

Davila, A., Fairén, A. G., Rodríguez, A. P., Schulze-Makuch, D., Rask, J., & Zavaleta, J. (2015). The Hebrus Valles exploration zone: Access to the Martian surface and subsurface. *First Landing Site/Exploration Zone Workshop for Human Missions to the Surface of Mars*. Retrieved February 6, 2017, from http://www.hou.usra.edu/meetings/explorationzone2015/pdf/1012.pdf.

Gazetteer of Planetary Nomenclature. (2016). USGS Astrogeology Science Center. Retrieved October 12, 2016, from http://planetarynames.wr.usgs.gov.

Hargitai, H. I. (2005). Planetary maps: Visualization and nomenclature. *Cartographica, 41*(2), 149–164.

Hill, J., Edwards, C. S., & Christensen, P. R. (2014). Mapping the Martian surface with THEMIS infrared global mosaics. In *Eighth International Conference on Mars*. Retrieved February 6, 2017, from http://www.hou.usra.edu/meetings/8thmars2014/pdf/1141.pdf.

Maki, J. N., Lorre, J. J., Smith, P. H., Brandt, R. D., & Steinwand, D. J. (1999). The color of Mars: Spectrophotometric measurements at the pathfinder landing site. *Journal of Geophysical Research, 104*(E4), 8781–8794.

Malin, M. C., Bell, J. F., Cantor, B. A., Caplinger, M. A., Calvin, W. M., Clancy, R. T., et al. (2007). Context camera investigation on board the Mars reconnaissance orbiter. *Journal of Geophysical Research E: Planets, 112*(5), 1–25.

Nagi, R. S. (2012). Maintaining detail and color definition when integrating color and grayscale rasters using No Alteration of Grayscale or Intensity (NAGI) fusion method. *AutoCarto*, September 16–18.

Nowicki, S. A., & Christensen, P. R. (2007). Rock abundance on Mars from the thermal emission spectrometer. *Journal of Geophysical Research, 112*(E5), E05007.

Rodriguez, J. A. P., Bourke, M., Tanaka, K. L., Miyamoto, H., Kargel, J., Baker, V., et al. (2012). Infiltration of Martian outflow channel floodwaters into lowland cavernous systems. *Geophysical Research Letters, 39*(22), 1–6.

Ruff, S. W., & Christensen, P. R. (2002). Bright and dark regions on Mars: Particle size and mineralogical characteristics based on thermal emission spectrometer data. *Journal of Geophysical Research, 107*.

Sacksteder, K. R., & Sanders, G. B. (2007). In-situ resource utilization for Lunar and Mars exploration. In *Reno, Nevada: 45th AIAA Aerospace Science Meeting and Exhibit*. Retrieved February 6, 2017, from http://arc.aiaa.org/doi/abs/10.2514/6.2007-345.

Schulze-Makuch, D., Davila, A., Fairén, A. G., Rodriguez, J. A. P., Rask, J., & Zavaleta, J. (2016). Ice Caves in Hebrus Valles: A target location for the first human mission to Mars. In *Reykjavik:*

Sixth Mars Polar Science Conference. Retrieved February 6, 2017, from http://www.hou.usra.edu/meetings/marspolar2016/pdf/6014.pdf.

Smith, D. E., Zuber, M. T., Frey, H. V., Garvin, J. B., Head, J. W., Muhleman, D. O., et al. (2001). Mars orbiter laser altimeter: Experiment summary after the first year of global mapping of Mars. *Journal of Geophysical Research: Planets, 106*(E10), 23689–23722.

Smith, D., Neumann, G., Ford, P., Arvidson, R. E., Guinness, E. A., & Slavney, S. (1999). *Mars global surveyor laser altimeter precision experiment data record.* NASA Planetary Data System. Retrieved February 6, 2017, from https://astrogeology.usgs.gov/search/map/Mars/GlobalSurveyor/MOLA/Mars_MGS_MOLA_ClrShade_merge_global_463m.

Thomson, B. J., & Head, J. W. I. (2001). Utopia Basin, Mars: Characterization of topography and morphology and assessment of the origin and evolution of basin internal structure. *Journal of Geophysical Research, 106,* 1–22.

Xu, Q., Zhang, Y., & Li, B. (2014). Recent advances in pansharpening and key problems in applications. *International Journal of Image and Data Fusion, 5*(3), 37–41.

XY Domain: A Sound Map Artwork for Communicating Big Data Characteristics

Antoni B. Moore and Charlotte Parallel

Abstract This paper details a sound map artwork with a visual map component in order to communicate the characteristics of Big Data, namely the volume, velocity, variety and uncertainty aspects of such datasets. It represents an attempt to address a part of the current research agenda on applying cartography to the visual representation and interaction with big spatial datasets. As an artist-scientist collaboration, the research concentrates on electrical signals and humans as transducers of those signals into data. We trace the evolution of the artwork, details construction, collection of representative signal data and context, leading to its exhibition as part of an "Art and Space" theme in a gallery in Dunedin, NZ. Finally, an explanation of and reaction to, the artwork is provided. Visitors reacted positively to the artwork, which conveyed the characteristics of Big Data through a mix of tactile and embodied interaction, sound representations and visual cartographic design/spatial data choice elements.

Keywords Sound map · Transduction · Electricity · Signals · Visual map

1 Introduction

In the modern age, information, signals and electricity surround us, flowing around and through us. Their presence can be detected, either visibly or invisibly. At some level, all can be represented by data. Our ability to generate and represent these data is unprecedented; indeed, we as humans could be depicted as transducers of objects and energy into data. This leads to more and more instances of confrontation with Big Data, where data comes to us in huge volumes, at great velocity and in a variety

A.B. Moore (✉)
School of Surveying, University of Otago, Dunedin, New Zealand
e-mail: tony.moore@otago.ac.nz

C. Parallel
Dunedin, New Zealand
e-mail: charlotteparallel@gmail.com

M.P. Peterson (ed.), *Advances in Cartography and GIScience*, Lecture Notes in Geoinformation and Cartography, DOI 10.1007/978-3-319-57336-6_12

of forms (Laney 2001). Much of this data is volunteered and of uncertain provenance. As we experience it in everyday life, these numerous "meetings" with data can lead to confusion. In cartography, it is one of the current major research challenges to be able to represent this data effectively, and a part of this is using the channel of art to add expressive power to data representations (Robinson et al. 2017).

We present *XY Domain* (Fig. 1), a sound artwork with a visual map that aims to aurally convey the well-known attributes of Big Data. A small urban environment dataset (electrical networks and signals—lighting, telecommunications, electrical sources, ATM) is represented in such a way as to artfully communicate the attributes of Big Data. The energies generated by these and other sources are all around us, like Big Data. As such, the artwork tangibly renders these often intangible energies that we transduce either through ourselves or through technology. Participants in the artwork walk on an urban map showing point sources of energy, triggering located and transduced sounds as they do so. The more participants there are on the artwork, the more sounds occupy the same channel, making for a confusion of noise, redolent of a large and confusing dataset. As in life, the human is the instigator of the data communication, transduction and release of energy. They are also bodily within the map environment, which is also aligned to and centred within the map referent (in this way, you are a part of the map).

The representational link to Big Data comes through that variable number of people walking on the map, entering and exiting the map area at different rates, triggering different sounds from the various types of phenomena depicted, that overlap and intermingle in unpredictable ways (respectively, volume, velocity,

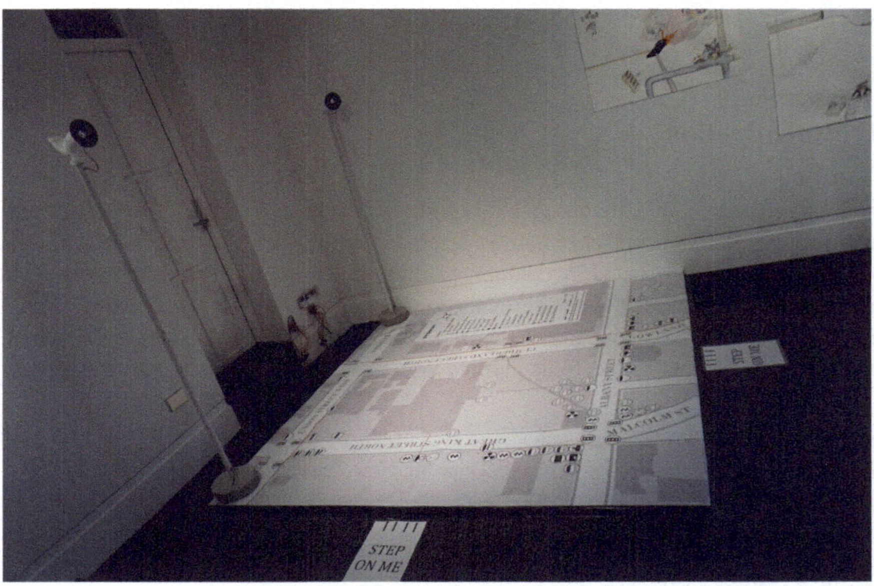

Fig. 1 XY Domain, a sound map artwork (*Photo credit* Pam McKinlay)

variety and veracity). That the data is collected through informal means, is unverified and is plotted visually on an incomplete context map layer, is a further statement on uncertainty.

This paper gives a full account of *XY Domain*. We provide a background to the process and entities that are being revealed (transduction, energies, humans and data) and how this translated into initial thinking for the artwork. We then trace the artwork's evolution, construction and context, leading up to its display in a public "Art and Space" exhibition.

2 Background

2.1 Transduction, Energy and Humans

As a research focus, the art practice of the second author is motivated by site-specific encounters with sound realised by the implicit processes of transduction. It follows that the artworks produced are spatially underpinned and map-like.

Listening is a form of attunement to any given place enabled by cochlear transduction. As a process, transduction can be discussed in two main ways. Firstly, as a concrete biological and acoustic tool and abstractly as an operation that allows for differing energetic realities to 'become,' articulating a field of relations between the likes of, human and non-human and the living and non-living.

Attunement is a focused attention to what creates an event, notably energies. Energy is a force, an all binding, obvious but largely invisible process. The interaction of mechanical and electromagnetic energies is located through transduction. As a living or powered mechanism, a transducer is the transformative interface that allows the change from one energy state to another. Every sound experienced in life has been transduced—connecting us as humans, the 'user' of mechanical and electromagnetic energy and all organic matter, to a larger energetic system.

The application of the physics of sound in art practices is a way of directing attention to the systems that we are in constant engagement with. Sound amplifies a larger field of energetics. It is a temporal flux of signals that reveals an ecological, social, political and geographic engagement with the lived-in world. The artistic use of energy, primarily electrical and electromagnetic, as "sound systems (consisting of bodies, technologies and acoustic vibrations)" is used in this project as a way of knowing and being in the world (Goodman 2010).

The process of transduction occurs in everything from the human ear to plants to viruses to cell phones, articulating a field of relations between one thing and another. This field of relations is a place of 'becoming' or the "intersection of diverse realities" that Adrian Mackenzie describes as a "knotting together of commodities, signs, diagrams, stories, practices, concepts, human and non-human bodies, images and places. They entail new capacities, relations and practices whose advent is not always easy to recognize" (Mackenzie 2003).

Attunement to a place or a situation is experienced through the whole of the sensorium. As individuals, the subjective senses embody signal as a materialist form, both concretely and abstractly. 'Materialist' is used as a term that refers to how waveforms travelling as forces within a medium, are site-specific indicators of a 'unique moment.'

Waveforms are considered as important artistic material that direct attention to the political, social, geographic and ecological milieu that are active in a situational experience with sound. The transducer as an object brings together a collection of forces. Each side of the transducer represents a different energy form and corresponding reality that are brought together in a transformative field, initiating a new set of relations.

'Transductive relations' (Adrian Mackenzie), 'lived electromagnetism' (Douglas Kahn) and 'technogenesis' (N Katherine Hayles) are all terms that describe the co-evolution of the human and non-human in an energetic continuum. What comes about cannot be predetermined but its 'coming to be' can be traced, mapped, sensed, detected and transformed through experience.

2.2 Big Data, Cartography and Art

The increasing ability to capture these 'energies' as data through technologically innovative data collectors, has led to an abundance of data. These data can be volunteered (collected by individuals with smartphones), directed (the result of professionally organized data collection), coming from a huge amount of remote and ground-based sensors (Kitchin 2013). Most of these data are also spatial (MacEachren and Kraak 2001), which means that visually, cartography is a natural partner with art to make these energies manifest. These data often adopt the characteristics of Big Data, where the size (volume), rate of growth (velocity) and different types (variety) of data (Laney 2001) is such that the computing power we currently have is insufficient to process it in a timely manner.

It follows that Big Data poses a significant challenge for cartography (Robinson et al. 2017), which forms a visual language for communicating spatial data, through representation and interaction. Furthermore, the role of art, having had a long history of being either a major or minor representational partner with maps and cartography (Casey 2005; Cosgrove 2005), is again starting to be recognized as an important channel of geospatial communication (Caquard and Taylor 2005).

Art has the potential to provide flexible geospatial visualisations and is a natural outlet for the narratives, emotions, underlying meanings, processes—indeed, energy flows—that are often found in Big Data (Robinson et al. 2017). As such, art lends itself to the variety, uncertainty, multi-scale and multi-dimensional aspects of big geospatial data. Through the *XY Domain* artwork, we will attempt to communicate the volume and velocity aspects of Big Data as well.

This artwork is an instance of an artist developing their own vision relative to maps and geospatial phenomena (one of the three categories of art-map collaborations stated by Caquard and Taylor 2005). It addresses a research challenge to "maximize creative contributions in expressing Big Data" which forms part of a research agenda for cartography and Big Data (Robinson et al. 2017) and a step towards the follow-up research opportunity to "facilitate engagement of artists with big data…"

2.3 *Cartography and Sound*

Sound adds an expressive communication channel to visual maps. Maps and sound have been co-investigated in various ways, from conventional visual maps of aural phenomena, such as noise (e.g., Arana et al. 2010), through embedded multimedia (e.g., narrative, music) in dynamic maps (e.g., Hu 2003) to ad hoc and considered use of sound as a cartographic variable. Krygier (1994) presented these sonic cartographic variables, an aural equivalent to Bertin's (1983) visual graphical variables. Variables such as volume and pitch were linked to quantitative data (applied to communicate uncertainty in the work of Fisher 1994), while timbre conveyed qualitative data.

Minard's Sound Map (2012) represented environmental (i.e., qualitative) sounds from 15 German cites, each city having 15 sound sites. The user had the ability to mix sounds from various points, forming a "collage of the city." Brauen and Taylor's Nunaliit Atlas (2008) also features located ambient sound. Müller and Scharlach (2001) present a noise map which uses volume as a quantitative variable.

In a survey of web-based sound maps, Brauen (2014, pp. 141–142) noted that "…the most common design paradigm currently used results in maps that provide access to acoustic media organized by location tags, playable through audio interfaces that are separated to greater or lesser degrees from the map itself." He notes that in most of the maps surveyed, the audiovisual system used has elements that distracts from or obfuscates the map (e.g., audio player on a separate webpage). He also notes that the examples surveyed generally emphasise user control over the located sound (i.e., in starting, stopping, altering volume etc., of the sound). See also Kornfeld et al. (2011) for guidelines on audio cartography.

XY Domain is an integrated interface (i.e., through the tactile button directly on the map, there is no separation between visual and audio interfaces), uses located qualitative sound data, triggered with minimal user control—the user is only able to start the broadcast of sound, potentially making for many sounds overlapping each other. These are all deliberate design choices in order to communicate the typical characteristics of Big Data and pervasive energies.

3 Context and Construction

3.1 Context

"Art and Space" was a thematic exhibition curated by Peter Stupples of the Dunedin School of Art, held in the Skinner Annex of Otago Museum. In its fourth iteration between the Dunedin School of Art and the University of Otago, "Art and Space" is colloquially referred to as the Science Series.

> The Science Series is a unique institution. A group of artists working in the city of Dunedin, most of whom are staff, students or graduates of the Dunedin School of Art, came together in late November 2015 with a similar number of research scientists at the University of Otago to talk about "art and space." (Stupples 2016)

This kind of art-science collaboration is a constant feature of SIGGRAPH and IEEE Vis conferences.

3.2 Idea Development

From a series of introductory emails the authors met for the first time at Charlotte's final MFA exhibition *Ecologies of Transduction* (Fig. 2a) at the Anteroom in Port Chalmers, a local town. In particular, "A sonic cartography of Venice" that formed part of the exhibition, featured detection, amplification and recording of Venetian canal and lagoon sounds. These sounds were accessed via located sound trigger buttons on a physical map exhibit (Fig. 2b). There was also a sonic map exhibit of Port Chalmers itself that featured sounds transformed from light using a repurposed solar panel (Fig. 3). These were to inform the development of *XY Domain*.

In discussion of mutual interests in cartography and how an artwork could be brought together, it was decided to create an interactive sound map. Initial thinking on this was that the entire exhibition space would form the extent of the sound map,

(a) **(b)**

Fig. 2 (**a**) Ecologies of transduction exhibition; (**b**) Sound trigger button on Venice exhibit

(a) (b)

Fig. 3 (**a**) The DIY light-to-sound transducer (repurposed solar panel) used for data collection; (**b**) the device being used to collect light data from a shop window as sound (data collection from the previous Venice exhibit)

which would have no visual element. The extent of the represented area varied in the discussions, from the immediate city block to the full extent of the city and environs, an area of some 50 km by 50 km.

In the hypothesized map, sounds would be triggered by motion sensors positioned at the entrances to the exhibition space. Therefore, the exhibition visitors would drive the characteristics of Big Data experienced through their number in the room (volume), the rate at which they enter the exhibition space (velocity) and the unpredictable ways in which sounds merge in the exhibition space, dependent on ad hoc conversations, other noises and environmental conditions (uncertainty). Variety would be expressed through the different sounds, first postulated to be sounds around Dunedin that could be clearly identified and mentally located (e.g., "nominal" type sounds as with Minard 2012).

In the end, due to there being other sound art exhibits in the gallery and issues with the sensitivity of motion detection sensors, it was decided to go with a contingency map that would deliver the immersive element of the gallery-as-map plan but through visitors walking on a visual map in a specific part of the gallery. The sound delivery would be similarly confined to the area of the map.

XY Domain is an electrical sound map that focuses on the transduced sound signals from electricity, a continuation in theme from the previous exhibition. *XY domain* references axis, data, code and area. A formal structure used to plot coordinates within a designated area. In this case, the area is the block surrounding the Skinner Annex.

3.3 Construction of the Sound Map

Within the chosen area, objects and signifiers of electrical networks and systems that are part of our day-to-day visual and operational landscape were mapped. Such objects included power poles, UHF aerials and ATM machines. These objects or signals in determined forms, transmit and receive electrical and electromagnetic signal to perform and power operations of our daily lives. The everyday use of electrical energy implies a relationship, a field of engagement between one thing and another. Mapping in this way is a kind of energetic forensics that observes and considers the socio-physical environment and its interaction between local signal ecologies.

To translate the electrical flickering of light on an ATM machine or the mechanisms of an electrical door, for example, a DIY light-to-sound transducer was used (Fig. 3). This was a solar panel wired to an audio jack which was then plugged into an amplifier. The solar-cell can convert a time-varying light signal into a time-varying electrical signal. This small electrical signal can then be fed to an audio signal amplifier and is then converted into sound.

Wandering about with a solar panel attached to a 2 m pole, holding an audio recorder, wearing headphones and wires drooping from the pole is not normal street attire and often verges on a performance of sorts. There were many curious passersby striking up conservations that often drew attention from security personnel!

An added element of *XY domain* is the human body within the map space to perform and activate the ongoing forces of a place. On one hand, the body directs attention to how we exist in a continuum of, and specific concentration within, the electromagnetic field through our capacity as a transducer. The other, where the activation of sound has the potential to relay intersecting and often unseen operations of geo-political, economic, ecological and social infrastructures.

The circuitry installed in the 2 × 2 m floor work was made possible by using two 16 channel electronic WAV trigger audio boards where each channel is connected to a button (see Fig. 2b) that when triggered activates the relevant sound. The WAV trigger boards are also polyphonic, enabling multiple sounds to be played at the same time. Underneath the floor map (see next section), 2 wires connected each button to the numbered track on the sound board. While the buttons and wiring system were concealed beneath the map, the wires all connected to the specified sound board that was plugged into the power point (Fig. 4). These were intentionally displayed, referencing networks and 'wired information.'

XY domain is therefore akin to an aural circuit that implies humans' interaction with an ongoing, invisible and often unheard cross pollination of electrical and electromagnetic signal.

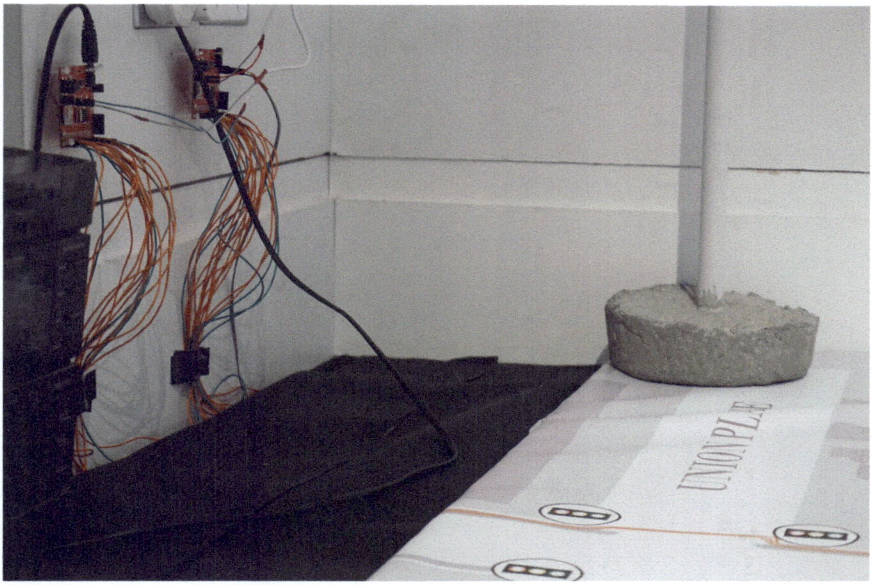

Fig. 4 Connection of the map to the soundboard, signifying networks (*Photo credit* Pam McKinlay)

3.4 Construction of the Visual Map

The visual map used as part of the artwork is shown in Fig. 5. The map (1:150 scale as exhibited—2 m × 2 m) was printed on PVC to withstand people walking on it, mounted on wood, with the sound trigger buttons standing proud of the floor surface under the PVC. Nominal point symbols were designed for the data collected with the transducer device, some drawing from symbols on electrical diagrams (antenna, radio station, transformer) in amongst the other types (ATMs, light from streetlamps and shop fronts, surveillance cameras, etc.). On the electrical theme, the point data was supplemented with underground electrical line data from Dunedin City Council. For context, road, path and parcel data was sourced from Land Information New Zealand (LINZ). The building contextual layer was derived from digital photogrammetry by the remote sensing classes at the School of Surveying (so it is incrementally added to every year). As such it is incomplete—a deliberate inclusion as a further commentary on Big Data. The map extent is one block around the exhibition space. As exhibited, it is aligned and oriented with the features that it represents. If you are facing north on the map, you are facing north in reality.

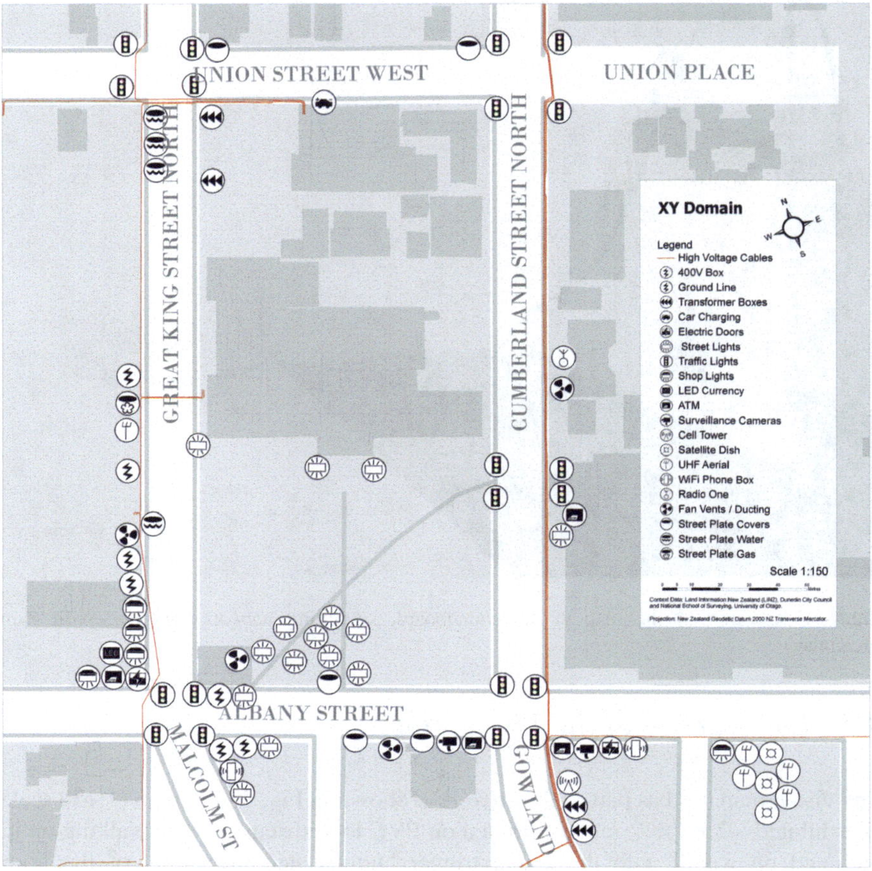

Fig. 5 The 2 m × 2 m visual map that forms the floor of the exhibit

4 Explanation and Reaction

XY Domain is sonically dormant until activated by a visitor to the map. Of course, the electronics are still active by their connection to a 240 V power source-drawing 5.5 V for each board—while also connected to speakers and a larger amplifier home audio system. The circuit is live but only realised when someone walks on it (Fig. 6a). The experiential nature of the work directly locates the participant in a techno-body relationship. As electric bodies within a circuit that in turn activates a wider constant electrical network in the surrounding streets.

A further aspect of the map is the egocentric element (the visitor is in the map) that is mirrored by the placement of self in the subject matter of maps seen and read through mobile devices. Each participant helps create the 'final work'. Some visitors choose to play it as an instrument, using their whole body to activate buttons

(b)

(a)

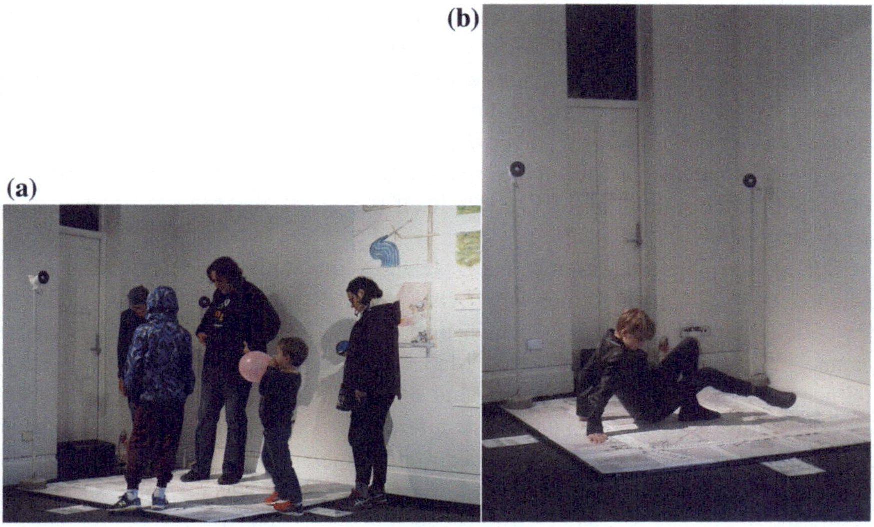

Fig. 6 (**a**) People entering and interacting with the artwork. Sound is emitted through the two visible speakers. (**b**) Example of a more playful approach to the artwork (*Photo credit* Pam McKinlay)

and the corresponding sounds at the same time (Fig. 6b). In some cases, children literally roll continuously over the map.

Most are more deliberate in their choices. By walking the map and referencing the key, they would then activate each symbol and consider its source. Two participants who interacted with the map in this way were eager to go back outside and scrutinize the surrounding streets through a new lens. Their experience created a fresh attention to their day-to-day electrical lives.

Not only are Big Data characteristics communicated by the array of sounds experienced, but elements of the visual map convey typical aspects of Big Data. The incompleteness of the building context layer has already been mentioned. In addition, a deliberate mistake has been left in the map—there is no car charging point on the map. These are intended to put across aspects of volunteered data, a large component of Big Data.

5 Conclusion

XY Domain is a tactile, visual and sound-based map artwork for communicating typical characteristics of Big Data. Map users engage with the map by walking on it, eliciting sound from a small transduced electrical signal dataset located as trigger foot buttons throughout the map. The intention is that the aural representation of data arrests the user through a *variety* of sounds from a variety of sources. With more users continually entering the artwork, come more sounds superimposed at *velocity*, eventually suggesting a dataset of *volume*. The cacophonous mix of sounds leads to confusion and *uncertainty*. These are all aspects of Big Data and so the artwork is meant to instill in the user some sense of what Big Data "feels like" (even though the sound dataset is not in itself Big Data), making this phenomenon more "knowable".

The artwork featured in this paper features qualitative and located sounds, instances of light converted into sound via a transducer. In this way, it is similar to the Sound Map (Minard 2012), which also features multiple sounds in the same channel. However, where the Sound Map intends a harmonious collage of environmental sound, *XY Domain* uses sound overlap for confusion to convey uncertainty. Confusion and uncertainty also references peoples ongoing and invisible cross pollination with electrical and electromagnetic signals.

Most previous examples of sound maps (including several in a survey of web-based sound maps conducted by Brauen 2014) emphasise the importance of user control over sound, delivered in an unambiguous single channel. The sounds in *XY Domain* are triggered by the participant stepping on a located button; they have no control over the sound after that moment. Another finding of Brauen's (2014) survey was the non-optimal interactivity of most of the sound maps featured (e.g., audio players getting in the way of visual web maps). *XY Domain* does not experience these problems, as it is an integrated map, a semi-immersive hard-copy map that uses tactile interactivity to trigger sound. The sound is delivered through non-obtrusive speakers, with no need for web browser windows to deliver content. As such, it employs a naturalistic form of interactivity.

Looking at it in a certain way, the data used for this artwork has quantitative aspects in that the number of people present on the artwork is directly represented in the amount of superimposed sounds. Also, between qualitative and quantitative representation, in future there should be exploration of 'sonic symbology'. In other words, is there an aural process analogous to map symbolization and generalization, making sounds simpler in waveform, but still recognizable? If successful, this may make sound a more efficient and clear communicator of spatial data, enhancing its value as part of an objective representation of a Big Data-set. As it is, however, sound may not be the best cartographic communicator of genuine Big Data.

Returning to the cartographic research agenda relating to Big Data, one of the main concerns is how to engage artists to address this representational challenge (Robinson et al. 2017). The form of artist-scientist collaboration embodied in the "Art and Space" exhibition and artworks such as *XY Domain* would seem an excellent way to do this, on current evidence. Although *XY Domain* is a one-time artwork (though it may be exhibited again in the future), the attempt it makes to represent the characteristics of Big Data should elicit more related investigations.

Acknowledgements We are grateful to Peter Stupples and Ruth Napper for organizing the "Art and Space" exhibition. Thanks to them and to fellow artists and scientists in the exhibition for their advice and support, particularly to Pam McKinlay for her photos. We acknowledge and thank Jack Tang at Dunedin City Council for the electrical cable data, Dr. Pascal Sirguey and successive SURV 309 (photogrammetry and remote sensing) classes for the buildings data (in progress) and Land Information New Zealand (LINZ) for the road, path and parcel data.

References

Arana, M., San Martin, R., San Martin, M. L., & Aramendia, E. (2010). Strategic noise map of a major road carried out with two environmental prediction software packages. *Environmental Monitoring and Assessment, 163,* 503–513.

Bertin, J. (1983). *Semiology of graphics.* Esri Press.

Brauen, G. (2014). Interactive audiovisual design for cartography: Survey, prospects, and example. In: D. R. F. Taylor & T Lauriault (Eds.), *Developments in the theory and practice of cybercartography* (2nd ed.). Amsterdam: Elsevier.

Brauen, G., & Taylor, D. R. F. (2008). Linked audio representation in cybercartography: Guidance from animated and interactive cartography for using sound. *Revista Brasiliera de Cartografia, 60*(3), 223–242.

Casey, E. S. (2005). *Earth-mapping: Artists reshaping landscape*. Minneapolis, MN: University of Minnesota Press.

Caquard, S., & Taylor, D. R. F. (2005). Chapter 12 Art, maps and cybercartography: Stimulating reflexivity among map-users. In D. R. F. Taylor (Ed.), *Modern cartography series* (pp. 285–307). Academic Press.

Cosgrove, D. (2005). Maps, mapping, modernity: Art and cartography in the twentieth century. *Imago Mundi, 57*(1), 35–54.

Fisher, P. F. (1994). Hearing the reliability in classified remotely sensed images. *Cartography and Geographic Information Systems, 21*(1), 31–36.

Goodman, S. (2010). *Sonic warfare: Sound, affect, and the ecology offear*. Cambridge & London: MIT Press.

Hu, S. (2003). Web-based multimedia GIS: Exploring interactive maps and associated multimedia information on the internet. In M. P. Peterson (Ed.), *Maps and the internet* (pp. 335–344). Elsevier.

Kitchin, R. (2013). Big data and human geography: Opportunities, challenges and risks. *Dialogues in Human Geography, 3*(3), 262–267.

Kornfeld, A.-L., Schiewe, J., & Dykes, J. (2011). Audio cartography: Visual encoding of acoustic parameters. In: A. Ruas (Ed.), *Advances in cartography and GIScience* (Vol. 1, pp. 13–32). Springer.

Krygier, J. B. (1994). Sound and geographic visualization. In A. MacEachren & D. R. F. Taylor (Eds.), *Visualization in modern cartography* (pp. 149–166). Oxford: Pergamon.

Laney, D. (2001). 3D data management: Controlling data volume, velocity, and variety. In *META*. Gartner.

MacEachren, A. M., & Kraak, M.-J. (2001). Research challenges in geovisualization. *Cartography and Geographic Information Science, 28*(1), 3–12.

Mackenzie, A. (2003). *Transduction: Invention, innovation and collective life.* Unpublished manuscript. Retrieved February 13, 2017, from http://www.lancaster.ac.uk/staff/mackenza/papers/transduction.pdf.

Minard, R. (2012). *Robin Minard/Sounding D/2010*. Retrieved February 13, 2017, from https://www.youtube.com/watch?v=NmT327ieEJg.

Müller, J.-C., & Scharlach, H. (2001). Noise abatement planning—Using animated maps and sound to visualise traffic flows and noise pollution. In *20th International Cartographic Conference,* Beijing, China (8 pp.).

Robinson, A. C., Demsar, U., Moore, A. B., Buckley, A., Jiang, B., Field, K., et al. (2017, in press). Big data and cartography: Research challenges and opportunities for making maps that matter. *International Journal of Cartography*.

Stupples, P. (2016). Introduction. In *Art and space: An exhibition, Otago Museum, H.D. Skinner Annexe*, September 16–October 2, 2016, Catalogue. Retrieved February 13, 2017, from https://issuu.com/dunedinschoolofart/docs/art_and_space_catalouge.

Reproducible Cartography

Timothée Giraud and Nicolas Lambert

Abstract This paper deals with the production of statistical maps as part of the wider reproducible research paradigm. The current and most widespread ways to produce statistical maps combine several software products in a complex toolchain that use a range of data and file formats. This software and diversity of formats makes it difficult to reproduce the same analysis and maps. The aim of this paper is to put forward a unified workflow that allows map production in a reproducible process. We suggest hereby the `cartography` package, an extension of the R software, that fills the need of specific thematic mapping solutions.

Keywords Reproducibility · Open-source · R · Statistical cartography · Map workflow

1 Introduction

Scientific claims have to be supported by evidence. The assessment of scientific results is possible through the availability of methods and data used by scientists, and reproducibility is a key element to validate studies. The main idea behind reproducible research is to release studies along with data and computer code that support scientific findings (Peng 2011).

This idea has been first addressed by Jon Claerbout, a geophysicist who established a standard toolchain to produce (and reproduced) each figure made in his department (Claerbout and Karrenbach 1992). The will to establish standards or to develop reproducibility is a key point in the reproducible research movement

T. Giraud (✉) · N. Lambert (✉)
Unité Mixte de Service RIATE - Centre National de La Recherche Scientifique, Paris, France
e-mail: timothee.giraud@ums-riate.fr

N. Lambert
e-mail: nicolas.lambert@ums-riate.fr

M.P. Peterson (ed.), *Advances in Cartography and GIScience*, Lecture Notes in Geoinformation and Cartography, DOI 10.1007/978-3-319-57336-6_13

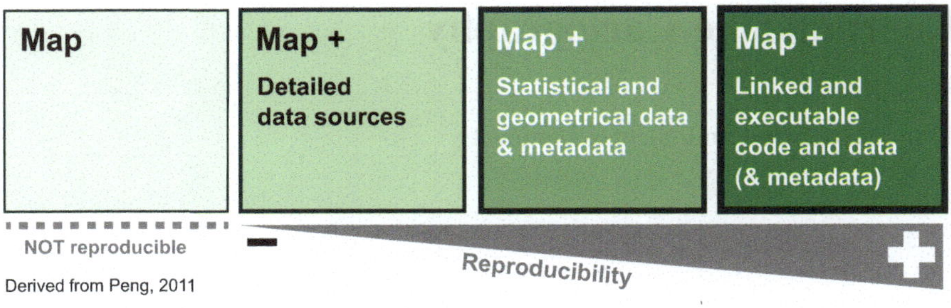

Fig. 1 The spectrum of map reproducibility

(Stodden and Miguez 2013) and scientific journals are increasingly keen on including datasets and code along with published articles (Peng 2009).

Discussions about reproducible research mostly take place in computational or statistical science fields. We argue that maps, as other graphics or statistical outputs, are part of scientific studies and should, as much as possible, be reproducible.

To be considered as evidence, a map should be open to debate and its construction should be transparent. Yet, maps produced in an academic context are currently made with a set of software products (spreadsheet, statistical software, GIS...) that slices the cartographic process. This multiplicity of tools and formats is an impediment to reproducibility. A fully reproducible map should be associated with the code and data used to produce it. Figure 1 describes what could be the spectrum of reproducibility applied to maps.

The first section of the paper demonstrates the relevancy of programming language solutions in mapping processes, especially in scientific contexts. In the second section, we put forward a unified workflow that allows map production in a reproducible process. We suggest hereby the `cartography` package, an extension of the R software that fills the need of specific thematic mapping solutions.

2 From GUI to Script

2.1 A Step Backward?

Most maps produced in an academic context are currently made with GIS or mapping applications that use graphical human-computer interactions. The use of graphical user interface (GUI) in computer science in general and cartography in particular is irrefutably a step toward more user-friendliness. But, this step may lead to the impossibility of reproducibility. To circumvent this weak reproducibility ability, most applications provide languages to build maps in their framework (e.g.,

Model Builder for ArcGIS or Python to create scripts in QGIS). However, these solutions hardly cover the full workflow from raw data to graphic representation and statistical findings.

Programming languages explicitly build on the idea of keeping traces of computations (data management, statistics, graphics and hence maps). Moreover, three main advantages to advocate the use of these scripting solutions are:

(1) The possibility of combining statistical operations before and after spatial operations within the same tool (i.e., unified toolchain). Building a map implies to implement a set of steps: data management, statistical analyses, geo-processing operations and eventually graphical display of results. Using two, three or more software products to conduct these operations introduces ruptures and a multiplication of data and file formats.
(2) The ability of using literate programming reports (i.e., reproducibility). Literate programming, introduced by Knuth (1992), strongly links analyses, statistical outputs and the code used to produce them in a single document. For example, R Markdown and Jupyter notebooks are respectively R and Python literate programming solutions based on the Markdown markup language.
(3) Most of scripting languages are open-source and the use of open-source frameworks provides full transparency on the methods and tools used to conduct analyses.

Scripting solutions can appear as a step back for cartographers who learned map making with languages such as ARC Macro Language (introduced by ESRI in 1986 or SAS macros. We argue that in the reproducible research framework, researchers must use literate programming solutions that enable the full traceability of their studies. A script that describes every step of the process, from raw data to comprehensive vector thematic maps that includes all statistical and geometrical transformations, is considered a full metadata document.

2.2 R as a Go-To Tool for Integrated Analysis

Several programming languages and software solutions are used to carry out studies in a unified workflow that integrates data handling and data processing (including spatial analyses procedures and spatial representations). Python and R emerge as the most prominent.

Python appears to be more versatile than R which is more focused on data. Spatial libraries exist for these two languages but the R spatial ecosystem is more dynamic and includes more spatial analyses methods. As statistical cartography practitioners, we have decided to invest in R development that allows a strong connection between statistics and cartography.

R is both a language and an environment for statistical computing and graphics. A large part of this open-source software popularity is due to its plug-ins, called packages. R packages are created by users and distributed through public repositories (Comprehensive R Archive Network, commonly referred to as CRAN). As of September 2016 there were 9387 packages. This system allows the development and sharing of software pieces in a transparent way.

Among its many packages, some are dedicated to the management, modification and display of spatial features. Three main packages are unavoidable: `sp` (Pebesma and Bivand 2005; Bivand et al. 2013), `rgdal` (Bivand et al. 2016) and `rgeos` (Bivand and Rundel 2016). `sp` is used to manage spatial data (vector and raster). To manipulate geographical projections and data import/export, `rgdal` provides bindings to the GDAL (Geospatial Data Abstraction Library, a translator library for raster and vector geospatial data formats) and PROJ (Cartographic Projections Library). `rgeos` is an interface to GEOS (Geometry Engine, Open Source, a geometrical operation library) aiming at geo-processing spatial features.

Several packages already allow to produce thematic maps in R. Most of them are described in the *Visualisation* topic of the "Analysis of Spatial Data" CRAN Task View (CRAN Task Views are guides that group and describe packages of specific domains). Nevertheless, none of them fits the need to combine easy mapping features with an elegant design of functions, a complete set of cartographic layout elements (arrow, scale, legends…) and a large set of representations.

With the `cartography` package, we offer an extension of the R ecosystem that enables fully reproducible cartography linked to data collection and data analysis. `cartography` is built upon and benefits from two of the main related spatial packages (`sp` and `rgeos`).

3 The Cartography Package

3.1 Design

The aim of `cartography` is to design high quality thematic maps similar to those designed with common mapping and GIS solutions. `cartography` targets two categories of users: cartographers new to R programming or R users new to cartography. Therefore, its functions have to be easy to use for cartographers and ensure compatibility with standard R workflows.

The package design follows some of the current practices in GIS workflows. Each function has two main arguments: a spatial object (e.g., GIS: vector layer/R: `sp` object) and an attribute table (e.g., GIS: csv/R: data frame). Each function focuses on a single cartographic representation (e.g., proportional symbols or choropleth representation) displayed on a georeferenced plot. This solution handles each representation as a layer and allows overlaying multiple representations on a map.

Functions have many parameters to fine-tune the cartographic representation. These parameters are similar to those used in GIS environments (e.g., width of the polygons boundaries, color palettes used in choropleth maps, customizable legend). Some additional functions are dedicated to layout design: customizable scale bar, a north arrow, a title, sources, and author information.

Following the common practice in R package development, each function in `cartography` is illustrated by a full example. Examples use datasets provided in the package (spatial and statistical datasets on European regions) and new users can easily reproduce them or adapt these scripts to their own data. A large range of export formats are available, in both vector and raster files, allowing final edition a graphic design software if needed (Giraud and Lambert 2016).

3.2 Main Features

`cartography` contains 35 functions that process and display spatial objects or map layout elements. Figure 2 gives an overview of the main features available in the package, indicating which functions are involved. Classical representations and some more advanced maps (that usually need geo-processing) are listed. The following figures are examples of maps solely built with `cartography`. The scripts used to generate them are available in a public repository (https://github.com/riatelab/ReproducibleCartography).

Figure 3 shows four maps produced with `cartography`. The first one uses the proportional symbols function (`propSymbolsLayer`) and displays the amount of population by country in Europe. The second one is a choropleth map (`choroLayer`) of the gross domestic product per inhabitant in euros.

The third and fourth maps combine two indicators. Symbol sizes are proportional to the amount of population and color scales are based on the distribution of gross domestic product per inhabitant (`propSymbolsChoroLayer`, `propSymbolsLayer + choroLayer`).

The three next maps are more complex. Their production involves geo-processing operations, usually scattered across multiple steps and intermediate files. `cartography` simplifies the process by wrapping all operations in its functions.

The “grid-cell” representation is a method that overcomes the arbitrariness and irregularity of administrative divisions. It highlights the main spatial trends by splitting the territory in regular blocks. Statistical values are distributed over a

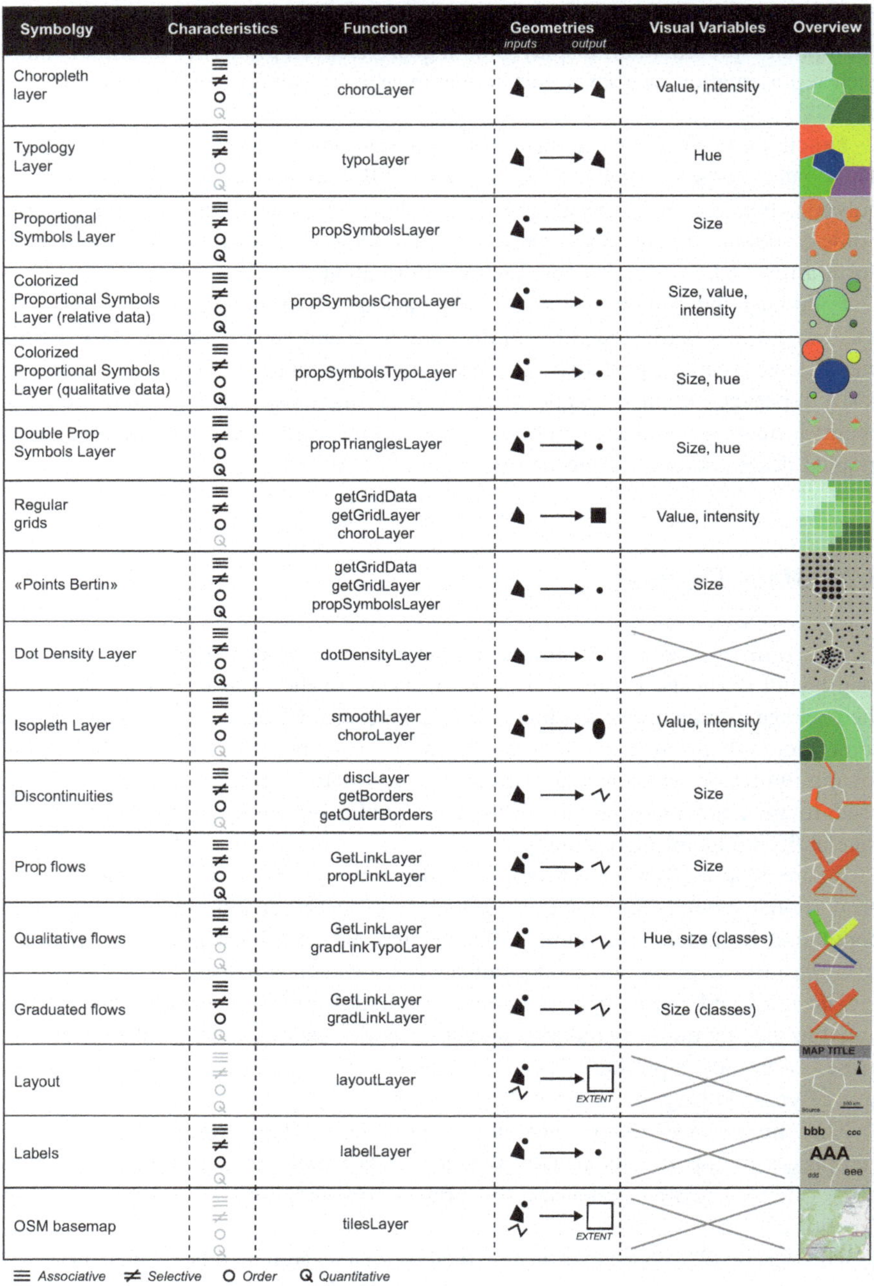

Fig. 2 Main features of `cartography`

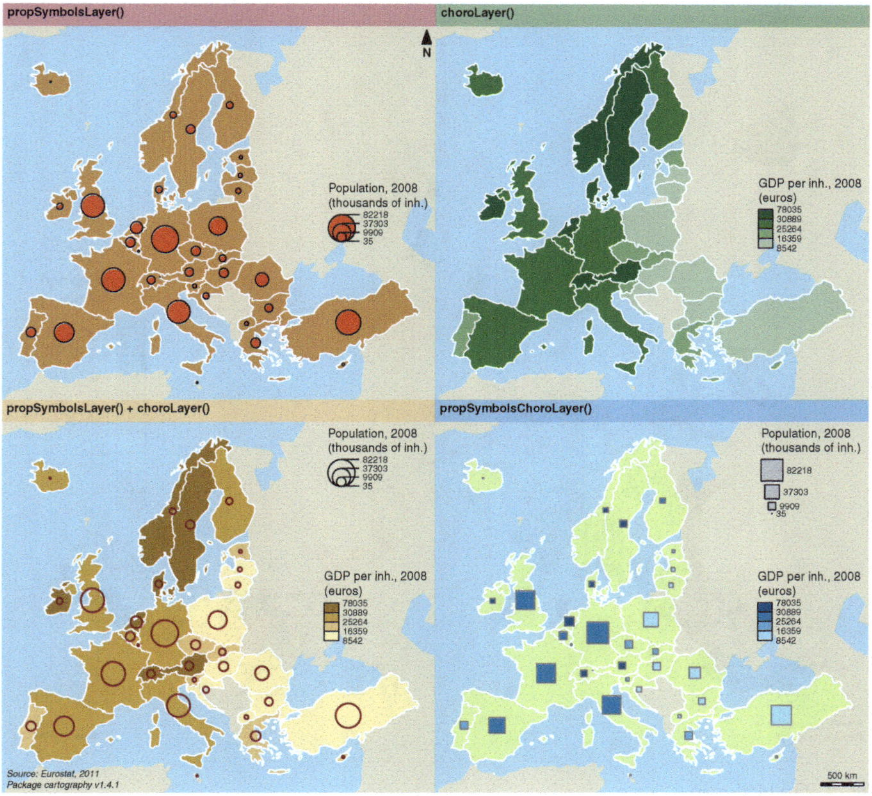

Fig. 3 Four classical maps

regular grid. Cell values are calculated taking into account the share of intersected areas between initial units and grid cells, and are eventually discretized and the grid is displayed as a choropleth map.

In `cartography`, this process relies on three functions. `getGridLayer` builds a regular grid (squares or hexagons) based on a spatial object and provides the shares of intersected areas. `getGridData` computes data that matches the grid layer. Finally, `choroLayer` displays the grid on a choropleth map (Fig. 4).

Discontinuity maps are based on the variation of a phenomenon in contiguous administrative units. This kind of representation does not focus on homogeneous regions, but rather on spatial breaks. Here, the thickness of the borders points out discontinuities intensities.

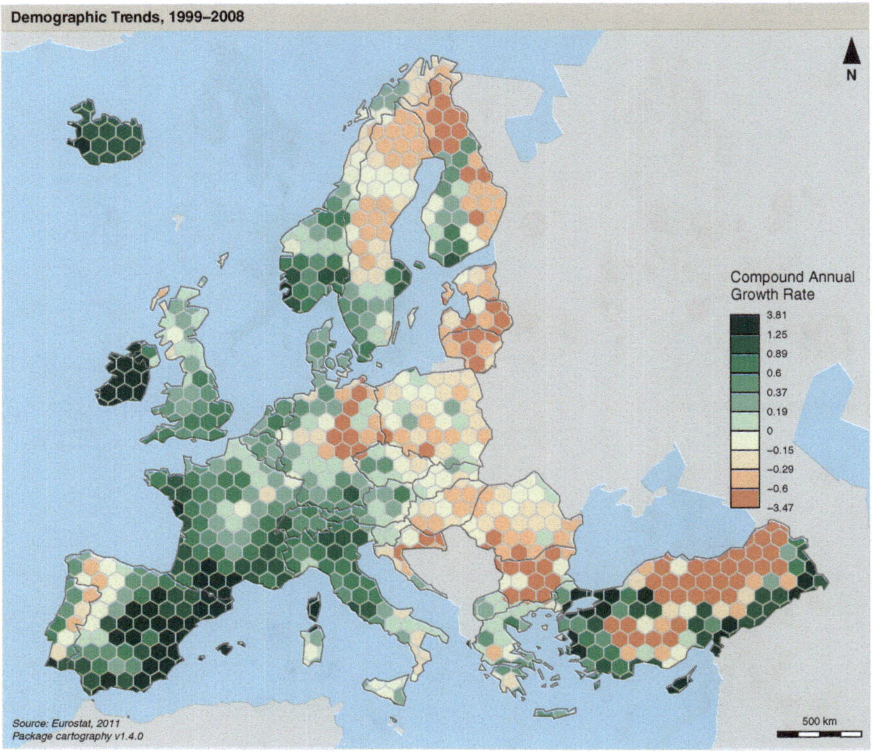

Fig. 4 Gridded map

The first step to build these maps is to extract borders from contiguous spatial units. The second step is to compute a discontinuity measure (either a ratio or an absolute difference). The third step consists in applying various values to the thickness of borders. Combining these discontinuities with a choropleth representation helps understanding the discontinuity direction (which one of two regions records the higher value).

In `cartography`, `getBorders` is used to extract borders between spatial units. `getOuterBorders` function can compute borders between non-contiguous units (maritime borders). `discLayer` function computes and displays discontinuities. A line's width reflects the discontinuity level between two neighboring units. The discontinuities layer can be supplemented by a choropleth layer (`choroLayer`). The resultant discontinuities representation is shown in Fig. 5.

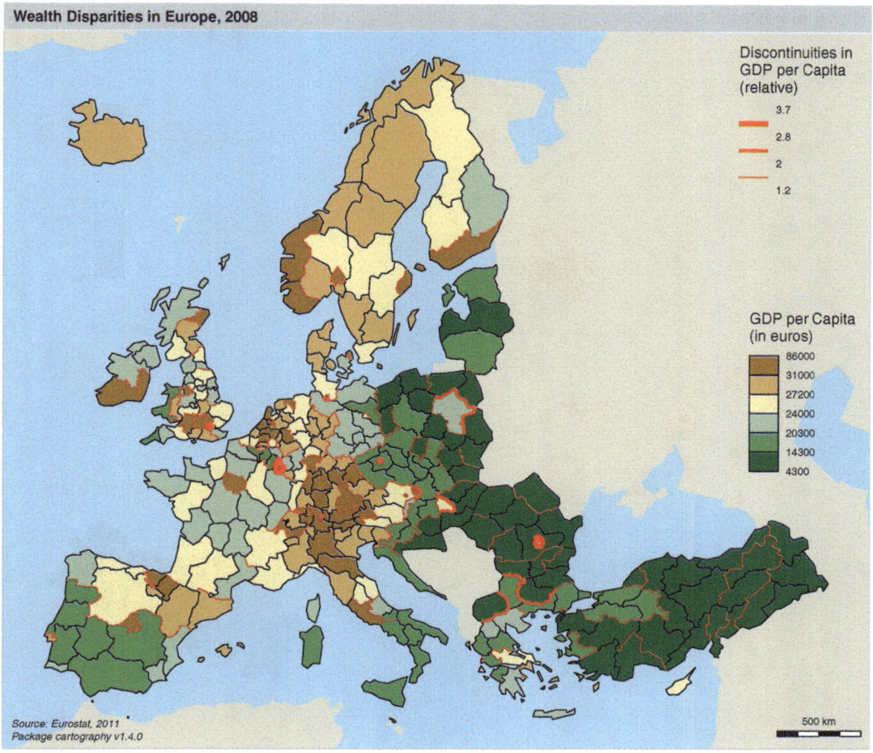

Fig. 5 Discontinuities map

Unlike choropleth maps that imply a phenomenon is spatially discrete, isopleth maps are based on the assumption that a phenomenon is spatially continuous. These maps use a spatial interaction modeling approach aiming at computing indicators based on stock values weighted by distances (Stewart 1942). This spatial representation is independent from the initial heterogeneity of the territorial division. The result is easy to read and can be considered as a bypass of the Modifiable Areal Unit Problem (MAUP) (Openshaw and Taylor 1979).

In `cartography` the `smoothLayer` function, that heavily depends on `SpatialPosition` (Giraud and Commenges 2016), allows a quick computation of these maps. The function takes a marked point layer and a set of parameters (a spatial interaction function and its parameters) as inputs and displays an isopleth map layer (Fig. 6).

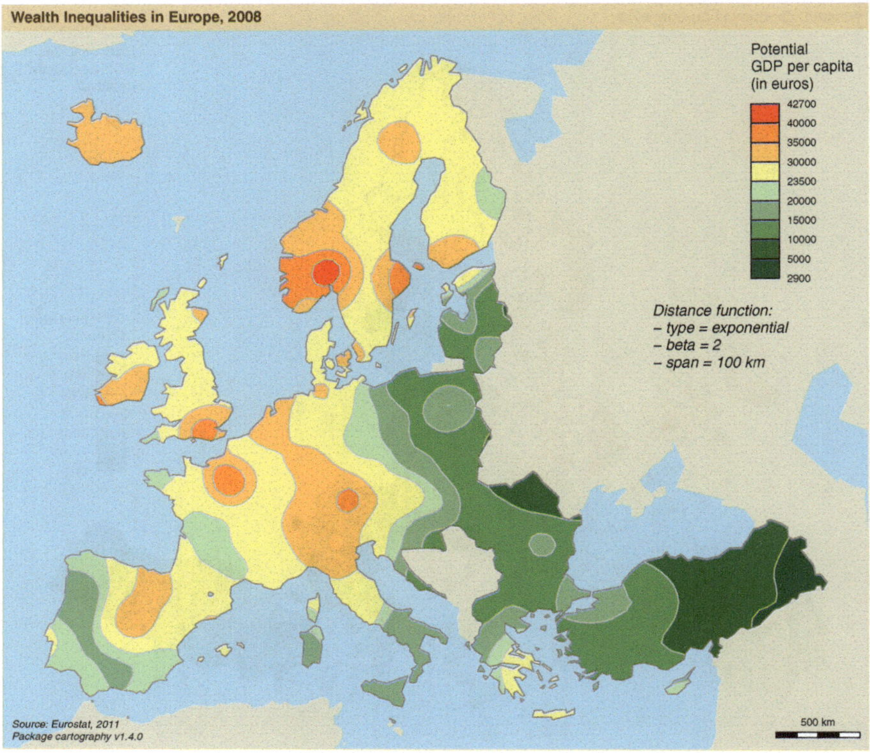

Fig. 6 Isopleth map

4 Conclusion

`cartography` was uploaded to CRAN in October 2015 and has been updated four times since (current version is 1.4.1). The development of the package happens on the online version control system GitHub (https://github.com/Groupe-ElementR/cartography). As of February 2017, the package has been downloaded 10,500 times from the RStudio CRAN. This number is an underestimate because CRAN is deployed across various mirror sites and only the RStudio site gathers this information. The number of downloads reveals an actual need of thematic mapping tools in reproducible research workflows.

Finally, we argue that maps should be more reproducible in order to obtain full acknowledgement as scientific productions. R and `cartography` (among other possibilities) could be relevant tools to ensure reproducibility in the academic context, hence promoting transparent scientific debates.

References

Bivand, R., Keitt, T., & Rowlingson, B. (2016). *rgdal: Bindings for the geospatial data abstraction library*. https://CRAN.R-project.org/package=rgdal.

Bivand, R., & Rundel, C. (2016). *rgeos: Interface to geometry engine—Open source (GEOS)*. https://CRAN.R-project.org/package=rgeos.

Bivand, R. S., Pebesma, E., & Gomez-Rubio, V. (2013). *Applied spatial data analysis with R* (2nd ed.). Springer, NY. http://www.asdar-book.org/.

Claerbout, J., & Karrenbach, M. (1992). Electronic documents give reproducible research a new meaning. In *Proceedings of 62nd Annual International Meeting of the Society of Exploration Geophysics* (pp. 601–604).

Giraud, T., & Commenges, H. (2016). *SpatialPosition: Spatial position models*. https://CRAN.R-project.org/package=/SpatialPosition.

Giraud, T., & Lambert, N. (2016). Cartography: Create and Integrate Maps in your R Workflow. *JOSS*, *1*(4). doi:10.21105/joss.00054.

Knuth, D. E. (1992). Literate programming. *CSLI lecture notes, Stanford, CA: Center for the study of language and information (CSLI), 1992*, 1.

Openshaw, S., & Taylor, P. J. (1979). A million or so correlation coefficients: three experiments on the modifiable areal unit problem. *Statistical Applications in the Spatial Sciences, 21,* 127–144.

Pebesma, E. J., & Bivand, R. S. (2005). Classes and methods for spatial data in R. *R News, 5*(2), 9–13.

Peng, R. D. (2009). Reproducible research and biostatistics. *Biostatistics, 10*(3), 405–408.

Peng, R. D. (2011). Reproducible research in computational science. *Science, 334*(6060), 1226–1227.

Stewart, J. Q. (1942). A measure of the influence of a population at a distance. *Sociometry, 5*(1), 63–71.

Stodden, V., & Miguez, S. (2013). Best practices for computational science: Software infrastructure and environments for reproducible and extensible research. *Available at SSRN 2322276.*

Part IV
Evaluating Map Quality

Effectiveness and Efficiency of Using Different Types of Rectangular Treemap as Diagrams in Cartography

Mengjie Zhou, Yan Cheng, Ning Ye and Jing Tian

Abstract Treemaps has been used as diagrams in cartography to visualize hierarchical, multivariate, and time series data through the construction of hierarchies. Several algorithms exist for generating Treemaps. Despite the popularity of the Treemap, few studies have analyzed the effectiveness and efficiency of different algorithms. A user study was conducted to evaluate the slice-and-dice, squarified, ordered, strip, and ordered squarified algorithms. In the user study, the accuracy and response time for completing tasks were recorded. The slice-and-dice algorithm was identified as most suitable to represent true hierarchy (intrinsic hierarchy), followed by pivot-by-size. Slice-and-dice was also identified as the most suitable algorithm for representing false hierarchy (imposed hierarchy). When representing time series, slice-and-dice, strip, and ordered squarified algorithms were all determined to be suitable.

Keywords Treemap · Cartography · User study · Effectiveness · Efficiency

---

M. Zhou · Y. Cheng · N. Ye · J. Tian (✉)
Key Laboratory of Geographic Information System, School of Resource and Environment Science, Ministry of Education, Wuhan University, Wuhan, People's Republic of China
e-mail: yutaka-2010@163.com

M. Zhou
e-mail: zmj.1990@163.com

Y. Cheng
e-mail: yancheng_0116@163.com

N. Ye
e-mail: Yuyuning0207@gmail.com

J. Tian
Key Laboratory of Digital Mapping and Land Information Application Engineering, National Administration of Surveying Mapping and Geoinformation, Wuhan University, Wuhan, People's Republic of China

M.P. Peterson (ed.), *Advances in Cartography and GIScience*, Lecture Notes in Geoinformation and Cartography, DOI 10.1007/978-3-319-57336-6_14

1 Introduction

Investigation on new representation methods has been one of the focuses of cartography. Treemap, a technique for visualizing hierarchical data has also been used in cartography to visualize socio-economic and demographic data. Treemap is suitable for detecting outliers, analyzing cause and effect, and discovering trends, and it has been successfully used to visualize various kinds of data, such as business information (Vliegen et al. 2006), electoral data (Wood et al. 2011), and social networks (Sathiyanarayanan and Burlutskiy 2015).

Because Treemap was originally designed to visualize hierarchical data, it can naturally be used to represent statistical data with spatial, temporal, or semantic hierarchies in cartography; for example, Jern et al. (2009) dynamically linked a Treemap to a choropleth map to facilitate the visualization of statistical data with a spatial hierarchy, while Slingsby et al. (2010a) used a Treemap to discover traffic patterns with a spatial hierarchy and visualize population data with a sematic hierarchy. Through construction of a false hierarchy (imposed hierarchy), Treemap can also be used to visualize multivariate statistical data and discovery correlation among different variables (Slingsby et al. 2010b), or to visualize flows and identify their spatial patterns (Wood et al. 2010).

Treemaps can be generated by several algorithms. These algorithms differ greatly in layout and may be suitable for different research questions or data types. Usability studies and controlled experiments are important in evaluating visualization techniques, because they help illustrate the potential benefits and limitations of visualization techniques as well as forming guidelines for their use (Plaisant 2004). Slingsby et al. (2009) proposed guidelines for a wide range of applications for designing Treemaps based on observations and experiences. Kong et al. (2010) proposed a set of design guidelines for creating effective rectangular Treemaps, concentrating on the design parameters of aspect ratio and luminance. The data represented in cartography is quite different from other kinds of data because it is spatially referenced and needs to be compared between different regions; therefore, the appropriate types of Treemap used as diagrams in cartography may vary in terms of the represented data and task types. Yet, little attention has been paid to it. Therefore, it is of interest to conduct a systematic evaluation on the effectiveness of different Treemap algorithms.

2 Treemap: A Brief Review

Treemaps can be generated by a variety of algorithms that can be classified into the three types listed in Table 1 based on their major design dimensions and salient characteristics.

Table 1 Algorithm classifications

Type	Algorithm
Keep input data order only	Slice-and-dice (Johnson and Shneiderman 1991)
Keep low aspect ratio only	Squarified (Bruls et al. 2000)
Keep balance between input data order and low aspect ratio	
Preserve one dimensional order	Cluster (Wattenberg 1999), ordered (Shneiderman and Wattenberg 2001), strip (Bederson et al. 2003), and spiral (Tu and Shen 2007)
Preserve two dimensional order	Ordered squarified (Wood and Dykes 2008), spatially ordered (Wood and Dykes 2008), and Nmap (Duarte et al. 2014)

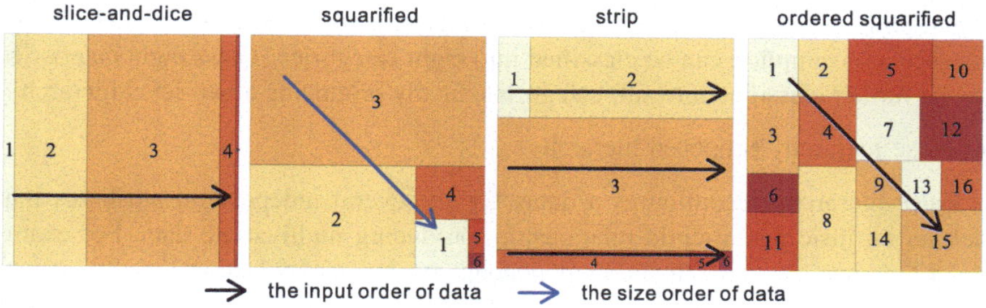

Fig. 1 An illustration of different Treemap types. The number of each rectangle indicates the order of data input

- **Keep input data order only.** Visual ordering in Treemaps is strictly consistent with one-dimensional ordering of the input data.
- **Keep low aspect ratio only.** A low aspect ratio is preserved without considering the mapping between input data order and visual ordering.
- **Keep balance between input data order and low aspect ratio.** The input data order (one-dimensional ordering or two-dimensional ordering) is preserved to some extent while maintaining a relatively low aspect ratio.

Our study concentrated on using Treemaps as diagrams to represent statistical data of individual regions. Geo-references can then be added through a point diagram map, a diagram map, a point grid map (Zhou et al. 2016) or a spatially ordered Treemap. This paper considers spatially ordered and Nmap algorithms to be methods for arranging nodes according to their geographic location rather than diagrams to represent statistical data of individual regions; therefore, these algorithms are not included in the user study. In the remaining types of Treemaps, we chose the algorithms that are used in existing Treemap software, such as treeMappa (http://www.treemappa.com/) and Treemap (https://www.treemap.com/). Therefore, **slice-and-dice**, **squarified**, **ordered (including pivot-by-middle, pivot-by-split-size, and pivot-by size)**, **strip**, and **ordered squarified algorithms** are evaluated here (see Fig. 1), and they cover all three types listed in Table 1. The ordered algorithm could not be illustrated by a simple graph, so it is not shown in Fig. 1.

3 Data Types Represented by Treemaps

Treemap algorithms differ greatly in the layouts, each suitable for different types of data. They are mainly used to represent true hierarchical, false hierarchical or spatio-temporal data, also called time series (Slingsby et al. 2009, 2010a, b).

- True hierarchy (intrinsic hierarchy)

True hierarchy expresses the notion of a subtype of another type. An instinctive parent-child relationship exists between superior and subordinate elements. For instance, consumption can be classified into eight categories. These eight categories can be further classified into sub-categories, finally producing a two-level hierarchy.

- False hierarchy (imposed hierarchy)

False hierarchy reconfigures a dataset with several independent attributes that lack an intrinsic parent-child relationship, producing multivariate data. For example, a Treemap can show the two independent attributes of consumption and income level, which have no intrinsic parent-child relationship and are therefore called a false hierarchy. This type of hierarchy can be used to represent potential relationships between two independent variables.

- Time series

Temporal data represent a distinct example of a true hierarchy. Increasing or decreasing trends or periodic changes can be intuitively conveyed by a Treemap.

4 User Study Design

A user study was conducted to evaluate the effectiveness and efficiency of different Treemap algorithms applied to different types of data. First of all, users must receive the transmitted information completely and correctly (effectiveness). Secondly, the result must be assessed in relation to the amount of time required to achieve full and correct information (efficiency). The user study was conducted as a between-subjects test rather than a within-subjects test. This is because the answers corresponding to the last Treemap will influence the answers for the next Treemap as the same group is asked to read seven types of Treemaps and answer their corresponding questions. In the between-subjects test, subjects were equally divided into seven groups. Each group was asked to read a type of Treemap and answer the same series of questions. The answers and response times were recorded and analyzed to evaluate the various algorithms.

4.1 Visual Tasks for Treemap Cartography

Presentation and analysis are usually the primary communication goals of a representation. In order to achieve these goals, users must perform several visual tasks. The user study identified six visual tasks to be tested, based on the visualization operations (Wehrend 1993) and visual tasks (Zhou and Feiner 1998) as well as the characteristics of the three data types (see Table 2).

- **Element locating** refers to finding special elements (rectangles) in Treemaps.
- **Visual comparison** describes a comparison of an attribute that is mapped to visual variables on Treemaps of different objects. The visual comparison tasks mainly ask for the differences between the sizes of rectangles in this study.
- **Visual ranking** requires the user to determine the order of elements (rectangles) according to special attributes. This task mainly asks for the differences between the sizes of rectangles in our study.
- **Characterize distribution** provides a set of data cases and a quantitative attribute of interest, requiring the user to characterize the distribution of that attribute's values over the set.
- **Correlation exploring** refers to identifying potential relationships among multiple variants represented by Treemaps.
- **Trend detecting** assesses the change patterns of objects over time, which can be described as increasing trends, decreasing trends, or periodic changes.

Table 2 Task types for Treemaps

Tasks	Types
Elements locating	1. Find two elements in one Treemap that are adjacent or not adjacent in the input order 2. Find same elements in two different Treemaps
Visual comparison	1. Make a comparison of sizes of two elements in a same Treemap 2. Make a comparison of sizes of same elements in two different Treemaps 3. Make a comparison of the proportions of the elements 4. Visually "add" the sizes of several elements and make a comparison
Visual ranking	Rank the elements by the sizes of rectangles
Characterize distribution	Characterize the distribution of that attribute's values over the set
Correlation exploring	Explore the correlation between two individual variables
Trend detecting	Detect trend (increasing, decreasing trend, or periodic change)

Element locating, *visual comparison*, *visual ranking*, and *characterize distribution* are the four basic visual tasks. The *characterize distribution* task may be initiated when the data contains a quantitative attribute. *Correlation exploring* is an important task for datasets characterized by false hierarchy. The final task, *trend detecting*, is used with time series.

4.2 Questionnaire Design

4.2.1 Dataset and Test Material

The questionnaire was divided into three parts (Part A, Part B, and Part C), corresponding to the three targeted types of data. When Treemaps are applied to cartography, different regions represented by Treemaps must be compared; therefore, the questionnaire used two represented regions to simulate comparisons. The Treemaps were made using treemappa (http://www.treemappa.com/), and the datasets used in the questionnaire and their corresponding Treemaps are as described below. The data came from 2011 Guangdong, Gansu, and Zhejiang Statistical Yearbook; and Data center of the Ministry of Environmental Protection of the People's Republic of China (http://datacenter.mep.gov.cn/).

1. Part A: True hierarchy (intrinsic hierarchy)
   - Dataset: Per capita annual consumption expenditure of urban households.
   - Treemap design: The proportions of the total expenditure of each province and consumption expenditure types are mapped by size and color, respectively (see Fig. 2).
2. Part B: False hierarchy (imposed hierarchy)
   - Part B (a): One ordinal variable and one nominal variable

- Dataset: Per capita annual consumption expenditure and construction of urban households by family income level.
- Treemap design: Consumption expenditure by family income level (ordinal) forms the first-level, followed by consumption purpose (nominal) in the underlying level. Size represents the proportion of the total expenditure of each province and color indicates consumption purpose (see Fig. 3).

- Part B (b): Two ordinal variables.

- Dataset: Average education level of rural families with different income levels.
- Treemap design: The proportion of total laborers by family income level (ordinal) constitutes the first-level, followed by education level (ordinal) in the underlying level. Size represents the proportion of total laborers of each province while color indicates the education levels (see Fig. 4).

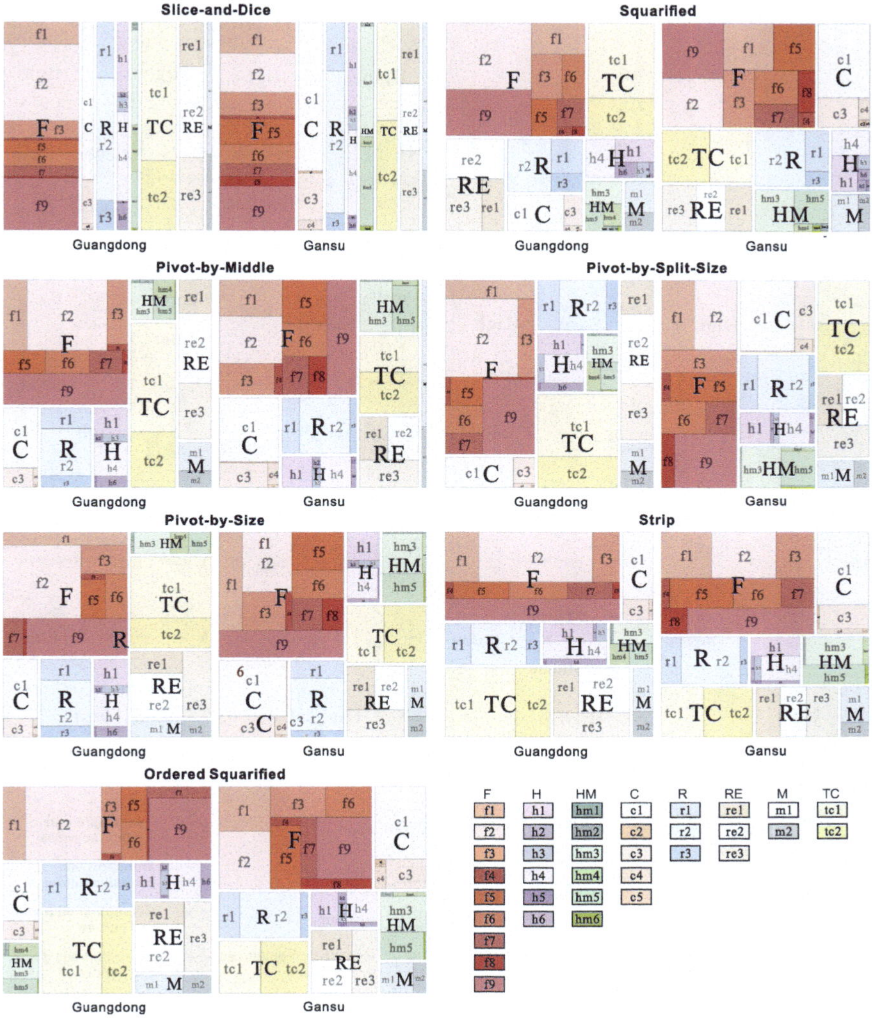

Fig. 2 Per capita annual consumption expenditure of urban households

3. Part C: Time series

- Dataset: Air quality levels.
- Treemap design: The year is divided by months at the first level and days at the second level. The air quality level is mapped to color, while the sizes of the sub-rectangles are fixed. However, the squarified algorithm arranges rectangles according to their sizes. Therefore, we did not apply the squarified algorithm to this dataset (see Fig. 5).

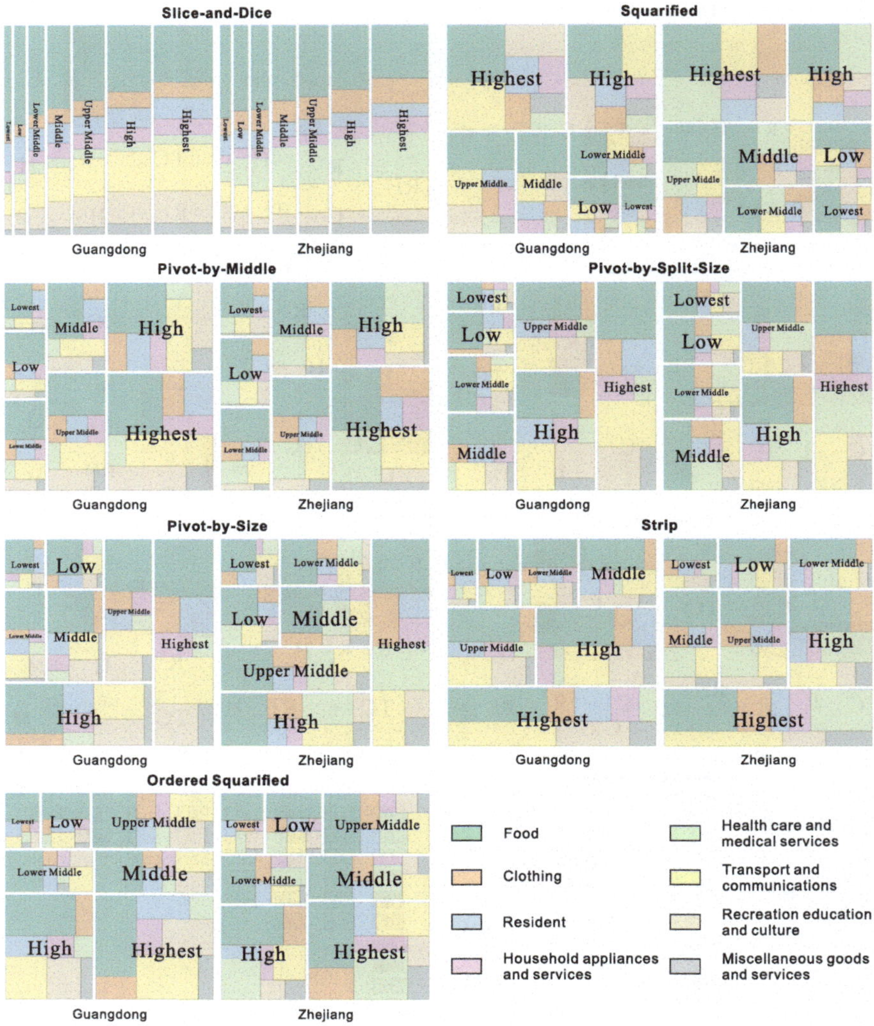

Fig. 3 Per capita annual consumption expenditure of urban households by family income level

4.2.2 Questions

We formulated several specific questions for each part based on the tasks listed in Table 2. When targeting rectangles in a visual comparison task, the difference between two rectangles cannot be obvious. To avoid subjects answering the questions based on existing knowledge without reading the Treemaps, some elements mentioned in questions and Treemaps were labeled using meaningless letters or numbers. Furthermore, several pilot trials were conducted before the user study.

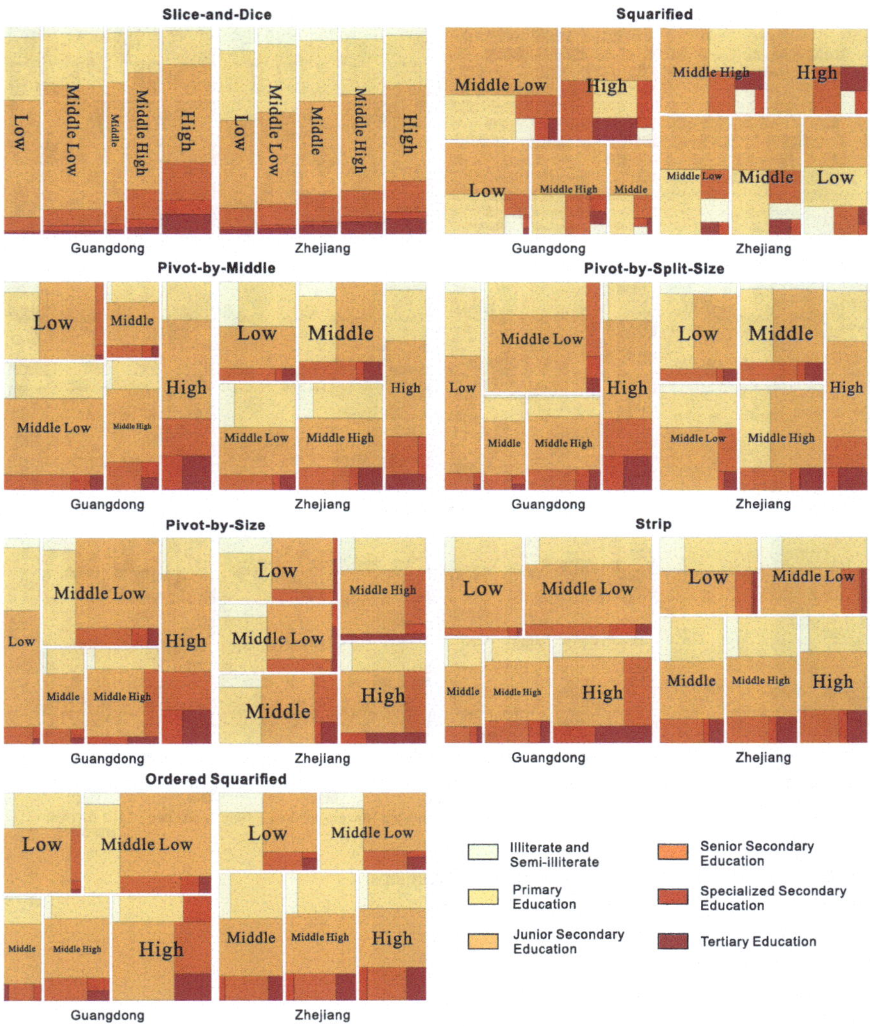

Fig. 4 Average education level of rural family's laborers by family income level

After the pilot trials, participants responded that the number of questions was too large and could cause visual fatigue and perfunctory psychology. Therefore, we deleted repeated tasks and their corresponding questions. Table 3 shows the final questionnaire. The complexity and difficulty of each part's questions grow stepwise.

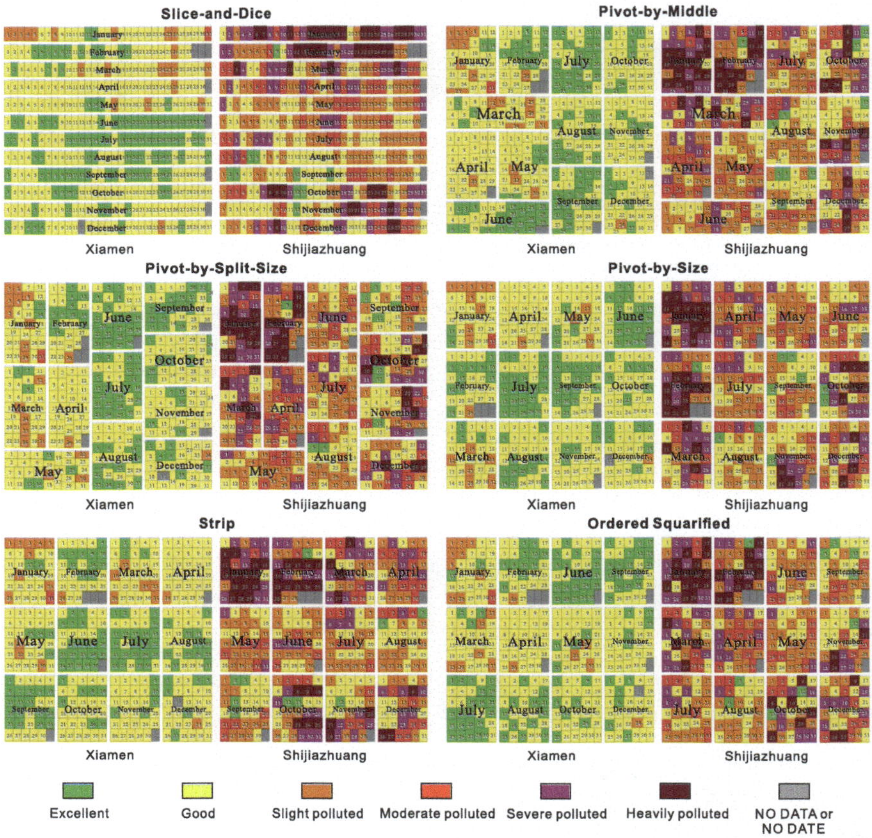

Fig. 5 Air quality levels as depicted through Treemaps

4.3 Procedure

The user study was conducted via a questionnaire on native HTML5 web pages. The subjects and questionnaires were equally divided into seven groups. Each subject was allowed to conduct the test independently before the deadline. The entire user study lasted for 14 days, from May 5, 2016 to May 19, 2016.

The first several pages of the web questionnaire contained detailed introductions to Treemaps and cartographic data, as well as instructions for answering. Each answer interface consisted of only one question with the corresponding Treemap placed under it. Subjects were asked to answer multiple-choice or free response without interruption. If subjects struggled to accomplish a task, they could choose the response "hard to tell or cannot tell." After completing one question, subjects clicked a button to jump to the next question, repeating the process until the entire questionnaire was complete. Subjects could not return to previous pages to change their selections.

Table 3 Questions for each type of statistical data

Task	Questions
Part A	
Elements locating	Q1. Find the f5 sub-rectangles in Guangdong and Gansu respectively Q2. Find the f6 and c7 sub-rectangles in Guangdong province respectively
Visual comparison	Q3. Compare the sizes of the f3 and f5 sub-rectangles in Gansu, and choose the bigger one Q4. Compare the sizes of the RE rectangles in Guangdong and Gansu, and choose the bigger one Q5. Compare the proportions of the f9 sub-rectangle in the F rectangle in Guangdong and Gansu, and choose the larger one
Visual ranking	Q6. Rank the F, C, R, TC, HM, H, RE and M rectangles according to their sizes in a descending order Q7. Rank the f1, f2, f3, f4, f5, f6, f7, f8, and f9 rectangles according to their sizes in a descending order
Part B(a)	
Elements locating	Q8. Find sub-rectangles numbered 5 in the "Lower Middle" rectangles in Guangdong and Gansu respectively
Visual comparison	Q9. Visually "add" the sub-rectangles numbered 6 in Guangdong and Gansu respectively, and choose the city that has the larger total area of the sub-rectangles
Correlation exploring	Q10. How does the proportion of the sub-rectangle numbered 1 in each first-level rectangle change with the family income increasing?
Part B(b)	
Characterize Distribution	Q11. Characterize the distribution of different education level people in middle-income families
Correlation exploring	Q12. In which province the sum of labor force whose education level is coded by 4, 5 and 6 changes more obvious with the family income increasing?
Part C	
Locate elements	Q13. Find the sub-rectangle representing the air quality of 7 June in Xiamen and the sub-rectangle representing the air quality of 21 August in Shijiazhuang respectively
Trend detecting	Q14. How did the air quality of Shijiazhuang change in November? Q15. How did the air quality of Shijiazhuang change in the whole year?

4.4 Subjects

The subjects were divided into seven groups. We invited subjects to participate in our study until the numbers of valid answer sheets from different groups were equal. Finally, the user study had 101 subjects in total, and we received 91 valid answer sheets from 54 females and 37 males. The subjects were found in school libraries randomly, each subject participated in the user study could get a gift, such as a notebook. Subjects had different academic backgrounds. They were asked to take a color vision test from Colblindor (http://www.color-blindness.com/color-arrangement-test/), and all reported normal color vision. In the brief background questionnaire, no participants reported using a Treemap before.

5 Results and Discussion

Response accuracy and response time—the indicators for evaluating the effectiveness and efficiency of Treemaps used as diagrams in cartography—were analyzed as follows.

We used one-way ANOVA test, Kruskal-Wallis test, and pairwise comparisons to analyze response time. Before analyzing response time, we conducted a Kolmogorov–Smirnov test between the response times of correct and incorrect answers for each question, concluding that there was no significant difference between the two groups in terms of response times. Therefore, we did not filter out the incorrect responses. As the response time distributions for most questions were skewed to the left, we first conducted lg-transformation to normalize the values (Ulrich and Miller 1993). If the P-value of the homogeneity test for variance was still less than 0.05 after lg-transformation, we used Kruskal-Wallis tests to analyze the group of response times; otherwise, we used the one-way ANOVA test. The entire analysis was executed in SPSS 20.0, with the results shown in Table 4. The significance tests reveal that the differences in response times between algorithms for most questions are statistically significant, except for Q2, Q6, Q7, and Q14.

Tables 5, 6, 7, 8, 9 and 10 show response accuracy and response time. The response times for Q2, Q6, Q7, and Q14 are not listed, because they lack statistical significance.

Element locating. The accuracy rate of the element locating tasks (Q1, Q8, and Q13) was 100% and is therefore not listed in Table 5. In Q1 and Q8, the results indicate that the slice-and-dice and strip algorithms took the least time. In Q1, the squarified algorithm took the longest time. And in Q13, pivot-by-split-size took the longest time. The capacity to retain the input data order and not change with data changes may influence the time required to accomplish elements locating tasks.

Visual comparison. As shown in Table 6, the slice-and-dice algorithm is most effective for completing visual comparison tasks, followed by the pivot-by-size and squarified algorithms based on the response accuracy. Overall, the evidence

Table 4 Values of statistical indicators

Question	P-value of homogeneity of variances	Method	F	H	P-value	Pairwise comparisons
Q1	0.068	One-way ANOVA	17.265	–	0.000	LSD
Q2	0.055	One-way ANOVA	2.136	–	0.058	–
Q3	0.006	Kruskal-Wallis test	–	20.549	0.002	Dunn-Bonferroni
Q4	0.017	Kruskal-Wallis test	–	21.238	0.002	Dunn-Bonferroni
Q5	0.184	One-way ANOVA	7.314	–	0.000	LSD
Q6	0.000	Kruskal-Wallis test		7.132	0.309	–
Q7	0.545	One-way ANOVA	0.360	–	0.902	
Q8	0.265	One-way ANOVA	6.577	–	0.000	LSD
Q9	0.187	One-way ANOVA	3.016	–	0.010	LSD
Q10	0.001	Kruskal-Wallis test	–	23.449	0.001	Dunn-Bonferroni
Q11	0.102	One-way ANOVA	5.161	–	0.000	LSD
Q12	0.000	Kruskal-Wallis test	–	11.375	0.000	Dunn-Bonferroni
Q13	0.075	One-way ANOVA	3.662	–	0.005	LSD
Q14	0.410	One-way ANOVA	2.046	–	0.082	–
Q15	0.079	One-way ANOVA	5.096		0.000	LSD

showing differences in response time is very weak. Results of response time in Q5 show that the slice-and-dice, strip, pivot-by-size, and ordered squarified algorithms take less time than pivot algorithms. Slice-and-dice takes the least time for Q9. However, the evidence is weak for the rest of the questions that address this task. The slice-and-dice Treemap fixes one side of each rectangle, making it much easier and more accurate to compare sizes. The pivot-by-size Treemap fixes one side of

Table 5 Response accuracy and response time (task type: Element locating)

Question	Response accuracy (%)	Response time (s)
Q1	N.A.	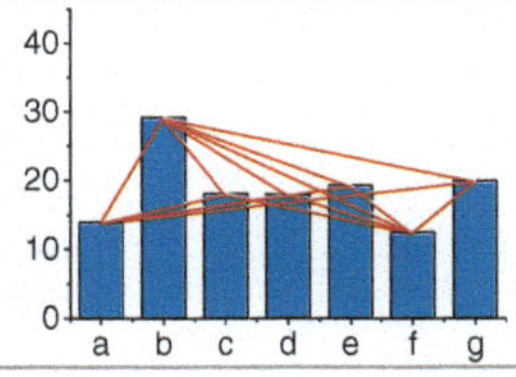
Q8	N.A.	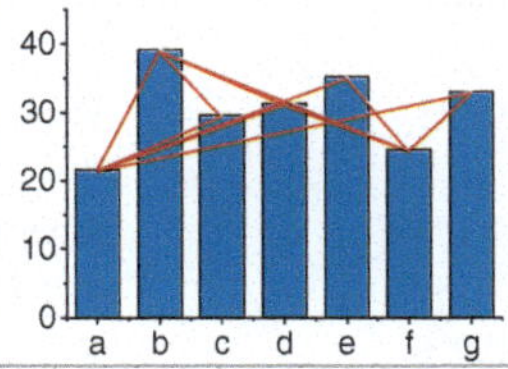
Q13	N.A.	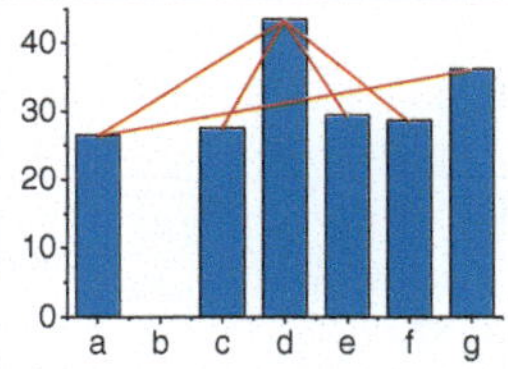 Statistically significant relationships
	a slice-and-dice, *b* squarified, *c* pivot-by-middle, *d* pivot-by-split-size, *e* pivot-by-size, *f* strip, *g* ordered squarified	

some rectangles, also making it easier to some extent. Although the squarified Treemap arranges sub-rectangles in order of size, which is helpful for recognizing the largest and smallest elements and making rough comparisons, accuracy is still lower than for the slice-and-dice Treemap.

Visual ranking. As shown in Table 7, results of Q6 indicate that the "F", and order of the "HM," "H," and "M" rectangles are easy to recognize across all Treemaps. However, a small number of subjects did not see the "M" rectangle in the slice-and-dice and pivot-by-middle Treemaps, perhaps because this rectangle is too narrow to be observed. The results of the pivot-by-split-size algorithm show that more than half of mistakes occurred among the "R," "TC," and "RE" rectangles. Mistakes in Q7 occurred in comparisons among the "f1," "f3," "f5," and "f6" rectangles, as well as in the comparison between the "f2" and "f9" rectangles and in the comparison between the "f7" and "f8" rectangles. Fixing one side of the sub-rectangles can help to complete the visual ranking task more easily and accurately. However, when very narrow sub-rectangles exist, the accuracy of the ranking task will decrease. Differences in response time for different algorithms

Table 6 Response accuracy and response time (task type: Visual comparison)

Question	Response accuracy (%)	Response time (s)
Q3	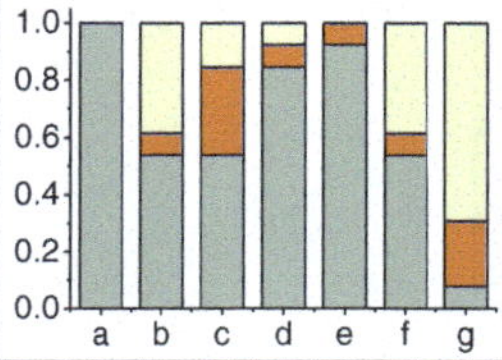	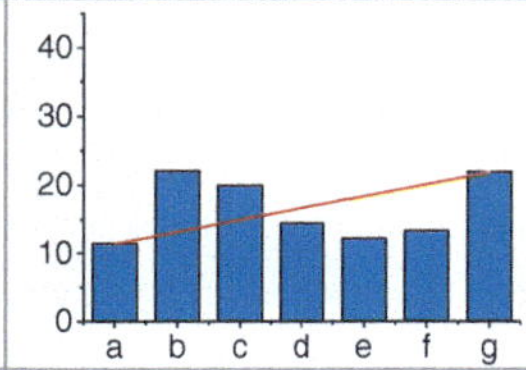
Q4	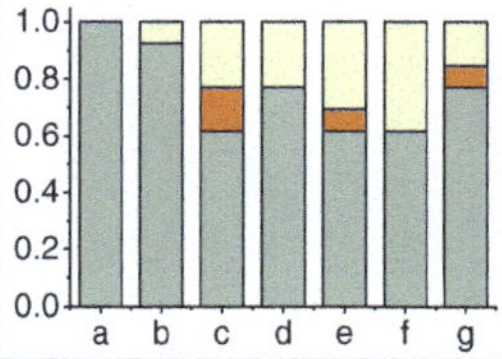	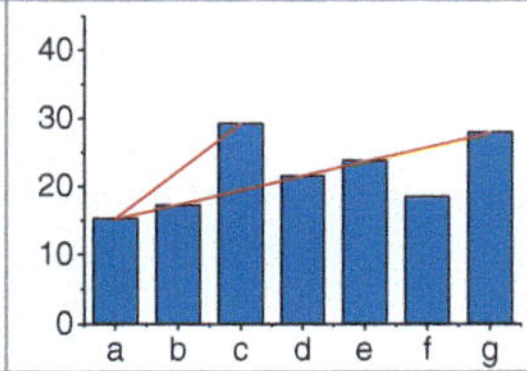
Q5	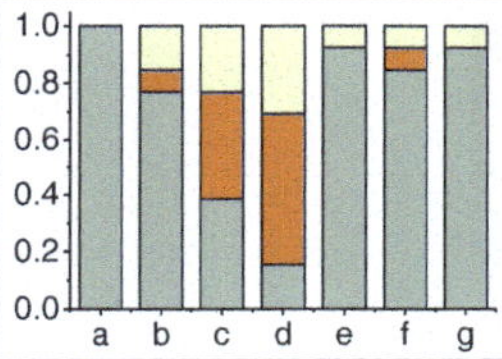	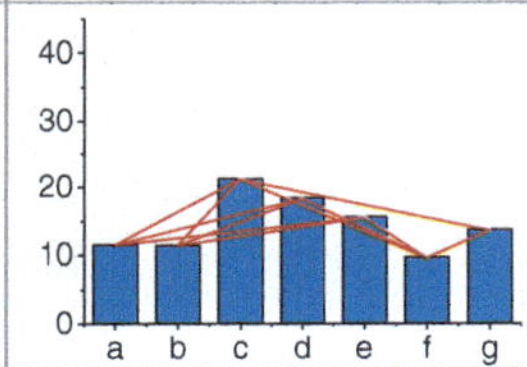
Q9	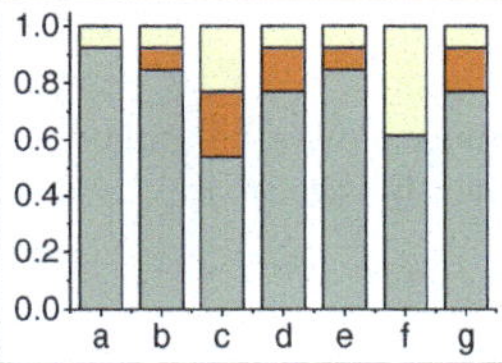	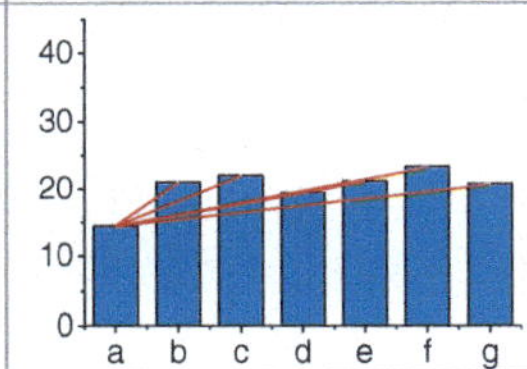
Total	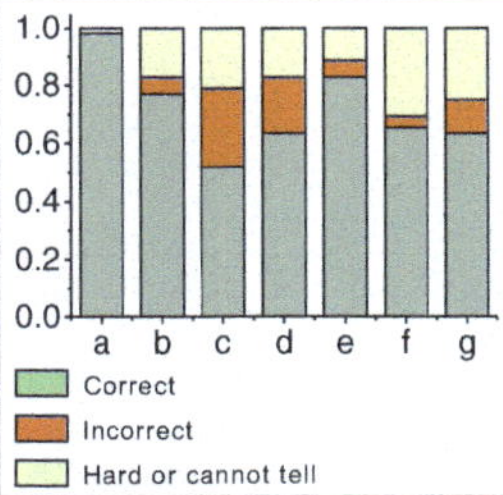	N.A. 
	a slice-and-dice, *b* squarified, *c* pivot-by-middle, *d* pivot-by-split-size, *e* pivot-by-size, *f* strip, *g* ordered squarified	

were not significant. Overall, the slice-and-dice and pivot-by-size algorithms performed best in the ranking task.

Characterize distribution. As shown in Table 8, results indicate that slice-and-dice is the most effective algorithm based on the response accuracy, followed by strip and pivot-by-size. The slice-and-dice, strip, and ordered squarified

Table 7 Response accuracy and response time (task type: Visual ranking)

Question	Response accuracy (%)	Response time (s)
Q6	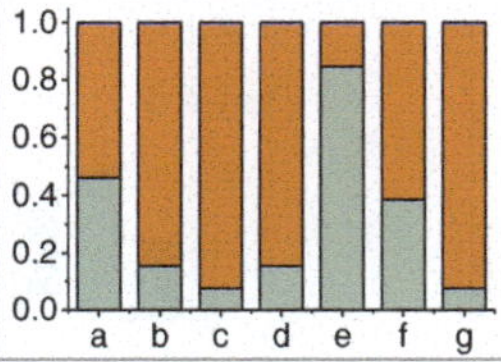	N.A.
Q7	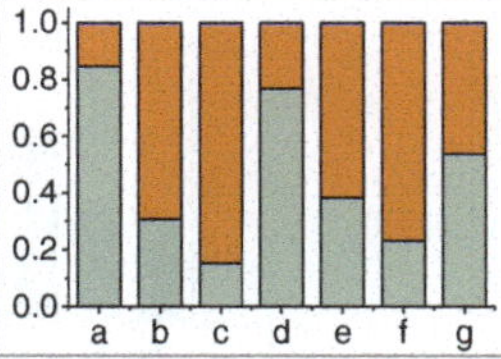	N.A.
Total	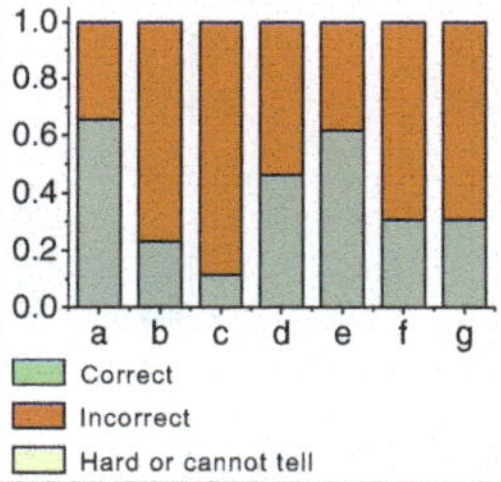	N.A.
	a slice-and-dice, *b* squarified, *c* pivot-by-middle, *d* pivot-by-split-size, *e* pivot-by-size, *f* strip, *g* ordered squarified	

Table 8 Response accuracy and response time (task type: Characterize distribution)

Question	Response accuracy (%)	Response time (s)
Q11	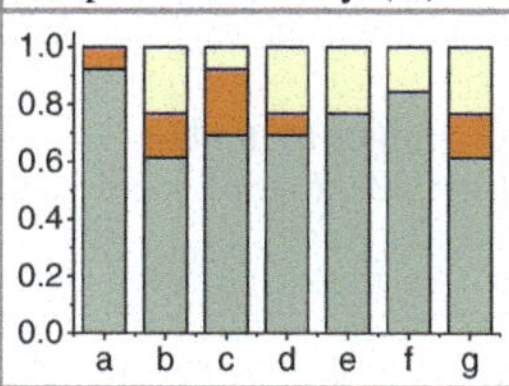	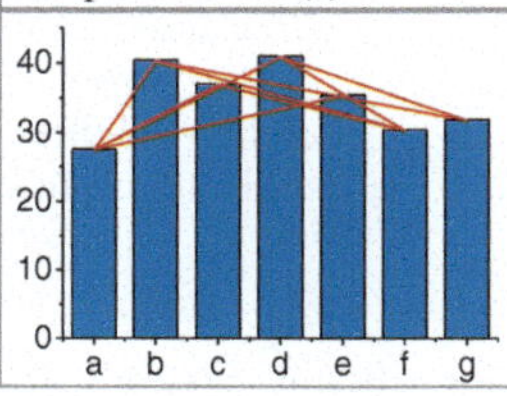

Table 9 Response accuracy and response time (task type: Correlation exploring)

Question	Response accuracy (%)	Response time (s)
Q10	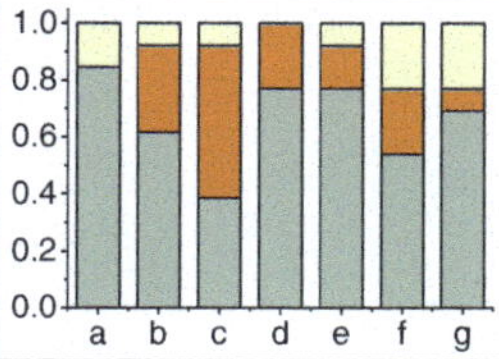	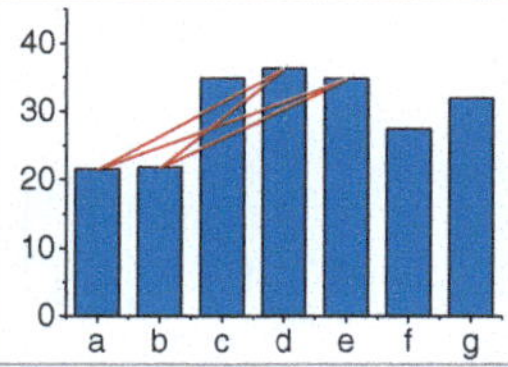
Q12	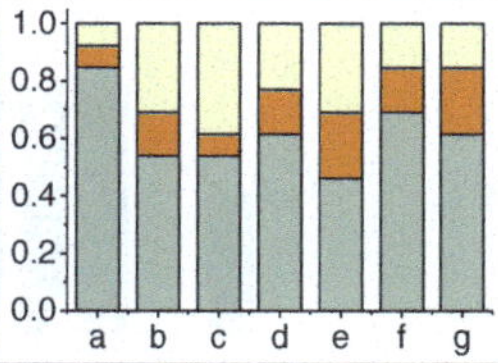	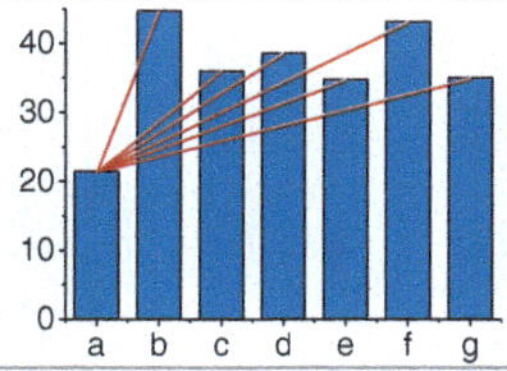
Total	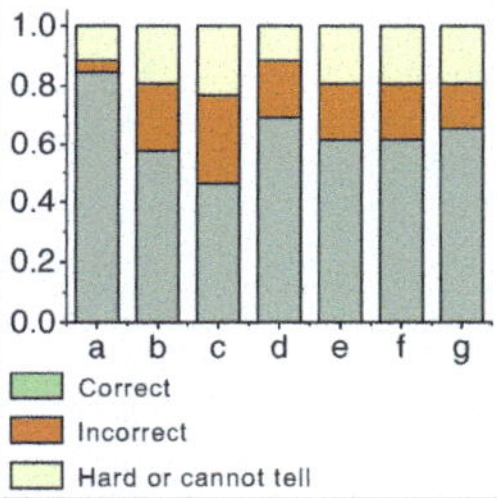	N.A. — Statistically significant relationships
	a slice-and-dice, *b* squarified, *c* pivot-by-middle, *d* pivot-by-split-size, *e* pivot-by-size, *f* strip, *g* ordered squarified	

algorithms are most efficient based on the response time. These results may mean that maintaining input order is crucial to the characterize distribution task.

Correlation exploring. As shown in Table 9, slice-and-dice is also the most effective algorithm for the correlation exploring task based on the response accuracy. In Q10, differences in response time among different algorithms were not significant, while in Q12, slice-and-dice took the least time. These results partially indicate that slice-and-dice is the most efficient algorithm. Slice-and-dice is suitable for detecting relationships between two independent variables. The reason could be that changes in attribute values do not cause changes in the structure of the slice-and-dice Treemap.

Trend detecting. As shown in Table 10, the slice-and-dice, strip, and ordered squarified algorithms performed better than the others in response accuracy. In Q15, the slice-and-dice and strip algorithms took less time. These results suggest that keeping relatively high consistency in input order and display order and displaying results in an order that users are accustomed to could be important to the trend detecting task.

Table 10 Response accuracy and response time (task type: Trend detecting)

Question	Response accuracy (%)	Response time (s)
Q14	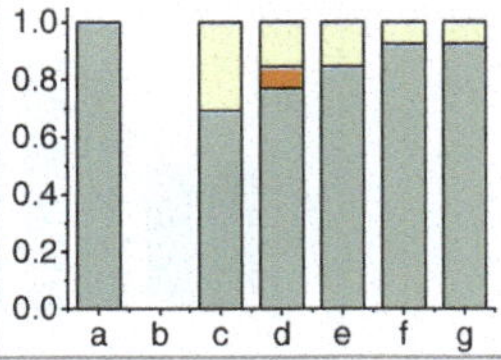	N.A.
Q15	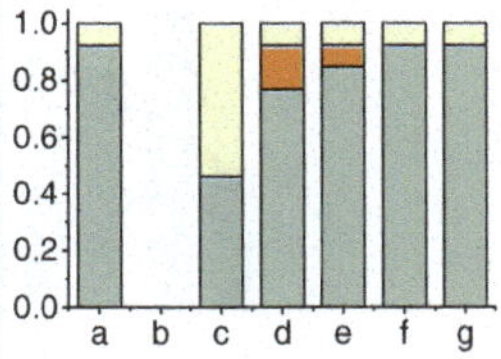	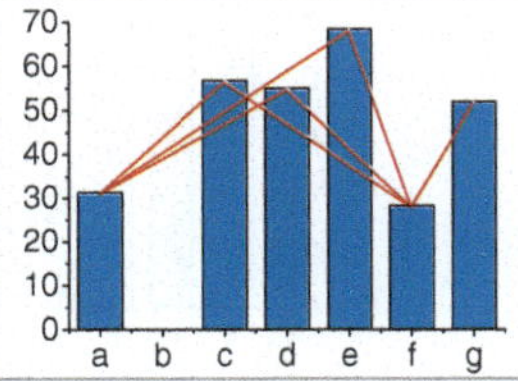
Total	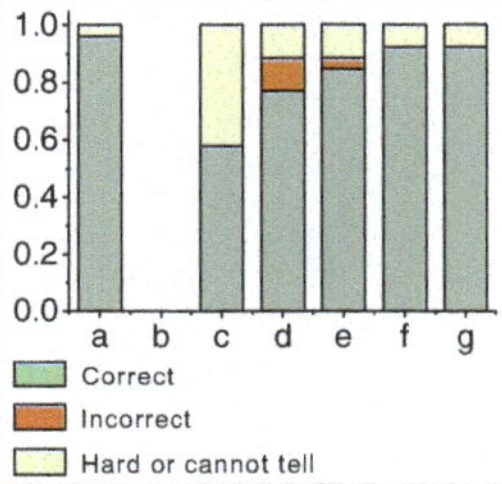	N.A. — Statistically significant relationships
	a slice-and-dice, *b* squarified, *c* pivot-by-middle, *d* pivot-by-split-size, *e* pivot-by-size, *f* strip, *g* ordered squarified	

6 Conclusion

This paper evaluated the effectiveness and efficiency of different Treemaps as diagrams in cartography. Considering both response time and response accuracy for each type of task, slice-and-dice is the most suitable algorithm to represent true hierarchy, followed by pivot-by-size; however, when very small rectangles occur in slice-and-dice, pivot-by-size works better. In representing false hierarchy, slice-and-dice is again the most suitable algorithm. When representing time series, the slice-and-dice, strip and ordered squarified algorithms are suitable. The evaluation results can be used as a guide for choosing Treemap algorithms.

The data that we evaluated contain only dozens of nodes, and the data are represented by static Treemaps. When the data volume increases, the data can be better represented by dynamic and interactive Treemaps. This study did not consider tasks for dynamic and interactive Treemaps. It is possible that both the effectiveness and efficiency of Treemap representation may change with the aid of good interaction, and this is the future work of our study.

Acknowledgements We appreciate the suggestions and comments from Michael Peterson and the anonymous reviewers. Our work was supported by the Special Program for Basic Science of China under Grant No. 2013FY112800.

References

Bederson, B. B., Shneiderman, B., & Wattenberg, M. (2003). Ordered and quantum treemaps: making effective use of 2d space to display hierarchies. *ACM Transactions on Graphics, 21*(4), 257–278.

Bruls, M., Huizing, K., & Van Wijk, J. J. (2000). Squarified treemaps. *Data visualization 2000* (pp. 33–42). Vienna: Springer.

Duarte, F. S., Sikansi, F., Fatore, F. M., Fadel, S. G., & Paulovich, F. V. (2014). Nmap: A novel neighborhood preservation space-filling algorithm. *IEEE Transactions on Visualization and Computer Graphics, 20*(12), 2063–2071.

Jern, M., Rogstadius, J., Åström, T. (2009). Treemaps and choropleth maps applied to regional hierarchical statistical data. In *2009 13th International Conference Information Visualisation* (pp. 403–410). IEEE.

Johnson, B., & Shneiderman, B. (1991). Tree-maps: A space-filling approach to the visualization of hierarchical information structures. In *IEEE Conference on Visualization, 1991. Visualization'91, Proceedings* (pp. 284–291). IEEE.

Kong, N., Heer, J., & Agrawala, M. (2010). Perceptual guidelines for creating rectangular treemaps. *IEEE Transactions on Visualization and Computer Graphics, 16*(6), 990–998.

Plaisant, C. (2004). The challenge of information visualization evaluation. In *Proceedings of the Working Conference on Advanced Visual Interfaces* (pp. 109–116). ACM.

Sathiyanarayanan, M., & Burlutskiy, N. (2015). Visualizing social networks using a treemap overlaid with a graph. *Procedia Computer Science, 58,* 113–120.

Shneiderman, B., & Wattenberg, M. (2001). Ordered treemap layouts. *In Proceedings of the IEEE Symposium on Information Visualization 2001* (Vol. 73078).

Slingsby, A., Dykes, J., & Wood, J. (2009). Configuring hierarchical layouts to address research questions. *IEEE Transactions on Visualization and Computer Graphics, 15*(6), 977–984.

Slingsby, A., Wood, J., & Dykes, J. (2010a). Treemap cartography for showing spatial and temporal traffic patterns. *Journal of Maps, 6*(1), 135–146.

Slingsby, A., Dykes, J., & Wood, J. (2010b). Rectangular hierarchical cartograms for socio-economic data. *Journal of Maps, 6*(1), 330–345.

Tu, Y., & Shen, H. W. (2007). Visualizing changes of hierarchical data using treemaps. *IEEE Transactions on Visualization and Computer Graphics, 13*(6), 1286–1293.

Ulrich, R., & Miller, J. (1993). Information processing models generating lognormally distributed reaction times. *Journal of Mathematical Psychology, 37*(4), 513–525.

Vliegen, R., Van Wijk, J. J., & Van der Linden, E. J. (2006). Visualizing business data with generalized treemaps. *IEEE Transactions on Visualization and Computer Graphics, 12*(5), 789–796.

Wattenberg, M. (1999). Visualizing the stock market. *In CHI'99 extended abstracts on human factors in computing systems* (pp. 188–189). ACM.

Wood, J., Badawood, D., Dykes, J., & Slingsby, A. (2011). BallotMaps: Detecting name bias in alphabetically ordered ballot papers. *IEEE Transactions on Visualization and Computer Graphics, 17*(12), 2384–2391.

Wood, J., & Dykes, J. (2008). Spatially ordered treemaps. *IEEE Transactions on Visualization and Computer Graphics, 14*(6), 1348–1355.

Wood, J., Dykes, J., & Slingsby, A. (2010). Visualisation of origins, destinations and flows with OD maps. *The Cartographic Journal, 47*(2), 117–129.

Wehrend, S. (1993). *Appendix B: Taxonomy of visualization goals* (pp. 187–199). Visual cues: Practical data visualization.

Zhou, M. X., & Feiner, S. K. (1998). Visual task characterization for automated visual discourse synthesis. *In Proceedings of the SIGCHI Conference on Human Factors in Computing Systems* (pp. 392–399). ACM Press/Addison-Wesley Publishing Co.

Zhou, M., Tian, J., Xiong, F., & Wang, R. (2016). Point grid map: A new type of thematic map for statistical data associated with geographic points. *Cartography and Geographic Information Science,* 1–16.

The Usability of a GeoVisual Analytics Environment for the Exploration and Analysis of Different Datasets

Irma Kveladze, Menno-Jan Kraak and Corné P.J.M. van Elzakker

Abstract A GeoVisual Analytics (GVA) environment is based on highly interactive and dynamic visualization techniques intending to reveal knowledge in complex and multivariate datasets. By depicting information, these techniques amplify human capabilities and facilitate the performance of cognitive tasks for pattern recognition, decision-making or analytical reasoning. However, the question is whether those visual representations are suitable for visualization of different types of data to perform similar tasks. The limited usability studies that have been done on interactive analytical environments have failed to yield a definite answer. Therefore, this paper presents an evaluation experiment on how an interactive GVA environment can be designed that will effectively support similar task execution processes for different use cases. In the GVA environment investigated, four graphic representations of the Space-Time Cube (STC), graphs and maps are interactively integrated to allow data manipulation from location, attribute and time perspectives for overview and detailed analysis. The results of the experiment revealed that the four visual representations in a GVA environment appear to be effective and efficient to explore different datasets with similar tasks, but require some sophistication.

Keywords GeoVisual analytics · Tools · Evaluations

I. Kveladze
Department of Civil Engineering, Aalborg University, Aalborg, Denmark

I. Kveladze (✉)
Department of Geoinformatics, Palacký University Olomouc, Olomouc, Czech Republic
e-mail: ik@civil.aau.dk

M.-J. Kraak · C.P.J.M. van Elzakker
Faculty of Geo-Information Science and Earth Observation, University of Twente, Enschede, The Netherlands
e-mail: m.j.kraak@utwente.nl

C.P.J.M. van Elzakker
e-mail: c.v.elzakker@utwente.nl

M.P. Peterson (ed.), *Advances in Cartography and GIScience*, Lecture Notes in Geoinformation and Cartography, DOI 10.1007/978-3-319-57336-6_15

1 Introduction

Modern technology has enabled the development of various interactive environments for visualization and analysis of complex datasets. These systems help users gain insight into complex relationships. However, it is required that such analytical tools should be evaluated to determine whether they meet that purpose (Thomas and Cook 2005).

We describe an experiment that investigated visual interaction strategies to measure the usability of a highly interactive environment for GVA. The constructed GVA was centered around a Space-Time Cube (STC) and interactively linked to graphs and a 2D map for the manipulation of different types of data for similar exploratory tasks. This is because GVA is often designed to solve only one particular problem. The question is whether the same environment can be made suitable to tackle similar problems with different datasets. Therefore, the main objective of this research is to understand how well the presented GVA tools are utilized by the test participants to execute similar tasks with different use cases. We want to know what makes them useful, and, if not, what are the issues to be considered.

In order to answer these questions, some background information on the evaluation of analytical environments will be presented. Then, the test set-up will be described in detail. Finally, the results will be presented and analyzed.

2 Usability of GeoVisual Analytics Environments

GVA environments are highly interactive complex systems that require systematic evaluation studies at different levels of their development. In those studies, potential users of the product may be asked to perform particular tasks (Nielsen 1993). Çöltekin et al. (2009) carried out a task based usability experiment related to interactive online analytical environments. In this controlled experiment, two differently designed interactive map interfaces operating on the same dataset were evaluated. The results revealed usability issues for both interfaces. One interface was more effective and the other was more efficient. In a follow-up research, Çöltekin et al. (2010) evaluated the users' visual analytics strategies based on the same interactive online interfaces. Results of the analysis show differences in efficiency metrics between different groups of users.

The use of GVA environments by two different groups, domain experts and non-domain experts, was also the objective of an experiment by Kveladze et al. (2015a) Their test results show that the domain experts are more effective and efficient in performing the analytical tasks because of their knowledge and motivation to investigate data than non-domain experts.

This study presents research that uses the same GVA environment but with different datasets. Similar to Kveladze et al. (2015a), we also deployed a combination of qualitative and quantitative methods to evaluate the exploratory environment and to identify usability issues.

3 Experiment Design

The interactive GVA environment for the experiment contained visual representations and functionalities for the analysis and manipulation of spatio-temporal data. The experiment involved two use case studies of simple datasets and complex annotations (Sect. 3.1). The GVA environment was evaluated with the same group of users in two sessions.

The test participants had to execute comparison and attribute-based tasks while following particular workflows (Kveladze and Kraak 2012). Similar to Kveladze and Kraak (2012) and Kveladze et al. (2015b), we also evaluated and defined data content design in terms of the cartographic design aspects for the research questions before constructing the GVA environments.

A number of qualitative (think aloud and user interviews) and quantitative (questionnaire, video/audio/screen/eye recordings) evaluation methods were used, and this made it possible to derive usability metrics on effectively completed tasks, the time required to complete the tasks, and user satisfaction. The test subjects also were expected to provide feedback on the usability of the GVA environment by discussing the different steps of their actions when completing given tasks. This helped to reconstruct their task execution strategy. In addition, based on the results, existing usability issues were identified and conclusions drawn regarding the future of the constructed GVA environment.

3.1 The Use Case Studies

The first use case study dealt with a simple dataset representing Napoleon's march to Moscow (Smith 2002; Tufte 1983). The data show details of Napoleon's Russian campaign with 15 different corps of his army. Each corps followed its own trajectory and information about time, passed populated areas, temperature, places of battles, and the number of available soldiers for each corps.

The second use case study is an annotated travel log of a trip in Estonia. The data set is considered complex because of the data variety: a path, collected by GPS is annotated by a large number of pictures, videos and documents.

3.2 Experimental GVA Environment

The GVA environment constructed for this usability experiment (Fig. 1) consisted of four different representations: a 2D map, a Space-Time Cube (STC) (Kraak 2008), a Time Wave (TW) and a Time Graph (TG) developed in the uDig program

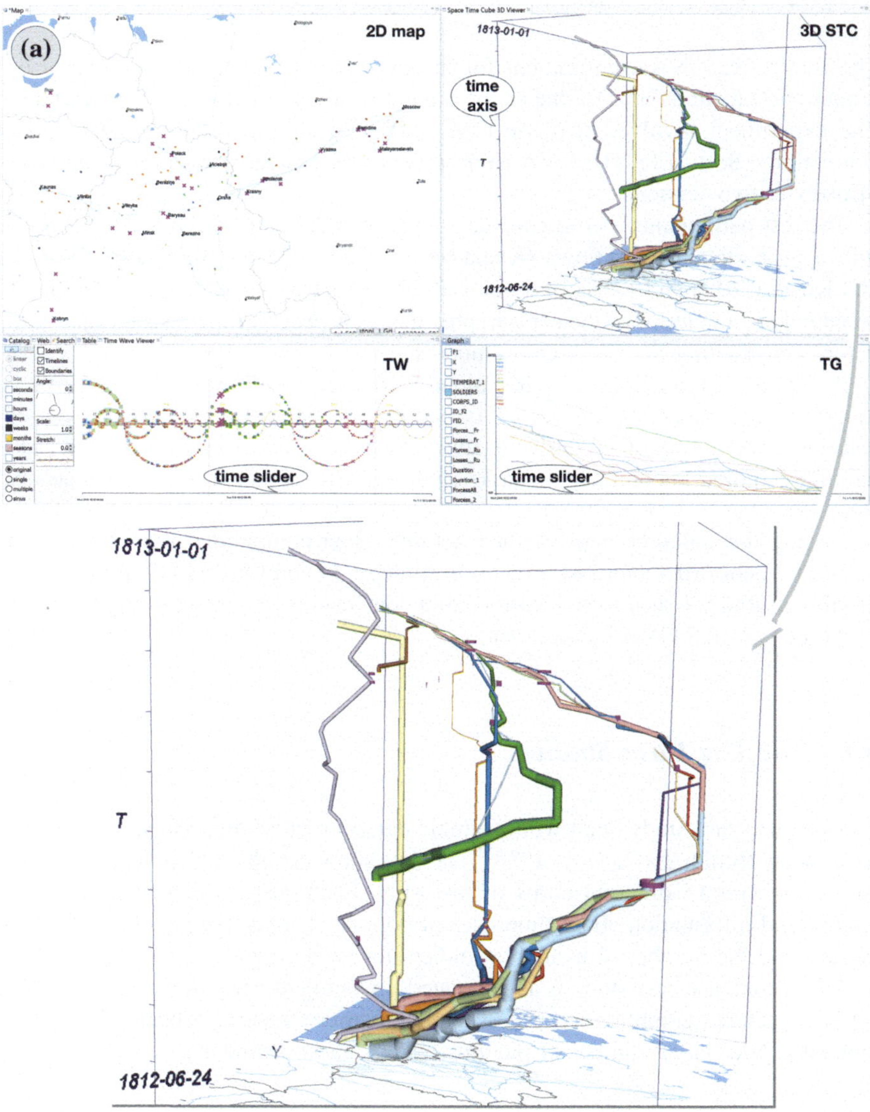

Fig. 1 Interactive visual environment for the experiment (**a**) simple dataset and (**b**) complex annotations

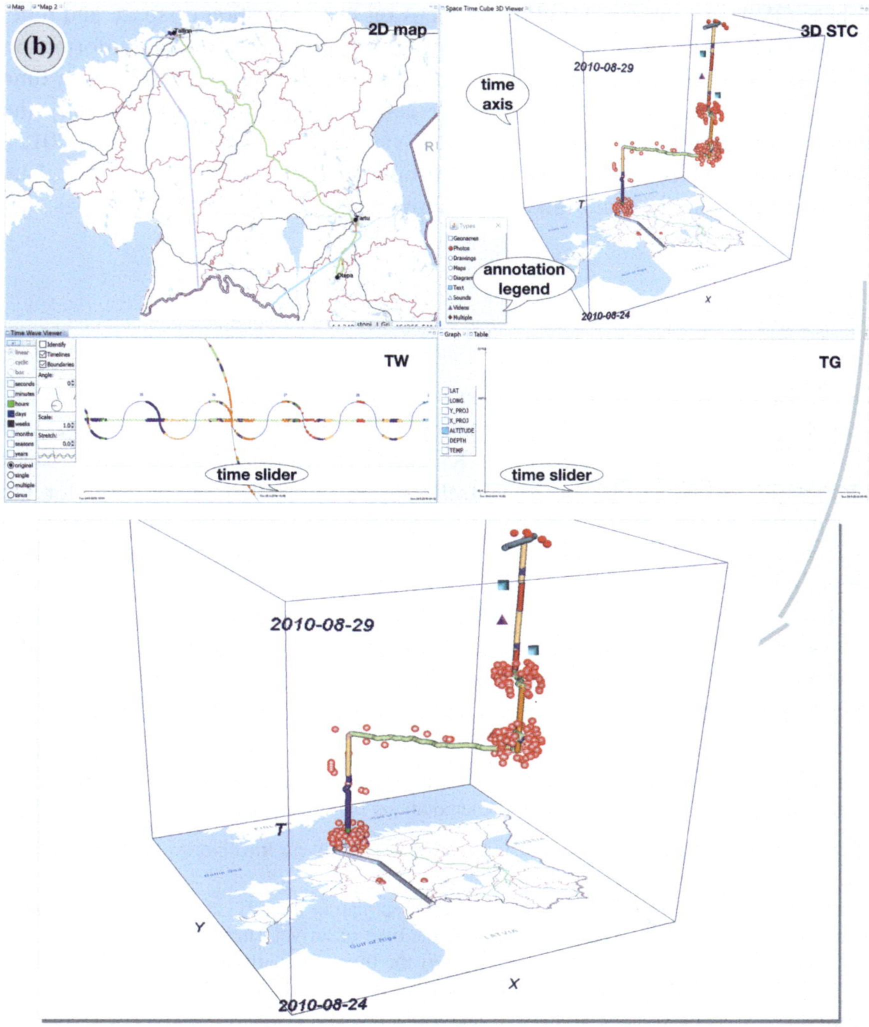

Fig. 1 (continued)

environment (Li 2010; Li and Kraak 2008). These visual representations allow users to analyse complex geographic datasets from spatial, temporal and attribute perspectives (Peuquet 2002).

The 2D maps allow the user to get a quick spatial overview of the whole dataset and show basic geographic information on administrative boundaries, hydrography and populated areas, and also use case related information, required for the exploration of the dataset. The STC was introduced in the 1970s to study individual lifestyles (Hägerstrand 1970). Over time, the popularity of this visual representation has grown and it became essential for the analysis of movement trajectories.

In these GVA environments it allows a focus on the integration of space and time, with an additional option for attribute information. It depicts the same information as the 2D map on its base map, while the use cases are represented from a time perspective as space-time paths. The TW and TG were introduced to study the temporal distribution of attribute values for various objects (Li and Kraak 2012). They enable a different, but integrated, attribute and temporal analysis of the data. The four representations are linked and allow query and interaction activities.

3.3 User Tasks

The Test Participants (TP's) were expected to manipulate the interactive environment and to execute four spatio-temporal tasks: *locate, identify, compare* and *characterize* (Zhou and Feiner 1998; Knapp 1995; Gotz and Zhou 2009). Their manipulation scenario or workflow itself was the object of observation for the experimenter, to understand how particular visual representations can be used in an interactive visual environment to support the data investigation process for the task execution.

The analysis of this task performance helped to obtain answers on the effectiveness, efficiency and user satisfaction for each visual representation in the GVA environment. The respective task and their expected execution scenarios are described for both case studies in Table 1.

Table 1 Tasks given during the experiment and expected execution scenarios for the simple datasets (task 1 and 2) and for the complex annotations (task 3 and 4)

Task 1	Please describe how long different corps were traveling to Moscow and back?
Execution scenario	TPs are expected to *locate* 2D base map of the STC at the time when corps arrived in Moscow and to name time of arrival Then continue moving base map up on time axis to *identify* the time of departure from Moscow and the end of sampaign
Task 2	Please look at the STC and TG and compare the difference between the beginning and the end of the campaign?
Execution scenario	TPs are expected to *compare* the number of solders at the beginning and at the end of the campaign through STC and TW. Then they should look at information on the temperature on the TG to *characterize* change
Task 3	Please describe how long the traveler stayed at different places? What do you think was the purpose of the stay/visit
Execution scenario	TPs are expected to locate the base map at the place of the long stops to identify time duration. then focus on TW TG and 2D map to characterize the purpose of the stay
Task 4	Please observe the types of annotations and find the area/place where the most diverse materials/information was collected?
Execution scenario	TPs are expected to compare different annotations in the STC to find the place of the most diverse information

3.4 Test Participants

A group of two females and five male participants participated in the experiment. They were of different age groups and were all scientists or PhD candidates with experience of working in the geo-domain. The majority of the participants indicated that they are mainly using 2D interactive maps and rarely 3D interactive maps. Six TPs claimed to be familiar with the STC, and five TPs appeared to be familiar with an interactive 2D map and STC in an analytical environment for investigating movement data. The majority of the participants rated themselves as very good or fair in being able to orient themselves in an analytical environment.

3.5 Experiment

Materials: The GVA environments as shown in Fig. 1.
Setup: The experiment was performed on a Windows 7 workstation running Tobii Studio 3.2. software and a Tobii X 60 Eye Tracker on 60 Hz resolution to capture the eye movements. The GVA environment, running in uDig, was displayed on a 24-inch LCD flat panel display with a 1680 × 1080 screen resolution.
Procedure: The experiment took place in a usability laboratory, where the participants were first informed about the functioning of the visual environment and the analytical tools required for the task execution. After familiarization with the environment, the participant's eyes were calibrated and the sessions started. The first two sessions were about the evaluation of the GVA environment, initially for the simple dataset and then the complex annotations. The third session was an interview to allow the participants to voice their opinion and experiences with the previous sessions. All three sessions were recorded and later processed for analysis.

4 Results

The recordings generated in Tobii Studio were divided into segments based on the task execution and user interview sessions for each TP. Ineffectively completed tasks were removed. For the effectively completed tasks, the Areas of Interests (AIOs) were identified and GazePlots generated for each TP for the quantitative analysis. For the qualitative analysis, verbatim transcription of recordings for each TP were made in ATLAS.ti.

4.1 Locate the Map

To perform the task *locate* with the simple dataset, most test participants moved the 2D base map of the STC along the time axis to find the point (Moscow) of spatio-temporal intersection and name the time when the corps arrived in Moscow. The majority performed this task on the 2D map and then continued with the exploration of temporal aspects in the STC. Such exploration behaviour indicates constant control and comparison of information in the GVA environment. This can be seen from the eye movement GazePlots of the participants for the task *locate* in Fig. 2.

The task execution strategy showed that the use of the 2D map for finding the spatial locations was more convenient than the STC, but the 2D map could provide only a spatial overview and was limited for seeing additional information. Another observation was that participants effectively used the interactive links between the base map of the STC and time slider of the TG to find spatio-temporal locations.

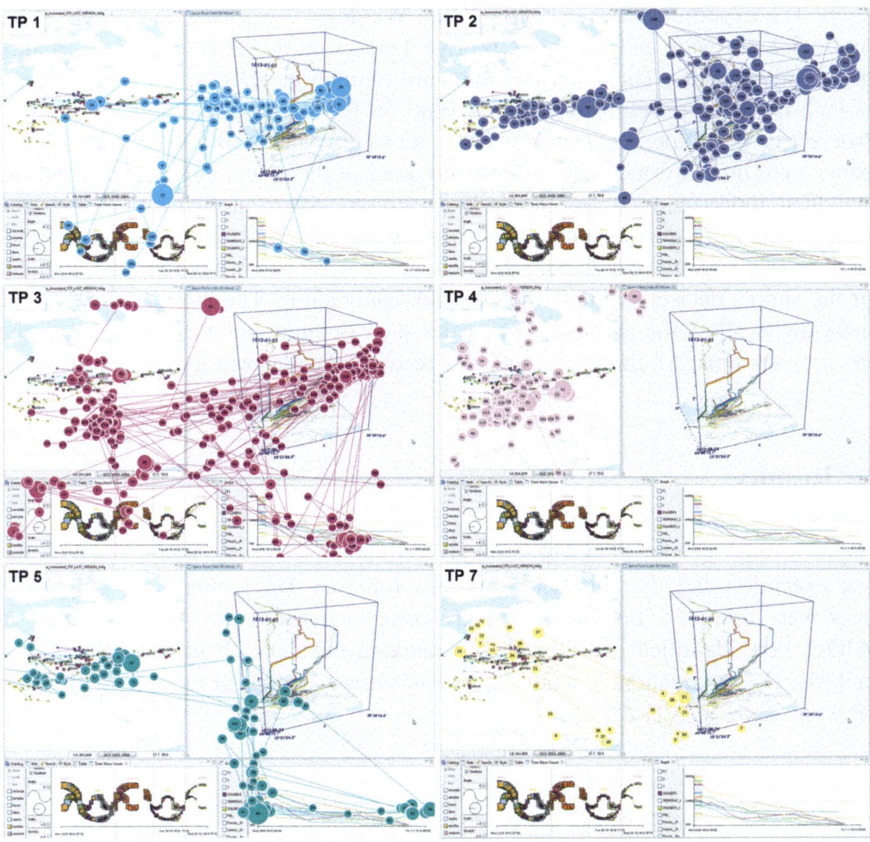

Fig. 2 GazePlots representing the use of the visual representations by the test participants during the execution of the task *locate* with the simple dataset. The small multiplies reveal the different task performance strategies of the test participants

With complex annotations, the majority of participants was synchronously observing other graphic representations to see the changes. A minority first looked at the temporal distribution of annotations in the STC, and then focused on the legend menu to understand map elements and annotation types. Thus, the execution of the task *locate* appeared to be easy for the exploration of the temporal component in the STC, especially based on the places of long stops, i.e., stations. Similar to the simple dataset use case, a minority of the participants completed the task *locate* using the time slider of the TG.

4.2 Identify Time

The task *identify* appeared to be easy to perform in the STC for the simple dataset. A minority of the test subjects preferred to complete the task with the TG and TW together. Several TP's found it easier to extract time related information from the TG and to see the duration of the event without selection activities. The selection function was found to be useful in such an interactive environment to *identify* information.

In order to *identify* the start and end time of the event, some participants used the time values shown on the time axis of the cube, while others focused on the time line of the TG and TW. Despite the multiple options to complete this task, the results revealed that the participants were somewhat confused about the overall functioning of the GVA environment when working with the simple dataset. An interesting observation was to see the participants' confusion in using both the STC and TG together. Most probably, this was caused by the nature of the (simple) dataset and the different ways of representing the data in both visualizations. The event in the TG is represented as a 'straight' line with its beginning and end on a linear time line. The end of the event shown in the TG was Moscow. In the cube and 2D map, Moscow is the most Eastern location of the round-trip event. This example shows that, some of test participants were not able to use the visual environment effectively, because either they did not understand the concepts behind the graphics or could not operate the environment.

For the majority of the participants the task *identify* was easy to perform with the complex annotations. When observing the data in the different representations, the majority of the users realized that overlapping annotations cover relevant information. Thus, the filtering option provided in the legend view was used (switch on/off) to observe the spatial and temporal distribution of required information in the STC and to *identify* time durations. Some used the time labels along the STC's time axis and 2D map, while others used the time slider of the TW and TG. The majority looked happy with the TW to perform this task, but were less happy with the TG.

4.3 Compare Differences

The task execution strategy for *compare* did not reveal any performance problems in the first case study, but showed a more effective use of visual representations than for the previous tasks. The majority of the participants was able to *compare* differences between the start and end of the event in one glance of the graphic representations. For comparison, some used only the STC and others explored the differences through the TG. A minority of participants studied the sizes of the corps in the STC, but for further details they used the TG and TW. While the majority preferred to move the time slider of the TG and observe appearing labels to follow the dynamic development of an event. During the execution of the task *compare,* one participant mentioned that "from a 1D graphical representation it is easy to see what happened, because the TW allows to compare battles and size of the corps together overlapped and makes it easier to draw conclusions."

For the annotations in the second use case, the task *compare* was mainly performed in the STC. To derive information on the diversity of annotations, the filtering option (switch on/off) available in the STC's legend menu was used effectively. The execution of this task revealed one minor issue related to the selection of annotations. In particular, selection of one particular annotation from many, was delayed due to the slow system response.

4.4 Characterize Change

All participants performed the task *characterize* less successfully and pointed to missing information in the first case study. It was observed that test subjects first identified the places between start and end of the event in the STC, and then used the TG and TW to see the difference in temporal duration. However, in order to *characterize* the factors influencing the campaign, the attribute information was required, and they could extract it from the TG.

Overall, the experiment revealed different exploration behaviours and ways of use. Test participants suggested improvements in the GVA environment's analytical functionality. They also suggested to implement a function for the STC that would help to differentiate by day and night, or by season.

Similar to the simple dataset, in the second case the participants started the task *characterize* from the STC by filtering annotations and observing the legend of the 2D map to match different activities. Some participants preferred working with the TW and the STC to investigate annotations at the places of the long stops. While participants were switching between the STC and 2D map, they discovered differences in colours shown on the 2D map and in the STC view for traveling activities due to the shading depth cue used for the space-time paths in the STC. This appeared to be confusing to at least one participant.

4.5 User Satisfaction

The majority of the participants considered the GVA environment to be useful because they could manipulate data from location, attribute and time perspectives, while a minority expressed the idea that the combination of the STC and TG was enough to answer questions of when, where and what. The orientation in this GVA environment did not appear to be a difficult task, although remarks were made to add extra geographic features for orientation in the detailed views.

The task *characterize* in the simple dataset was named as relatively difficult, although finding the places of long stops was an easy task. The majority of participants indicated that the STC is a useful representation to target time related questions, but visual clutter was highlighted as a disturbing issue. A minority emphasized that they had difficulties manipulating the STC. Besides, they suggested to make the base map of the STC interactive to allow the selection of objects that could be seen in the 2D map view.

The 2D map was found to be useful to start the data exploration and to understand the dynamic process as a whole. The TW was seen as too cluttered for the simple dataset because of the 15 trajectories, while with the complex annotations it could represent traveling activities for the single trajectory in an effective way. The participants could derive general sequential trends for both use cases and see time periods or cycles. The TG was useful for the exploration of the simple dataset, but useless for the complex annotations since the information on heights of the terrain was not considered useful for exploration. The convenient interactive link with the STC allowed to control time in an easy way in the GVA environment, but the test persons would like to see more options to zoom in/out.

The most appreciated aspect of the GVA environment was interactivity and synchronous relocation of the base map and time slider between four visual representations that allows different information from each to be derived. The majority of the participants used a combination of visual representations while executing the given tasks and the interactivity between the STC, TW and 2D map was not even considered to be sufficient.

4.6 Task Performance

In the experiment with the simple dataset, six participants out of seven could perform tasks effectively (Fig. 3), while for the complex annotations, five participants out of seven were able to complete tasks. The results also showed some difference in the performance of similar tasks. For instance, five participants out of seven were able to complete the task *characterize* effectively with complex annotations, while for the simple datasets six participants could complete it. In general, the task *characterize* was identified as the most ineffective task, and the tasks *identify* and *locate* as the most effective tasks for the complex annotations.

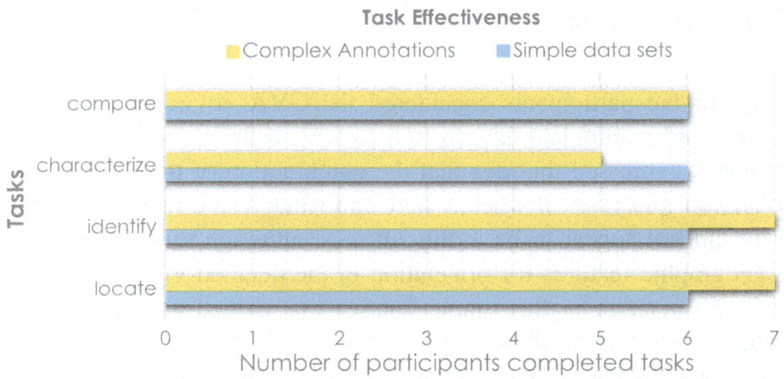

Fig. 3 The task effectiveness of participants during the experiment

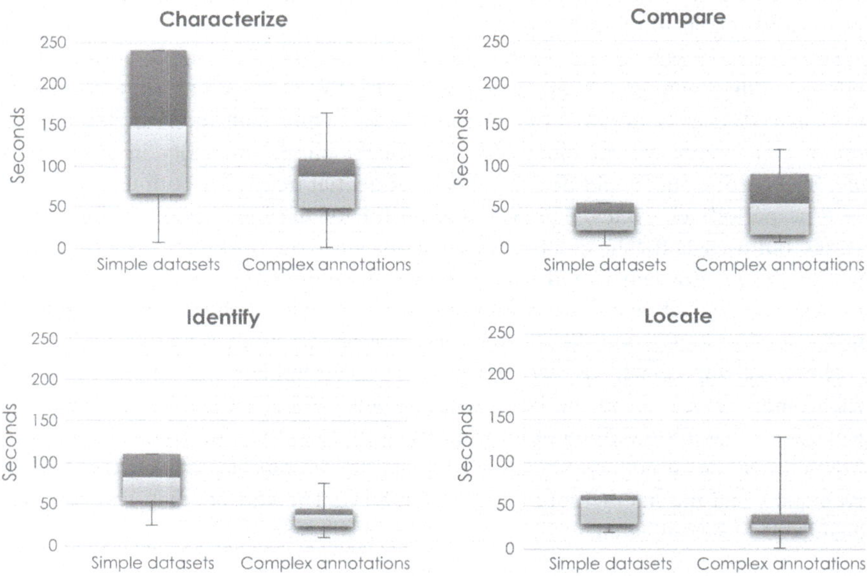

Fig. 4 Average task efficiency of participants during the experiment

Compared to the other tasks, the task *characterize* required the combined use of the visual representations since each view was representing at least one attribute. For instance, in the case of the simple dataset, TPs had to analyse the combination of qualitative (names, locations) and quantitative (temperature, size of the corps) information to provide answers while, for complex annotations, it was only qualitative (transport mode, locations, activities/annotations) information. As for the task *compare,* it was performed well by six test subjects for both use case studies.

The efficiency also revealed some differences (Fig. 4). The statistical analysis shows that the tasks *locate, identify* and *characterize* on average were performed in a relatively shorter time for complex annotations than for the simple dataset. This can be explained by familiarization and experience gathered with the GVA environment during the first session with the simple dataset.

4.7 Use of the Visual Representations in the GVA Environment

To analyse the use of the visual representations during the task performance, the recordings were revisited and new AOIs identified in terms of the use of visual representations. The efficiency metrics derived in seconds were transferred to percentages to reveal the relative use of each visual representation during the performance of each task. It appeared that the STC was the most used visual representation (Fig. 5). For instance, in the complex annotations case study, the tasks *identify* and *locate* were completed for 64% in the STC, while for *characterize* and *compare* this reaches more than 70%. In the case of simple datasets, it was obvious that the major visual representation used was the STC again, but in total it did not exceed 50%.

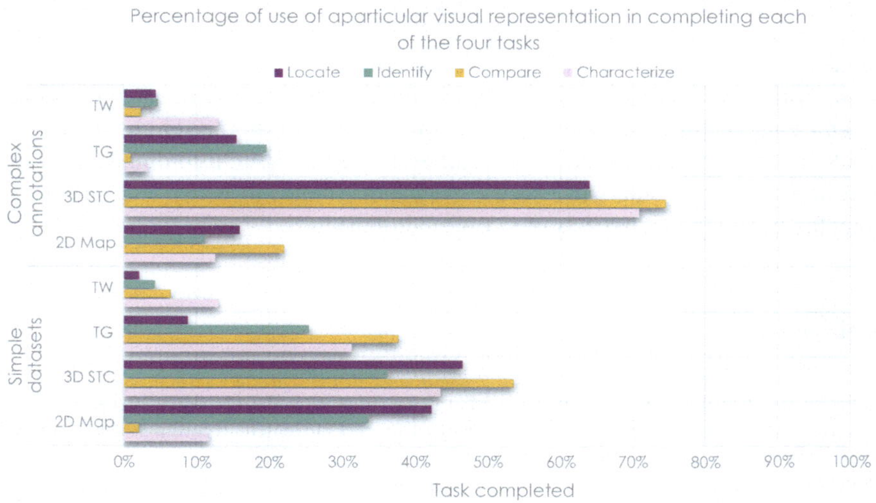

Fig. 5 The graph represents the percentage of use of a particular visual representation in completing each of the four tasks in the two use cases during the experiment

5 Conclusions

The purpose of this study was to determine how GVA environments can be improved, with a focus on interaction principles and usability. The experiment helped define whether selected graphic representations created for one use case can be used to perform similar tasks in different use cases. In other words, how universal is the application of certain forms of graphic representation, despite the influence of the nature of the data characteristics and tasks.

The coordination between views appeared to be very useful for exploration. However, the experiments in general revealed some usability issues regarding the coordination between visual representations.

Representations implemented for one dataset are not completely useful for other datasets. For instance, in case of the simple datasets the TG performed well, but with complex annotations it was less useful due to the lack of the user interest in data characteristics represented. Despite this, the experiment showed that the GVA environment was effective and efficient to tackle similar problems for both datasets, although representation and their GVA functionality needed adaption.

Another finding was derived from the usability experiments' video/audio/screen recordings. They show obvious advantage for working on actual tasks, i.e., the participants could use tools and manipulate the GVA environment better at every new task due to the learning effect during the experiments. This confirms that tool use experience and familiarization with GVA environments are more effective with actual tasks since they require cognitive thinking to answer given questions.

As with other empirical research, some usability aspects and research questions on efficiency metrics with a bigger group of users would still need to be done. The results of this research would inform further testing design. Accordingly, we suggest and encourage others to execute evaluation studies to better understand the complex user-interface aspects of GVA environments.

References

Çöltekin, A., Fabrikant, S. I., & Lacayo, M. (2010). Exploring the efficiency of users' visual analytics strategies based on sequence analysis of eye movement recordings. *International Journal of Geographical Information Science, 24*(10), 1559–1575.

Çöltekin, A., Heil, B., Garlandini, S., & Fabrikant, S. I. (2009). Evaluating the effectiveness of interactive map interface designs: A case study integrating usability metrics with eye-movement analysis. *Cartography and Geographic Information Science, 36,* 5–17.

Gotz, D., & Zhou, M. X. (2009). Characterizing users' visual analytic activity for insight provenance. *Information Visualization, 8*(1), 42–55.

Hägerstrand, T. (1970). What about people in regional science? *Papers in Regional Science, 24*(1), 6–21.

Knapp, L. (1995). A task analysis approach to the visualization of geographic data. In T. L. Nygers, D. M. Mark, R. Laurini, & M. J. Egenhofer (Eds.), *Cognitive aspects of human-computer interaction for geographic information systems* (pp. 355–371). The Netherlands: Kluwer Academic Publishers.

Kraak, M. J. (2008). Geovisualization and Time—new opportunities for the space–Time Cube. In M. Dodge, M. McDerby, & M. Turner (Eds.), *Geographic visualization* (pp. 293–306). UK: Wiley.

Kveladze, I., & Kraak, M. J. (2012). What do we know about the space—time cube from cartographic and usability perspective? In *Proceedings of AutoCarto 2012: The International Symposium on Automated Cartography, Columbus, Ohio, USA, 2012*. Cartography and Geographic Information Society.

Kveladze, I., Kraak, M. J., & Van Elzakker, C. P. J. M. (2015a). The space-time cube as part of a GeoVisual analytics environment to support the understanding of movement data. *International Journal of Geographical Information Science, 29*(11), 2001–2016.

Kveladze, I., Kraak, M. J., & van Elzakker, C. P. J. M. (2015b). Evaluation of the cartographic design aspects for the space-time cube data content. In *27th International Cartographic Conference, ICC 2015*. Rio de Janeiro, Brazil: ICC.

Li, X. (2010). *The time wave in the time space, a visual exploration environment for spatio-temporal data*. University of Twente, ITC.

Li, X., & Kraak, M. J. (2008). The time wave. A new method of visual exploration of geo-data in timespace. *The Cartographic Journal, 45*(3), 193–200.

Li, X., & Kraak, M. J. (2012). Explore multivariable spatio—Temporal data with the time wave: Case study on meteorological data. In A. G. O. Yeh, W. Shi, Y. Leung, & C. Zhou (Eds.), *Advances in Spatial Data Handling and GIS: 14th International Symposium on Spatial Data Handling* (pp. 79–22). Berlin: Springer.

Nielsen, J. (1993). *Usability engineering*. Boston: Morgan Kaufmann Publishers Inc.

Peuquet, D. (2002). *Representations of space and time* (p. 394). New York: Guilford.

Smith, D. G. (2002). *Armies of 1812: The grand Armeé and the armies of Austria, Prussia, Russia and Turkey*. Spellmount: Staplehurst.

Thomas, J. J., & Cook, C. A. (2005). *Illuminating the path: The research and development agenda for visual analytics*. Washington, USA: IEEE Computer Society.

Tufte, E. R. (1983). *The visual display of quantitative information*. Cheshire: Graphic Press.

Zhou, M. X., & Feiner, S. K. (1998). Visual task characterization for automated visual discourse synthesis. In *Proceedings of the SIGCHI Conference on Human Factors in Computing Systems, LA, USA, 1998*. Addison-Wesley Publishing Co.

Characterizing Maps from Visual Properties

Catherine Domingues and Laurence Jolivet

Abstract Customized mapping needs to rely on a strategy to help mapmakers in designing their maps. Our approach is based on analogy between existing sample maps and the wished customized map. The challenge consists in being able to offer the appropriate maps according to each mapmaker's specifications. We have focused on the perception of seven selected visual properties of maps. To qualify this perception, a user test was set up. A sample map database was built in order to collect assessment of participants about visual properties. Test results were investigated through property and by statistical features. Analyses show that among properties, some are scored as more *extreme* and/or *unanimous* than others. Correlations between properties were also highlighted, which clarifies the characterization of sample maps. We then suggested a way to discover the database in a custom-made map process by using relevant queries based on statistical features of properties. Recommendations on exploiting a sample map database are finally formulated.

Keywords Custom-made map • Visual property • Color • Visual property statistical feature • User test

1 Introduction

A number of present cartographic services offer maps which aim at matching users' specific needs. These services enable cartographers and non-cartographers to view or make custom-made maps. They adjust visualization rules to map content or they fulfill specific visualization features. Such tools are provided by national mapping agencies (NMAs), private companies, personal or company blogs, research projects.

C. Domingues (✉) · L. Jolivet
Institut national de l'information géographique et forestière, Saint-Mandé, France
e-mail: catherine.domingues@ign.fr

L. Jolivet
e-mail: laurence.jolivet@ign.fr

M.P. Peterson (ed.), *Advances in Cartography and GIScience*, Lecture Notes in Geoinformation and Cartography, DOI 10.1007/978-3-319-57336-6_16

For instance, the Ordnance Survey (OS) and IGN, respectively the British and French NMAs, enable Web users to make custom-made maps for paper printing or online publication. These maps contain added value in regards to topographic reference data produced by the NMAs. OS Maps offered by the Ordnance Survey is an online mapping and route planning service that can be used to visualize 1:25 and 1:50 k maps, to find, create and share routes and to access keys or advanced features. Web service "Carte à la carte" from IGN, the API OSM CloudMade, maps of London by the blog mappinglondon.co.uk and more broadly GPS maps proposed by car navigation companies, are some other examples of custom-made maps. These services can be clustered according to their purpose, with some being more oriented toward exploring data than making a map for presentation. A common challenge is dealing with different types of data (for instance, adding the mapmaker's thematic data on blank maps provided by the service) and to apply specified visualization rules (colorful range, generalization, etc.).

The various approaches to make custom-made maps lead to an abundant production of maps which do vary in both content and appearance (Kent and Vujakovic 2011). Several approaches also require maps—data and symbolization—to lean on in order to guide users in map design process. Existing maps can be classified to be more easily handled by tools and accessed by mapmakers.

Existing maps may also be a source of inspiration during the design process. Indeed, map design is a process which crosses art and science (Krygier 1995; Christophe and Hoarau 2012). Keates (1984), quoted by Kent (2005), suggested: "the aesthetic properties of maps are significant to both map users and cartographers." Beautiful maps are attractive and their contents are more easily remembered. Even if extremely subjective and open for debate, a consensus seems to exist that enables us to qualify some maps as "beautiful".

Kent (2009) and Field and Demaj (2012) report that beautiful maps are likely to attract more confidence: readers spend more time reading them, and thus, better understand the information. The notion of style can be used to characterize maps. Style includes specific cartographic constraints as readability, efficiency, and aesthetics (Mapstyle project, Ory et al. 2015). Excellence in map design may be assessed by using criteria explained in Demaj and Field (2012). Nevertheless, map assessment is a personal matter which explains why tests are necessary to test maps (Pickle 2003).

The concern here is with making custom-made maps which are suitable for users' needs and preferences. The global approach is to support mapmakers in a design process through the analogy between maps—those presented as examples and imagined by mapmakers—instead of a list of principles in cartography. There must be enough sample maps to offer a wide range of possibilities and sources of inspiration. Consequently, samples are too numerous to be shown to the mapmaker all at once and need to be characterized.

Our purpose is to organize a sample map database to support analogical reasoning. To this end, visual properties about maps are defined and studied on the basis of user test results. The approach is described in Sect. 2. The visual properties which ensure qualification and characterization of the sample maps are explained in

Sect. 3. Assessment of sample maps is based on a test presented in Sect. 4. In Sects. 5 and 6, the test results are examined to determine visual properties for each map and then to analyze visual properties through their statistical features. Proposals to organize the sample map database based on the test are detailed in Sect. 7. Conclusions and perspectives are proposed in the last section.

2 Approach to Make Custom-Made Maps

Map design can be defined as the process of conceptualization and production of a map within a formalized goal (Slocum et al. 1999). It has to take into account data availability, human perception constraints and aesthetic considerations.

We have tackled the design of custom-made maps by analogical reasoning using sample maps. Sample maps enable showing how a map may look like if sample symbolization is applied to a mapmaker dataset. Our approach is to make and characterize a database of sample maps in order to use analogy between one or several samples and the map the mapmaker wishes to make.

Analogy is the common aspect several elements share. Analogy between maps can be defined from the description of their components considered as quantitative criteria: number of key lines, number of hierarchical levels in the key, or symbolization features. Analogy can also be based on more subjective criteria. It depends on individual perception which includes personal tastes, culture and the context in which a person visualizes a map. Specifically, perception of visual attributes of maps mentioned as visual properties is the main criterion we exploit.

We have focused on topographical maps as they are well-known (Kent 2009). Map symbol reading requires little interpretation and the topographical landscape elements can easily be recognized (road network, coastline, river system, etc.). This intuitiveness makes it possible to offer a wide range of colors, even unrelated to usual drawing codes, if not to usual connotations of colors. It would not be the same for statistical maps in which colorful ranges have to be related to values and their repartition (Brewer et al. 2003).

To support mapmakers in the customization process, we have decided to present map samples as a global picture and not only map keys. Map keys are generic and can be applied to different datasets. They are less dependent on the spatial configuration of mapped features (Piket 1972 cited in Kent 2009). The problematic is similar in image processing when analyzing an image using a color histogram that does not take into account spatial distribution and neighborhood effect (Neumann and Neumann 2005).

In the scope of cartography, we have assumed that presenting sample maps instead of map keys enables the mapmaker to imagine more easily how his/her dataset would look like (Kent 2009). Lastly, as assessment of visual properties can vary according to map readers, a user test was set up to collect individual perceptions. The test is based on qualitative words that describe visual properties and can be seen as inputs to the sample map database during the custom-made map process.

3 Visual Properties of Sample Maps

In the chosen analogy process, each sample map is seen as a whole. Evaluation of sample maps is conducted from qualitative words which apply to global sample maps. Those words are referred to as properties. Those properties focus on the visual variable color, leaving aside the other six visual variables (according to Bertin 1983). The notion of color is mentioned as a means to organize and represent distinct features (Keates 1996 cited in Kent 2009).

Choice of colors, as it can be extended to the other cartographic visual variables, is generally made within the limits of readability of the map and within conventions linked to the culture of one country. For example, in most of NMA's European standards, shades of blue are used for water, green for vegetation, etc., even if each country brings changes (Forrest and Kinninment 2001).

Among those criteria, color is identified as having a central role in the symbolization because it takes an important part in the appearance of the map (Bertin 1983; Robinson et al. 1995). A number of seven visual properties (not to be confused with the seven visual properties of Bertin 1983) were defined mainly in relation to color range and so to the colors displayed on maps. They are issued from an initial longer list of qualitative words (Dominguès 2008). The list is as follows: *warm* (opposed to *cold*: in the test, maps that are scored as *not warm* are considered as *cold*); *pastel*; *bright*; *sober*; *rich*; *original*; *realistic*.

The properties apply to sample maps and their assessment by participants is left to individual perception of the maps. Participants can score a map as more or less compliant with a given property. If the score is very low, the map would be considered as being assessed by the *non-<property>* where *<property>* is one of the seven properties. For each map, scores given by all participants enable us to study properties one-by-one and to compare them. Maps are also analyzed together with the others from properties characterized by statistical features:

- A property may be assessed as the most ***typical*** if the score average is the most distant from the mid-value (i.e., the neutral rate: 0) compared to the other properties. Only one property corresponds to that criterion except when several properties have the same rate. In most cases, each sample map has a single *typical* property;
- A property may be assessed as ***extreme*** if the average of participants' scores is close to the limit of the assessment interval. The property is considered as having a strong significance in a positive meaning of the property, or in the opposite (negative). Several properties can be assessed as *extreme* per map. This statistical feature gives information about the properties assessed in most radical ways;
- A property may be assessed as ***unanimous***: it applies to a property rated with similar values by the test participants. This statistical feature corresponds to a low standard deviation (StD) among participants' scores.

4 Test Protocol

The process aims to qualify a sample maps database on which tools of custom-made map would be based. It relies on specific research hypotheses concerning maps and participants. Those hypotheses condition the design and the content of the sample map database as well as the test implementation.

4.1 Research Hypotheses About Database Design and Test

The design procedure of the sample map database is based on the properties. The test aims to give hints to classify the sample maps from these properties. On the other hand, assessing maps from the properties is not only descriptive; it is also linked to individual perception of maps. Finally, evaluation of visual map properties also depends on criteria that are not trivial to quantify. Therefore, the research hypothesis must be detailed:

- For each sample map, it is possible to identify properties that would best characterize the map;
- Properties are not supposed to be correlated;
- For all the sample maps, property scores vary among participants but it is possible to extract trends (in the concordance or divergence of opinions) in the qualification of maps by test participants.

Those hypothesis need to be discussed from the test results.

4.2 Sample Map Database

The customization process needs to show adequate number of sample maps to mapmakers. The design approach aims to make a database with enough varied samples so they can serve as a model and inspire mapmakers. Then, the sample map database is built in order to be evaluated according to two variables: (a) symbolization with a focus on the color; (b) visualization scale impacting the selection of data and the geographical extent of the maps. Only the symbolization (a) is tackled in the article. The sample maps vary depending on different criteria:

- The number of colors (hue and shade) of the range of colors;
- Saturation of colors: from grayed to more saturated colors;
- Emphasis on themes: one or several themes might be highlighted;
- The visualization scales: 1:25 and 1:50 k.

One specific area has been chosen to make samples as specified by its built-up density: "urban" and its landscape: "coast/sea" (Jolivet 2009). The corresponding dataset contains four main themes: buildings, roads, wooded areas and hydrography. The dataset for the two scales are quite similar. Changes happened to the building and road themes in which the geometry was generalized for 1:50 k and with fewer classes. The extent is 1.95 km^2 for scale 1:25 k and 12.25 km^2 for scale 1:50 k. Finally, 52 keys were defined from qualified colorful ranges used in design and decoration (Sawahata 2001) to create 104 (= 52 colorful ranges × 2 scales) sample maps.

4.3 Implementation of the Test

The test was done using only paper maps. The sample map database was divided in four sets presented in different orders so as to minimize the fatigue effect (Van Elzakker and Griffin 2013). Participants could choose when and where they would take the test. There was no time limit (observed average duration: 40 min). The panel of participants was mainly people who use maps or work on cartographic matters and would be interested in custom-made maps. 16 people took the test. It is a limited number of participants though quite homogeneous in mapping skills.

During the test, maps were presented one after the other to the participants who were asked to score the seven properties with a grade included in the normalized interval [−2; +2]. +2 means high agreement, −2 disagreement (thus, it means the agreement to the inverse meaning of the property, e.g., *not warm* means *cold*), and 0 means a neutral opinion. Only integer values were allowed which entails five potential grades organized symmetrically as agreement or disagreement with the property. Test exploitation deals with the score of each property given by each user about all the maps.

5 Characterizing Maps with Visual Properties from the User Test

Analyzing the test results aims to bring to light how sample maps are perceived. We deal with the first research hypothesis: every map can be significantly characterized by some visual properties (henceforth called property). The operating method is based on the statistical features of the samples: *typical*, *extreme*, *unanimous*.

5.1 Typical Property per Map

The *typical* property is defined among the seven properties when the average of participants' scores is the closest to the limits of the interval of potential values, i.e.,

2 corresponding to the positive meaning of the property, or −2 corresponding to the opposite meaning. Each map has a unique *typical* property, except in the case of equality. The aim is to determine if the *typical* property of a map is really *typical*. If the *typical* property score is closer to the following one(s), the notion of *typical* property is not as relevant. Thus, distance is calculated between the *typical* property average score and the average score for the property that is the second closest to the limit of the interval.

That last property is called the second *typical* property. Distances only take into account the absolute value of average scores. They range from 0 to 2. Only 7% of the maps have a distance equal to 0 which means that those maps have at least two *typical* properties. 48% of the maps have a distance smaller than 0.2 between the two properties (this threshold represents 10% of the interval extent in absolute value) and 74% smaller than 0.4. Thus, very few maps have a *typical* property which is very distant from the 6 other properties. The average distance between the *typical* property and the second *typical* property is equal to 0.3. The corresponding standard deviation is equal to 0.2. No map has a distance between the most *typical* properties higher than 1.2.

5.2 Extreme Property(-ies) per Map

A property corresponds to the statistical feature *extreme* if the average value among participants is close to the limit of the interval. A threshold value of 1 was set meaning that average scores should be below −1 or above 1. The threshold is purposely quite discriminative: it includes 29% of the average values. The question is to determine if all sample maps have at least one *extreme* property.

Considering the threshold set to 1, 21 maps have no *extreme* property. There are similar, though decreasing, counts of maps that have 1, 2 or 3 *extreme* properties, respectively 26, 18 and 16. Few maps are concerned by 4 or 5 *extreme* properties, respectively 4 and 3, and no map has 6 or 7 *extreme* properties.

The group of maps with no *extreme* property is quite numerous: 20% of the maps. This result contradicts the possibility that each map has at least one *extreme* property. That group contains maps that are visually quite diverse in terms of colors, within each map and between maps. Various colors coexisting in a map may be an explanation of the fact that participants find it difficult to give *extreme* and so discriminative grades. The largest group gathers maps with one property scored as *extreme*: 25%. The fact that the number of maps decreases dramatically from 4 *extreme* properties suggests that the defined properties can be discriminative but not for all the maps, or that not all properties are discriminative.

5.3 Unanimous Property(-ies) per Map

Statistical feature *unanimous* depends on the range of scores given by test participants. Standard deviation is used as an indicator to qualify distances between participants' scores of positive or opposite meaning. A threshold value on standard deviations was set in order to determine if the property can be characterized as *unanimous*. The value was fixed on the base of the range of all standard deviation values which verify a normal distribution centered between 0.8 and 1. A relevant discriminative value is 0.8 as it concerns around 30% of the series.

There are few maps (6) that have no *unanimous* properties. There are also a few maps with more than 5 *unanimous* properties and none of the maps above 5. 33% have one unique *unanimous* property. Corresponding maps have diverse color ranges: colored-grey maps as well as maps with a wide colorful range. The main part of the counts concerns maps that have 1 and 2 *unanimous* properties (67%). As a brief conclusion, we can say that almost all maps (94%) have at least one *unanimous* property which may be interesting for the exploration of the DB.

6 Analysis of Visual Properties Through Statistical Features

The test deals with maps through defined visual properties and leads to acquiring knowledge on those properties. Knowledge includes statistical features on properties and correlations between properties and between statistical features of properties. The knowledge represents metadata on the DB of sample maps. Maps are selected and displayed according to statistical features, giving the opportunity to illustrate a statistical feature with representative maps and to be precise on the interpretations of the test. Finally, it will be useful to choose the best sample maps corresponding to the user specifications.

6.1 Statistical Feature: Typical

Occurrences of sample maps were computed according to their *typical* property. Percentages are presented in Table 1.

The property which meets the most statistical feature *typical* is *warm*, the second most *typical* property is *realistic* and the third is *pastel*. In Fig. 1, each *typical* property is presented divided in the opposite meanings. Displayed maps have been selected on the base of the highest difference between the score of the most *typical* property and the score of the second most *typical* property.

Table 1 Percentages of sample maps distributed between the seven properties meeting statistical feature *typical* (properties are seen in both meanings). The total reaches a little over 100% since 6 maps have 2 *typical* properties and 1 map has 4 *typical* properties. From dark orange to light orange: the highest number of maps, the second highest and the third highest

Statistical feature *typical*	*warm*	*pastel*	*bright*	*sober*	*rich*	*original*	*realistic*
Percentage of maps	47%	13%	5%	8%	4%	8%	29%

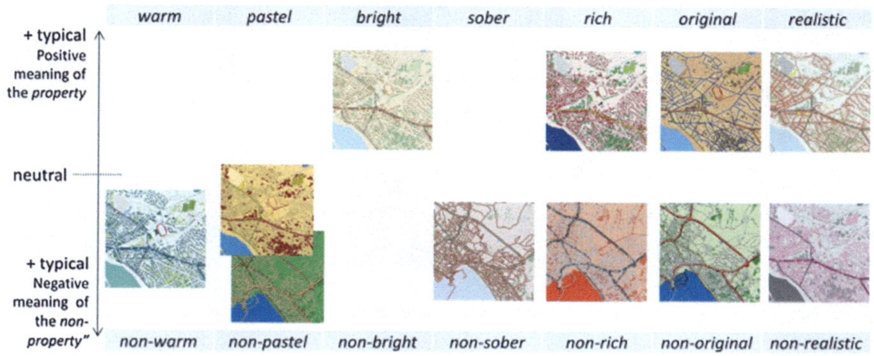

Fig. 1 Statistical feature *typical* is illustrated here by maps with the highest distance between the 1st and 2nd most *typical* properties. All the properties verify this statistical feature though only three properties—*rich*, *original* and *realistic*

If we take a closer look to the distances between 1st and 2nd *typical* properties, one can notice that all distances higher than 0.5 correspond to non-properties: *non-warn*, *non-realistic* and *non-pastel*. The most frequent is *non-warm*. 87% of the maps have a non-property scored as *typical* so few maps can be characterized with a *typical* property in the positive meaning. A discriminative use for DB exploration by this statistical feature *typical* would be through the non-property meaning.

6.2 Statistical Feature: Extreme

The goal is to determine which properties are concerned by the statistical feature *extreme* with the selection threshold still equal to 1. The first result is the number of sample maps scored as *extreme* for each property (Table 2). The property *warm* verifies the statistical feature *extreme* for the highest number of maps (54%). So, this property can be considered as the most *extreme* in terms of occurrence of maps. The second property which was scored the most as *extreme* by participants is *realistic* and the third one is *original*. The least scored property as *extreme* is *bright* (8%). Those results are similar to the characterization like *typical*, though the property *original* has much greater occurrence for *extreme*.

Table 2 Percentages of sample maps that correspond to the statistical feature *extreme* for each of the seven properties. The total is higher than 100% as several properties can be *extreme*

Statistical feature *extreme*	*warm*	*pastel*	*bright*	*sober*	*rich*	*original*	*realistic*
Percentage of maps	54%	14%	8%	15%	13%	36%	48%

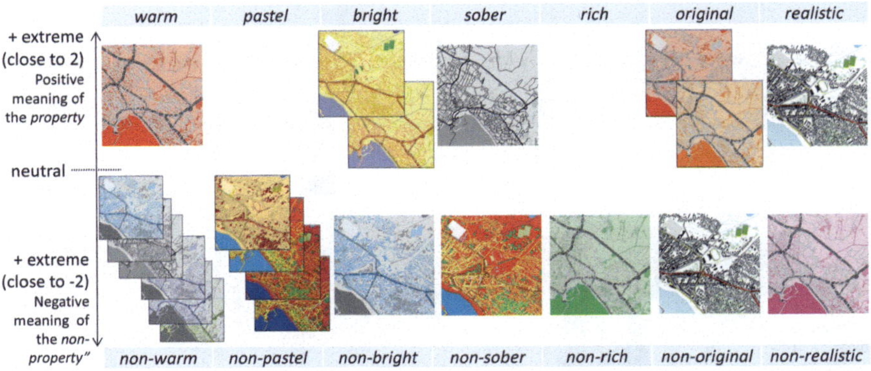

Fig. 2 The most representative maps scored as *extreme* in a positive and a negative meaning for each property. The displayed maps for the *extreme* properties are thus included in the first line of Table 2. *Pastel* and *rich* have no map verifying the statistical feature in the positive meaning

Several elements of interpretation can be highlighted. First, the property *warm* can be assumed as quite clear and significant for participants. *Warm* and its opposite meaning *non-warm* (or *cold*) can be seen as objective properties as they can be directly associated with the colors displayed on maps. Even if some colors can be ambiguous like purple or some green hues, most of the colors are commonly classified by people as *warm* or *cold* (Itten 1985). What can vary among users' perceptions is the considered extents and salience of *warm* and *cold* colors on maps.

The second most *extreme* property *realistic* can be interpreted by the fact that participants are quite used to recognizing classical styles for topographic maps. A part of tested sample maps does not have a classical symbolization. Thus, the property *realistic* along with the property *original* is extremely scored by users. Besides, relatively few spaces with no color at all (blank spaces) are left. This leads to a complete cover of maps by colors, which strengthens the (*non-*)*realistic* or the *original* aspect. Concerning the other properties that are considered as less *extreme*, we can interpret that the properties may not be as easy to understand or that they are not as discriminative as the previous ones.

The second result deals with identifying maps that are representative for a property qualified as *extreme*, either positively (e.g., extremely *warm*) or negatively (e.g., extremely *non-warm*). Notice that in Sect. 5, occurrences were not divided between the two meanings. Figure 2 represents maps that are the most representative for each property scored as *extreme* positively or negatively. The representativeness is associated to the distance to the interval limits. The shortest the distance is (and inferior to 1), the more *extreme* the property is for the map and so the more discriminative the property is for the map.

For the interpretation of the results partly displayed in Fig. 2, all of the maps and scores were exploited. The first remark is that representative maps for statistical features *extreme* can be the same for different properties. For instance, *original* and *realistic* are two properties that invert the most representative maps at opposite meanings. A second aspect is that statistical feature *extreme* is more supported in the opposite meaning of the property and with shorter distances to interval limits. Only the most representative maps for *extreme* properties, *bright* and *original,* are in the positive meaning. It could imply that maps tend to be more discriminated by properties when opposite in meaning. Knowledge on statistical feature *extreme* can be summarized by the fact that mainly three properties are concerned: *warm*, *realistic* and *original*, and that the other properties tend to be less discriminative, though maybe more uncommon and accurate, to discover the sample database.

6.3 Statistical Feature: Unanimous

The statistical feature *unanimous* enables us to identify maps scored similarly among participants. It is interesting to consider properties which tend to reach a consensus and properties which arouse discussion. Percentages of maps are given in Table 3 by properties with a threshold on standard variation still equal to 0.8.

The proportion of maps that have the *warm* property that have been scored unanimously is the highest. Then, the property *realistic* corresponds to the second highest percentages of maps and *original* to the third. The property that concerns the least number of maps is *rich*. When averages of standard deviations are put into perspective, those results are quite coherent. Standard deviation averages indicate that *warm* is the most *unanimous* property and *sober* the least one. It may be assumed that *warm* is a property that is the most recognizable by people and that is commonly apprehended. To represent each property when scored as *unanimous*, maps corresponding to the smallest and the highest standard deviation between participants are selected in Fig. 3.

Same maps can be scored as the most *unanimous* for one or several properties in the same or opposite meaning. For instance in Fig. 3, the map scored least unanimously for (*non-*)*pastel* is also scored less unanimously for (*non-*)*original*, but most unanimously for (*non-*)*warm*. The less *unanimous* map for *bright* is the most

Table 3 Percentages of maps by properties when these properties were scored unanimously

Statistical feature *unanimous*	*warm*	*pastel*	*bright*	*sober*	*rich*	*original*	*realistic*
Percentage of maps	66%	22%	17%	14%	12%	27%	39%

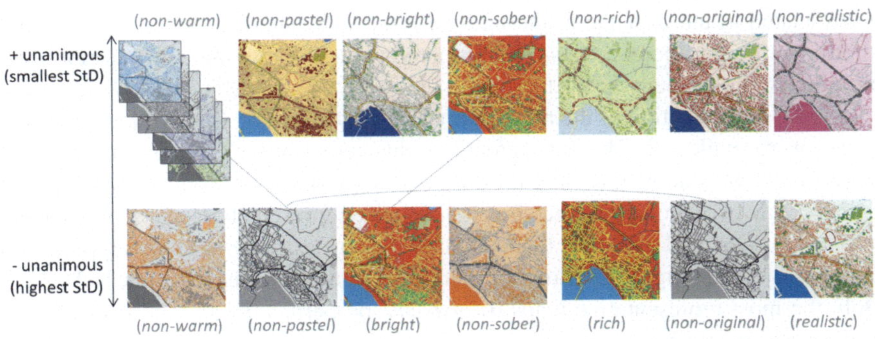

Fig. 3 For each property, maps that are the most (and with a StD < 0.8) and the least *unanimous* are represented. Positive or negative meaning is indicated. *Dotted lines* link identical samples

unanimous for (*non-*)*sober*. The statistical feature *unanimous* is a good way to take into account the different perceptions of maps and properties as all properties correspond to several unanimously scored maps or on the contrary to several not unanimously scored.

6.4 Correlations Among Properties and Among Statistical Features

Two types of correlations may be sought to better understand raw scores and statistical features of properties: between properties (i.e., are the properties statistically independent?) and between statistical features (i.e., do statistical features manifest themselves simultaneously? and what does it mean?).

Firstly, the goal is to determine if significant relations of correlation exist between properties and if so in what meaning (positive or negative). If several properties are correlated, it might be relevant not to use all of them to qualify sample maps. Correlations are calculated from the average of scores of participants based on Pearson's formula in R with a confidence interval set to 95% (Table 4).

The strongest correlation is between the property *realistic* and the property *original* in an opposite direction. The second strongest one is between *warm* and *bright* in a positive way. The property *warm* is also quite strongly correlated positively with *rich* and oppositely with *sober*. The other significant correlations are lower (values between −0.7 and 0.7). Property *warm* is the most correlated with the others; only *realistic* is not significantly correlated with *warm*. The less correlated property is *realistic*, only with *rich* and *original*. Globally, there are 6 positive correlations and 8 negative ones. An explanation is that correlations are due to the meaning of the properties themselves in participants' perception; in that sense *bright* and *warm* might be close, or in opposite meaning *original* and *realistic*,

Table 4 Pearson's correlation values between properties. Only the significant correlations are indicated (i.e., p-value below 0.05): positive correlations in magenta, negative correlations in blue. Strong value > 0.7 or < −0.7 are in dark hues. Medium values in [0.5; 0.7] and [−0.7; −0.5] are in medium hues. Low values in [0.5; −0.5] are in light hues. Hatched cells are not relevant

Pearson's correlation	*warm*	*pastel*	*bright*	*sober*	*rich*	*original*	*realistic*
warm							
pastel							
bright							
sober							
rich							
original							
realistic							

and *sober* and *rich*. A second explanation might come from the fact that participants may associate tested properties (Hinz et al. 2007).

Secondly, we have seen that some maps highly scored as *extreme* (6.2) and *unanimous* (6.3) could be the same depending on the properties. The goal is to determine if correlations can be found not only between properties but also between their statistical features. It would mean that statistical features simultaneously manifest themselves for certain properties. Pearson's correlation values were computed with a confidence interval at 95%. All significant values are positive and under 0.7. Hence only the relations higher than 0.5 were considered between:

- *Extreme* and *unanimous* are correlated for *sober*; it means that statistical feature *extreme* for *sober* visual property is shared by most participants;
- *Typical* and *unanimous* are correlated for *realistic*; when a map is unanimously *realistic*, this property obviously detached from all the others;
- *Extreme* and *typical* are correlated for *warm* and *pastel* which indicates that those properties take precedence.

7 Exploring and Increasing the Sample Map Database

The synthesis of test results enables to offer the most adequate sample map to mapmaker query and to guide new sample map design.

The challenge was to determine how to select and order the sample maps in individual map-making process. First of all, the precedent analysis (see Sect. 6.4) shows that the visual properties are consistent. Only *original* and *realistic* are very correlated (but independent of the others) and may be merged in a unique property. Then, the point is to select the best sample maps to answer mapmaker queries. We consider that a sample map may legitimately represent a visual property if this property includes the statistical feature *unanimous*. Once the property *unanimous* realized, we propose two ways to choose samples.

Table 5 Occurrences of maps corresponding to the exploration of the database from statistical features of the visual properties. *Original* and *realistic* were in opposite meaning. There are only samples available for *extreme* and *unanimous*, and not for *typical* and *unanimous*. *Pastel*, *sober*, and *non-bright* have no samples whereas *non-warm* and *non-realistic* have a great number of samples

	warm	*pastel*	*bright*	*sober*	*rich*	*realistic*
No. of samples	5	0	4	2	0	1
	non-warm	*non-pastel*	*non-bright*	*non-sober*	*non-rich*	*non-realistic*
No. of samples	40	4	0	9	1	48

As first intended, we consider that the best samples to illustrate a given visual property have to join both statistical features: *extreme* and *unanimous* for this property. Another way to meet mapmakers' queries consists in looking for sample maps which have both *unanimous* and *typical* properties. To demonstrate feasibility and relevance of these intentions, Table 5 indicates the count of sample maps available for the exploration based on *extreme* and *unanimous*, and potentially *typical* and *unanimous*, statistical features.

Lastly, if the mapmaker is not satisfied by the selected sample maps, it can be seen as an unusual or innovative map-making design. Then, an alternative selection way may propose samples with *extreme* or *typical*, but not *unanimous* property in order to differ from average preferences.

The map-making design process calls for varied and numerous sample maps. On the other hand, the analysis of statistical results highlights that sample maps are missing for several visual properties. For example, Table 5 shows that, concerning the properties *pastel*, *rich* and *non-bright*, there is no sample with both *unanimous* and *extreme* or with both *unanimous* and *typical* statistical features. This lack is coherent with the fact that for these properties there is no correlation between *unanimous* and *extreme* (or *unanimous* and *typical*). This pitfall should be taken into account in new sample map design.

8 Conclusions and Perspectives

In the context of custom-made mapping, we proposed to show sample maps to mapmakers in order to define his/her map's visual properties by analogy. Then, the delicate issue was how to organize the sample map database and to define a method to select relevant samples. We proposed to use statistical features of visual properties. The significant correlations observed between the defined visual properties make it possible to reduce their list (second research hypothesis). The propositions in Sect. 7 show that visual properties enable to structure the sample map database (first research hypothesis).

Lastly, trends in participants' opinions have been identified from the statistical feature *unanimous* (third hypothesis). The major perspective plans to enrich the sample map database without implementing a new user test. In order to achieve this, additional metadata concerning key and spatial distribution of geographic themes would be required (as in Domingués and Karchi 2013); these metadata would be inputs for a training model which would predict visual properties of a map. They could also be exploited for providing tools to navigate in the properties by sliding from maps that verify more or less that property (Christophe 2012). Enrichment of the metadata would ideally lead to an automatic insertion of maps covering various types of landscapes in the database and thereby to expand the number of sample maps proposed to mapsmakers.

References

Bertin, J. (1983). *Semiology of graphics. Diagrams, networks, maps* (trans.: Berg, W. J.). Madison, Wis.: University of Wisconsin Press.

Brewer, C. A., Hatchard, G. W., & Harrower, M. A. (2003). ColorBrewer in print: A catalog of color schemes for maps. *Cartography and Geographic Information Science*, *30*(1), 5–32.

Christophe, S. (2012). Cartographic styles between traditional and original (towards a cartographic style model). In *Proceedings of AutoCarto 2012*, Colombus, Ohio, USA, September 16–18.

Christophe, S., & Hoarau, C. (2012). Expressive map design based on pop art: Revisit of graphic semiology? *Cartographic Perspectives, 73,* 61–74.

Demaj, D., & Field, K. (2012). Reasserting design relevance in cartography: Some examples. *The Cartographic Journal, 49*(1), 77–93.

Domingués, C. (2008). Description de cartes géographiques. In *Proceedings of EGC 2008, 8èmes journées francophones "Extraction et Gestion des Connaissances"—Workshop on "User modeling and Web interface customization"*, Sophia-Antipolis, France, January 29–February 1 (pp. 59–68).

Domingués, C., & Kharchi, T. (2013). Map classification according to their visual global properties. In *Proceedings of ICC 2013*, Dresden, Germany, August 25–30.

Field, K., & Demaj, D. (2012). Reasserting design relevance in cartography: Some concepts. *The Cartographic Journal, 49*(1), 70–76.

Forrest, D., & Kinninment, E. (2001). Experiments in the design of 1:100 000 scale topographic mapping for Great Britain. *The Cartographic Journal, 38*(1), 25–40.

Hinz, A., Michalski, D., Schwarz, R., & Herzberg, P. Y. (2007). The acquiescence effect in responding to a questionnaire. *GMS Psycho-Social-Medicine*, *4*. Retrieved June 20, 2007, from https://www.ncbi.nlm.nih.gov/pmc/articles/PMC2736523/.

Itten, J. (1985). Art de la couleur. Réédition en 2004. Dessain et Tolra.

Jolivet, L. (2009). Characterizing maps to improve on-demand cartography—The example of European topographic maps. In *Proceedings of 17th Conference on GIScience and Research in UK (GISRUK'09)*, Durham, UK, April 1–3.

Keates, J. S. (1984). The cartographic art. *Cartographica, Monograph, 31,* 37–43.

Keates, J. S. (1996). *Understanding maps* (2nd ed.). Harlow: Longman.

Kent, A. J. (2005). Aesthetics: A lost cause in cartographic theory? *The Cartographic Journal, 42* (2), 182–188.

Kent, A. (2009). Stylistic diversity in European state 1:50 000 topographic maps. *The Cartographic Journal, 46*(3), 179–213.

Kent, A. J., & Vujakovic, P. (2011). Cartographic language: Towards a new paradigm for understanding stylistic diversity in topographic maps. *The Cartographic Journal, 48*(1), 21–40.

Krygier, J. B. (1995). Cartography as an art and a science? *Cartographic Journal, 32*(6), 3–10.

Neumann, L., & Neumann, A. (2005). Style transfer techniques using hue, lightness and saturation histogram matching. In *Proceedings of Computational Aesthetics in Graphics, Visualization and Imaging*, Girona, Spain (pp. 111–122).
Ory, J., Christophe, S., Fabrikant, S. I., & Bucher, B. (2015). How do map readers recognize a topographic mapping style? *The Cartographic Journal, 52*(2), 193–203.
Pickle, L. W. (2003). Usability testing of map designs. In *Proceedings of the 35th Symposium on the Interface*, Salt Lake City, USA, March 12–15.
Piket, J. J. C. (1972). Five European topographic maps: A contribution to the classification of topographic maps and their relation to other map types. *Geografisch Tijdschrift, 6,* 266–276.
Robinson, A. H., Morrison, J. L., Muehrcke, P. C., Kimerling, A. J., & Guptill, S. C. (1995). *Elements of cartography*. New York: Wiley.
Sawahata, L. (2001). *Color harmony workbook: A workbook and guide to creative color creations.* Rockport Publishers.
Slocum, T. A., McMaster, R. B., Kessler, F. C., & Howard, H. H. (1999). *Thematic cartography and geovisualization* (3rd ed.). Pearson.
Van Elzakker, C. P. J. M., & Griffin, A. L. (2013). Focus on geoinformation users. *GIM International, 27,* 20–23.

How Hard Is It to Design Maps for Beginners, Intermediates and Experts?

Gáspár Albert, Virág Ilyés, Csaba Szigeti, Dávid Kis and Dávid Várkonyi

Abstract An online map reading test was done with 859 subjects to statistically measure the efficiency of information retrieval from three different cartographic images of the same area. The differences between the maps were defined by the graphic variables of size, color, pattern, etc., for six map data categories: linear features, hydrography, land cover, elevations, point-like objects, geographic names. The subjects solved map reading tasks related to these categories. Cartographic images were designed for each of the three map reader groups: beginners, intermediates and experts. The design method and the grouping were based on the results of previous studies, and the grouping was done with a competency test prior to the map reading task. The results showed the effectiveness of information retrieval from the three different cartographic images. Conclusions about the efficiency were done concerning the age, gender and level of expertise of the subjects.

Keywords User-oriented maps · Online maps · Map data types · User studies

---

G. Albert (✉)
Department of Cartography and Geoinformatics, Eötvös Loránd University, Budapest, Hungary
e-mail: albert@ludens.elte.hu

V. Ilyés
Department of Survey Statistics, Eötvös Loránd University, Budapest, Hungary
e-mail: ilyesvirag@gmail.com

C. Szigeti
Doctorate School of Earth Sciences, Eötvös Loránd University, Budapest, Hungary
e-mail: szgtcsaba@map.elte.hu

D. Kis · D. Várkonyi
Faculty of Informatics, Eötvös Loránd University, Budapest, Hungary

M.P. Peterson (ed.), *Advances in Cartography and GIScience*, Lecture Notes in Geoinformation and Cartography, DOI 10.1007/978-3-319-57336-6_17

1 Introduction

Cartographers have known for a long time that map users understand the same map with different efficiency (Petchenik 1977; Thorndyke and Stasz 1980). We've also know that differently displayed maps can help unskilled users to understand spatial information (Board 1978). To understand the rules and cognitive processes behind these phenomena, user oriented maps were designed, and user studies were conducted following the work of similar studies (Montello 2002; Allen et al. 2006; Ooms et al. 2014). It was shown that the amount of information comprehended by a user depends on various cognitive competences, such as sign recognition, sense of directions, distance- coordinate- and scale-reading, and recognition of morphology (Muir 1985; Clarke 2003). Using tests targeting these skills in map related experiments, the role of expertise was clearly demonstrated concerning the level of map-data comprehension as was found by Clarke (2003). According to Gerber (1981), the difference between an "experienced" and an "unexperienced" map reader lies in their long-term memories: while the first group has previous knowledge about various map symbols, the second one has to "learn" the map during the process of reading it. Engaging in tasks like consulting the legend consumes capacity of the working memory, so the amount of information extracted from the map will be less for beginners than for the experts. This phenomenon is explained by the cognitive load theory (Bunch and Lloyd 2006), and was observed many times in map user studies (Harrower 2007; Ooms et al. 2014). The amount of displayed information, thus, will be a crucial element in the legibility of a map.

2 Thoughts of the Map Maker and the Map Reader

Petchenik (1977) was probably right: it is unlikely that the thoughts of the map maker and the map reader will be exactly the same. However, map reading competency studies regarding the various map data types have already produced many clues to reduce the difference. Basically two or three user categories are distinguished in these studies: novices/experts (Deeb et al. 2012; Ooms et al. 2012) and beginners, intermediates and experts (Gerber 1981; Clarke 2003; Albert et al. 2016). Concerning the six map data types that are necessarily distinguished in a topographic map database (Thompson 1979; Buckley et al. 2004; Usery et al. 2009), we present some examples of which type of user is sensitive to which map content:

(1) Linear features (road network, transportation, and tourist trails). These are usually the structuring elements of a map and they draw immediate attention of map readers (Ooms et al. 2014). The interpretation of linear structures is the most challenging for all map readers; however, it has been found that beginners make five times more mistakes than experts (Albert et al. 2016).
(2) Hydrography (creeks, rivers, lakes, springs). Hydrographic symbols do not vary greatly by time (Albert 2014) and help to interpret the surface morphology

(Potash et al. 1978). Linear and area features of the hydrography are "easy to understand" features for all map readers.

(3) Land cover (forest, meadow, vineyard, etc., built-in area). Interpretation of natural and artificial coverage can be facilitated with familiar patterned symbols (Barkowsky and Freksa 1997).

(4) Hypsography (elevations, contours, shading, scarps, peaks). Intermediate and beginner map readers have difficulties in interpreting hypsography (Eley 1992; Wakabayashi 2013). Contour line interpretation requires both navigation and symbol recognition skills (Montello et al. 1994; Murakoshi and Higashi 2015).

(5) Point-like objects (symbols, significant man-made or natural objects). It was shown by Albert et al. (2016) that 15% of experts, 38% of the intermediate and 66% of the novice map users make mistakes in map symbol interpretation. It was also shown by Easterby and Hakiel (1981) that a symbol is comprehended more easily if it has descriptive details.

(6) Geographic names (name attributes of all objects displayed as labels). Map labeling rules (Imhof 1975) are implemented into algorithms where the preference of orientation, shape and texture varies according to expertise (Deeb et al. 2012; Ooms et al. 2012). It was shown that both experts and beginners prefer sans-serif font types, but beginners are more tolerant to diversity.

So, if such remarkable differences exist, how is it possible to design maps which transmit the same meaning from the map maker to the differently skilled map readers? To answer this question, we created three different maps of the same area, and with an online test, we examined the effectiveness of the three different cartographic images in the case of three user categories: beginners/intermediates/experts. In the map making process, the "Bertinian" graphic variables (Bertin 1967) were used to enhance certain map data types together with the usual generalization methods (McMaster and Shea 1992).

3 What Questions Can Be Answered with the Experiment?

The research was aimed to answer several questions about the usability and effectiveness of the produced cartographic images. It was expected that the results would help to design automatically generated online maps for those individuals who would reveal their map reading skills (i.e., by filling out a test). These questions were as follows:

- Do the different maps help to achieve better results on map related tasks?
- Were the average results better when the participants used their own groups' map?
- Are there any significant differences in the percentage points of correct answers among the map reading groups?

- Are there significant differences in the completion time of different user groups?
- Are there any differences in the completion time of each task among the map reading groups?
- Are there any relations between the age and completion time?
- Are there any differences between the applied map scales of each task?
- Are there any relationships between age/gender and applied map scales?

In order to answer our research questions, we gave each participant maps designed for a specific map reading group—half of the participants got maps designed for their own map reading level, and 25% of them received maps prepared for another group.

4 Categorization of Map Readers

The first component of the map reading test measured the competency. For this, an online questionnaire was presented for each participant, and measured seven competences:

1. interpretation of topographic elements,
2. interpretation of hypsography,
3. interpretation of map symbols,
4. interpretation of geographic names,
5. orientation and mental rotation skills,
6. application of scale bar,
7. distance and travel time estimation.

The tripartite categorization of the participants followed the same method which was used by Albert et al. (2016). Each of the competences was measured with one multiple-choice question with four answers, and the "do not know" option. The assignments were linked to four simple maps containing only those map-data types which were necessary to solve the questions relating to them. This reduction of content was necessary to lessen the cognitive load of the participants during the test.

Online tests that are too long are often left unfinished by participants so the overall time should not take longer than 15–20 min. The duration of the competency test was planned to take 5–10 min, so the second part of the test could be similarly long. The sequence of both the maps and the questions was randomized.

5 Differently Designed Cartographic Images and the Test Questions

For the second component of the test, we created three online maps (level-B for beginners, -I for intermediates, and -E for experts) of the same area. The maps were embedded in a web page that contained the test questions (Szigeti et al. 2017—a parallel paper in this volume). The varying design and generalization level of the maps are based on the results of various studies (see Sect. 2).

The test area had been chosen to have a large variety of topographic elements (multiple road types, a large city, villages, etc.) and diverse hypsography with both hills and plains. The source of the map data was OpenStreetMap (OSM) and SRTM 3 arc second elevations. The data was processed with the QGIS 2.14 open source desktop and server-side applications.

This part of the test involved seven questions (Q1–Q7); two of them were interactive, and the other five were multi-choice questions with four answers. Measuring the same competencies as in the first component of the test, the questions were aimed to examine whether the maps were adequate for the targeted map reader group. Additionally, the answering time and the zoom level at the time of the answer were also recorded. The sequences of the questions and the cartographic images were randomized again, but in this case an online map server provided the cartographic images from the same (but differently visualized) database for all participants. The user interface in the browser made it possible to change the zoom level dynamically between 1:1,000 and 1:250,000, which modified the map content quite similarly to the OSM zoom levels (Table 1).

The interactive Q1 question tested the orientation skills. Participants clicked along a city route following a written description. The results showed how well the maps helped the participants define relative (right/left) or cardinal (North/South) directions.

The Q2 question tested the interpretation of linear objects. The task was to choose a route that did not intersect with any main road. Four options were listed, and each of them showed up on the map when the participants clicked on their radio

Table 1 The zoom levels of the online maps. Multiple entries of the same object types in a column indicate a different design and/or generalization

<table>
<tr><th>Zoom levels</th><th colspan="6">Polygons</th><th colspan="7">Lines</th><th colspan="9">Points</th></tr>
<tr><td>Till 1:250,000</td><td rowspan="3">Field</td><td rowspan="3">Vineyard</td><td rowspan="3">Forest</td><td rowspan="5">–</td><td rowspan="11">Built-in</td><td rowspan="2">Hypsographic name</td><td rowspan="3">Tourist route</td><td rowspan="8">Highway</td><td rowspan="8">Primary road</td><td rowspan="8">Secondary road</td><td rowspan="2">–</td><td rowspan="2">–</td><td rowspan="5">–</td><td rowspan="3">–</td><td rowspan="6">–</td><td rowspan="8">–</td><td rowspan="10">–</td><td rowspan="6">City name</td><td>–</td><td rowspan="2">–</td><td rowspan="3">–</td><td rowspan="5">–</td></tr>
<tr><td>1:150,000</td><td rowspan="5">Village name</td></tr>
<tr><td>1:75,000</td><td rowspan="2">Hypsographic name</td><td rowspan="4">Bridleway, track</td><td rowspan="4">Path</td><td rowspan="4">District name</td></tr>
<tr><td>1:70,000</td><td rowspan="3">Field</td><td rowspan="3">Vineyard</td><td rowspan="3">Forest</td><td rowspan="3">Tourist route</td><td rowspan="8">Railway station; Gas station</td><td rowspan="8">Peak name</td></tr>
<tr><td>1:50,000</td><td rowspan="7">–</td></tr>
<tr><td>1:35,000</td><td>Bu.</td><td rowspan="2">Street name</td><td rowspan="6">Railway station name</td></tr>
<tr><td>1:15,000</td><td rowspan="5">Field</td><td rowspan="5">Vineyard</td><td rowspan="5">Forest</td><td rowspan="3">Building</td><td rowspan="5">Tourist route</td><td rowspan="5">Bridleway, track</td><td rowspan="5">Path</td><td rowspan="5">Shop; Doctor; Guesthouse; Cave; Shelter; Tower; Water; Fireplace; Radio tower; Bridge</td><td rowspan="5">–</td><td rowspan="5">–</td><td rowspan="2">District name</td></tr>
<tr><td>1:10,000</td><td rowspan="4">Street name</td></tr>
<tr><td>1:8,000</td><td rowspan="3">Highway</td><td rowspan="3">Primary road</td><td rowspan="3">Secondary road</td><td rowspan="3">Restaurant; Sight, Monument</td><td rowspan="3">District name</td></tr>
<tr><td>1:5,000</td><td rowspan="2">Building</td></tr>
<tr><td>1:4,000 to 1:1,000</td><td>Pub; Cross; Bench;</td></tr>
</table>

Table 2 The extent of generalization on the maps designed for the three groups of participants. The numbers indicate the total numbers of items of the listed feature types on a given scale for the whole test area

	Beginner (number of displayed objects)				Intermediate (number of displayed objects)				Expert (number of displayed objects)			
Scale level	1: 75,000– 1: 35,000	1: 35,000– 1: 15,000	1: 15,000– 1: 5,000	1: 5,000– 1: 1,000	1: 75,000– 1: 15,000	1: 15,000– 1: 1,000	1: 35,000– 1: 15,000	1: 5,000– 1: 1,000	1: 75,000– 1: 15,000	1: 15,000– 1: 1,000	1: 35,000– 1: 15,000	1: 5,000– 1: 1,000
Fields	372		476		393		476		476		476	
Forest	243		277		250		277		277		277	
Buildings	–	165	1893	3570	–	165	4439	18945	–	165	14432	18945
Bridleway, track	502		3345		1091		3345		1981		3345	
Path	188		1182		351		1182		582		1182	

button. Road types varied in color and thickness on all maps. On smaller scale (<1:15,000) versions of the maps, the unpaved roads (bridleways and tracks) were generalized using the refinement method (McMaster and Shea 1992). The level-B map had the highest, and the level-E the lowest extent of generalization (Table 2).

The Q3 question tested the interpretation of land cover by selecting the correct answer out of four possibilities (farmland, suburban, urban, village). The map contained both man-made and natural objects, and buildings only appeared on large scales (>1:35,000). The beginners' map was the most, and the experts' map was the least generalized. On larger scales (>1:75,000), the beginners had detailed, and the intermediates had schematic tree symbols and green background on the forest polygons to facilitate the recognition (Fig. 1).

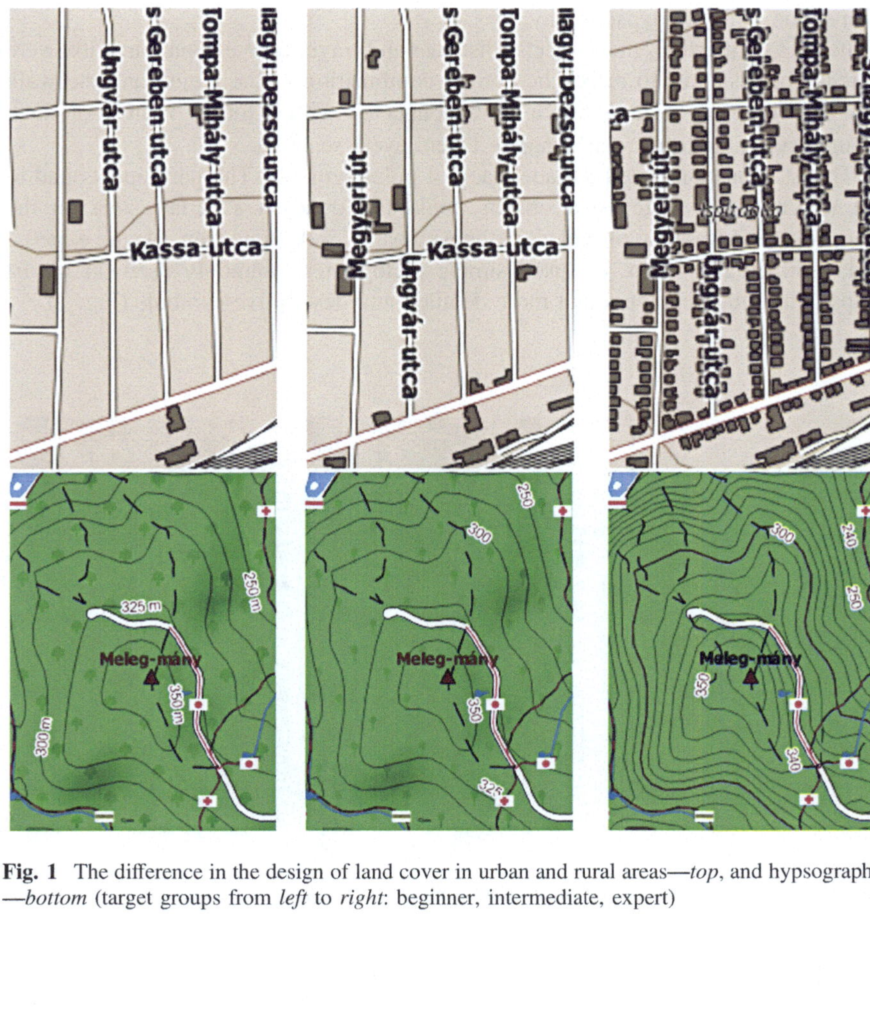

Fig. 1 The difference in the design of land cover in urban and rural areas—*top*, and hypsography —*bottom* (target groups from *left* to *right*: beginner, intermediate, expert)

The Q4 question tested the interpretation of hypsography interactively. Here the task was to click on the steepest slope of a four-sided hill. The judgement of landform steepness, however, is different from the usual contour reading tasks (Sholl and Egeth 1982), so the results were treated accordingly. The contour line intervals on the experts' map were 10 m, while on the level-I and B maps they were 25 m. The latter two also contained hill shading. In order to help the beginners' comprehension, we also added measurement units (m) to the values of the contour lines (Fig. 1).

The Q5 question measured the interpretation of geographic names. Subjects had to tell from four possibilities what village a specific road crosses. The road crossed all, but only one of them was a village. By clicking on them, the map zoomed to their locations. All types of settlement names (towns, villages, districts) had the same fonts, but different sizes and typefaces. The colors of hypsographic names were brown on the beginners' and intermediates' maps to emphasize the semiotic connection with the hypsography.

In the Q6 question, the subjects' distance and travel time estimation skills were tested. The task was to select the correct combination of the length and the walk time between two points, which were connected with a tourist route. All three groups were able to use the dynamic OpenLayers scale bar.

The Q7 question tested the adequacy of point symbols. The participants had to tell which of the four listed objects could be found in a certain area on the map. A legend was also provided on the web page for this question. On the experts' and intermediates' map, we used simple pictograms commonly used on tourist maps, while the beginners had more detailed and descriptive symbols (Fig. 2).

Fig. 2 The difference in the design of point symbols (from *left*: beginner, intermediate, expert)

6 Database—Sampling and Weighting

The data was collected between July and September 2016, and the survey was distributed online in Hungarian language. There were a total of 937 participants, but only 859 tests were evaluated. Discarded were those that were unfinished and those for which the testing time was less than 5 min in the first or second part of the test. In addition, results were also left out for those that did not give their precise educational qualification. The sample is representative with respect to gender and age categories of the 15–79-year-old Hungarian population. (A sample is said to be representative with respect to a variable if the relative distribution in the sample is equal to its relative distribution in the population.) Based on the first test's results, our sample consists of 257 beginner, 305 intermediate and 297 expert map users.

7 Proportion of Good Answers

The differences between map reading groups by questions show that the proportion of the good answers in the beginner group is significantly lower compared to the same score in the intermediate and expert group (except Q2, the interpretation of topographic objects). Regarding the intermediate map readers' performance in the case of Q1, Q3 and Q7, the proportion of their good answers is similar to that of the experts, while it falls behind in the other tasks (Table 3).

It was expected that the respondents' average percentage points would depend on the maps they received. In the case of the intermediates, this tendency was well represented on the average percentage point results. However, the map designed for the beginners worked oppositely to our preliminary expectations—it did not close

Table 3 Proportion of good answers by questions and map reading groups

	Map reading category		
	Beginner	Intermediate	Expert
(Q1) Orientation skills	2.9%$_a$	7.6%$_b$	9.8%$_b$
(Q2) Interpretation of topographic objects	45.7%$_a$	53.2%$_a$	72.2%$_b$
(Q3) Coverage	70.4%$_a$	79.8%$_b$	87.4%$_b$
(Q4) Interpretation of hypsography	67.4%$_a$	90.4%$_b$	98.4%$_c$
(Q5) Interpretation of geographic names	43.7%$_a$	63.9%$_b$	83.9%$_c$
(Q6) Distance and travel-time estimation	54.0%$_a$	74.0%$_b$	85.2%$_c$
(Q7) Interpretation of map symbols	34.3%$_a$	62.9%$_b$	70.0%$_b$

Note Values in the same row that don't share the same subscript are significantly different at $p < 0.05$ in the two-sided test of equality for column means (t-tests). The dark blue cells are significantly higher than the light blue ones in absolute value. Tests assume equal variances

Table 4 Proportion of good answers by question, map reading groups and maps used for the second test

	Beginner map readers			Intermediate map readers			Expert map readers		
	Type of maps used in the second test			Type of maps used in the second test			Type of maps used in the second test		
	Beginner	Intermediate	Expert	Beginner	Intermediate	Expert	Beginner	Intermediate	Expert
(Q1) Orientation skills	0.7%$_{a}$	8.5%$_{b}$	2.4%$_{a,b}$	5.6%$_{a}$	9.1%$_{a}$	7.4%$_{a}$	11.5%$_{a}$	8.0%$_{a}$	10.0%$_{a}$
(Q2) Interpretation of topographic objects	56.0%$_{a}$	45.9%$_{a}$	60.5%$_{a}$	55.2%$_{a}$	73.4%$_{a}$	56.3%$_{a}$	75.9%$_{a}$	86.6%$_{a}$	70.6%$_{a}$
(Q3) Coverage	74.7%$_{a}$	69.7%$_{a}$	79.6%$_{a}$	67.4%$_{a}$	79.3%$_{a,b}$	90.3%$_{b}$	86.4%$_{a}$	93.4%$_{a}$	84.7%$_{a}$
(Q4) Interpretation of hypsography	65.1%$_{a}$	64.8%$_{a}$	81.4%$_{b}$	86.1%$_{a}$	88.3%$_{a,b}$	97.2%$_{b}$	98.9%$_{a}$	97.9%$_{a}$	98.9%$_{a}$
(Q5) Interpretation of geographic names	43.0%$_{a}$	50.3%$_{a}$	51.2%$_{a}$	53.9%$_{a}$	75.8%$_{b}$	51.4%$_{a}$	76.7%$_{a}$	97.9%$_{b}$	80.2%$_{a}$
(Q6) Distance and travel-time estimation	60.7%$_{a}$	55.3%$_{a}$	51.5%$_{a}$	76.2%$_{a}$	76.5%$_{a}$	67.6%$_{a}$	91.6%$_{a}$	84.6%$_{a}$	82.1%$_{a}$
(Q7) Interpretation of map symbols	35.0%$_{a}$	39.0%$_{a}$	37.8%$_{a}$	55.3%$_{a}$	64.5%$_{a}$	72.2%$_{a}$	73.7%$_{a}$	68.2%$_{a}$	69.2%$_{a}$
Average points [max. 7]	3.27$_{a}$	3.36$_{a}$	3.54$_{a}$	3.89$_{a}$	4.58$_{b}$	4.31$_{a,b}$	5.04$_{a}$	5.26$_{a}$	5.00$_{a}$

Note for explanation of subscripts and colors, see Table 3

the gap between the map reading groups. For the better understanding, it is worth examining the second test's questions step-by-step with special attention to the beginners' and intermediates' performance (see Table 4).

In the case of Q1 (orientation skills), the beginners' average score was significantly higher when they used level-I maps compared to the one created for them. The result suggests that, for this kind of task, the very low density of the built-in area could negatively affect the respondents' results.

Regarding Q2 and Q3, there is no significant difference in the proportion of good answers, but remarkable differences can be found again in the interpretation of hypsography (Q4): those beginners who used level-E maps performed better than those who worked on level-B or I. This result could be explained with the denser contour intervals and the lack of hill shading on the experts' map.

Answers to questions Q5, Q6 and Q7 showed no significant difference, but there are observable tendencies towards one or another groups' map, which may be accounted for many parameters (such as the contour intervals, land cover, the unpaved roads' generalization, the coloring of the peak names and labels, etc.).

Intermediate map readers who worked on level-B or E maps generally performed worse than those who used their own level's map. The only significant exceptions were the Q3 (coverage) and Q4 (hypsography). In these cases, the proportion of good answers was 9–11% higher when the respondents used level-E maps, suggesting that the result was affected positively by the higher level of details on the experts' map.

8 Completion Time

The time interval between clicking on the "next" buttons on two subsequent web pages to proceed in the test was recorded and analyzed for every participant (both who answered correctly and incorrectly). Since the test was accessible from any Internet-capable devices, the response time could be affected by the speed of the connection. The observed tendencies remained generally the same, when only the correct answers were accounted. During the analysis, the focus was mainly on intragroup differences (Table 5). Based on the completion times, the Q1 proved to be the hardest of all. It was followed by the Q2 and the Q7.

Beginners, who got maps designed for their own level, spent more time with the test on average (15 min.) compared to those who used the intermediates' (12.14 min.) and experts' (12.33 min.) maps. A clear tendency shows that the beginners' map was not adequate for beginners in the sense that it was too simplified. This is revealed mainly in four types of tasks: (1) orientation skills, (2) interpretation of hypsography, (3) distance and travel-time estimation, (4) interpretation of map symbols. The positive exception was the Q5 task, where the beginners were faster when they worked on their level's maps.

The intermediate map readers were quickest with the maps designed for their level (14.67 min.), which is followed by the case when they received beginner (15.22 min.) and expert maps (16.53 min.). For them, only the Q4 task was problematic on their own map, which was solved significantly faster by those who received the expert maps.

Table 5 Average response time in minutes by question, map reading groups and maps used in the test

	Beginner map readers			Intermediate map readers			Expert map readers		
	Type of maps used in the second test			Type of maps used in the second test			Type of maps used in the second test		
	Beginner	Intermediate	Expert	Beginner	Intermediate	Expert	Beginner	Intermediate	Expert
(Q1) Orientation skills	5.53_{a}	3.90_{b}	$4.85_{a,b}$	5.08_{a}	5.84_{a}	8.05_{b}	5.10_{a}	5.11_{a}	6.12_{a}
(Q2) Interpretation of topographic objects	2.39_{a}	2.56_{a}	1.96_{a}	4.21_{a}	2.50_{b}	$2.62_{a,b}$	2.47_{a}	2.40_{a}	2.16_{a}
(Q3) Coverage	1.28_{a}	1.38_{a}	0.82_{a}	0.88_{a}	0.90_{a}	1.07_{a}	0.76_{a}	0.94_{a}	0.98_{a}
(Q4) Interpretation of hypsography	1.28_{a}	0.67_{b}	$0.79_{a,b}$	0.81_{a}	$0.70_{a,b}$	0.56_{b}	0.80_{a}	0.81_{a}	0.60_{a}
(Q5) Interpretation of geographic names	1.32_{a}	1.49_{a}	1.69_{a}	1.38_{a}	1.43_{a}	1.05_{a}	1.32_{a}	1.08_{a}	1.39_{a}
(Q6) Distance and travel-time estimation	1.20_{a}	0.86_{b}	$0.95_{a,b}$	1.07_{a}	1.53_{a}	1.15_{a}	1.16_{a}	1.21_{a}	3.24_{a}
(Q7) Interpretation of map symbols	2.00_{a}	1.27_{b}	1.27_{b}	1.79_{a}	1.78_{a}	2.04_{a}	1.37_{a}	1.44_{a}	1.42_{a}
Total [minutes]:	15.00_{a}	12.14_{b}	12.33_{b}	15.22_{a}	14.67_{a}	16.53_{a}	12.98_{a}	12.99_{a}	15.90_{a}

Note for explanation of subscripts and colors, see Table 3

The experts, although they could have answered the questions correctly on any maps, completed the test the slowest when they worked with level-E maps (15.9 min.); the use of other maps resulted in a 3-min shorter time on average. This tendency showed up in the cases of more than half of the questions, except Q2 (interpretation of topographic objects), and Q4 (interpretation of hypsography).

A slight correlation (0.1) was observed between the participants' age and their response time, which means that solving the tasks took slightly longer with an increase in age.

9 Map Scale

To find out the preferred scale levels of different map reading tasks, the website logged the maps' zoom level relative to the default value, after the participants left the web page. The default zoom level for each task is denoted by 0 by the OpenLayers plugin, while the level of zooming is represented with positive integers, and zooming out with negative integers. Each zoom level halves or doubles the map scale.

The general tendencies show that participants at the Q1 task mainly used larger scales, and experts used zooming more than beginners and intermediates. Smaller scale maps were used more frequently at the Q2, Q3 and Q7 questions, and the experts used smaller scales than the beginners (Table 6). In these cases, the scale level has a low inverse correlation (−0.3) with the correct answers, implying that more correct answers were given with smaller scales.

No strong relationship was found between the participants' age and the applied map scales, and the participants' gender only caused slight differences regarding the application of map scales with women respondents using zooming less frequently than men.

Table 6 Average zoom level by questions and map reading groups calculated from the integers recorded with the OpenLayers plugin. Positive numbers indicate zooming-in, while negative values indicate a tendency to zoom-out

	Map reading category		
	Beginner	Intermediate	Expert
(Q1) Orientation skills	0.047_a	0.029_a	0.336_b
(Q2) Interpretation of topographic objects	-1.236_a	-1.487_b	-1.727_c
(Q3) Coverage	-0.007_a	-0.049_a	-0.066_a
(Q4) Interpretation of hypsography	-0.006_a	0.008_a	0.111_a
(Q5) Interpretation of geographic names	-0.041_a	0.015_a	-0.064_a
(Q6) Distance and travel-time estimation	0.009_a	0.084_a	0.110_a
(Q7) Interpretation of map symbols	-0.310_a	-0.425_b	-0.554_b

Note for explanation of subscripts and colors, see Table 3

10 Summary

A previous study carried out by Albert et al. (2016) noted that group-specific mistakes are hard to be found among beginner map readers due to the diversity of people within the group. Since they have different strengths and weaknesses in cognitive skills, their mistakes vary. The present paper's results show that while it was achieved to design good maps for the intermediate map readers—in a way that their average performances approach the experts' results, in the beginners' case it is a more complicated task. In measuring the beginners' performance with different tasks, it is clear that their results are highly affected by the density of the land cover symbols and contour lines. Concerning these map data types, they usually performed better with maps that were designed for intermediates or experts. This suggests that the object density of the built area was unusually low in the beginner maps' case (over-generalized), and the use of smaller contour intervals (and the avoidance of the hill-shading) was welcomed by the beginners, and it affected the results positively. Although in all tasks, the tendency is clear that the proportion of good answers was higher if they used intermediate and expert maps, in the case of the geographic names, response times show that they still performed more quickly on maps designed for their own level.

From the aspect of response time, a clear tendency shows up that emphasizes the conclusions derived from the correct answers. Beginners answered the questions the slowest, and those intermediates the fastest who responded to maps designed for their own map reading level. Expert map readers solved the tasks more quickly if they used beginner or intermediate maps. A weak relationship between the respondents' age and the completion time was also observed.

Regarding the zoom levels used by the participants, there are differences between the map reading groups: the experts and the beginners used slightly smaller maps scales while intermediates used slightly larger map scales. Only small differences were observed by gender.

The study presented here can help future experiments and map applications in two ways: partly as a methodological paper and partly by providing data for dynamic online maps. The latter belongs to the wider discipline of adaptive map creation (Sarjakoski and Sarjakoski 2008), which emerged with the technology of online maps. Adaptive map design based on the results of our research is discussed from the aspects of utilizing purely open source data and tools (Szigeti et al. 2017). The primary aim of our research was to statistically measure the usability of the designed maps, and the settings of the experiment made it possible to achieve this aim. Although the effective map design was only partly achieved, the numerical results, categorized by the basic map data types are informative. Online map applications—aiming to produce different maps for different people—will have to define explicitly the level of generalization, and the optimal map keys for the individual users. The findings presented here may serve as a stepping stone for this process.

Acknowledgements The study was carried out by the Research Group for Experimental Cartography and was financed from the grant of the Council for Talent Management at Eötvös Loránd University for multidisciplinary research.

References

Albert, G. (2014). What does an archive map tell the contemporary map readers? In E. Livieratos & M. Pazarli (Eds.), *9th International Workshop on Digital Approaches to Cartographic Heritage* (pp. 288–298). Budapest, Hungary: International Cartographic Association.

Albert, G., Ilyés, V., Kis, D., Szigeti, Cs., & Várkonyi, D. (2016). Testing the map reading skills of university students. In T. Bandrova & M. Konecny (Eds.), *6th International Conference on Cartography and GIS* (pp. 188–199). Albena, Bulgaria: Bulgarian Cartographic Association.

Allen, G. L., Cowan, C. R. M., & Power, H. (2006). Acquiring information from simple weather maps: Influences of domain-specific knowledge and general visual–spatial abilities. *Learning and Individual Differences, 16*(4), 337–349.

Barkowsky, T., & Freksa, C. (1997). *Cognitive requirements on making and interpreting maps.* Paper presented at the International Conference COSIT, Pennsylvania, USA.

Bertin, J. (1967). *Sémiologie graphique: les diagrammes–les réseaux–les cartes*. Paris: Mouton.

Board, C. (1978). Map reading tasks appropriate in experimental studies in cartographic communication. *Cartographica: The International Journal for Geographic Information and Geovisualization, 15*(1), 1–12.

Buckley, A., Hurni, L., Kriz, K., Patterson, T., & Olsenholler, J. (2004). Cartography and visualization in mountain geomorphology. In M. P. Bishop & J. F. Shroder (Eds.), *Geographic information science and mountain geomorphology* (pp. 253–287). Chichester, UK: Springer, Praxis.

Bunch, R. L., & Lloyd, R. E. (2006). The cognitive load of geographic information. *The Professional Geographer, 58*(2), 209–220.

Clarke, D. (2003). Are you functionally map literate. Cartographic renaissance. In *Proceedings 21st International Cartographic Conference (ICC 2003)*, Durban, South Africa.

Deeb, R., Ooms, K., & De Maeyer, P. (2012). Typography in the eyes of Bertin, gender and expertise variation. *The Cartographic Journal, 49*(2), 176–185.

Easterby, R. S., & Hakiel, S. R. (1981). Field testing of consumer safety signs: The comprehension of pictorially presented messages. *Applied Ergonomics, 12*(9), 143–152.

Eley, M. G. (1992). Component processing skills in the interpretation of topographic maps. *Cartographica, 29*(1), 35–51.

Gerber, R. V. (1981). Competence and performance in cartographic language. *The Cartographic Journal, 18*(2), 104–111.

Harrower, M. (2007). The cognitive limits of animated maps. *Cartographica, 42*(4), 349–357.

Imhof, E. (1975). Positioning names on maps. *The American Cartographer, 2*(2), 128–144.

McMaster, R. B., & Shea, K. S. (1992). *Generalization in digital cartography*. Washington: Association of American Cartographers.

Montello, D. R. (2002). Cognitive map-design research in the twentieth century: Theoretical and empirical approaches. *Cartography and Geographic Information Science, 29*(3), 283–304.

Montello, D. R., Sullivan, C. N., & Pick, H. L. (1994). Recall memory for topographic maps and natural terrain: Effects of experience and task performance. *Cartographica, 31*(3), 18–36.

Muir, S. P. (1985). Understanding and improving students' map reading skills. *The Elementary School Journal, 86*(2), 206–216.

Murakoshi, S., & Higashi, H. (2015). Cognitive characteristics of navigational map use by mountaineers. *International Journal of Cartography, 1*(2), 210–231.

Ooms, K., De Maeyer, P., & Fack, V. (2014). Study of the attentive behavior of novice and expert map users using eye tracking. *Cartography and Geographic Information Science, 41*(1), 37–54.

Ooms, K., De Maeyer, P., Fack, V., Van Assche, E., & Witlox, F. (2012). Investigating the effectiveness of an efficient label placement method using eye movement data. *The Cartographic Journal, 49*(3), 234–246.

Petchenik, B. B. (1977). Cognition in cartography. *Cartographica, 14*(1), 117–128.

Potash, L. M., Farrel, J. P., & Jeffrey, T. E. (1978). An approach to assessment of relief formats for hardcopy topographic maps. *The Cartographic Journal, 15*(1), 28–35.

Sarjakoski, L. T., & Sarjakoski, T. (2008). User interfaces and adaptive maps. In *Encyclopedia of GIS* (pp. 1205–1212). Springer.

Sholl, M. J., & Egeth, H. E. (1982). Cognitive correlates of map-reading ability. *Intelligence, 6,* 215–230.

Szigeti, Cs., Albert, G., Ilyés, V., Kis, D., & Várkonyi, D. (2017). On the way to create individualized cartographic images for online maps using free and open source tools. In M. P. Peterson (Ed.), *Advances in Cartography and GIScience—Selections from the International Cartographic conference 2017*. New York: Springer.

Thompson, M. M. (1979). *Maps for America: Cartographic products of the U.S. Geological Survey and others*. Reston, VA, Washington: USGS.

Thorndyke, P. W., & Stasz, C. (1980). Individual-differences in procedures for knowledge acquisition from maps. *Cognitive Psychology, 12*(1), 137–175.

Usery, E. L., Finn, M. P., & Starbuck, M. (2009). Data layer integration for the national map of the United States. *Cartographic Perspectives, 62,* 28–41.

Wakabayashi, Y. (2013). Role of geographic knowledge and spatial abilities in map reading process: Implications for geospatial thinking. *Geographical Reports of Tokyo Metropolitan University, 48,* 37–48.

Interaction Problems Found Through Usability Testing on Interactive Maps

Elaine Gomes Vieira de Jesus, Patricia Lustosa Brito and Vivian de Oliveira Fernandes

Abstract Due to the technological changes beginning in the 1970s, significant advances were seen in the field of cartography, of which the move from analog maps to digital and online ones stands out. This context highlights the importance of obtaining greater efficacy in cartographic communication, permitting the interaction between the user and the content of the map. For interfaces to be developed in an effective way, different approaches must be taken into consideration, one of these is related to the concept of usability. The objective of this study is to analyze the usability of the interactive Brazilian maps Geobahia and Geopolis with relation to the use of the data and information they present. To achieve this objective, usability tests were carried out with the aim of evaluating the quality of use of the interfaces, measured by the ease with which the users concluded their proposed tasks. The tests showed certain difficulties confronted by these users in the realization of proposed tasks that indicate the possible presence of usability problems with these interfaces.

Keywords Interactive maps · Usability · Interfaces

1 Introduction

In the modern world, maps have progressively become a greater part of people's lives. However, it is clear that cartographic products are not always able to meet the user's needs. This study examines interactive online maps, computational environments on the web that permit the display and manipulation of geospatial data

E.G.V. de Jesus (✉) · P.L. Brito · V. de Oliveira Fernandes
Department of Transport Engineering and Geodesy, Salvador, Bahia, Brazil
e-mail: lanegeografia@hotmail.com

P.L. Brito
e-mail: britopatricia@hotmail.com

V. de Oliveira Fernandes
e-mail: vivian.fernandes@ufba.br

M.P. Peterson (ed.), *Advances in Cartography and GIScience*, Lecture Notes in Geoinformation and Cartography, DOI 10.1007/978-3-319-57336-6_18

and information through the computational interface (available set of computational tools) and map interface (mapped region).

In order for these interfaces to be developed in an effective manner, different perspectives should be taken into consideration. One of these is the concept of usability as defined by NBR 9241 in part 11 (Usability Orientation), a collection of technical Brazilian standards. Cybis (2003) points out that in this guideline that the usability of systems is defined as those characteristics that permit the user to achieve their objectives and satisfy their needs, within a certain context of utilization. Given this, performance and user satisfaction are specified and measured based on the level or achievement of the sought objectives in the interaction (efficacy), by the resources allocated to achieve these objectives (efficiency), and by the degree of acceptance of the product by the user (satisfaction).

Based on many studies (Nielsen 1993; Tóbon 2002; Adrienko et al. 2003; Falat 2007; Nivalla 2007; Mendonça 2009), it was observed that one of the most effective methods of usability evaluation is usability testing, carried out with system users. Falat (2007) and Delazari (2004) studies emphasizes that this testing makes possible the identification of critical situations, as well as the construction of proposals for improvements.

2 Theoretical References

2.1 Interactive Maps

Interactive maps are computational environments that, according to Maziero (2007), permit the propagation of cartographic information through the interactions of users with the interface. For the author, in these environments, the interactions of the user can be divided into two moments. One is the interaction with the computational interface, which is related to the set of available computational tools. The other refers to the interaction with the interface itself, which represents the mapped region.

Due to the enormous variety of themes, applications, and purposes of an interactive map, the functions it integrates can range from basic ones up to functions of spatial analysis, using the tools present in GIS applications, that constitute the *computational interface*. Understanding this, different authors Maziero (2007), Kramer (2007) and Mendonça (2009), have carried out studies with an emphasis on the functions that compose these environments.

According to Mendonça (2009), some tools are considered basic in an interactive map, such as those of navigation (*zoom in*, *out* and *pan*). Others, like distance measurement, overlapping features, selection of multiple features and queries, simple or by attribute, are relevant, but not so essential, although they make the completion of simplified spatial analyses possible.

The navigation tools (*zoom* and *pan*), layer selection, and acquisition of information and attributes of a specific feature were considered by Dent et al. (2009) to be common as much in interactive maps as in geographic information systems (GIS). These authors call attention to the fact that not all tools need be labelled or visible at all times. Because of this, it is necessary to be careful in the selection of the correct scale and the degree of generalization for the presentation of this information.

In a study developed by Kramer (2007), it was shown that users had greater difficulties in understanding what tools of interactivity were necessary and how they should work effectively. Interviews with focus groups showed that users considered the following tools as most needed in order of priority: zoom in and zoom out; map printing; legend display; moving the map (pan); selection of a specific characteristic in the map and obtaining information about it.

2.2 Usability and Evaluation of Interfaces

Usability is a term that can be applied to the use of web interfaces. It is possible to measure it by the efficient and effective completion of tasks by the user, by the minimum quantity of errors, and by user satisfaction. As Wincler and Pimenta (2002) note, it is necessary to consider issues of usability for the development of web interfaces in order to minimize problems encountered during the interaction with the user.

An interface can be evaluated with methods and techniques with or without the presence of a user. Among the methods and techniques without the presence of a user that are more frequently used in scientific works, inspections by guidelines and checklists, heuristic evaluations, and ergonomic criteria stand out. The most well-known method of evaluation is usability testing, which has the objective of identifying critical points in the interaction of the user with the system that has problems of usability. Techniques that can be utilized in the carrying out of testing are: asking the user to verbalize their actions in the method known as "think aloud," measurement by performance, application of questionnaires and interviews, and the use of surveillance systems that record the commands of the user during the interactions. Using these methods, problems of usability are identified when users are observed to have difficulties in concluding their tasks.

According to Wincler and Pimenta (2002), the observation of the evaluator during testing can occur directly, such as in watching users complete tasks in the same room or through one-way mirrors that do not allow the user to see the evaluator. The information can be registered through the evaluator's notes or through cameras or audio recorders. These records permit the evaluator to identify and discuss the usability problems found in the interaction. Two cameras are suggested, one to record the reactions of the user and the other to register the commands given by the user on the computer screen.

Some of the techniques that can be used to perform these tests are: stimulating the user to verbalize their actions, known as Think Aloud (as used by Andrienko et al. 2003), performance measurement, application of questionnaires and interviews, and the use of spy systems, which record user commands during interactions.

3 Methodology

3.1 Participants

The usability tests were carried out in December 2014 and January 2015 with 24 users of the Federal University of Bahia, including: six higher education teachers, six master's degree students (three of whom were from the graduate program in Geography and three from the postgraduate program in Urban Environmental Engineering), six freshman undergraduate students (three from the Geography and three from the Surveying and Cartographic Engineering program) and six seniors undergraduate students (three in Geography and three from Surveying and Cartographic Engineering).

3.2 Stimuli and Apparatus

The following were used for the completion of the testing: one laptop; a testing form created in Google docs; the laptop's video recorder; the "screenshot captor" application, which recorded the screen during the interactions of the user; and the interactive maps Geobahia (Fig. 1) and Geopolis (Fig. 2).

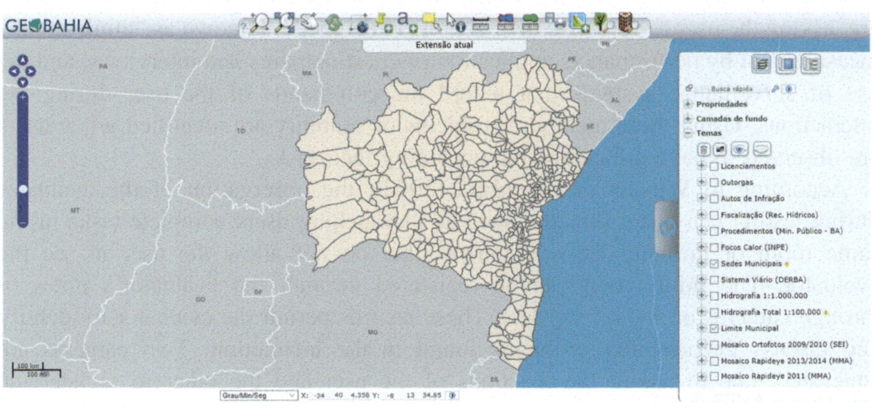

Fig. 1 Geobahia interface

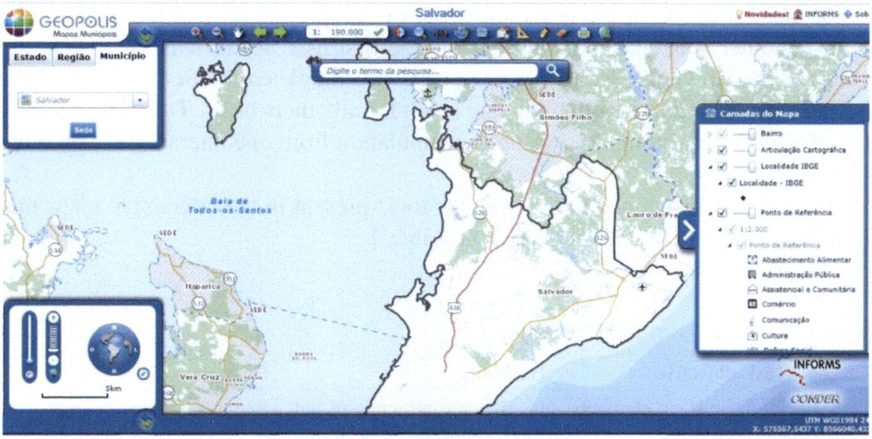

Fig. 2 Geopolis interface

The interactive map Geobahia is a product of the Institute of the Environment and Water Resources (INEMA), encompasses the whole state of Bahia, and presents data and information connected to the environmental field.

The institution responsible for the interactive map Geopolis is the Company of Urban Development of the State of Bahia (CONDER), which covers the municipality of Salvador and the Metropolitan Region and includes basic information about these areas as well as data from the 2010 Census.

Despite the fact that the interactive maps analyzed in the usability test cover different areas and themes, the tools and the method for completing tasks is very similar in these interfaces, making only slight adaptations in the tasks necessary.

3.3 *Procedures*

The test was created in an online form in Google docs, and the link was sent to the participants and responded to during the test, in the presence of the evaluator. This form contained brief information about the interactive maps (responsible organization, theme, and objectives for the Geobahia and Geopolis maps). The participants were told that the test was aimed at evaluating the interface and not the user and that they should, as they completed the tasks, say what they were thinking and what they intended to do (think aloud). It is worth highlighting that the evaluator stayed with the user during the whole test (with the intention of motivating them) but did not indicate the steps that should be followed. The user had to find ways to complete the tasks without the evaluator's help. Three tests lasted on average an hour and a half for each individual user.

For the realization of the tasks, the order of the interactive map tested from one user to another was changed. Thus, while the first user began with Geobahia, the second user with Geopolis, and so on. The users completed the first task with both and then the second task with both, and so on for all others tasks. This methodology was adopted to not permit knowledge accumulation from one interactive map to the other map.

The usability test aimed to evaluate the tools present in the interactive maps that were grouped in tasks and described in Table 1.

Table 1 Tasks proposed for the usability test

Tasks	Task objective	Geobahia	Geopolis
Task 1	Find the basic elements and utilize the navigation tools	Identify the presence of the basic elements. Indicate the scale in which the original map is represented. Utilizing the *zoom in, out,* and *pan* tools, identify the site of the municipality of Salvador and indicate what scale you are observing	Same as Geobahia
Task 2	Use the distance measurement tool	An individual is located in Salvador and needs to buy equipment that is sold in three municipalities (São Francisco do Conde, Salinas das Margaridas, and Dias D'Ávila). What municipality would be closest to the individual?	An individual is located in the neighborhood Barra in Salvador and wants to go to a green space. The options are the Zoological Garden and the City Park. Which of the two would be closer to the individual?
Task 3	Select layers/themes, find the toponymy, and utilize the feature identification tool	Enable the hydrographic layer 1:1,000,000, enable the label, and ask for information about the Capivara River that passes by the municipality of Camaçari. Are the names of the rivers easy to identify?	Enable the neighborhood layer, ask for information about the neighborhood Ondina in Salvador and find the labels. Are the labels easy to view? What is the resident population and demographic density of Ondina?
Task 4	Select layers/themes, represent the symbology, and utilize the data output tools	Enable the road system layer (DERBA), view the map and the legend and save in PDF. What did you think of the symbology of the map? Is it possible to interpret the information using the legend?	Produce a thematic map with the total population—resident population by neighborhood, view the map and the legend, and save in PDF. What did you think of the symbology of the map? Is it possible to interpret the information using the legend?

(continued)

Table 1 (continued)

Tasks	Task objective	Geobahia	Geopolis
Task 5	Use the query by attribute tool	Complete an attribute query with the theme of municipal limits, and identify the municipalities that belong to Salvador's greater metropolitan area. What are the municipalities?	Complete an attribute query on the reference points layer and theme of education. Identify the Centro Mutliplo Oscar Cordeiro and view it on the map. What is the address and neighborhood?

Source de Jesus (2015)

4 Results and Discussion

The results are evaluated and discussed according to the group of tasks presented in Table 1.

Task 1—Basic Elements and Tools of Navigation

From the usability test, it was noted that in task 1 (Table 1), on Geobahia, the users were able to find the majority of the basic elements with the exception of the legend, which represented 68% of the difficulties encountered. The next most frequent problem, finding the indication of north, represented 9% of all difficulties. On Geopolis, the greatest difficulty was also related to the legend, representing more than 50% of the difficulties. This is due to the fact that few users noticed that there was no icon with a legend and that they needed to open the layers to view the representation.

Next, finding the title represented nearly 25% of the difficulties, since many users did not notice that changing between the municipal, regional, and state options would change the title automatically. Again, the north arrow presented an obstacle, this time with 7% of all difficulties. There were also problems in mastering the navigation tools *zoom in* and *out* in the toolbar, which represented 3% (Fig. 3).

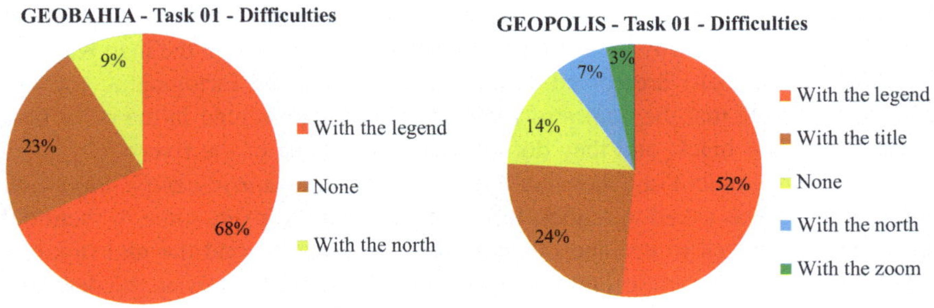

Fig. 3 Difficulties with Task 01 on the interactive maps. (*Source* de Jesus 2015)

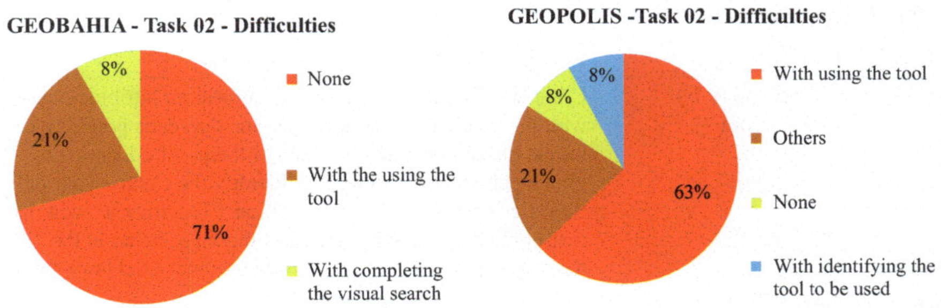

Fig. 4 Difficulties with Task 02 on the interactive maps. (*Source* de Jesus 2015)

Task 2—Distance Measurement Tool

The usability tests show that users did not have many problems with completing task 2 with the Geobahia map (Table 1). The most frequently cited difficulty was mainly the lack of practice in manipulating the tool, which represents more than 20%. The second difficulty was finding the municipalities (8%). This difficulty came up primarily because the search tool present on the interactive map was not working, so the users have to search for the municipalities visually.

With Geopolis, the users had a great difficulty completing this task. The greatest difficulty encountered was with manipulating the tool, both in understanding how to end the measurement, which is ended with a double-click, and in identifying the location that would do the measurement utilizing the tools of *pan* and distance simultaneously, totaling 63%. They also had difficulty in identifying the option to be utilized (freehand distance or straight line, which represented 8%) expressed in Fig. 4. Other reasons were the visual search of locations for measurement (16%) and the fact of not knowing how to erase one measurement to complete another (5%).

Task 3—List of Layers, Toponymy, and Feature Identification Tool

Essentially two problems were identified with the toponymy on Geobahia. The first has to do with the way in which the options are described on the chart of attributes that can be shown on the map. No editing of these labels was done, making understanding more difficult for the user who has no way of knowing what some of those texts mean. The way in which these labels can be shown on the map was also a problem, since many are copied. Labels do not follow the cartographic conventions that indicate that the names of rivers should be represented in the color blue and not the color black, and they do not follow the shape of the rivers.

The second problem on Geobahia has to do with the *feature identification tool.* When the user was asked to request information about the river, some of them did not know which tool to use and others took a long time to find the tool (totaling 24%). It was also difficult to find the layers (21%), in order to activate the desirable one, since the list of layers is located on a hidden flap on the right side (Fig. 5).

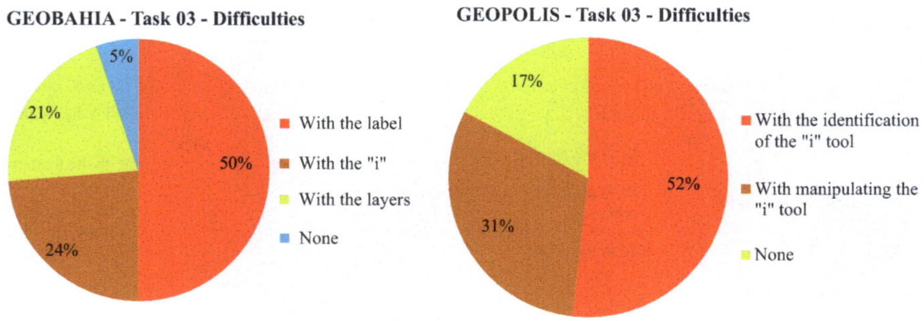

Fig. 5 Difficulties with Task 3 on the interactive maps. (*Source* de Jesus 2015)

There is no location for enabling the toponymy in Geopolis. It is activated in the same way as the layers, by changing the scale. However, it was identified that on the smaller scales, unnecessary information appears in detriment of other information. For example, the label of some avenues in Salvador appear, but the names of neighborhoods do not.

For task 3 in Geopolis, the main problems encountered were understanding which tool should be used, including problems in finding the tool, totaling 52%, and the next most frequent problem was perceiving how the tool functioned (with options of different layers) and mastering it, due mainly to the response time (representing 31%), as shown in Fig. 5. Two factors contribute to this difficulty. The feature identification tool is located on the toolbar, dividing an icon with the buffer option. This way of grouping the tool creates some difficulty in finding it, given that, the "i" is not very visually perceptible. The other factor is related to the excess of steps between clicking and obtaining the desired information.

Task 4—Layers, Symbology, and Tools for Data Output

On Geobahia, the difficulties encountered by the users for task 4 (Table 1) were: finding the legend (28%), finding the save in the .pdf option (24%), and getting access to the layers (7%), as shown in Graphic 4. These difficulties are due the fact that to access the legend they need first to access the list of layers on the tab on the right side of the screen, enable the road system layer—DERBA, and use the icon marked legend on the upper part of the tab or on the layer properties on the legend option (symbology).

The usability test revealed that on Geopolis, task 4, the greatest difficulty involved finding the legend on the thematic map (41%), encountered by 17 users; next were observed problems related to the production of the thematic map, encountered by 12 users, and representing 32% of difficulties, being difficulties as much with finding a thematic option as with manipulating it (since the tab cannot be closed). If this occurs, the map loses the classification of the data. Those problems occur on Geopolis, because the main layers are enabled to be viewed in a certain scale automatically. In the choosing of layers, priority was given to enabling in a

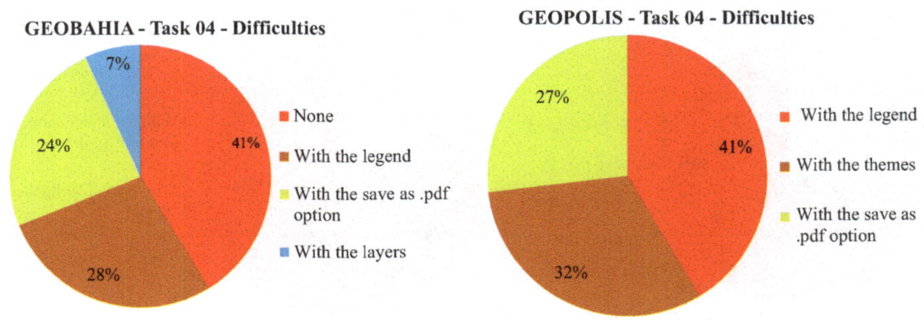

Fig. 6 Difficulties with Task 04 on the interactive maps. (*Source* de Jesus 2015)

conventional way. For this, the user was asked to find one of the thematic layers, which, in this interactive map, are located on the toolbar, on the option "show themes." The user was to enable the layer "Demographic Census 2010—resident population by neighborhood," located on the show themes icon, to view the symbology of the map in comparison with the legend. After selecting "resident population by neighborhood," the user needed to choose the option "total population."

And finally, saving the map as a .pdf was also a difficulty (27%) because the users expected to find an option "save map as .pdf" separate from the printing option, and few had knowledge of the option to save by selecting .pdf as a printer, as shown in Fig. 6.

Task 5—Query by Attributes

On Geobahia, it was observed that the users completed task 5 in two ways (as much by the attributes chart as by the tool itself). However, there were problems with both options, since many users knew neither how to complete a query by attributes nor which tool to utilize.

The query, through the tool, presented itself as a more practical and fast option, but the users had great difficulties with this process, mainly in identifying the tool to be utilized (33%), which is inserted in the option selection. There were also problems with the manipulation of the tool, especially in the selection of the expression and viewing of the search result (totaling 39%). The use of an expression to find the municipality was also a problem because the name of the territory could be typed. An extremely small check box must be selected for the territory options to appear.

When the query was completed by attributes chart, the difficulties consisted in finding the municipalities one by one, viewing all the pages (20%), since they did not place the information in order, a fact which prevented four users from finding some municipalities, as shown in Fig. 7.

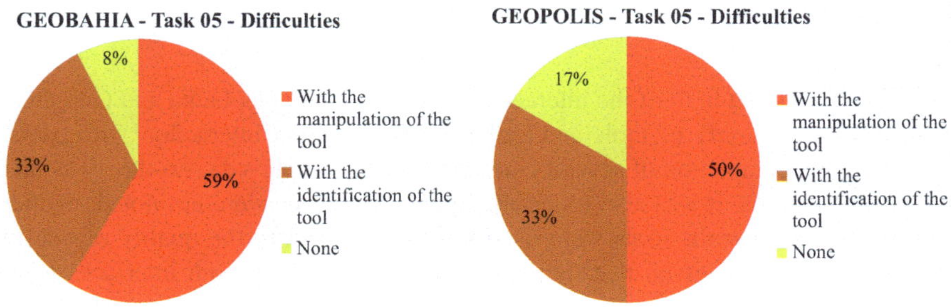

Fig. 7 Difficulties with Task 05 on the interactive maps. (*Source* de Jesus 2015)

On Geopolis, practically all users performed the query by attributes through the tool, and even those who tried to perform by the layer "reference points'' ended up giving up because they did not know the scale of representation.

Therefore, the main difficulties faced by the users in this task were in relation to the identification of the tool (because they did not know which tool they would use and took time to find it), representing 33% of the difficulties and difficulties in manipulating the tool, involving problems to assemble the expression and to identify the page that was the requested site, with 50%, as shown in Fig. 7.

These problems occurred because all of the points of reference with the theme of education in Salvador appear on the user request, so to find the Centro Múltiplo Oscar Cordeiro it is necessary to search the pages on the lower part of the window. After finding it, clicking on the magnifier brings the location closer. The interactive map goes directly to the location, but if there are several nearby points, the user must pass the cursor of the mouse over them to show their names.

The analyses made it possible to verify that great differences were neither perceived with regard to the utilization of the tools nor with relation to the quality of one interactive map as compared with the other, since the number of difficulties in both interfaces was similar. In tasks 1, 2, and 4, the users encountered more difficulties with Geopolis, and on tasks 3 and 5 they encountered more difficulties with Geobahia, as shown in Table 2.

Table 2 Number of users who presented some difficulty in completing the tasks

Tasks	Difficulties with Geobahia	Difficulties with Geopolis
1	19	20
2	19	21
3	22	19
4	12	24
5	21	19

Source de Jesus (2015)

5 Conclusions

It was observed that both of the interactive maps analyzed, Geobahia and Geopolis, present a great variety of tools and make possible a good interaction with users, permitting the completion of several common tasks in people's day-to-day lives that use geospatial data. It was also found that great differences were not noted, neither with respect to how the tools were used nor with respect to the quality of one in relation to the other, given that the number of difficulties on both interfaces were similar.

Based on the proposal to evaluate the aspects of the interface and interactivity, it became clear that the greatest usability problems of these interactive maps correspond with the difficulties encountered by the users. Given this, three major usability problems were identified. The first is with regards to the legend, which integrates the basic elements, and the second is related to the query by attributes, problems occurring with both interactive maps, and the third is related to the data output to create the map in .pdf, a problem occurring only on Geopolis. Other tools that presented a great number of difficulties on Geobahia were the toponymy (labels) and the final layout, and on Geopolis: the tool for layer identification and the final layout.

Therefore, it is recommended that on Geobahia: (i) The legend icon may be accompanied by the name "legend" or directly display the representations, as layers are enabled/disabled; (ii) The query by attribute tool, once performed by the table, could be able to sort the information that the user wants to see, allowing the user to see fewer municipalities to check which territory it belongs to. If it is performed by the tool query by attribute, it should separate the option selection from the option attributes. It would also be interesting to have a window indicating what the user should do when the tool is enabled, to explain how to assemble the expression; (iii) the toponymy could be activated by the option "enable toponymy" next to the layer name, or shortly after opening it. As for the label that will be enabled, only names that mean something concrete for the user, such as "feature name," should be listed.

On Geopolis it is recommended that: (i) The user could have access to the maps representations by clicking on an icon for the legend or on the name "legend"; (ii) In the data output, it is essential that the interactive map presents the option of saving in .pdf, as well as printing (as in Geobahia), because the user may not have a .pdf printer installed on the computer; (iii) The query-by-attribute tool could allow to search other columns other than field and theme, such as neighborhood or other attributes, as this would reduce the size of the list that the user has to look for, even though it is in alphabetical order. And, it should also identify the requested feature directly, emphasizing it in relation to the others, this way it would not be necessary for the user to move the mouse over the feature to see the result; (iv) The identification tool must also be redesigned, in order to take less time to respond to the user's command; the text "click on the map to identify points" could be removed, because it also identifies lines and polygons leading de user to doubts about the tool functions. Finally, the result of the identification tool could show all layer options, so the user would not have to click the arrow to see them.

Adaptations on Geobahia and Geopolis map layouts interface are also recommended so that the final map presents all the basic elements (title, scale, legend, north, coordinates and reference map), and not just the title and the legend. It is also important that these elements are editable.

It is known that it is a great challenge to develop maps that achieve a perfect cartographic communication between the interface and the user. For that, it is necessary to know needs and difficulties encountered by the target public of an interface. However, this study represents one more advance in the analysis of usability of interactive maps in the access of geospatial data and information, since it permits the identification of serious usability problems found on the interfaces that can be adapted to better meet user needs.

References

Andrienko, G., Andrienko, N., & Voss, H. (2003). GIS for everyone: The CommonGIS project and beyond. In M. P. Peterson (Ed.), *Maps and the internet* (pp. 131–146). Elsevier.

Cybis, W. A. (2003). Engenharia de Usabilidade: uma abordagem ergonômica. Apostila LabIUtil, Universidade Federal de Santa Catarina. Retrieved March 23, 2016.

Companhia de desenvolvimento urbano do Estado da Bahia (CONDER). Geopolis. Retrieved February 12, 2016.

Delazari, L. S. (2004). *Modelagem e implementação de um Atlas Eletrônico Interativo utilizando métodos de visualização cartográfica*. USP: São Paulo. Tese doutorado.

Dent, B. D., Torguson, J. S., & Hodler, T. W. (2009). *Cartography: Thematic map design* (6th ed.). McGrawHill Higher Education, New York.

Falat, D. R. (2007). *Avaliação de Mapas na Web: Questões relativas à interface e à interatividade*. Curso de Pós-graduação em Ciências Geodésicas, UFPR: Dissertação de Mestrado.

Instituto do meio ambiente e recursos hídricos (INEMA). Geobahia. Retrieved November 07, 2016.

de Jesus, E. G. V. (2015). Análise de Usabilidade de Mapas Interativos nas Atividades de Ensino Superior. Dissertação de Mestrado – Mestrado em Engenharia Ambiental Urbana, Universidade Federal da Bahia, Salvador.

Kramer, R. E. (2007). Interaction with maps on the internet—It's all about the user. In: *Proceedings of the ICA Commission on Maps and the Internet*, Warsaw, Poland.

Maziero, L. T. P. (2007). *Influência dos aspectos das interfaces na comunicação dos mapas interativos e a proposição de diretrizes para o design dessas interfaces*. Tese de Doutorado - Curso de Pós-graduação em Ciências Geodésicas: Universidade Federal do Paraná, Curitiba.

Mendonça, A. L. A. (2009). Avaliação de interfaces para mapas funcionais na web. 2009. 189f. Dissertação de Mestrado em Ciências Geodésicas - Departamento de Geomática, Universidade Federal do Paraná, Curitiba.

Nielsen, J. (1993). *Usability engineering*. EUA: Ap Professional.

Nivalla, A.-M. (2007). Usability perspectives for the design of interactive maps. Tese de Doutorado, Department of Computer Science and Engineering, Helsinki University of Technology, Finlândia.

Tobón, C. (2002). Usability testing for improving interactive geovisualization techniques. *Centre for Advanced Spatial Analysis*. Working Paper Series.

Winckler, M., & Pimenta, M. S. (2002). Avaliação de usabilidade de sites web. Retrieved September 20, 2016, from http://pt.scribd.com/doc/73087845/2002-Winckler-Pimenta-Eri-2002-Cap3.

The Apprehension of Overlaid Information in a Web Map

Dzenan Dumpor and Terje Midtbø

Abstract Map mashups are now common when presenting information in digital media. This paper explores the relation between the background information (basemap) and the information that is displayed over the basemap. Will the user be disturbed by the background when solving tasks related to the overlaid information? To answer this and other questions, a web-based experiment was prepared. The participants in the experiment were introduced to different tasks based on combinations of data primitives in the top layer. Three different basemaps were tested. In a focused part of the experiment, tasks related to the combination of area and points in the top layer were explored more closely. The experiment indicated that the tasks related to different coloured areas in combination with points of different colours were more time consuming to solve on a topographic map with many details.

Keywords Web maps · Basemap · Web experiment

1 Introduction

Detailed maps at multiple scales are available today through the World Wide Web (WWW) and mobile devices. The barrier of entry for creating and using map applications has been lowered by the introduction of services such as Google Maps, OpenStreetMap and Bing Maps. Whether a user is interested in sharing geographic information with others, or finding the shortest route from A to B, an easy accessible map application can be used to help them achieve their goal.

D. Dumpor · T. Midtbø (✉)
Norwegian University of Science and Technology, Trondheim, Norway
e-mail: terjem@ntnu.no

D. Dumpor
e-mail: Dzenan.Dumpor@geodata.no

M.P. Peterson (ed.), *Advances in Cartography and GIScience*, Lecture Notes in Geoinformation and Cartography, DOI 10.1007/978-3-319-57336-6_19

A map mashup, the most popular mashup on web (Fichter 2016), imparts thematic information on top of a background layer. The background layer, or basemap, acts as a reference layer for the thematic information. Basemaps contain a varying degree of information, depending on the degree of generalization applied at the actual zoom level. Although a basemap is a vital part of any map mashup, relatively little research has been done in examining how basemaps compare to each other.

In this paper, three basemaps used for web maps by ESRI are explored. A web application has been created to test the performance of each basemap. Conducting a web experiment introduces several challenges. It is necessary to develop a sophisticated web-application including a high degree of usability.

2 Background and Related Work

Mashup is a term that originated in the music industry (Roth and Ross 2009). A mashup results when mixing a vocal track over an instrumental tune. Likewise, a map mashup emerges when two different map layers are melted together. Usually this comes into being when a layer of a particular theme is visualized over a basemap.

There are some challenges in using map mashups. The basemap normally supports zooming and panning. In a simple map mashup, the overlaid symbols then tend to clutter, in particular when the map is zoomed out to get an overview. This is a classic cartographic generalization problem and it has been explored by many authors (Neun et al. 2008; Su et al. 1997; Ware and Jones n.d.; Tauscher and Neumann 2016; Harrie 1999).

A map becomes less legible when it is overloaded with details. Stigmar and Harrie (2011) introduced a set of measures to make it possible to give numerical values to different grades of legibility. This is tested by introducing different areas in a map in different levels of detail (LOD). Bereuter and Weibel (2013) and Korpi and Ahonen-Rainio (2013) discuss how to reduce the cluttering problem in digital media. When imposing a new set of symbols on the top of a basemap, it is desirable that the symbols appear in balance with the background. Since the imposed symbols, in this case, represents the "main theme" of the map, they need to show some prominence. To make this possible, without making a glaring presentation, one needs to involve less emphasized basemaps.

The most common way to make a map mashup is to include the thematic data layer through an Application Programming Interface (API) provided by web services for general maps. Peterson (2009) points out that more than 25 different providers of web services have APIs for background maps available. Many of these offer various basemaps, based on different styles and ready to be used as backgrounds. In Google Maps, for example, it is most common to select between the standard Google Map and a satellite photo. In contrast to most other online map providers, Google provides the option to style the map by, for example, changing to more muted colors.

There might be different criteria for selecting basemaps. Nivala and Sarjakoski (2007) introduce adaptive maps for mobile units where both the base map and the icon layer are adapted to the users' situation and interests. Different map colours are used to illustrate the variation in seasons of the year. Though they are investigating the users experience with different backgrounds, they are not doing objective usability tests for how different backgrounds influence the legibility of the map.

Relatively little research has been done when it comes to how data layers are influenced by a change of the basemap in a web/mobile application. There are, however, many works that are related to the problem. Gibin et al. (2008) uses Google Maps as basemap for their "London profiler" web application. Their idea is to put transparent choropleth maps on top of the basemap from Google. They claim that Google Maps "is ideal for this use" without any further evidence. In their application, the influence of the basemap is adjusted by using different degrees of opacity on the top layer. Dong et al. (2014) use eye tracking to explore how people identify areas of interest (AOI) on imagery basemaps. Their study compares the use of an unmodified orthographic image to a digitally enhanced image. They conclude that the image enhancements improved both the effectiveness and efficiency in identifying areas of interests. Most map mashups are based on static information on top of a static basemap. Lai and Yeh (2004) studied how dynamic symbols work on a static basemap. They concluded that high amount of contrast between the base-map and symbols for the foreground data was more important when both of them were static. By using dynamic (blinking) symbols there were improved visibility even with lower contrast.

Contrast between the overlaid symbols and the basemap is a key factor for map mashups, and Chesneau (2011) establishes the importance of contrast in colours between symbols in the map and the background. Colour contrasts are used to enable the map viewer to distinguish features from each other and from the background of a map (Peterson 2009). A study where participants were looking for a target among many distractors displayed on a map, found time spent on the task decreased with increased contrast between the colour of the target and the colours of the distractors. Contrast among distractor colours did not significantly influence time spent on the tasks (Lloyd 2005). By utilizing the ability of contrast to distinguish features from one another, we can ensure that our most important data is brought to a viewer's attention by providing adequate contrast in colour choices, independent of a dark or light background (Peterson 2009).

Whether the background is dark or light does not matter as much as providing adequate contrast. Evolutionary, humans' ability to see contrast is much more important than the ability to judge darkness. Darkness of an object depends on its surroundings and the context (day/night etc.) (MacEachren and Mistrick 1992). Eyes can adapt to light or dark surroundings, and still be able to see contours because of adequate contrast between objects. A well-known rule, which is related to contrast, is 'five-shades-are-enough'. This is often used on choropleth maps and states the human eye can only match five shades of the same hue (Peterson 2009).

3 Web-Based Experiment

The focus of this paper is to explore how different basemaps influence legibility of the "main theme" in a map mashup. A basemap prepared for good legibility will support the data layers presented on top, rather than taking attention away from them. The paper aims to answer the following questions:

1. How do basemaps influence how fast participants solve tasks related to the overlaid data?
2. How will the participants experience in using maps impact their performance?
3. Are there any differences between genders when solving the same tasks?

To answer these questions, a web-based experiment was prepared. A web-based experiment may have more inconsistent results than an experiment in controlled environments. However, the ability to reach a large number of participant in their natural environment is a considerable benefit (Midtbø and Nordvik 2007). The validity of web-based experiments is confirmed by Krantz and Dalal (2000) where matching results were found between web-based and controlled experiments. For the experiment, ASP.NET was selected as the back-end tool, and Dojo Toolkit was used on the front-end.

The experiment itself was tailored to show how background maps influence the apprehension of an additional data layer. The idea was to use the basemaps as distractors, while the participants focus should be on the foreground information. Three basemaps created by ESRI (ESRI 2016) were selected for the experiment. These are shown in Fig. 1 and consist of:

- **Topographic Map** is based on light colours and includes features such as cities, water features, physiographic features, roads and railways, airports, landmarks and shaded relief imagery for added context.
- **The World Imagery** basemap consists of satellite images.

Fig. 1 The same map extent shown with the three basemaps: topographic, world imagery and *dark grey* canvas

- **The Dark Grey Canvas** is created with a dark and neutral background providing only key information. The use of light or prominent colours on a dark background may also establish a visual hierarchy where the users' data is put in the foreground.

Two locations in Mexico, Tijuana and Colima, were selected as areas for the experiment. The top data layers were generated from public shapefiles supplied by the National Institute of Statistics and Geography in Mexico, containing point features of medical-centres, plazas, markets and schools. Since the participants in the experiment were expected to be located in Norway, the distant locations would help to avoid bias caused by local knowledge.

Three types of tasks were created. These are differentiated by the combination of geometric primitives. The questions in the experiment were based on the comparisons of point/point, point/line and point/area (Fig. 2).

A qualitative colour scheme was used to separate different features in the map. The colours were defined by ColorBrewer (2016) (Fig. 3). These colours seemed to work well across all the selected basemaps.

The web application created for the experiment consists of two main components, a web map embedded on the left side of the screen, and a toolbar on the right containing all the information the participants need to know to solve the tasks (Fig. 4).

The toolbar contains the *question, three alternative answers, next button, button to re-establish the start view of the map, legend and progress bar*. The questions were evenly distributed between the three geometrical combinations and had a structure like these examples:

Point/point: Which point is closest to a school? Which point is closest to a medical-centre? Which point is in total closest to both a school and a medical-centre?

Fig. 2 Combination of different geometric entities: point/point, point/line and point/area

Fig. 3 The selected colours for the data layers

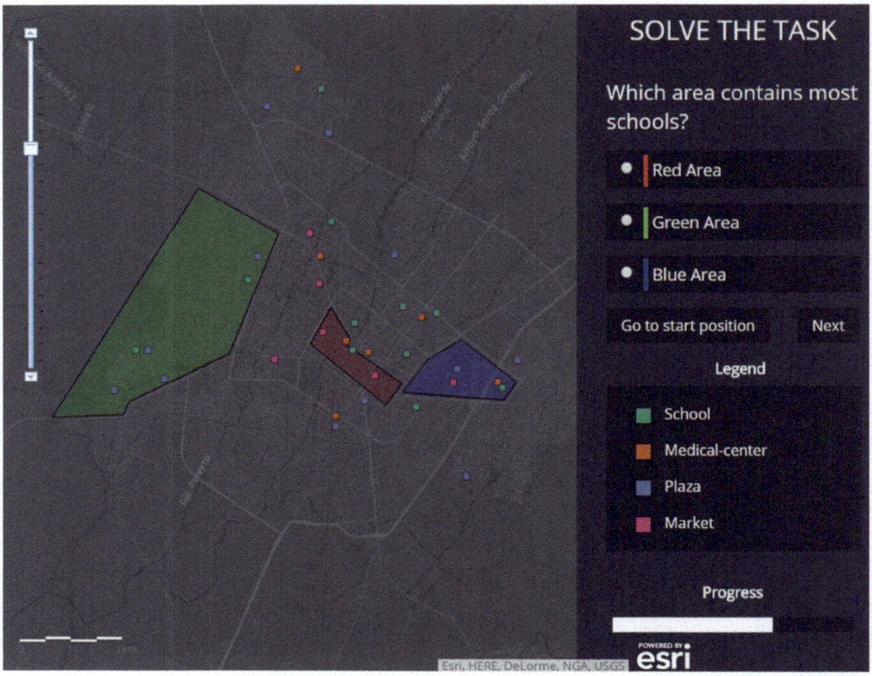

Fig. 4 The applications as seen by the participants in the experiment

Point/line: Which path starts at a school and crosses the river? Which path goes from a plaza to a market? Which path passes the largest number of markets on its way?
Point/area: Which area contains six schools and only one medical-centre? Which area contains most markets? Which area does not contain a plaza?

Figure 4 shows the final application for the experiment after a pilot study was accomplished. The pilot study involved seven participants who gave valuable feedback that helped to make the experiment more applicable.

The procedure for the final experiment went like this:

1. A welcome page containing information about the experiment.
2. A tutorial where the participants are exposed to tasks, one for each geometrical combination.
3. The participants complete the experiment based on three basemaps for each geometrical combination.
4. The participants answer general questions.

The participants' performance was timed during the experiment. Consequently, it was important to familiarize the participants with the task through a tutorial.

Since we were testing the influence of the basemaps, the order of the basemaps was regulated to avoid bias. As there are three basemaps in the experiment, six sequences of the basemaps are possible. The used sequences were logged, and every time a participant started the experiment, he/she was exposed to the current least used combination.

During the experiment, we recorded *time elapsed on task, all map operations (e.g., zoom/pan), the basemap used for the task and if the answer to the task was correct or not.* In addition, general information about each participant was requested. This included *sex, age, the used device (tablet or computer), experienced/non-experienced map users* and *their concentration during the experiment.*

4 Results

The web-based experiment was divided into two parts. First, an initial experiment, where the participants were exposed to questions related to all the three geometrical combinations, followed by a focused experiment that concentrated on the most interesting geometrical condition. The second part was shorter and involved a larger group of people.

Two groups of people were involved in the initial experiment. An invitation was sent through e-mail to students at Programme for Engineering and ICT at the Norwegian University of Science and Technology (NTNU). The invitation was also sent to employees at Kristiansand municipality. In this part, 60 male and 32 female participants were recorded. Two of the female participants were skipped because of an abnormality in testing. Hence, the recordings from 90 participants were used in the studies.

The focused experiments included 270 valid submissions, 190 males and 80 females. In this group, students from Programme of Civil and Environmental Engineering were invited, in addition to a group of geomatics professionals in Norway.

In the initial experiment, the participants did solve three tasks for each geometrical combination. The tasks themselves were easy to solve, so it was decided only to study the time each participant used to find an answer. These three tasks used the same basemap, and the time for each of these tasks were summarized before analysing the data. Table 1 gives a brief overview of the elapsed time that was recorded during the experiment.

To meet the requirement for normal distribution, the data sets were transformed by the function $f(x) = \log_{10}x$. Further analysis by using Anderson-Darling tests and Bartlett's tests showed that the point/point dataset were eligible for analysis of variance (ANOVA), while the point/line and point/area datasets were eligible for Kruskal-Wallis. The hypotheses for both the ANOVA and the Kruskal-Wallis are:

$$\text{H0}: m1 = m2 = \cdots = mk$$

$$\text{H1}: \text{Minimum two of the mk's are different}$$

Table 1 Results from the initial experiment

	Topographic	World imagery	Dark-grey
Point/point tasks			
No. of observations	29	31	30
Sample mean (s)	48.3	45.7	42.7
Standard dev. of x (s)	21.1	18.8	14.8
Point/line tasks			
No. of observations	32	28	30
Sample mean (s)	57.0	66.9	63.0
Standard dev. of x (s)	19.9	33.1	27.0
Point/area tasks			
No. of observations	29	31	30
Sample mean (s)	50.5	63.8	49.7
Standard dev. of x (s)	18.0	39.5	30.6

Table 2 Results for the analysis of differences between basemaps, initial experiment

ANOVA for point/point tasks	F-value: 0.67	H_0 accepted
	Critical value: 3.10	No significant differences
Kruska-Wallis for point/line tasks	H-value: 0.89	H_0 accepted
	Critical value: 5.99	No significant differences
Kruska-Wallis for point/area tasks	H-value: 4.75	H_0 accepted
	Critical value: 5.99	No significant differences

where, mk is the rank mean of group k. Table 2 shows the results from the analysis with a confidence interval of 95%.

As we can see, there was no significant difference when comparing the three basemaps. The point/area task was, however, close to the critical value (significant with a confidence interval of 90%) and was selected for a closer study in the focused experiment.

Statistical analysis was executed for some other combinations of the data sets. Two-sampled t-test indicated a significant difference in time spent between males and females when all the tasks were added together.

One small change was implemented for the focused experiment. The area covering symbols were made slightly more transparent in order to increase the influence of the basemaps (Fig. 5). The results are showed in Tables 3 and 4. Since the data seemed to indicate a small difference between female and male participants, these are listed separately.

The participants indicated if they were experienced or non-experienced map users. Since one of the distribution channels of the experiment was through a web site for geomatics professionals, the share of experienced users was quite high. 40 females and 100 males considered themselves as experienced, while 40 females and 90 males declared the status non-experienced map users.

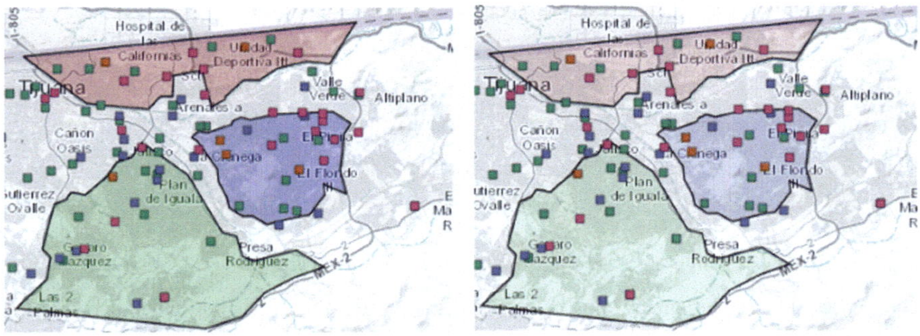

Fig. 5 Transparency was slightly altered from the initial experiment (*left*) to focused experiment (*right*)

Table 3 Results from the focused experiment

	Total	Male	Female	With experience	Without experience
No. of observations	270	190	80	139	131
Sample mean (s)	60.0	61.6	56.1	65.8	53.8
Standard dev. of x (s)	30.8	33.5	22.7	35.4	23.4

Table 4 Detailed results from the focused experiment

	Topographic	World imagery	Dark grey
Point/area tasks			
No. of observations	90	89	91
Sample mean (s)	68.7	56.1	55.2
Standard dev. of x (s)	38.8	26.4	22.9
Male participants only			
No. of observations	63	63	64
Sample mean (s)	73.4	55.4	56.1
Standard dev. of x (s)	42.2	27.1	25.2
Female participants only			
No. of observations	27	26	27
Sample mean (s)	57.5	57.8	51.7
Standard dev. of x (s)	26.3	24.3	16.1

A Mann-Whitney U test was used to check differences between genders and experience levels. This test assumes that the samples have approximately the same variances and distribution, and is used to examine if the median of the two samples differs. The hypotheses of the Mann-Whitney U test are

Table 5 Analysis of difference between genders and experience

Differences between genders	Z-value: 0.54	H_0 accepted
	Critical value: 1.96	No significant differences
Differences between experienced and non-experienced map users	Z-value: 3.39	H_0 rejected
	Critical value: 1.96	Significant differences

Table 6 Analysis of difference between basemaps for different groups

Differences between basemaps, all participants	H-value: 10.0	H_0 rejected
	Critical value: 5.99	Significant differences
Differences between basemaps, male participants only	H-value: 13.9	H_0 rejected
	Critical value: 5.99	Significant differences
Differences between basemaps, female participants only	H-value: 0.24	H_0 accepted
	Critical value: 5.99	No significant differences
Differences between basemaps, non-experienced map users	H-value: 14.32	H_0 rejected
	Critical value: 5.99	Significant differences

$$H_0: median_A = median_B$$
$$H_1: median_A \neq median_B$$

Table 5 shows the results from the analysis with a confidence interval of 95%.

Next, we used Kruskal-Wallis one-way analysis of variance to see how different basemaps were influencing different groups. Results from this analysis are presented in Table 6 (95% confidence interval).

A multiple comparison test confirms that the topographic basemap is significantly different from the others. This is in accordance with the expectations when looking at the numbers in Table 4.

5 Discussion and Conclusions

The initial experiment, which included visual analysis of different combinations of data primitives on top of a basemap, showed no significant differences between any of the basemaps. We did notice one significant difference. Female participants spent more time completing the experiments than the male participants. The focused experiment did not verify this finding. Since the largest group of participants in the initial experiment was a group of students close to the authors, some of the students were interviewed about their contribution in the experiment. Several female participants stated that they often double-checked their answers before proceeding to the next task. This additional step would add to the total time elapsed for each task set for female participants.

The focused experiment revealed that the experienced participants used a longer time than the non-experienced (Table 3). This might seem counter-intuitive, as one would expect experienced participants to be familiar with most basemaps presented as well as with web maps in general. Other similar studies have shown that experienced participants achieve better results in tasks where symbol recognition and other relatively simple perception tasks are involved (Konečný et al. 2011). The tasks in this experiment do not give experienced participants any advantages with respect to symbol recognition as all features and choices do not use often-utilized cartographic symbols, but rely instead on simple geometries differentiated by colour.

Although no significant differences were found in the initial experiment, there is a tendency for experienced participants to use more time than non-experienced participants for both experiments. No hard evidence is found as to why this is the case but there may be several contributing factors. Being more familiar with zoom and/or pan operations, experienced participants might use this more than non-experienced participants. By zooming and/or panning to confirm an answer for a task would use more time. If unfamiliar with these operations, one might rely on an immediate judgement based on a single map. By using their knowledge about web maps, experienced users are able to get more out of the information in the map.

The focused experiment showed differences in the time used for solving tasks on the three different basemaps. The time used was significantly higher when using the topographic map as the background. Further analysis showed that this was significant even at the 99% confidence interval. Since the area feature on top of the basemap was more transparent, the figure-ground relationship between the areas and the basemap may have changed. By letting more of the background through the area symbols, it could be expected that the basemap richest in details ended up as more disturbing. By reducing the opacity, the visual hierarchy might have been reduced to a level where there were less clear differences between the text labels presented on the topographic basemap and the areas for the three choices (see Fig. 5).

The world imagery is also rich in detail. However, just like the dark grey canvas, the world imagery is darker in nature and the reduced opacity may have improved the colour contrast between the background and the coloured point symbols in the foreground. This may have made the overlaid data more legible and the tasks easier to solve. This is in accordance with the studies of (Lai and Yeh 2004; Chesneau 2011; Lloyd 2005). In a further investigation, it might be interesting to take a closer look at how different grades of opacity influences similar tasks.

The experiment revealed that the topographic map was significantly less efficient both for the male participants and for the non-experienced map users. Experienced participants may have been able to filter out text labels on the topographic basemap as they were of no importance when completing the tasks, while non-experienced participants might have spent more time for this "mental filtering process." It is, however, difficult to explain why the male participants spent more time on the tasks involving the topographic basemap and the females did not.

Throughout both experiments, the dark grey basemap showed promising and consistent results. If no reasons exist for using the topographic or world imagery basemaps, and data layers are being presented using simple geometries differentiated by colour value and hue, the dark grey basemap appears to be a good choice.

References

Bereuter, P., & Weibel, R. (2013). Real-time generalization of point data in mobile and web mapping using quadtrees. *Cartography and Geographic Information Science, 40*(4), 271–281.

Chesneau, E. (2011). A model for the automatic improvement of colour contrasts in maps: Application to risk maps. *International Journal of Geographical Information Science, 25*(1), 89–111.

ColorBrewer. (2016). ColorBrewer: Color advice for maps. Retrieved July 13, 2016, from http://colorbrewer2.org/.

Dong, W., Liao, H., Roth, R. E., & Wang, S. (2014). Eye tracking to explore the potential of enhanced imagery basemaps in web mapping. *The Cartographic Journal, 51*(4), 313–329.

ESRI. (2016). ArcGIS online, ready to use maps. Retrieved July 12, 2016, from http://www.esri.com/software/arcgis/arcgisonline/features/maps.

Fichter, D. (2016). What is a mashup? In N. C. Engard (Ed.), *More library mashups*. Retrieved March 18, 2016, from http://books.infotoday.com/books/More-Library-Mashups.shtml.

Gibin, M., Singleton, A., Milton, R., Mateos, P., & Longley, P. (2008). An exploratory cartographic visualisation of London through the Google maps API. *Applied Spatial Analysis and Policy, 1*(2), 85–97.

Harrie, L. E. (1999). The constraint method for solving spatial conflicts in cartographic generalization. *Cartography and Geographic Information Science, 26*(1), 55–69.

Konečný, M., Kubíček, P., Stachoň, Z., & Šašinka, Č. (2011). The usability of selected base maps for crises management—Users perspectives. *Applied Geomatics, 3*(4), 189–198.

Korpi, J., & Ahonen-Rainio, P. (2013). Clutter reduction methods for point symbols in map mashups. *The Cartographic Journal, 50*(3), 257–265.

Krantz, J. H., & Dalal, R. (2000). Chapter 2—Validity of web-based psychological research A2—Birnbaum. In H. Michael (Ed.), *Psychological experiments on the internet* (pp. 35–60). San Diego.

Lai, P.-C., & Yeh, A. G.-O. (2004). Assessing the effectiveness of dynamic symbols in cartographic communication. *The Cartographic Journal, 41*(3), 229–244.

Lloyd, R. E. (2005). Attention on maps. *Cartographic Perspectives, 52,* 28–57.

MacEachren, A. M., & Mistrick, T. A. (1992). The role of brightness differences in figure-ground: Is darker figure? *The Cartographic Journal, 29*(2), 91–100.

Midtbø, T., & Nordvik, T. (2007). Effects of animations in zooming and panning operations on web maps: A web-based experiment. *The Cartographic Journal, 44*(4), 292–303.

Neun, M., Burghardt, D., & Weibel, R. (2008). Automated processing for map generalization using web services. *GeoInformatica, 13*(4), 425–452.

Nivala, A.-M., & Sarjakoski, T. L. (2007). User aspects of adaptive visualization for mobile maps. *Cartography and Geographic Information Science, 34*(4), 275–284.

Peterson, G. (2009). GIS cartography: A guide to effective map design. CRC Press. http://www.crcnetbase.com/doi/book/10.1201/9781420082142.

Peterson, M. P. (2009). The brave new world of online mapping. In *Proceedings of ICC*. http://icaci.org/files/documents/ICC_proceedings/ICC2009/html/nonref/13_16.pdf.

Roth, R. E., & Ross, K. S. (2009). Extending the Google maps API for event animation mashups. *Cartographic Perspectives, 64,* 21–40.

Stigmar, H., & Harrie, L. (2011). Evaluation of analytical measures of map legibility. *The Cartographic Journal, 48*(1), 41–53.
Su, B., Li, Z., Lodwick, G., & Muller, J.-C. (1997). Algebraic models for the aggregation of area features based upon morphological operators. *International Journal of Geographical Information Science, 11*(3), 233–246.
Tauscher, S., & Neumann, K. (2016). A displacement method for maps showing dense sets of points of interest. In G. Gartner, M. Jobst, & H. Huang (Eds.), *Progress in cartography* (pp. 3–16). Cham: Springer International Publishing.
Ware, J. M., & Jones, C. B. (n.d.). Conflict reduction in map generalization using iterative improvement. *GeoInformatica, 2*(4), 383–407.

Visualization of Environment-related Information in Augmented Reality: Analysis of User Needs

Kateřina Chmelařová, Čeněk Šašinka and Zdeněk Stachoň

Abstract The main motivation for this study was to enable paragliding pilots to use Augmented Reality features during flight. For orientation in the air, paragliding pilots use a map as the main source of information. It is therefore important that the information is presented in the most easily accessible way. The aim of the present study was to investigate the needs of paragliding pilots with respect to maps and the way maps provide them with information about the environment. The participants were three professional paragliding instructors with different amounts of experience. The methods employed involved a semi-structured interview and a practical, real-flight test involving AR glasses; the test was done by only one of the pilots. Two categories of data were identified, with the first involving data related to geographical terrain and the other to the pilot's immediate surroundings. Our results suggest that an optical see-through display would be more suitable for use in the air than the video-based glasses we utilized. If an optical see-through display were used, the terrain could be seen through the glasses, with terrain-related information displayed in the map. AR could be used to present flight-related information as well. Our findings suggest that the way in which information is presented to pilots should be chosen based on the category of the information, namely, whether it is terrain-related or flight-related. In addition, it was found that the small display in the right corner of the pilot's field of view is rather distractive and insufficient to fulfill the pilots' needs. Our results also showed that future AR applications should be designed to be transferable between different platforms: glasses, mobile phones, and tablets.

K. Chmelařová (✉) · Č. Šašinka
Department of Psychology, Center for Experimental Psychology and Cognitive Sciences, Masaryk University, Brno, Czech Republic
e-mail: sprinarova.katka@gmail.com

Č. Šašinka
e-mail: ceneksasinka@gmail.com

Z. Stachoň
Laboratory of Geoinformatics and Cartography, Department of Geography, Masaryk University, Brno, Czech Republic
e-mail: zstachon@geogr.muni.cz

M.P. Peterson (ed.), *Advances in Cartography and GIScience*, Lecture Notes in Geoinformation and Cartography, DOI 10.1007/978-3-319-57336-6_20

Keywords Augmented reality · Human-computer interaction · User test · Map-related information · Paragliding · Optical see-through display

1 Introduction

The importance of maps is constantly increasing. In cartography, more and more Information and Communication Technologies (ICT), including Geographical Information Systems (GIS) and Global Positioning Systems (GPS), are utilized. In addition, Inertial Measurement Units (IMUs), such as accelerometers, gyroscopes and magnetometers are employed in outdoor environments to track the position and rotation of a body.

AR glasses determine their position according to GPS and IMUs (Barbour and Schnidt 2001; Bichler et al. 2012; Elarebi et al. 2016). The challenge is in selecting an appropriate information visualization method for the user.

Our research aims to establish the most effective ways of presenting the information contained in a map by means of Augmented Reality (AR) glasses. AR glasses are a type of Head-mounted Display (HMD), a device that is worn on the head. It utilizes a small screen to present images or videos. Augmented reality is achieved through overlaying real-world elements with computer-generated sensory input. The information is presented in 2D; alternatively, it is presented as a 3D image that can be studied from multiple directions. AR is a technology that superimposes a computer-generated input on a users' view of the real world, thus providing a composite view.

Several studies are available in which AR is used in the context of cartography and environmental information representation (Behzadan et al. 2015; Chung et al. 2016; Fajnerova et al. 2016); however, in most of them, AR systems are used on the ground. The system we used in our study aims to enhance the pilots' view of the environment during a paragliding flight.

The objective of the present study is twofold: first, it is to gather information about what data are available to paragliding pilots before and in-flight; second, it is to determine what information is needed and what the limitations of currently used systems are.

2 Background

Kraak and Brown (2003) describe a cartographic visualization process as a translation or conversion of geospatial data from a database into maps and map-like graphics. Geospatial data handling consists of the acquisition, storage, manipulation and visualization of geospatial data in the context of particular applications. According to Larkin and Simon (1987), two types of equivalence of visual representations can be distinguished: informational and computational. When two visual representations are informationally equivalent, all of the information contained in one of them can be drawn from the other, and vice versa. Computational equivalence, then, refers to the fact that the same amount of 'computation' (i.e., effort) is needed to extract the information given explicitly in each of two informationally equivalent representations. In earlier surveys on AR, many types of AR displays were identified (Azuma 1997; Billinghurst et al. 2015). Military aircraft and helicopters use head-up displays (HUD) and helmet-mounted displays (HMD) to present graphic without requiring the users to divert their attention from their usual viewpoint. For working with maps, two types of displays can be used: video see-through AR displays and optical see-through AR displays. Video see-through displays combine closed-view HMD with one or two head-mounted cameras; they are limited by the camera(s)' frame rate(s). Optical see-through HMDs work similarly to HUDs, but they employ an optical combiner attached to the head. Optical combiners typically reduce the amount of light that the user sees from the real-world environment. The displaying capabilities of devices with optical combiners are not limited by the properties of the video cameras employed. One of the advantages of AR HMDs consists in allowing communication between multiple users. According to Ellis et al. (1991), collaborative interaction can take place either at the same time/in the same place, or in different times/at different places; thus, in accordance with Table 1, meeting room technologies and wall projectors would fall into the top left category, real-time document editors and video conferencing into the bottom left category, group calendars into the top right category, and emails and blogs into the bottom right category.

A number of projects have focused on virtual space collaboration. One of these was WearCom, a project presented by Billinghurst and Kato (1999), which set out to explore the possibility of using wearable computers in remote collaboration. Using fiducial markers, the authors of the project succeeded in making a remote call through audio-visual connection. Hölleler et al. (1999) developed a system enabling to connect indoor and outdoor AR devices. Any action performed by an indoor device was captured by the outdoor device; therefore, the system could be used as a navigation device. Poelman et al. (2012) presented an AR system that allows remote crime scene investigation. The project consisted of the following phases: (1) establishing the requirements for supporting crime scene analysis; (2) creating a mock-up design; (3) mock-up design evaluation; (4) designing a mediated reality system; and (5) evaluation of the mediated reality system. The above project served as an inspiration for the present study, with the initial part of the project being reproduced in detail.

Table 1 Time space matrix (Ellis et al. 1991)

	Same time	Different times
Same place	Face-to-face interaction	Asynchronous interaction
Different places	Synchronous distributed interaction	Asynchronous distributed interaction

Kamilakis et al. (2016) used two-dimensional zoomable maps to visualize points of interest in close surroundings of the user. Their results suggest that most participants spent more time attending to the mobile AR environment than to the map interface. The authors emphasized that the above should be interpreted as a sign of experimentation with an unfamiliar environment, rather than an indication of user preferences.

The present study aims to understand paragliding pilots' informational needs as well as their operational environment, and to investigate the limitations of currently used systems.

3 Methods

Our research sample consisted of three professional paragliding instructors with different amounts of experience. All of them had an instructor's license. Pilot 1 had three years experience; pilot 2 and pilot 3 had above 15 years of experience.

First, a semi-structured interview focusing on AR visualization possibilities was administered. In addition to involving a predefined set of questions, the interview allowed the pilots to talk freely about their paragliding experience and to identify the difficulties they faced.

Second, a user test involving the Vuzix M100 AR glasses was conducted in real-flight conditions. After the paragliding flight, the pilots were asked about how the flight differed from their usual flying conditions. Vuzix M100 Smart Glasses have a small video see-through display in the upper right corner of the right eye. The display is opaque, with only the video camera's image of the real world shown. The device has an aspect ratio of 4:3, with a field-of-view of 15°, which is equivalent to looking at a 4-inch display from the viewing distance around 35 cm. An integrated head-tracker allows 3 Degrees-of-Freedom (3 DOF) head-tracking; in addition, the system is equipped with a 3 DOF gesture engine. Three means of control are available: button control (via dedicated buttons on the glasses), voice control and two-dimensional gesture control.

Finally, the pilots were asked to provide feedback on the AR glasses and suggestions concerning the use of AR during a paragliding flight.

4 Results

In the first subsection, a brief overview of general information related to paragliding is included. The information was provided by three paragliding pilots comprising our sample. The second subsection contains results of the semi-structured interview; the third subsection deals with the results of the user test. Only one of the pilots underwent the user test.

4.1 General Background

The highest reported altitude for paragliding was approx. three kilometers (around 10,000 feet), the average altitude being approx. 6,500 feet. Paragliding pilots use a radio for communication. Typically, the radio is attached to the harness shoulder strap in order to decrease the influence of aerodynamic noise. Below is a list of the necessary instrumentation and information the pilots need in flight:

- Map + GPS
- Forward speed
- Gliding angle (+ distance)
- Vario (rate of climb and descent indicator)
- Altitude (a paraglider's height over sea level)

One of the paragliding pilots from our sample provided the following comment:

- *"The most important information to know in flight includes current altitude, whether I am ascending or descending, and forward speed."*

4.2 Areas of Application of Augmented Reality in Paragliding

In the semi-structured interview, several areas of application of AR glasses in paragliding were identified by the paragliding pilots:

1. Performance flying
2. Training
3. Safety purposes

The use of AR glasses during performance flying and competition flying will provide the pilots with georeferenced information, thus giving them an edge over other pilots. Another possible area of application is "air caching," a game where the players follow a pre-defined set of geographical points.

Of particular interest would be the application of AR glasses in paragliding training. The three professional instructors in our sample suggested they would find it useful to be able to distinguish "their" group from other paragliding pilots. One of the pilots provided the following comment:

- *"I want to know where the other people are."*

AR would allow paragliding beginners to locate the other trainees and to collaborate with them. Information about latitude and longitude will help the trainee to

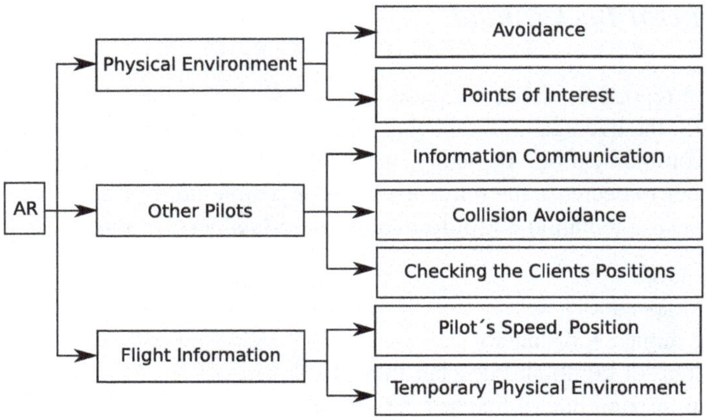

Fig. 1 Benefits of AR in paragliding

determine their own position, too. Another useful feature would be flight-path/altitude tracking, which would make it possible to:

- Watch the other people in training
- Locate people quickly (for the instructor)
- Provide help in unexpected situations

In order to serve the above purposes, the tracking would obviously need to be operative during the entire flight of each of the trainees.

Of the three areas of application listed above, safety reasons are the most important. Safety in paragliding includes avoiding dangerous (or potentially dangerous) objects such as high-voltage power lines, wind plants and chimneys.

An overview of the benefits of AR use in paragliding is presented in Fig. 1.

In the semi-structured interviews, three categories of information were identified that could be presented through augmented reality:

- Static/dynamic information
- Textual/graphical information
- Non-interactive/interactive visualization

While static information (e.g., the position of turning points) is available pre-flight, dynamic information includes flight-related data such as current position, weather and communication with other paragliding pilots.

Textual information is used where precise numerical/textual values are needed, whereas graphics may serve the purpose of representing real-world objects.

Non-interactive presentation of information is suitable for when the information is needed to solve a particular task but the pilot does not have free hands to interact with the device. Interactive presentation, on the other hand, is needed for interaction with AR glasses.

4.3 User Test

After the real flight with AR glasses, one of the pilots from our sample was asked to answer several questions. The first question concerned the pilot's opinion on the use of AR in a paragliding flight. *"The glasses were not helpful in any respect. Wearing them during the flight was aggravating. It was left to my imagination what benefits the glasses might actually bring."*

Then, the pilot was asked to suggest any improvements that would make the glasses more amenable with a paragliding pilot's needs. *"Rather than a display in front of the glasses, I would appreciate a see-through display so that I could see the terrain, overlaid with terrain-related data."*

Another question concerned possible uses of AR in paragliding. *"For me, the only reason to look at the display instead of the environment would be to acquire important data that would help me in tactical decision making; this contributes to better quality as well as higher safety of the flight. What I find especially attractive is that by rotating my head I could see the terrain behind the glasses. As regards information related to my air surroundings, I would prefer having it displayed in front of me; however, I would need to try this in practice to be able to decide for sure."*

Then, the pilot was asked about his experience with the M100 glasses during the paragliding flight. His task was to evaluate the glasses from the viewpoint of a Subject Matter Expert (SME) who needs to reach a given destination. AR should be able to make it easy.

"The glasses that I tried are not suitable either for sport or flying; they are too light and unstable. They cannot be used with any of the helmets paragliding pilots use. Also, the display is too small; I had to focus a lot when searching for a particular content. The display took up a major portion of my field of view. Yet, sight is the main sense in flying. The small display attracted my attention more than I would wish. This can be a very dangerous distraction, especially when several paragliding wings are rotating in a thermal. However, when exploiting lifts, we typically have 5–10 min to correct our decisions, so attending to the display was not a problem."

The last question was related to interaction with the AR map layer. *"Since we are flying with our gloves on, controlling the device would be difficult. The system settings should be set up prior to the actual flight. In the air, only the screen switch should be used. The button for switching between displays should be attached to the driving rope or to the helmet."*

The above suggests that it might be useful to create a visualization system that would be compatible not only with AR glasses, but also with mobile phones and tablets so that those not having (or not willing to wear) AR glasses could make use of the enhanced view capabilities.

5 Discussion

Our findings suggest that it might be useful to create a universal visualization system that would be compatible not only with AR glasses, but also with smart-phones or similar devices. The main conclusion that follows from the user test in the real environment is that the information that will be presented in AR should be divided as follows:

- The information that is better to visualize in relation to terrain
- The information that are visualized in relation to pilot (data important for tactical decisions)

One of the possible explanations for the above findings is that paragliding pilots need to be thoroughly familiar with flight-related information; in addition, they need to understand how these characteristics are changing towards the environment.

If the Vuzix M100 AR glasses are not suitable to be used in paragliding, a question remains of what glasses *are* suitable. A plausible alternative might be the use of binocular glasses capable of displaying flight-related as well as environment-related information. For safety reasons, it is essential that the device can be switched off when need be.

6 Conclusions

The results of our study showed that there are multiple ways to investigate, calculate and visualize data in AR. The semi-structured interview we used revealed several concepts that would be worth exploring in future research. In our study, three possible areas of application of AR in paragliding were identified: performance flying, paragliding training (e.g., aiding location of the trainees both in the air and on the ground); and safety reasons (e.g., helping the pilots to avoid trees, chimneys, wind power plants and high-voltage lines).

Our results also showed that AR can be used to display several different types (modes) of information:

- Static versus dynamic
- Text versus graphics
- Information visualization only versus information visualization with interaction

The user test results indicate a need for developing a universal system for both AR glasses and mobile phones. The need stems from the fact that all paragliding pilots cannot be expected to be equipped with AR glasses; therefore, mobile phones could be used instead, providing these are able to present the same amount of information. However, an in-depth research is needed here. The pilots we interviewed reported that a significant drawback of mobile phones when it comes to AR and other systems, e.g., GPS, is their low battery power.

The results of the user test indicate poor suitability of the Vuzix M100 AR glasses for the purposes of either sport or flight (where the user is attending to a map). The glasses are too fragile to be used in changeable environmental conditions. Finding a suitable alternative is the objective of future research.

Terrain-related information should be presented depending on changes in the altitude of the terrain. It is important that the displayed information is highly accurate and not prone to changes due to head movement. The results of our research indicate the necessity to choose an appropriate way of data visualization, based on whether the information to be presented is terrain-related or flight-related.

Following where we left off, future research could focus, for instance, on evaluating the suitability of different types of AR glasses with an optical-see through technology for paragliding purposes. Also, using a larger sample of participants would be advisable, along with testing the resulting data for statistical significance.

Acknowledgements This research has been supported by funding from the project of Masaryk University under the grant agreement No. MUNI/M/0846/2015, 'Influence of Cartographic Visualization Methods on the Success of Solving Practical and Educational Spatial Tasks'. We gratefully acknowledge the support and willingness of Dalibor Carbol and his paragliding school El Speedo s.r.o. without which the present study could not have been completed.

References

Azuma, R. T. (1997). A survey of augmented reality. *Presence: Teleoperators and Virtual Environments, 6*(4), 355–385.

Barbour, N., & Schmidt, G. (2001). Inertial sensor technology trends. *IEEE Sensors Journal, 1*(4), 332–339. Retrieved October 17, 2016, from http://ieeexplore.ieee.org.ezproxy.techlib.cz/stamp/stamp.jsp?arnumber=983473.

Behzadan, A. H., Dong, S., & Kamat, V. R. (2015). Augmented reality visualization: A review of civil infrastucture system applications. *Advanced Engineering Informatics, 29*(2), 252–267.

Bichler, S., Ogris, G., Kremser, V., Schwab, F., Knott, S., & Baca, A. (2012). Towards high-precision IMU/GPS-based stride/parameter determination in a outdoor runners' scenario. *Procedia Engineering, 34,* 592–597.

Billinghurst, M., & Kato, H. (1999, March). Collaborative mixed reality. In *Proceedings of the International Symposium on Mixed Reality* (pp. 261–284). Retrieved October 20, 2016, from https://www.vs.inf.ethz.ch/edu/SS2005/DS/papers/ar/billinghurst-ar.pdf.

Billinghurst, M., Clark, A., & Lee, G. (2015). A survey of augmented reality. *Foundations and Trends® Human–Computer Interaction, 8*(2–3), 73–272.

Chung, J., Pagnini, F., & Langer, E. (2016). Mindful navigation for pedestrians: Improving engagement with augmented reality. *Technology in Society, 45*, 29–33. Retrieved October 20, 2016, from http://www.sciencedirect.com/science/article/pii/S0160791X16000142.

Elarebi, T., Suprem, A., & Deep, V. (2016). Orientation and displacement detection for smartphone device based IMUs. *IEEE Access*. Retrieved October 17, 2016, from http://ieeexplore.ieee.org.ezproxy.techlib.cz/document/7748482/.

Ellis, C. A., Gibbs, S. J., & Rein, G. (1991). Groupware: some issues and experiences. *Communications of the ACM, 34*(1), 39–58. Retrieved October 10, 2016, from http://papers.cumincad.org/cgi-bin/works/Show?39a6.

Fajnerova, et al., (2016). Human cognitive enhancement tested in virtual reality environments. In *Proceedings of the 11th International Conference on Disability, Virtual Reality and Associated Technologies*, Los Angeles, California, USA, 20–22 September 2016. Retrieved February 02, 2017, from http://centaur.reading.ac.uk/66645/8/ICDVRAT2016_Full_Proceedings_11th%20_Conf.pdf.

Höllerer, T., Feiner, S., Terauchi, T., Rashid, G., & Hallaway, D. (1999). Exploring MARS: Developing indoor and outdoor user interfaces to a mobile augmented reality system. *Computers and Graphics, 23*(6), 779–785. Retrieved October 17, 2016, from http://www1.cs.columbia.edu/~drexel/research/Hollerer-1999-CandG_Elsevier.pdf.

Kamilakis, M., Gavalas, D., & Zaroliagis, C. (2016, June). Mobile user experience in augmented reality vs. maps interfaces: A case study in public transportation. In *International Conference on Augmented Reality, Virtual Reality and Computer Graphics* (pp. 388–396). Springer International Publishing.

Kraak, J. M., & Brown, A. (Eds.). (2003). *Web cartography*. CRC Press.

Larkin, J. H., & Simon, H. A. (1987). Why a diagram is (sometimes) worth ten thousand words. *Cognitive Science, 11*(1), 65–100.

Poelman, R., Akman, O., Lukosch, S., & Jonker, P. (2012, February). As if being there: mediated reality for crime scene investigation. In *Proceedings of the ACM 2012 Conference on Computer Supported Cooperative Work* (pp. 1267–1276). ACM. Retrieved October 10, 2016, from http://www.arlab.nl/manual-uploads/poelman-et-al-as-if-being-there-final2012.pdf.

Part V
Geographic Analysis

Analysis and Visualization of the Urban Residents' Income-Related Happiness Index in China

Ying Song, Yang Yu, Yanfang Liu, Zixi Liu, Qianyi Li and Guoguang Xu

Abstract Using the Theil index and correlation analysis, this paper analyzed the Urban Residents' Income-Related Happiness Index from multiple perspectives, including the current situation, regional disparities, spatial–temporal variations, and the mutual relationships of the indicators based on data from 289 prefecture-level cities in China from 2006 to 2015. The results indicated the following: (1) Overall, the Urban Residents' Income-Related Happiness Index in China was lower in 2015 and was unevenly spread across space. Several cities with higher Income-Related Happiness Index were primarily situated in the southwest region; (2) The within-province disparities were higher than the between-province disparities; (3) For the past 10 years, the Urban Residents' Income-Related Happiness Index has exhibited an overall downward trend, and the cities that had slight increases in their Income-Related Happiness Index were primarily situated in North China; and (4) The correlation between the Urban Residents' Income-Related Happiness Index and the per capita gross regional product is not significantly positive. Most cities in China can be categorized into three types: strong happiness–small economic development, weak happiness–large economic development, and weak happiness–small economic development scales. In addition, this study visualized the index analysis results. All of these results are expressed using different maps.

Keywords Urban Residents' Income-Related Happiness Index · Analysis and visualization · Regional disparities · Spatial–temporal variations · Indicator correlations

Y. Song (✉) · Y. Yu · Y. Liu · Z. Liu · Q. Li · G. Xu
School of Resource and Environmental Science,
Wuhan University, Wuhan, Hubei Province, China
e-mail: ysong818@126.com

Y. Song · Y. Liu (✉)
Key Laboratory of Geographic Information System, Ministry of Education,
Wuhan University, Wuhan, Hubei Province, China
e-mail: yfliu610@163.com

M.P. Peterson (ed.), *Advances in Cartography and GIScience*, Lecture Notes in Geoinformation and Cartography, DOI 10.1007/978-3-319-57336-6_21

1 Introduction

National happiness has been a very popular topic since the 1970s. Multiple scholars have focused on "well-being" and the "happiness index" (Bradburn 1969; Easterlin 2001; Kahneman et al. 2005; Xing 2006; Han and Zhou 2012; Li and Wang 2013; Pan 2015). One study found that the higher the gross national product of a country, the lower the correlation between individual well-being and income (Veenhoven 1991).

Urban centers propel China's economic development. Therefore, determining the current level of the Urban Residents' Income-Related Happiness Index of prefecture-level cities is an interesting topic. Regional disparities in the Urban Residents' Income-Related Happiness Index over the past 10 years should also be examined. The characteristics of the spatial–temporal distribution as well as the relevance of the connection between the Urban Residents' Income-Related Happiness Index and the urban economic development scale are also important. Therefore, exploring and revealing the spatial–temporal distributions of the Urban Residents' Income-Related Happiness Index is the focus of this paper.

2 Research Region and Data Sources

The data analysis focused on data from prefecture-level cities. Of these cities, Chaohu in Anhui Province was adjusted to a county-level city in 2011 and was subsequently removed from the analysis. Bijie and Tongren in Guizhou Province became prefecture-level cities in 2011, and Haidong in Qinghai Province became a prefecture-level city in 2012. We used the original administrative district as the analysis unit before the year when the administrative level changed to compare data from multiple years. In total, we analyzed 289 prefecture-level cities.

Data were obtained primarily from the China City Statistical Yearbook 2006–2015. Part of the data also came from China's provincial statistical yearbooks and municipal statistical bulletins.

3 Research Method

3.1 Analysis of the Current Situation

This paper uses the calculation method for the gold content of the GDP (the Income-Related Happiness Index) proposed by Yang (Li and Wang 2013), which is expressed as follows:

$$GCGDP = \frac{PCDI}{PCGDP} = \frac{(PCUDI + PCRNI) \times UR}{PCGDP} \tag{1}$$

where *GCGDP* refers to the gold content of the GDP (%), *PCDI* is the per capita annual disposable income in Yuan, PCGDP is the per capita Gross Domestic Product (Yuan), PCUDI is the per capita annual disposable income of urban residents (Yuan), PCRNI is the per capita net income of farmers (Yuan), and UR is the proportion of the urban population in percent. Based on the above equation, this paper proposes an equation to measure the Urban Residents' Income-Related Happiness Index as follows:

$$URIRHI = \frac{per\ capita\ annual\ disposable\ income\ of\ urban\ residents\ (yuan)}{per\ capita\ Gross\ Regional\ Product\ (yuan)} \tag{2}$$

Data used in the analysis of the current situation were obtained from the 2015 yearbook, and the population used in the calculation is the urban resident population.

3.2 *Analysis of Regional Disparities*

The analysis of regional disparities is one of most important objectives of regional economic research. This paper uses a second-order decomposition of the Theil index to analyze the within-province and between-province disparities of the Income-Related Happiness Index.

The Theil index is an important index for measuring regional economic disparities, because the Theil index can divide the disparities into between-region disparities and regional disparities and is therefore widely used to study regional disparities. In this paper, the Theil index is calculated by weighting the income proportion:

$$T = \sum_{i=1}^{N} y_i \log_2 \frac{y_i}{p_i} \tag{3}$$

where N is the number of regions, Y_i is the proportion of the GRP in the region i, and P_i is the proportion of the population in the region i. A smaller Theil index indicates smaller regional disparities in economic development and greater disparities between regions.

The equation of the first-order decomposition of the Theil index is as follows:

$$T = T_{WR} + T_{BR} = \sum_i \left(\frac{Y_i}{Y}\right) \sum_j \left(\frac{Y_{ij}}{Y_i}\right) \log_2 \left(\frac{Y_{ij}/Y_i}{P_{ij}/P_i}\right) + \sum_i \left(\frac{Y_i}{Y}\right) \log_2 \left(\frac{Y_i/Y}{P_i/P}\right) \tag{4}$$

where the provincial administrative unit refers to the basic space unit, T_{WR} refers to regional disparities, T_{BR} refers to between-region disparities, and Y_{ij} and P_{ij} are the GRP and population, respectively, of the provincial administrative unit J in the region i. The second-order decomposition of the Theil index is implemented with the prefecture-level administrative unit as the basic spatial unit. The equation for the second-order decomposition of the Theil index is as follows:

$$\begin{aligned} T &= T_{WP} + T_{BP} + T_{BR} \\ &= \sum_i \left(\frac{Y_i}{Y}\right) \sum_j \left(\frac{Y_{ij}}{Y_i}\right) \sum_k \left(\frac{Y_{ijk}}{Y_{ij}}\right) \log_2 \left(\frac{Y_{ijk}/Y_{ij}}{P_{ijk}/P_{ij}}\right) \\ &\quad + \sum_i \left(\frac{Y_i}{Y}\right) \sum_j \left(\frac{Y_{ij}}{Y_i}\right) \log_2 \left(\frac{Y_{ij}/Y_i}{P_{ij}/P_i}\right) + \sum_i \left(\frac{Y_i}{Y}\right) \log_2 \left(\frac{Y_i/Y}{P_i/P}\right) \end{aligned} \tag{5}$$

where T_{WP} refers to the within-province disparities, T_{BP} refers to the between-province disparities, Y_{ijk} and P_{ijk} are the GRP and population, respectively, of the prefecture-level administrative unit k in the provincial administrative unit j of the region i, and Y and P are the GRP and population, respectively, of all regions in China.

Based on the characteristics of the Theil index, that measures regional disparities, the Theil index's value of the Income-Related Happiness Index can be calculated and the regional disparities of the Urban Residents' Income-Related Happiness Index in China can be analyzed and expressed on a map of China. According to the principle of the Theil index, y_i in Eq. (3) is replaced by the proportion of the Income-Related Happiness Index in region i, Y_{ijk} in Eq. (5) is replaced by the proportion of the Income-Related Happiness Index of the prefecture-level administrative unit k in the provincial administrative unit j of the region i, and Y is the Income-Related Happiness Index of all regions in China.

3.3 *Analysis of the Spatial–Temporal Variations*

3.3.1 Analysis of the Temporal Variation

In this section, the top 20 cities ranked according the Urban Residents' Income-Related Happiness Index from 2006 to 2015 are counted, the changes of the indicators over time are plotted, and the change trend is analyzed.

3.3.2 Average Annual Growth Rate

The average annual growth rate of the Urban Residents' Income-Related Happiness Index over 10 years can be calculated as follows:

$$average\ annual\ growth\ rate = \left(\frac{this\ period}{the\ past\ N\ years}\right)^{\frac{1}{N-1}} - 1. \quad (6)$$

The cities are divided into four types according to the annual growth rate: increased significantly, increased slightly, declined slightly, and declined significantly. The categories are used to analyze the speed of development and the stability of the Urban Residents' Income-Related Happiness Index.

3.4 Analysis of Indicator Correlations

GRP is an important indicator to measure the scale of regional economic development, and one of the most important macroeconomic indicators. To determine whether income well-being is correlated with the scale of economic development, the following equation is used for the correlation analysis between the Urban Residents' Income-Related Happiness Index and the GRP:

$$\overline{X} = \frac{1}{n}\sum\nolimits_{i=1}^{n} X_i, \quad \overline{Y} = \frac{1}{n}\sum\nolimits_{i=1}^{n} Y_i.$$

Four types of associations can be used:

$$\begin{cases} \frac{X_i}{\overline{X}} \geq 1 \cup \frac{Y_i}{\overline{Y}} \geq 1\ strong\ happiness - large\ economic\ scale \\ \frac{X_i}{\overline{X}} \geq 1 \cup \frac{Y_i}{\overline{Y}} < 1\ strong\ happiness - small\ economic\ scale \\ \frac{X_i}{\overline{X}} < 1 \cup \frac{Y_i}{\overline{Y}} \geq 1\ weak\ happiness - large\ economic\ scale \\ \frac{X_i}{\overline{X}} < 1 \cup \frac{Y_i}{\overline{Y}} < 1\ weak\ happiness - small\ economic\ scale \end{cases} \quad (7)$$

where $\overline{X}$ is the average of the national Income-Related Happiness Index, $\overline{Y}$ is the average of the GRP, X_i is the Income-Related Happiness Index of city i, and Y_i is the GRP of city i.

According to Eq. (7) and based on the Income-Related Happiness Index and the GRP in 2015, Chinese cities can be divided into four types: strong happiness–large economic development (high Income-Related Happiness Index–high GRP), strong happiness–small economic development (high Income-Related Happiness Index–low GRP), weak happiness–large economic development (low Income-Related Happiness Index–high GRP), and weak happiness–small economic development (low Income-Related Happiness Index–low GRP) scales and can be expressed on a map of China using the quality base method.

4 Analysis and Expression

4.1 Current Situation of the Urban Residents' Income-Related Happiness Index

The current situation of the Urban Residents' Income-Related Happiness Index is analyzed using data from 2015. The Urban Residents' Income-Related Happiness Index and the GRP are classified statistically using the natural split point method and divided into five levels, which are then grouped based on the natural aggregation of the data to maximize the differences among levels. On a map of China, the Urban Residents' Income-Related Happiness Index is expressed on the first visual level using the statistical chart method (with a red heart-shaped symbol). The gross regional product is expressed as the second visual level using the hierarchical statistical method shown in Fig. 1.

In 2015, the number of cities belonging to the different scales in terms of the Urban Residents' Income-Related Happiness Index of China's 289 prefecture-level cities ordered from highest to lowest were 53, 10, 72, 90, and 64. Of these cities, the city with the highest Index was Longnan in Gansu Province at 1.67, followed by Dingxi and Zhaotong, both of which had an index of more than 1.60. The city with

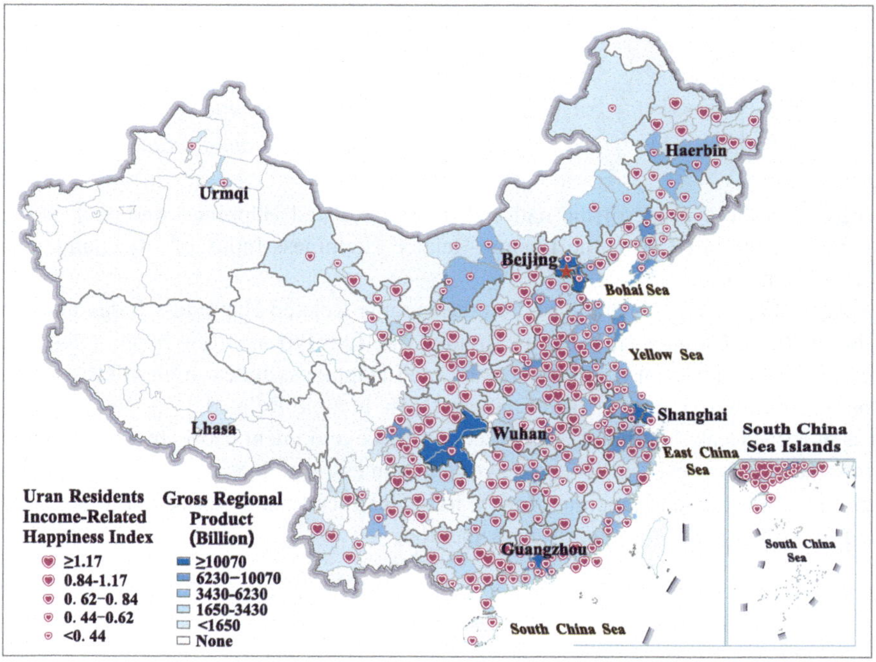

Fig. 1 Urban Residents' Income-Related Happiness Index in 2015

the lowest index was Ordos in the Inner Mongolia Autonomous Region with 0.17, followed by Daqing and Dongying with less than 0.25.

The data show that the number of cities with higher Income-Related Happiness Index, with less strong-happiness, or with the sum of the first two levels being ≥ 0.84 accounted for only 21.8% of the total number of cities. The sum of the first three levels (≥ 0.62) also accounted for only 46.7% of the total.

Figure 1 shows that spatially, cities with higher Income-Related Happiness Index (≥ 0.84) are primarily situated in Southwest China and in parts of Central China, in particular the Gansu, Guangxi, Sichuan, and Yunnan Provinces. The number of cities in these provinces with a higher Income-Related Happiness Index was 27, accounting for 42.9% of the total number of cities with a higher Income-Related Happiness Index. Cities with a middle-level Income-Related Happiness Index (0.62–0.84) were spread across the country, and no obvious clustering phenomenon was observed. However, cities with lower Income-Related Happiness Index (<0.62) were concentrated primarily in eastern regions, especially in southeast coastal cities, such as Fujian, Guangdong, Zhejiang, and Jiangsu. In addition, the Income-Related Happiness Index of provincial capitals are generally lower than those in other cities in the same province. However, the distribution of the GRP and the Income-Related Happiness Index is very different, and cities with higher GRP (623 billion Yuan) are primarily coastal cities and provincial capitals.

4.2 Regional Disparities in the Urban Residents' Income-Related Happiness Index

4.2.1 Multi-level Disparities

To facilitate the analysis, China was divided into six regions, namely, North China, Northeast China, East China, South China, Southwest China, and Northwest China. The analysis shows the results of the second-order decomposition of the Theil index and indicates that, compared to 2006, the total disparity of the Urban Residents' Income-Related Happiness Index, the between-region disparity, the between-province disparity, and the within-province disparity, all decreased in 2015. The decrease in the within-province disparity was more obvious, showing a reduction of 0.0321, while the between-region disparity only changed by 0.0108. The rank of the contribution of the three disparities in the two-year data to the total disparity (the proportion of each disparity to the total disparity) did not change, and the order from large to small disparity is as follows: within-province disparity, between-province disparity, and between-region disparity. This order confirms that the smaller the spatial unit, the greater the degree of regional economic disparity (Feng et al. 2015). The between-region disparity in the two years did not change and remained stable at approximately 16.0%. The between-province disparity was reduced from 28.3% in 2006 to 24.4% in 2015, whereas the within-province disparity increased from 55.6% in 2006 to 59.6% in 2015.

4.2.2 Between-Province Disparities

Between-province disparities refer to disparities in indicators between provinces in a geographic area. The analysis shows that the between-province disparities increased slightly in Northeast China, whereas the degrees in the other five regions decreased, with a significant decrease in the southwest. This phenomenon is directly related to the decrease in the Income-Related Happiness Index as well as the increase in the population in the Guizhou Province and the decrease in the Income-Related Happiness Index in the Yunnan Province.

4.2.3 Within-Province Disparities

Within-province disparities refer to index differences between prefecture-level cities in the province. In 2015, the within-province disparities could be divided into five levels using natural breaks in the classification, namely: ≥ 0.29, 0.19–0.29, 0.14–0.19, 0.11–0.14, and <0.11, corresponding to 2, 6, 7, 6, and 5 cities, respectively. Similar to those in 2015, the within-province disparities in 2006 could be divided into five corresponding levels with 6, 5, 6, 6, and 3 cities, respectively. To facilitate the comparison below, the three middle levels were together referred to as the middle level; thus, the three levels were the highest level (≥ 0.29), the lowest level (<0.11), and the middle level.

In terms of quantity, the analysis shows that the number of cities in the middle and lowest levels of the within-province disparities increased by 2 in 2015 compared to the number of cities in 2006, whereas the number of cities in the highest level decreased by 4: the Guizhou, Heilongjiang, Sichuan, Jilin, Yunnan, and Zhejiang Provinces in 2006 reduced to just the Sichuan and Heilongjiang Provinces in 2015.

The 27 provinces were classified according to the within-province disparities as follows:

- Increased significantly (≥ 0.05): Xinjiang, Sichuan, Heilongjiang, and Anhui;
- Increased slightly (0–0.05): Gansu, Inner Mongolia, Shaanxi, Hebei, and Guangxi;
- Declined slightly (−0.06–0): Guizhou, Jiangsu, Zhejiang, Jilin, Fujian, Guangdong, Shandong, Hubei, and Hunan; and
- Declined significantly (<−0.06): Liaoning, Shanxi, Henan, Hainan, Qinghai, Ningxia, Jiangxi, and Yunnan.

The analysis shows only nine provinces with increasing within-province disparities, with significant increase in Xinjiang Uygur Autonomous Region. The disparities in the other 17 provinces experienced different degrees of reduction, especially Guizhou Province whose within-province disparity decreased significantly.

Using the rank of the within-province disparities in 2006 and 2015, a conclusion can be reached that the largest increases were in the Guizhou and Jiangsu provinces whereas the largest declines were in the Anhui Province and the Xinjiang Uygur Autonomous Region.

Concerning the two aspects above, the Guizhou Province experienced the most significant reduction in within-province disparities and the most significant increase in rank. However, note that a reduction in disparities does not indicate an enhancement in the residents' income well-being. Instead, a reduction can show that the residents' income well-being is based on the scale of population. The Residents' Income-Related Happiness Index in Guizhou Province decreased in 2015 compared to that in 2006, whereas the Urban Residents' Income-Related Happiness Index in Liupanshui, Zunyi, Anshun, and Bijie decreased markedly. The population of Bijie and Tongren in 2015 increased significantly and accordingly the two cities were promoted to prefecture-level cities, whereas the population of Zunyi decreased significantly.

The above changes caused the gap in the ratio of the Urban Residents' Income-Related Happiness Index to the population in the cities of Guizhou Province to decrease, that is, the share that Urban Residents' Income-Related Happiness Index of each city occupied for the province was closer to the share that population of each city occupied for the province, resulting in a decrease in the Theil index.

The series of phenomena mentioned above are the reason for the significant decrease in the within-province disparities in Guizhou Province. The Xinjiang Uygur Autonomous Region's within-province disparities increased and its rank decreased significantly. The Urban Residents' Income-Related Happiness Index of Urumqi and Karamay did not develop synchronously with a significant change in the population; however, the former increased and the latter decreased, resulting in the gap in the ratio between the Urban Residents' Income-Related Happiness Index and the population becoming larger.

4.3 Analysis of the Spatial–Temporal Changes in the Urban Residents' Income-Related Happiness Index

4.3.1 Temporal Changes

Limited diversity was observed in the names of the top five cities in the annual list of the Urban Residents' Income-Related Happiness Index during the period 2006–2015. These cities include Dingxi (10 years), Zhaotong (9 years), Longnan (6 years), Guyuan (6 years), Bijie (5 years), and Fuyang (5 years). Dingxi consistently took the first or second place in terms of the index. Two of the above cities are located in Gansu Province and the remaining in Yunnan Province, Ningxia Hui Autonomous Region, Guizhou Province, and Anhui Province. As such, it appears that the cities leading in terms of ranks with respect to the index are not located in the most developed areas.

Among the top 20 cities in the list, we see Dingxi, Longnan, Guyuan, Zhaotong, Fuyang, Bozhou, Luan, and Bijie, with no evident changes noted over the examined time period. However, Bozhou exhibited an upward trend in its ranking on an annual basis, and Bijie together with Lu'an, displayed an abrupt downward movement in 2015, which could be attributed to the plummeting per capita gross domestic product of Bijie in 2014 and the plunging urban per capita disposable income of Luan in 2015. Longnan ranked first for the second time on the 2015 list of the Urban Residents' Income-Related Happiness Index. A greater divergence could be observed in the index data between the top 5 cities and the 15 cities that had been lagging, where only minimal variance was observed in the index. Economically voluminous cities such as Beijing (ranked 226), Shanghai (ranked 197), Guangzhou (ranked 275), and Shenzhen (ranked 283) failed to win over other cities for that year.

4.3.2 Characteristics of the Spatial Distribution of the Annual Growth Rate

The Urban Residents' Income-Related Happiness Index exhibited an overall downward trend during the past 10 years. In 2006, the average was 0.84, whereas the average in 2015 was only 0.65, indicating a decrease of 0.27 compared to that of 2006. The average annual growth rate was −0.028. Of the 289 cities, 233 cities exhibited a decreasing trend, of which 53 exhibited a significant decrease (<−0.050). Baishan in Jilin Province and Bijie in Guizhou Province exhibited the largest decrease at −0.078, followed by Sanming in Fujian Province at −0.077 and Zunyi and Liupanshui in Guizhou Province, both of which stood at −0.074. Only 56 cities exhibited increasing Urban Residents' Income-Related Happiness Index, accounting for 19.4% of the total. Only two cities experienced significant increases: Zhengzhou in Henan Province at 0.28 and Karamay in the Xinjiang Uygur Autonomous Region at 0.054.

In the spatial distribution, cities with slight increases in the Urban Residents' Income-Related Happiness Index were distributed primarily in the northeastern and northern regions. Cities with a slight decline were spread across the country, whereas cities with a significant decline were concentrated primarily in the southwestern regions and in Jiangsu, Fujian, and Inner Mongolia.

4.4 Correlation Analysis of Indicators

The correlation analysis of the Urban Residents' Income-Related Happiness Index and the city economy scale demonstrated that most cities in China can be categorized into strong happiness–small economic development, weak happiness–large economic development, and weak happiness–small economic development scales. The cities categorized into each type numbered 119, 70, and 91, accounting for 41.2%, 24.2%, and 31.5%, respectively, of the total number of cities. Cities with the strong happiness–small economic development scale were distributed primarily in the central and western regions. Cities with the weak happiness–large economic development scale were primarily coastal cities and provincial capitals, as well municipalities directly under the central government. Cities with the weak happiness–small economic development scale were distributed in all provinces. Only nine cities were included in the strong happiness–large economic development scale, and they were distributed in the Shandong, Hebei, Henan, Hunan, and Zhejiang Provinces (see Fig. 2).

The correlation analysis of the Urban Residents' Income-Related Happiness Index and the GRP shows that the residents' income well-being and the economic development scale of 189 cities (accounting for 65.4% of the total number of cities) in China were synchronized and that the larger the economic scale, the lower the Urban Residents' Income-Related Happiness Index. By contrast, their income happiness index is generally and relatively higher. Ideally, a city's economy and the people's livelihood should develop simultaneously. The government should focus on increasing the disposable income of residents and improving people's livelihood while the economy is developing rapidly. Cities falling in the weak happiness–large economic development scale should be more people-oriented and consider the

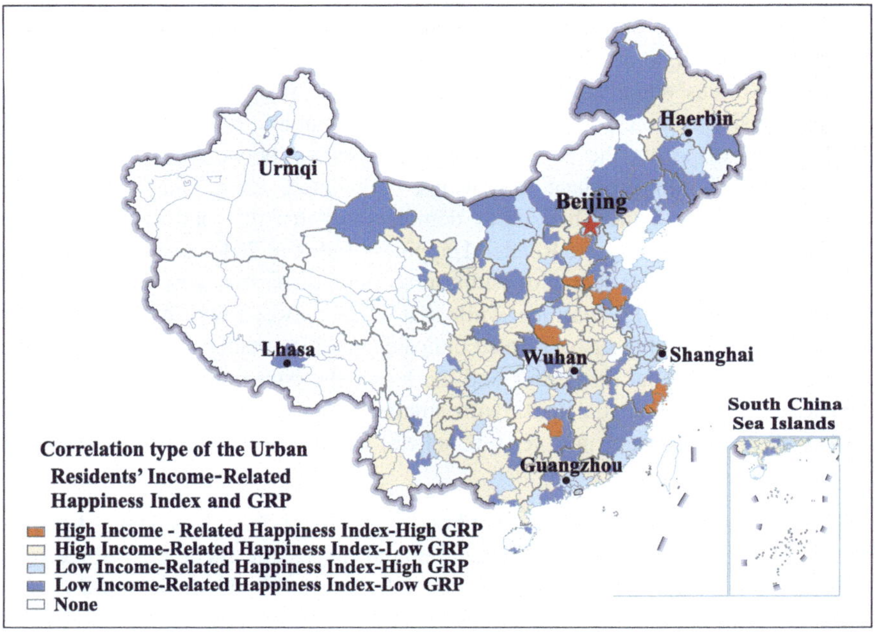

Fig. 2 Correlation type of the Urban Residents' Income-Related Happiness Index and GRP

enhancement of the residents' income well-being as a prerequisite for economic growth. Meanwhile, cities falling in the strong happiness–weak economic development scale should develop their economy rapidly but ensure that the sense of well-being of their people is not weakened. To promote ideal cities in China, the income well-being and the economy should grow simultaneously.

5 Conclusions

Using data on 289 prefecture-level cities in China over the period 2006–2015, this paper conducted an in-depth study of the Urban Residents' Income-Related Happiness Index by analyzing the current situation, regional disparities, spatial–temporal variation, and correlations of the economic indicators. The following conclusions were reached:

(1) Overall, the Urban Residents' Income-Related Happiness Index in China is low and is unevenly spread across space, with the higher Income-Related Happiness Index cities concentrated in the southwest region where the economy is relatively backward.
(2) For the past 10 years, the Urban Residents' Income-Related Happiness Index appears to have an overall downward trend. Although the economy has been developing rapidly, the income-happiness of residents has not increased, and instead the average decreased from 0.84 to 0.65—a decrease of 22.6%. People's livelihood and economic development are obviously not synchronized, and both aspects in some cities are in a backward stage. Government departments need to focus equally on the economy and the people's livelihood to synchronously promote people's livelihood and economic development.
(3) The total disparity, the between-region disparities, between-province disparities, and within-province disparities, calculated using the second-order decomposition of the Theil index demonstrates that the total disparity of the Urban Residents' Income-Related Happiness Index in 2015 is smaller than that in 2006 and that the degree of within-provincial disparities was higher than the between-provincial disparities, while that of the between-region disparities was the lowest. The total disparity also indicates that the smaller the spatial unit, the greater the regional disparities and the larger the contribution to the total disparity. However, spatial units with smaller disparity are mostly in low-level balances, which are not very beneficial to the development of people's livelihoods.

To further improve people's livelihoods and the Urban Residents' Income-Related Happiness Index, the per capita annual disposable income of urban residents should increase appropriately while the economy develops rapidly. The government should also focus on the synchronous development of the economy and people's livelihoods, implement different development strategies in different types of cities, and promote city development toward the strong happiness–large economic development scale. Simultaneously, regional disparities should be reduced and cooperative regional development should be encouraged by promoting each region toward the high-level development of people's livelihoods and the economy.

Acknowledgements This research was supported by the "Twelfth Five-Year Plan" basic work project of national science and technology, the Research and Compilation of the National Atlas (Project No. 2013FY112800).

References

Bradburn, N. M. (1969). The structure of psychological well-being. *American Sociological Review*, (3).

Easterlin, R. A. (2001). Income and Happiness: Towards a Unified Theory. *Economic Journal, 111*(473), 465–484.

Feng, C. C., Zeng, Z. R., & Cui, N. N. (2015). The economic disparities and their spatiotemporal evolution in China since 2000. *Geographic Research, 34*(2), 234–246.

Han, B., & Zhou, G. W. (2012). A review of recent research on China's income vs. GDP radio. *Journal of Shenzhen University* (*Humanities and Social Sciences*), 29(5), 95–99.

Kahneman, D., Krueger, A. B., Schkade, D. A., et al. (2005). A survey method for characterizing daily life experience: The day reconstruction method. *Science, 306*(5702), 1776–1780.

Li, Y., & Wang, J. R. (2013). The gold content of GDP ranking of 31 provinces in 2012. *China Economic Weekly*, (8).

Pan, J. H. (2015). Spatiotemporal pattern of gold content of GDP at prefecture level or cities above in China. *Scientia Geographic Sinica, 35*(12), 1502–1510.

Veenhoven, R. (1991). Is happiness relative? *Social Indicators Research, 24*(1), 1–34.

Xing, Z. J. (2006). The index system construction and tracking research of happiness index. *Data Magazine, 8,* 10–12.

Displaying Voter Gains and Losses: Local Government Elections in South Africa for 2011 and 2016

Peter M.U. Schmitz

Abstract Every five years, the South African population goes to the polls to elect local government councilors. These elections are fought along party lines, however, independent candidates are also allowed to participate. This paper looks at the change of voter behavior between two subsequent local government elections from 2011 and 2016. The percentage voter gains and losses for the top three political parties in South Africa are shown. In the election of 2016, the ruling African National Congress (ANC) party suffered its biggest loss since 1994. The Democratic Alliance (DA) showed an increase in votes between the two elections and, the new kid on the political block, the Economic Freedom Fighters (EFF), showed its biggest gains in the same areas where the ANC showed its largest losses. Three methodologies were used to show these voter behaviors, namely cartograms, 3D maps and thematic cartogram. The latter was used to show both losses and gains in the same cartogram.

Keywords Voters · Cartogram · 3D maps · Thematic maps · Elections

1 Introduction

The author has had the privilege since 2004 of being involved in the South African national, provincial and local government elections. His involvement has been as a member of South Africa's Council for Scientific and Industrial Research (CSIR)

P.M.U. Schmitz (✉)
CSIR Built Environment, Meiring Naude Rd, Brummeria, Pretoria, South Africa
e-mail: pschmitz@csir.co.za

P.M.U. Schmitz
Department of Geography, Geoinformatics and Meteorology, University of Pretoria, Lynnwood Rd, Hatfield, Pretoria, South Africa

P.M.U. Schmitz
Fakultät für Vermessung, Informatik und Mathematik, Hochschule für Technik, Stuttgart, Schellingstrasse 24, 70174 Stuttgart, Germany

M.P. Peterson (ed.), *Advances in Cartography and GIScience*, Lecture Notes in Geoinformation and Cartography, DOI 10.1007/978-3-319-57336-6_22

election night analysis team that forecasts the final results of an election when between 5 and 10% of the voting districts have been declared. He witnessed firsthand how the fortunes of the political parties have changed after each election, most notably between the 2011 and 2016 local elections where the dominant African National Congress (ANC) went from 62% in 2011 to 54% in 2016 at the national level, and lost three of the major metropolitan areas to the opposition. The biggest winners were the Democratic Alliance (DA) and the Economic Freedom Fighters (EFF). This paper describes, through the use of maps, where these three major parties won and lost votes in these two elections.

2 Mapping Political Voting Results

The mapping of political voting results in the USA dates back to 1880s when Scribner's Statistical Atlas of the United States was published showing various presidential elections and the 1880, 1884 and 1888 popular vote (Hewes and Gannet 1883). It seems that the 1884 and 1888 popular vote were inserted in later editions of the original 1883 publication since they are not listed in the table of contents of the atlas nor had they been given plate numbers. These two maps used different colors to denote Republicans (grey) and Democrats (red). The 1880 popular vote map is shown in Fig. 1. The blue hue denotes percentage Republican

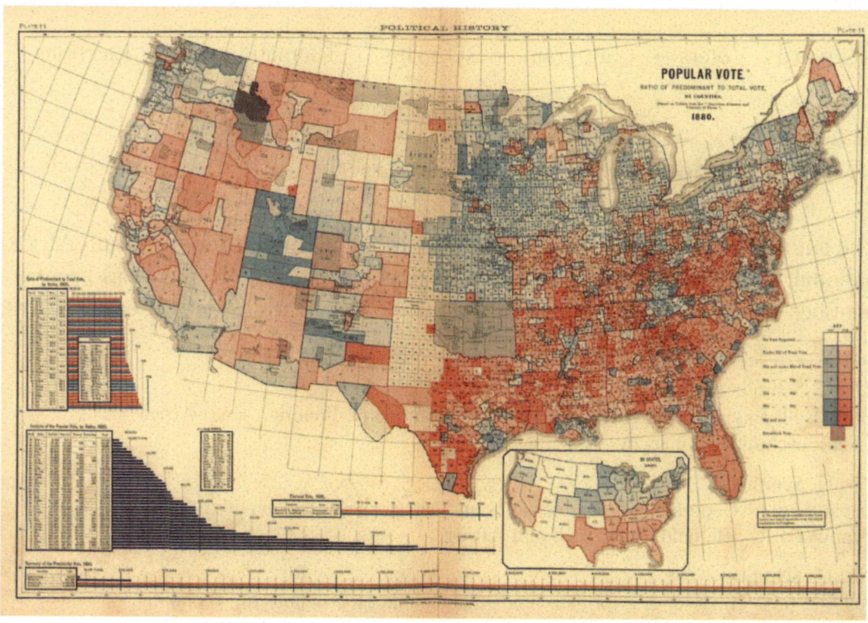

Fig. 1 The 1880 US presidential election. *Source* Plate 11, Hewes and Gannet (1883)

vote per county and the red hue, the Democrat vote percentage per county. The color indicates 50% or more votes for the party in a county.

Other scholars such as Turner and Libby advocated studies of voting results to understand voting behavior in the US in the late 1800s (Bogue 1968). Various methodologies besides the normal thematic map showing the winning party were utilized to represent voting results. Stoffel et al. (2012) used a technique called gradient fill of polygons where the polygon is proportionally shrunk to illustrate the number of votes obtained. The winning party uses the original constituency polygon and the party that came second uses a smaller version of the same polygon and in a different color to the winning party. The system can be applied to elections where two or more parties participate.

Vanderbei (n.d.) used a color triangle to map US presidential voting patterns. Red is used to represent the Republicans, blue for the Democrats and green for any other minor party. The mixture of the various parties' color intensities indicated the proportion of votes garnered. If a county voted 100% Republican, it appeared red, or blue if the county was 100% Democrat. If both garnered 50% of the vote in a county, the county was colored purple. If the county was split three ways, the color was grey. Figure 2 shows an example of Vanderbei's map for the 1968 US presidential election.

Another form for displaying election results is the cartogram. A cartogram is a map where geographical regions such as states, provinces or wards are scaled to

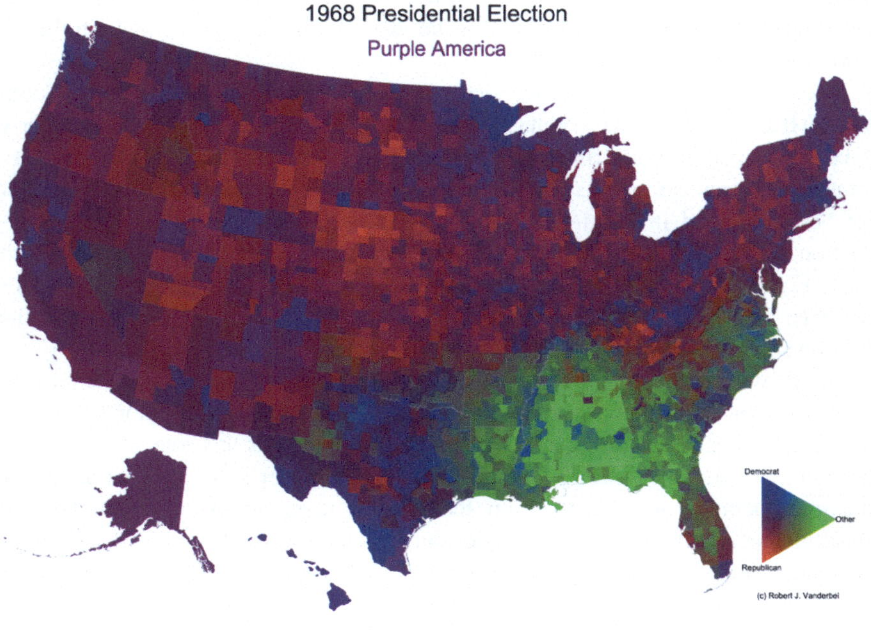

Fig. 2 Results of the 1968 US presidential election. *Source* Vanderbei (n.d.), 5 Copyright, 2017, Vanderbei

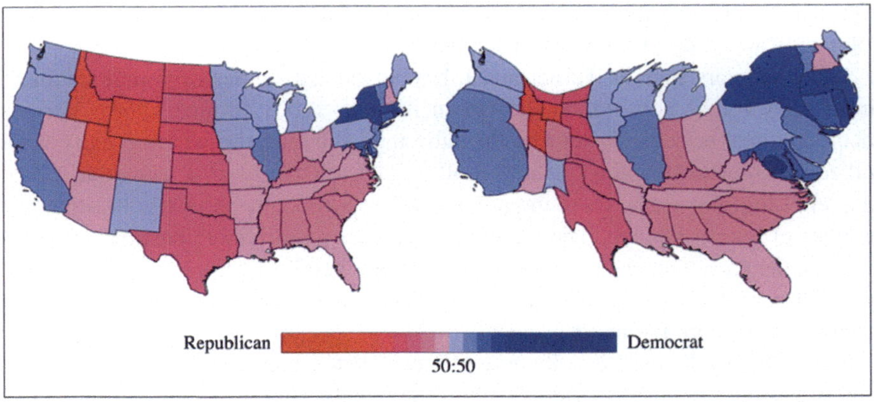

Fig. 3 Maps of the 2000 US presidential elections. *Source* Gastner and Newman (2004: 7502, Fig. 3c, f). Copyright (2004) National Academy of Sciences, U.S.A.

some statistic such as voting results, population and income (Nusrat and Kobourov 2016; Gastner and Newman; 2004; Gastner et al. 2005). Figure 3 shows the 2000 US election results as a standard thematic map (on the left) and the results proportional to the number of votes each state has in the Electoral College (Gastner and Newman 2004).

Maps are currently used to show election results in various countries around the world. Sometimes innovative methods are utilized to show the results such as the Lego maps showing election results in the 2016 federal elections in Australia by Fairfax Media (Fairfax Media 2016).

Since the 1999 South African elections, the Electoral Commission of South Africa (IEC) has used maps on its election results board to show the winning party at various levels ranging from voting districts to the provinces. The IEC is constitutionally mandated to ensure free and fair elections in South Africa. The author has been involved in South African elections since 2004 as part of the CSIR's election prediction team and has firsthand experience of the IEC's election result maps. Figure 4 shows an example of the IEC's result maps in one municipality for the 2016 local government elections that could be accessed by the public on the IEC's website.

The News 24 website for the 2016 local government elections published a map to show voter shifts between the two elections where the purple color shows the voter shift away from a political party and the green colors the voter shift towards the political party (News 24 2016). The aim of this paper is to demonstrate a few mapping methodologies apart from the general thematic maps shown here to illustrate voter shifts in South African elections. Three methods will be discussed in the methodology section namely cartograms, 3D maps and thematic maps combined with a cartograms.

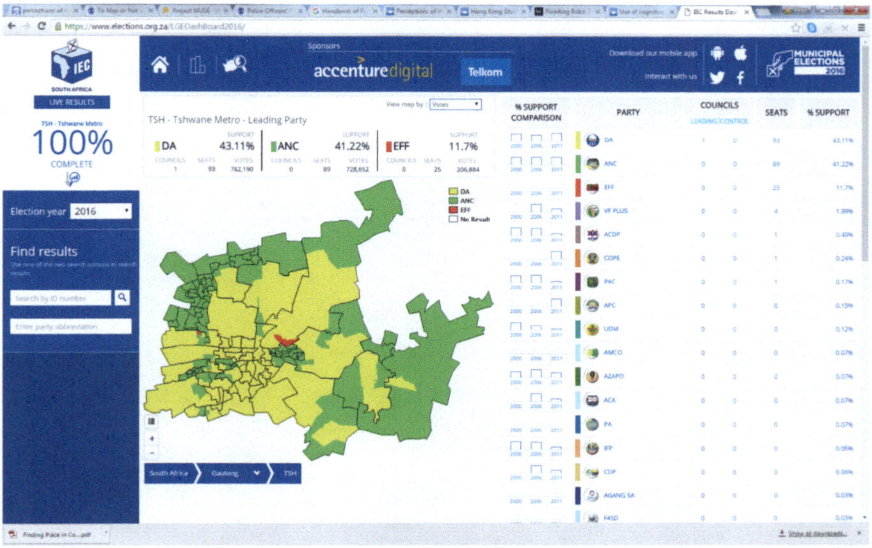

Fig. 4 LGE 2016 results for the City of Tshwane. *Source* (IEC 2016), used with permission)

3 Methodology

This paper looks at three methods to illustrate voter gains and losses between two successive local government elections, namely the 2011 and 2016 local government elections in South Africa. Figure 5 shows the local governments, municipalities, in South Africa. There are myriad of methods to illustrate voter gains and losses, but we decided to use cartograms and 3D mapping since both show changes by magnitude. These changes also illustrate where the gains and losses were in South Africa. Several maps show hues to illustrate voter gains and losses like those in Figs. 1 and 2.

3.1 Cartograms

Cartograms have been around since 1870 and are maps where the regions are scaled proportionally to a selected statistic such as populations, party votes and income (Tobler 2004; Nusrat and Kobourov 2016). Nusrat and Kobourov (2016) list several methodologies employed to draw cartograms ranging from the rubber map method, Dorling cartograms, and mosaic cartograms to diffusion-based cartograms. Diffusion-based cartograms were developed by Gastner and Newman (2004) shown in Fig. 3. The two authors subsequently developed an ArcGIS tool that could generate these cartograms. This tool was used to for the cartograms showing voter

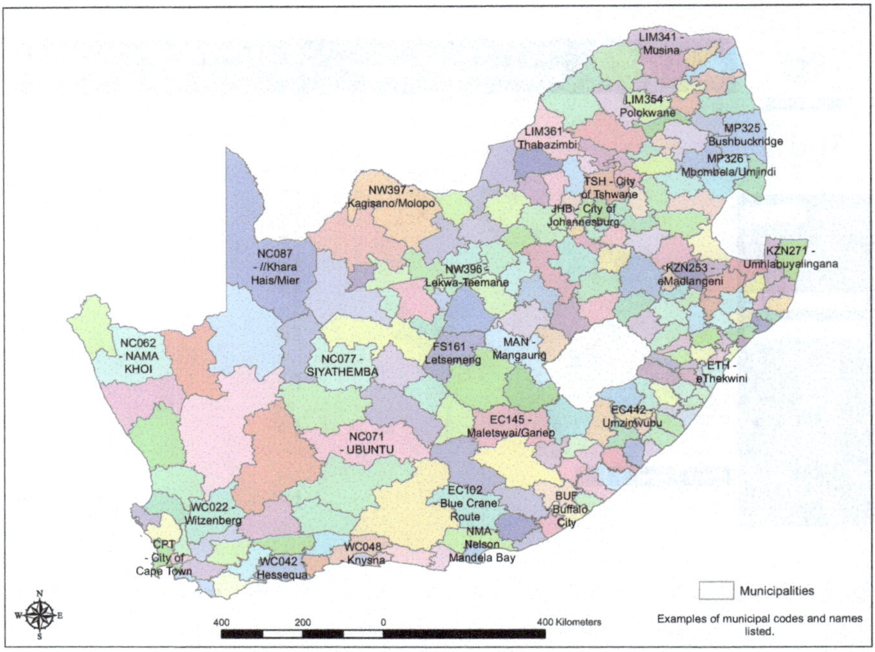

Fig. 5 Municipalities in South Africa

gains and losses for selected political parties in the South African 2016 local government elections.

The selected parties are the African National Congress (ANC), Democratic Alliance (DA) and the Economic Freedom Fighters (EFF). The EFF only came into existence in 2013 and participated in the 2014 national and provincial elections, and the 2016 local government elections. Thus, only the voter gains will be shown in the cartograms and other maps. Tobler (2004) indicated that a standard map should be used as a companion to understand the cartogram using the "brushing technique" whereby the map reader brushes (highlights) the area of interest on a standard map and it automatically highlights the same area on the cartogram. For this paper, an inset consisting of a standard thematic map was used to guide the map reader in the interpretation of the cartogram that illustrates voter shifts. The next section discusses 3D mapping.

3.2 Three-Dimensional (3D) Mapping

Three-dimensional mapping has been in use in GIS since the early 1980s (Raper and Kelk 1991). The use of three-dimensional mapping has a wide range of applications ranging from mapping noise levels in urban areas (Stoter et al. 2008);

3D buildings for the planning of telecommunication networks in a city (Coors 2003) to creating 3D models from LIDAR data (Schwarz 2010) and 3D terrain models (Raper and Kelk 1991). Vanderbei (n.d.) made a 3D rendering of voting data of the 2004 US presidential election by combining it with population density. The population is rendered in 3D with the voting data draped over it. In this paper, the percentage voter gain or loss is presented in 3D using ESRI's ArcScene software. The 3D maps are created for all three parties: ANC, DA and EFF.

3.3 Thematic Map Combined with Cartogram

The Gastner and Newman cartogram uses absolute values, meaning only positive numbers. Losses, normally shown as a negative number, i.e., −13.4% were converted to positive numbers for mapping as a cartogram. To combine both voter gains and losses into a single cartogram, the negative values of voter losses were converted into positive values. The resultant cartogram was shown as a thematic map showing gains and losses. This is similar to the Gastner et al. (2005) example where the cartogram was created based on the population per state and the voter

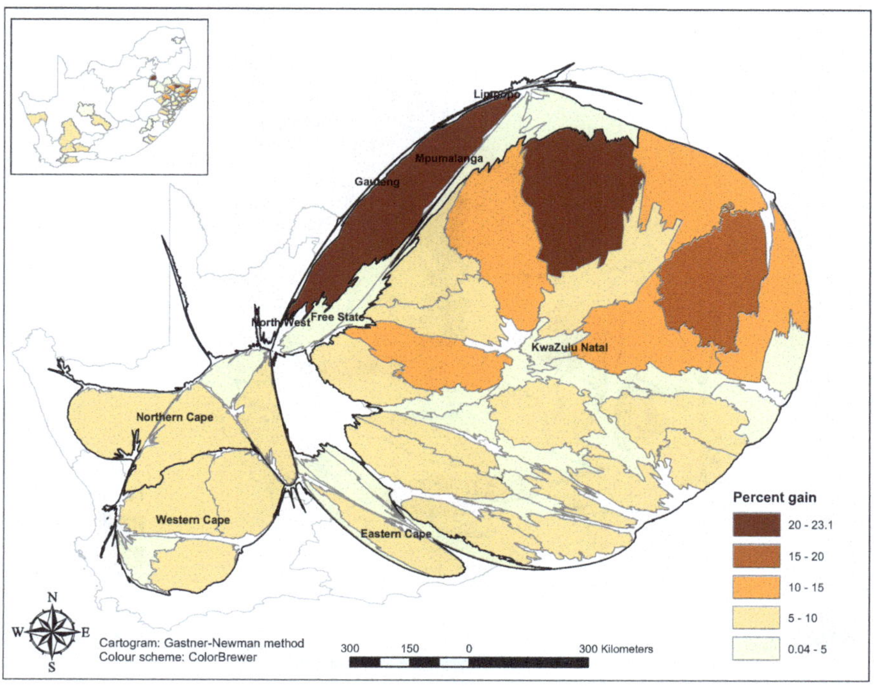

Fig. 6 Cartogram showing ANC gains

results of the 2004 US presidential elections a choropleth map. The color schemes for all the maps generated were based on those given by Brewer (2016).

4 Discussions

As mentioned in the previous section, we decided to use cartograms to show the extent of voter gains and losses for the three main parties. Each cartogram, based on the Gastner and Newman (2004) method, will be discussed in more detail. In this section, the gains and losses were mapped separately to show their extent. Figure 6 shows the ANC gains at the municipal level between the two local government elections.

Figure 6 indicates that the ANC had the most gains in KwaZulu Natal province followed by a few municipalities in the Eastern Cape, Northern Cape and Western Cape. On the other hand, the ANC suffered voter losses in the other local government areas in South Africa as shown in Fig. 7. As can be seen, the largest losses suffered by the ANC were in Limpopo province, North West province and Gauteng. The cartogram shows the largest distortions in these provinces. The province with the smallest number of losses is KwaZulu Natal.

Although the ANC lost votes in most of the local governments (municipalities), they did not lose the municipality to the opposition. However, the losses were

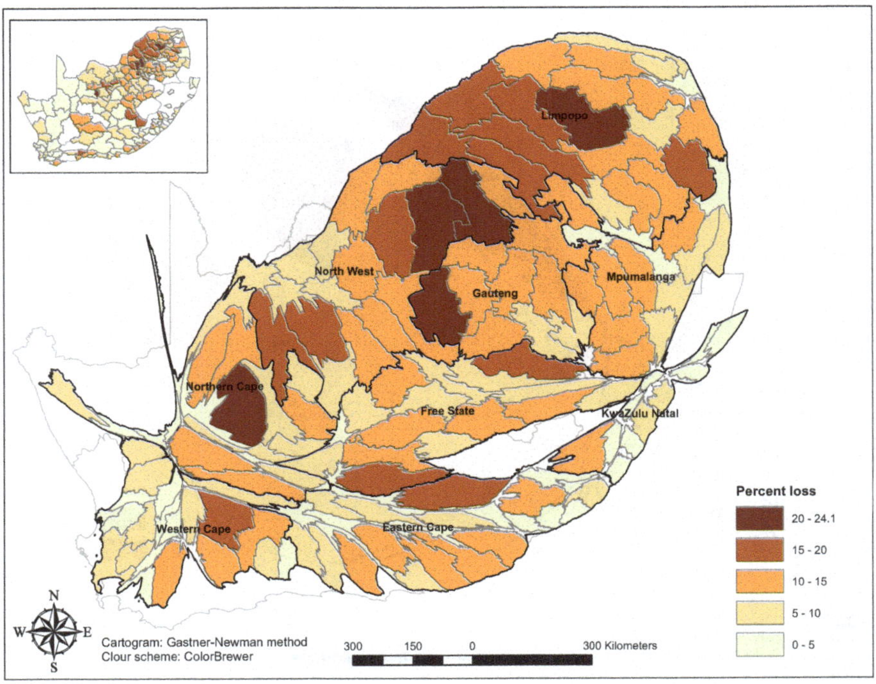

Fig. 7 ANC losses in 2016 local government elections

significant and the party had to form coalitions to remain in control of three of the local governments. The three metropolitan municipalities lost were Nelson Mandela Bay (Port Elizabeth), Tshwane (Pretoria) and the City of Johannesburg. In addition, the ruling party lost 21 local governments (non-metropolitan municipalities).

The 3D maps show these trends from a different perspective; the main difference is that the distortion is volumetric, keeping the geographical shape of the local government intact. Figure 8 shows clearly where the most dramatic shifts in voter behavior occurred in South Africa.

The biggest drawback in using the cartograms based on the Gastner and Newman (2004) method is that it cannot process negative numbers which, in this example, was used to show percentage voter losses. To accommodate this shortcoming, the data set was split into two separate datasets, one showing the gains and the other the losses. The negative values in the latter have been changed to absolute values. Municipalities with no data in these two data sets were assigned the value "0".

This approach distorts the zero (0) value municipalities into very small polygons and this leads to the distortions as shown in Figs. 6 and 7. To overcome this problem, the voter losses have been converted into absolute values with a tag indicating that these represent voter losses and the others are voter gains. Fortunately, the Gastner and Newman (2004) method retains the attributes of the original Shapefile and thus it is possible to combine thematic mapping with the cartogram showing the gains and losses in one cartogram as illustrated in Fig. 9.

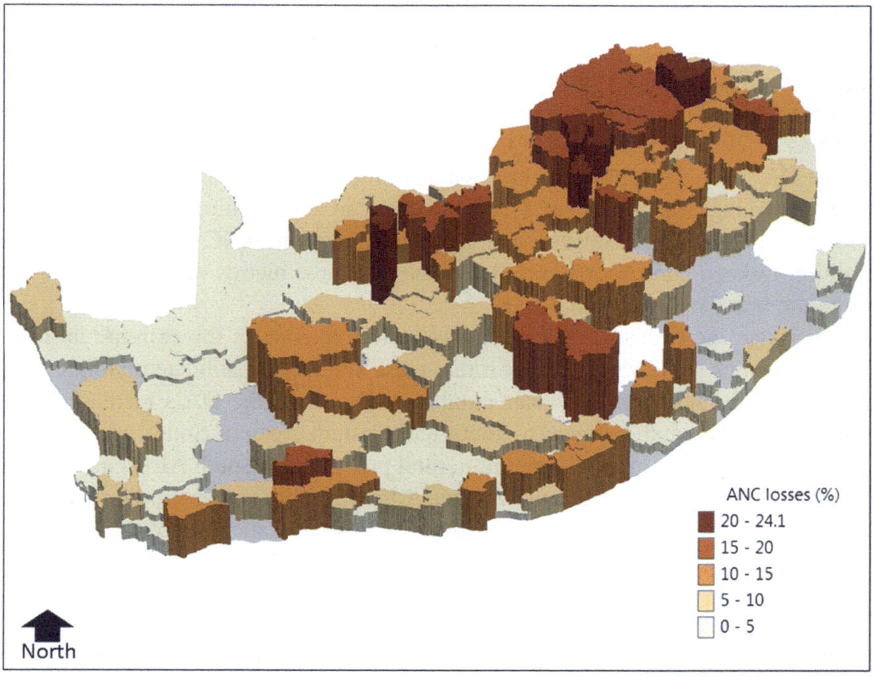

Fig. 8 3D map showing ANC losses in the 2016 local government elections

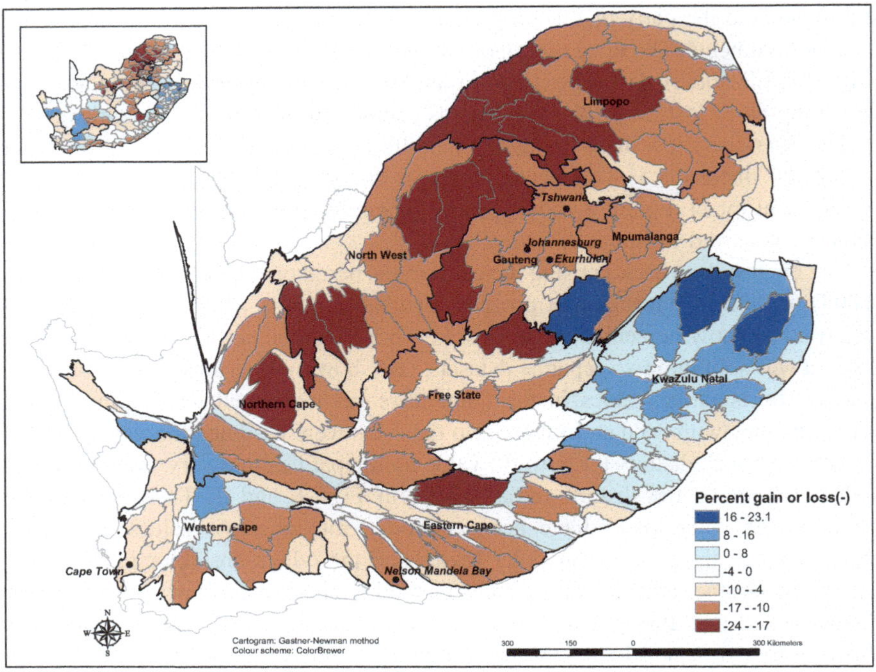

Fig. 9 Cartogram indicating both voter gains and losses by the ANC

Figure 10 shows the gains made by the Democratic Alliance (DA), currently the largest opposition political party in South Africa. The DA strengthened its majority in the Western Cape Province when compared to the previous local government elections in 2011 as illustrated in Fig. 10.

The DA made the most significant gains in the metropolitan areas, albeit with a small percentage but sufficient to enable them to take control of three metros: Nelson Mandela Bay, Tshwane and Johannesburg. The ANC was forced into a coalition to keep control of Ekurhuleni. The last three metros are in Gauteng, the economic heartland of South Africa.

A surprise increase in DA voter gains was in Musina in the extreme north of South Africa as well as some other rural areas. The DA consolidated its hold over Cape Town by increasing its voter numbers. In general, the DA's increases were prominent in the south-eastern, southern and western areas of South Africa.

The DA lost votes in four of the more rural provinces in South Africa namely the Free State, Northern Cape, North West and Mpumalanga. However, the rural voter loss for the DA was less in the provinces of Eastern Cape, KwaZulu Natal and Limpopo. The voter gains and losses are also clearly indicated by the thematic cartogram in Fig. 11. Figure 12 shows the DA voter gains in 3D.

The EFF is a young political party on the South African political scene and garnered 8.3% of the vote at a national level (IEC 2016). The cartogram in Fig. 13 clearly shows the voter percentages for the EFF. The cartogram shows similarities

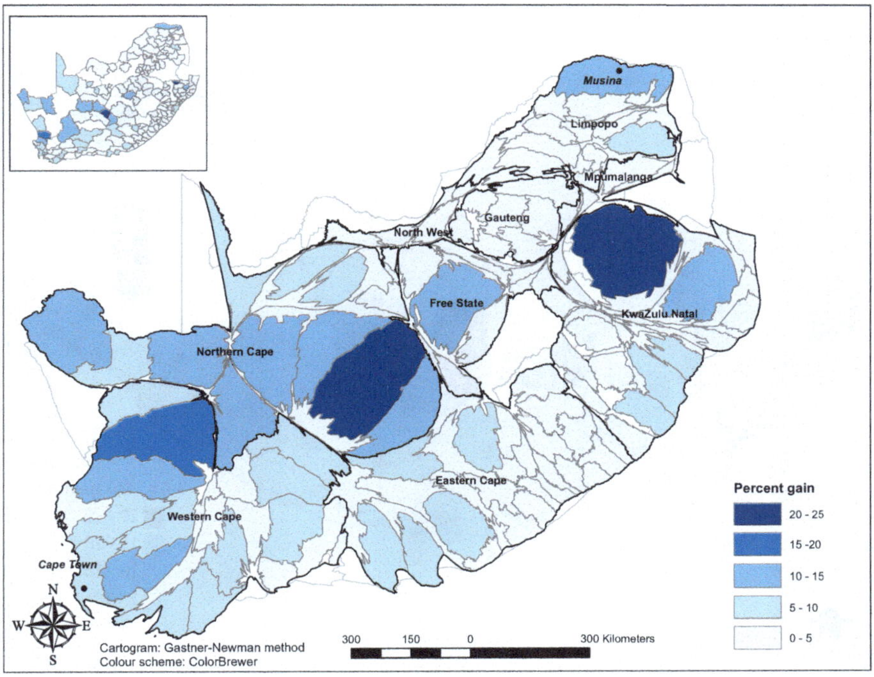

Fig. 10 DA percentage voter gains in the 2016 local government elections

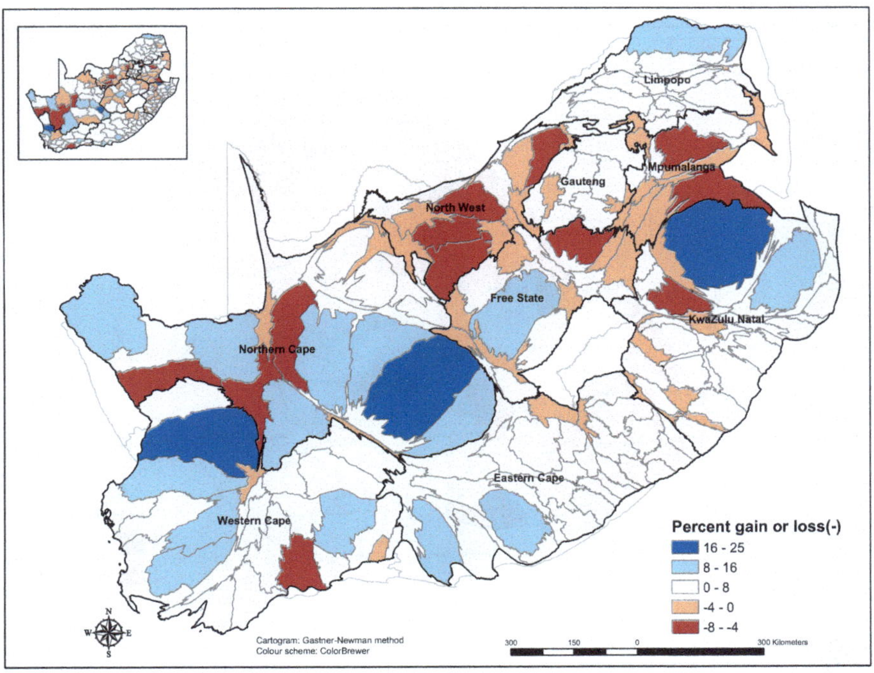

Fig. 11 Thematic cartogram showing DA voter gains and losses

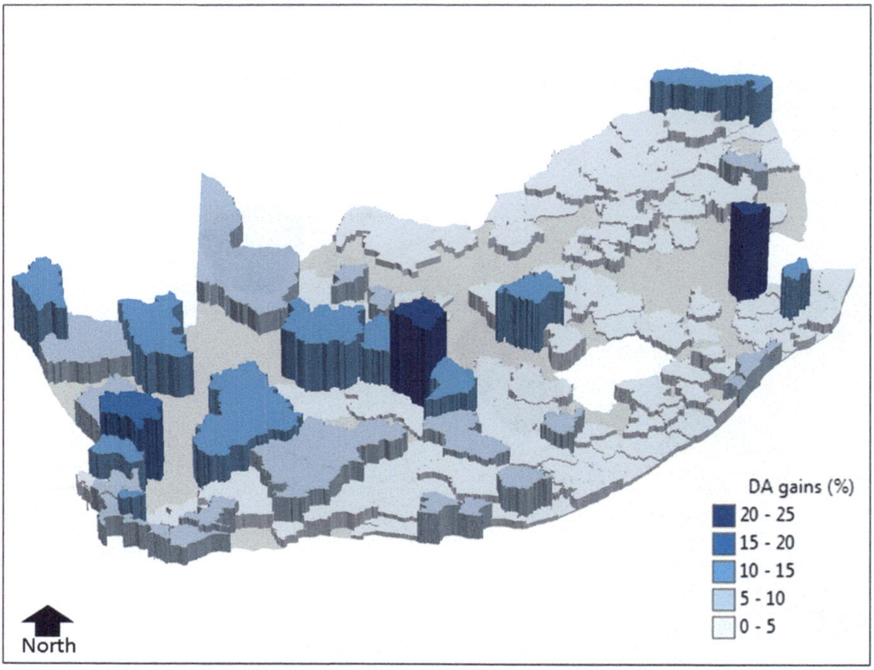

Fig. 12 3D map showing DA voter gains

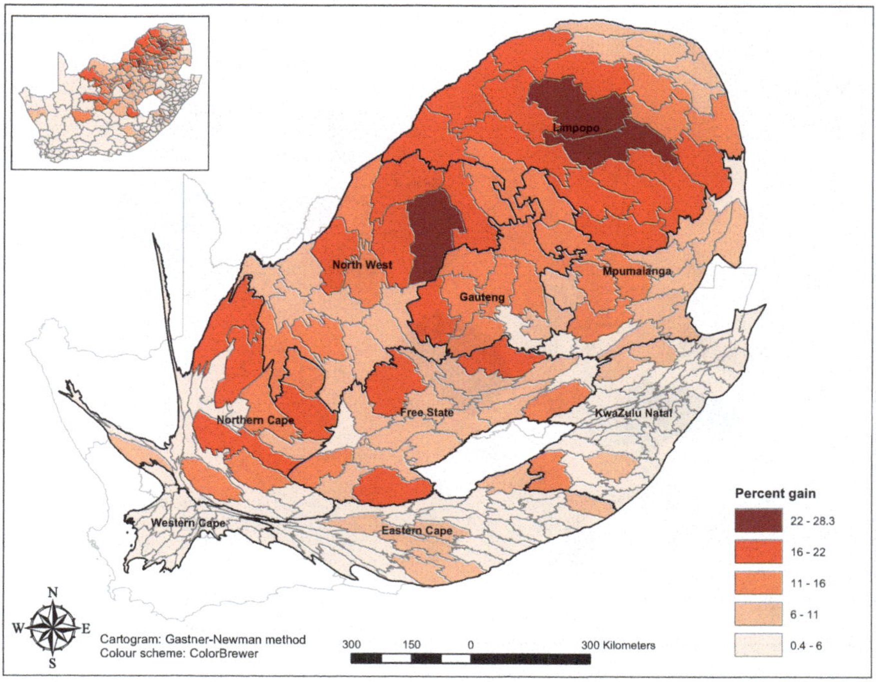

Fig. 13 EFF voter gains in the 2016 local government elections

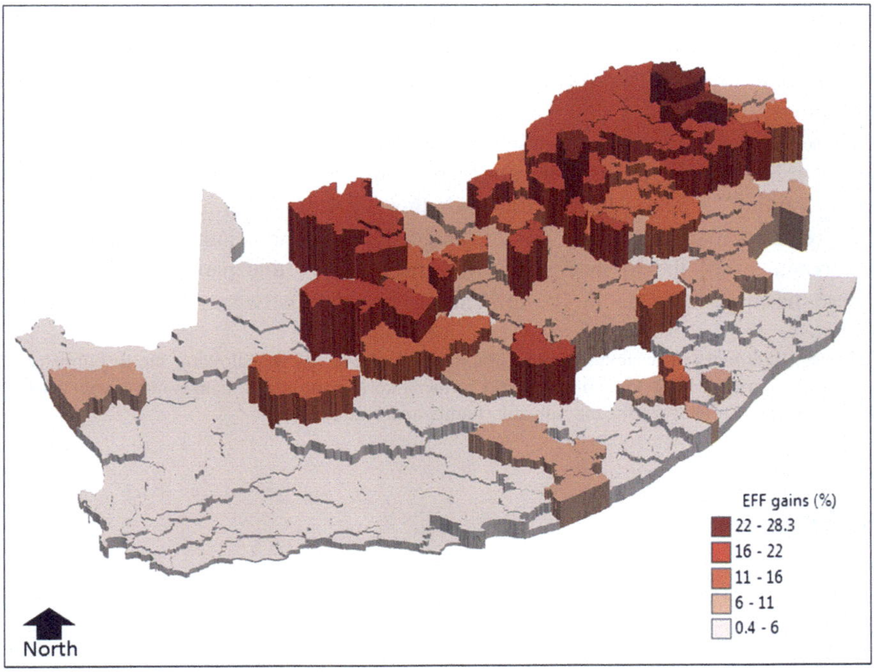

Fig. 14 3D representation of EFF voter gains in the 2016 local government elections

to voter losses of the ANC in Fig. 6. Figure 14 shows the 3D map of voter gains for the EFF. It is true that the EFF did take more votes from the ANC than the DA did, but a significant number of voters did not go to the polls in the traditional ANC areas as a protest to ANC's political conduct over the last few years (Kimmie 2016).

5 Conclusions

This paper has proposed three methods to display voter gains and losses between two local government elections. The three methods were cartograms, 3D maps and thematic cartograms. The last was used to show both gains and losses in voter percentages.

Both cartogram methods illustrate to the map reader the size of the loss or gain in voter percentages, not simply a color code to show the magnitude as in a choropleth map. However, the cartogram may confuse the map reader not familiar with South African about the location of a municipality in South Africa. It is, therefore, necessary to include a small map to guide the reader. This is based on Tobler's (2004) idea of having a standard map that can be "brushed" to indicate to the map reader which is the corresponding polygon in the cartogram. The cartogram is, however, a novel way of representing voter behavior.

The 3D map shows the changes differently. A 3D map may be more intuitive for a map reader since it keeps the original shape of the polygon. It requires less "training" of the map reader to read and understand the map compared to reading a cartogram.

It would be interesting to sample a general population to determine how they would react to these maps. The research should look at a "before training" and "after training" scenario to determine whether a cartogram or a 3D map would be the preferred map to display voter change. Despite the relative advantages and disadvantages of the methods, all three discussed in this paper clearly indicate where the major changes in voting behavior occurred between the two elections.

Acknowledgements The author would like to thank the CSIR for allowing me to publish and present the paper at ICC. The author would further like to convey his thanks and appreciation to the independent reviewers for the valuable inputs and commentary.

References

Bogue, A. G. (1968). United States: The 'New' political history. *Journal of Contemporary History, 3*(1), 5–27.

Brewer, C. A. (2016). *Designing better maps: A guide for GIS users* (2nd ed.). Redlands, California, USA: ESRI Press.

Coors, V. (2003). 3D-GIS in networking environments. *Computers Environment and Urban Systems*.

Fairfax Media. (2016). *The Lego Election*. Retrieved August 8, 2016, from http://www.smh.com.au/interactive/2016/election_lego/#/home.

Gastner, M. T., & Newman, M. E. J. (2004). Diffusion-based method for producing density equalizing maps. *Proceedings of the National Academy of Sciences (USA), 101*, 7499–7504.

Gastner, M. T., Shalizi, C. R., & Newman, M. E. J. (2005). Maps and cartograms of the 2004 US presidential election results. *Advances in Complex Systems, 8*(1), 117–123.

Hewes, F. W., & Gannet, H. (1883). *Scribner's Statistical Atlas of the United States, showing by graphic methods their present condition and their political, social and industrial development.* New York: C. Scribner's Sons. Atlas. Retrieved September 8, 2016, from https://www.loc.gov/item/a40001834/.

Independent Electoral Commission of South Africa (IEC). (2016). Results summary—all ballots. http://www.elections.org.za/content/LGEPublicReports/402/Detailed%20Results/National.pdf Downloaded PDF on 10 October 2016.

Kimmie, Z. (2016). *Personal communication during election night forecasting 5 August 2016.* Pretoria: CSIR.

News 24. (2016). Local Government Elections. Retrieved August 8, 2016, from http://www.news24.com/elections/results/lge#year=2016&map=compare&election=pr.

Nusrat, S., & Kobourov, S. (2016). The State of the Art in Cartograms. *EuroVis 2016, 35*(3). Computer Graphics Forum, The Eurographics Association and Wiley.

Raper, J. F., & Kelk, B. (1991). Three-dimensional GIS. In D. J. Maguire, M. F. Goodchild, & D. W. Rhind (Eds.), *Geographical information systems, volume 1: Principles*. New York, USA: Lonman Scientific and Technical, Wiley.

Schwarz, B. (2010). Mapping the world in 3D. *Nature Photonics, 4*, 429–430. http://www.nature.com/naturephotonics.

Stoffel, F., Janetzko, H., & Mansmann, F. (2012). Proportions in categorical and geographic data: Visualizing the results of political elections. In *Advanced visual interfaces*, Capri Island, Italy. ACM 978-1-4503-1287-5/12/05.

Stoter, J., de Kluijver, H., & Kurakula, V. (2008). 3D noise mapping in urban areas. *International Journal of Geographical Information Science, 22*(8), 907–924.

Tobler, W. (2004). Thirty five years of computer cartograms. *Annals of the Association of American Geographers, 94*(1), 58–73.

Vanderbei, R. J. (n.d.). *On Graphical Representations of Voting Results.* Retrieved September 5, 2016, from http://s3.amazonaws.com/academia.edu.documents/30799282/10.1.1.140.7759.pdf?AWSAccessKeyId=AKIAJ56TQJRTWSMTNPEA&Expires=1471859134&Signature=CTpHxDmDryzt31xe9sXL1aRNTCg%3D&response-content-disposition=inline%3B%20filename%3DOn_graphical_representations_of_voting_r.pdf.

Mapping Community Vulnerability to Poaching: A Whole-of-Society Approach

Peter M.U. Schmitz, Duarte Gonçalves and Merin Jacobs

Abstract Wildlife crime, especially rhinoceros poaching, is a major problem in South Africa. Approaches to reduce wildlife crime in South Africa range from very sophisticated, near-military interventions to engaging with nearby communities. A method investigated here is the use of a whole-of-society approach. This paper investigates the community analysis part of the whole-of-society approach that was used to reduce levels of wildlife crime. Indicators of vulnerability are discussed based on research conducted by several authors and selected on the basis that data can be obtained from generally available sources such as the census. Indicators linked with high levels of risk are identified for communities next to game parks. The goal is to reduce the vulnerability of individuals in these communities to be coerced into wildlife crime by criminal syndicates.

Keywords Wildlife crime · Indicators · Whole-of-society · Communities · Game parks

P.M.U. Schmitz (✉)
CSIR Built Environment, Meiring Naude Rd, Brummeria, Pretoria, South Africa
e-mail: pschmitz@csir.co.za

P.M.U. Schmitz
Department of Geography, Geoinformatics and Meteorology, University of Pretoria, Lynnwood Rd, Hatfield, Pretoria, South Africa

P.M.U. Schmitz
Fakultät Für Vermessung, Informatik und Mathematik, Hochschule Für Technik, Schellingstrasse 24, 70174 Stuttgart, Germany

D. Gonçalves · M. Jacobs
CSIR Defence, Peace, Safety and Security, Meiring Naude Rd, Brummeria, Pretoria, South Africa

M.P. Peterson (ed.), *Advances in Cartography and GIScience*, Lecture Notes in Geoinformation and Cartography, DOI 10.1007/978-3-319-57336-6_23

1 Introduction

Rhinoceros poaching, other wildlife poaching, and plant harvesting are international problems. In most instances, there are links to trans-national crime syndicates (Williams and Godson 2002). Initiatives are currently in place to involve communities and private landholders in the sustainable use of wildlife in South Africa. Other examples to promote conservation include Zimbabwe's CAMPFIRE initiative and the National Community-Based Natural Resource Management Programme of Namibia (Child 2012). However, as with other economic activities, these only employ a limited number of people from communities. The remainder of the community, although benefitting indirectly, may still be vulnerable, due to their socio-economic circumstances, to the lure of high prices offered by criminal syndicates.

The aim of this paper is to pinpoint vulnerabilities to criminal syndicates in a whole-of-society approach. Vulnerabilities to be examined include inadequate education, poverty, tenure issues, employment status and proximity to game reserves. All of these make community members vulnerable to poaching as a way to improve their livelihoods. The identification of these criteria was based previous research by Abbott (2008), Bezuidenhout (2014) and Warchol (2007) that describe why community members could get involved in poaching. These criteria were augmented by Williams and Godson (2002) and Duffy et al. (2013).

The purpose here is to identify and map these vulnerabilities within 20 km from existing game parks and reserves in South Africa. This distance was chosen because it is the distance that a person can walk in one day or night. The final result was a risk map showing areas of high vulnerability. Once these areas have been identified, then it is possible to investigate the various vulnerabilities and to develop strategies to reduce these. The paper is organized as follows: The next section discusses the whole-of-society approach and the identified vulnerabilities. The methodology section discusses the acquisition of data or proxies and mapping the geographical distribution of the identified vulnerabilities. Finally, there is a discussion of the results, ending with conclusions and future research.

2 Whole-of-Society

The whole-of-society approach indicates the inclusion of society, organisations and government takes an active role in conservation, disaster resilience and security, to list a few examples Lindberg and Sundelius (2012). Internationally, there has been a trend to reconsider the security chain that consists of four generic tasks, namely analysis, prevention, response and evaluation (de Spiegeleire 2011). Over time, efforts and spending have tended to focus on responding to incidents. However, the outcome of a response is never certain and can have profound consequences.

There is a shift in other countries to prevention and analysis. Such analysis is referred to as foresight in this paper. The return to the security chain suggests that there is more focus internationally on systemic approaches to security. It is critical to note that the security chain must be carried out faster than the problems are changing. The consequence of this is that the ability of organisations to respond to changes in the environment (dynamic or uncertainty) requires high levels of integration (Hoogervorst 2009). When multiple organisations are involved, this integration must extend across them.

It has been argued that all that is required is a tool (such as software) so that information is integrated and the relevant people can have access to the information. Then, once this is achieved, everyone will collaborate and common goals are not required. However, based on insights from the international community, sharing information is necessary but not sufficient—a cross-organization strategy is also required (International Federation of Red Cross and Red Crescent Societies 2000).

Information collaboration does not create a common approach and different organisations may still work against each other. Also, information collaboration may improve tactical and operational level situations but does not address problems which require strategic interventions. It does not create deep change nor does it re-align paradigms. For complex problems, all levels need to be addressed simultaneously, and the structure or the paradigm underlying the problem must be changed. Furthermore, information collaboration ignores power (also commonly ignored in organisational literature) which could disrupt, stall or undermine any effort.

The broader focus of a whole-of-society approach is important because, with the complexity of wildlife crime, different aspects of the problem are interconnected. If stakeholders are tempted to address the problem, new problems may result. For example, while law enforcement is understandably frustrated by its inability to get ahead of wildlife crime, it cannot refer to its efforts as a "war on poaching" or refer to poachers as "insurgents," and then hope to work with communities. It is important to understand that while parks are necessary for conservation, they fragment land and people.

Including all stakeholders in the development of interventions increases the chance that these interventions are successful. When some stakeholders are excluded, interventions may be rejected and/or important stakeholder issues left unaddressed.

In order to deal with the complexity of wildlife crime a vertically and horizontally integrated approach is proposed, as in Fig. 1, where the context of the problem and interventions are indicated on the left-hand side and methodology on the right-hand side. For wildlife crime, there are a number of hierarchical levels that are of concern:

- The first level is the context level and includes the end-users of wildlife products in Southeast Asia, the illicit product distribution network, the syndicates involved in transnational organised crime, poaching networks in southern Africa, wildlife parks, and communities around parks. This paper is aimed at

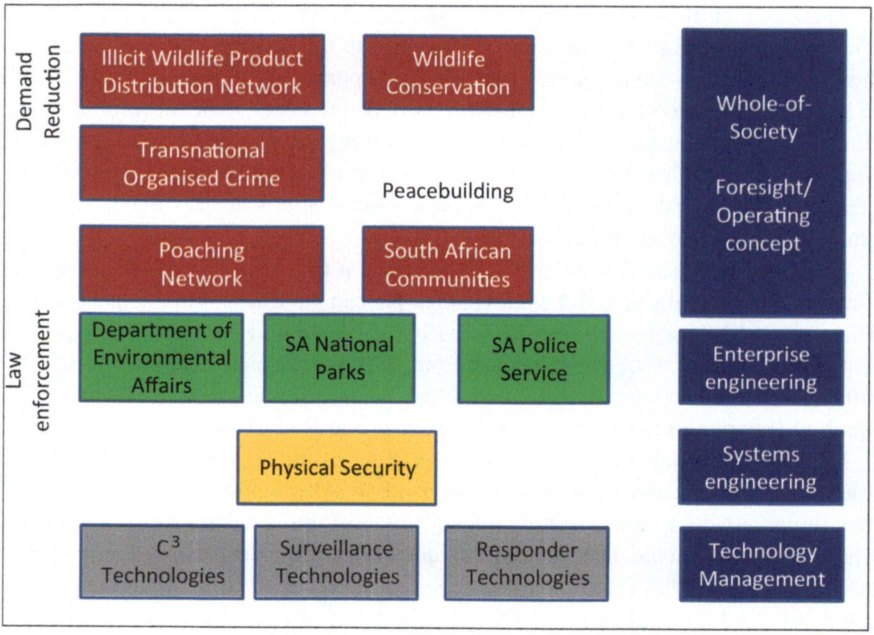

Fig. 1 *Vertically* and *horizontally* integrated approach: Wildlife crime example

this level and investigates a methodology to determine vulnerabilities in communities. These are identified and addressed as part of the whole-of-society approach.

- The second level includes the government departments and other organisations involved in interventions. One of the interventions is law enforcement which includes the South African Police Service (SAPS), the national Department of Environmental Affairs (DEA), and South African National Parks Board (SANParks), amongst others, the NGOs which support law enforcement, and Onderstepoort Veterinary Genetics Laboratory (VGL). This level provides capabilities for the intervention. Other interventions might include demand reduction or management, or community interventions.
- The third level is a particular capability within an organisation such as physically securing the Intensive Protection Zone.
- The fourth level includes the technologies for a particular capability, such as command, control and communications, surveillance technologies and responder technologies.

One level should not be assumed to be more important than another and activities should occur on all levels simultaneously.

The integrating methodologies on the right-hand side of Fig. 1 are at various levels:

- Whole-of-society—A whole-of-society is concerned with developing interventions outside of departmental mandates, for individual organisations and communities, based on foresight and drawing on different disciplines.
- Enterprise engineering (single departments and interoperability between these) —capability evaluation and assessment against strategic requirements (the realm of enterprise engineering). This is not just about ICT but includes organisational design (leadership, rewards and recognition, and culture).
- Systems engineering—technical systems and capabilities (the realm of systems engineering).
- Technology management—developing, acquiring and maturing new technologies.

Each integrating methodology level has discipline specific methodologies that are used. The focus of this work is the whole-of-society level. The approach is problem focused as opposed to discipline focused or organization-centric and vertically integrated with enterprise engineering/architecture and technical systems. The timeous flow of information in both directions on the vertical and horizontal is critical for appropriate responses to security risks.

The whole-of-society process is depicted using a linear sequence to communicate the approach in Fig. 2, but it is understood that it is not as simple as this in practice and requires iteration. There are two main areas: Foresight and

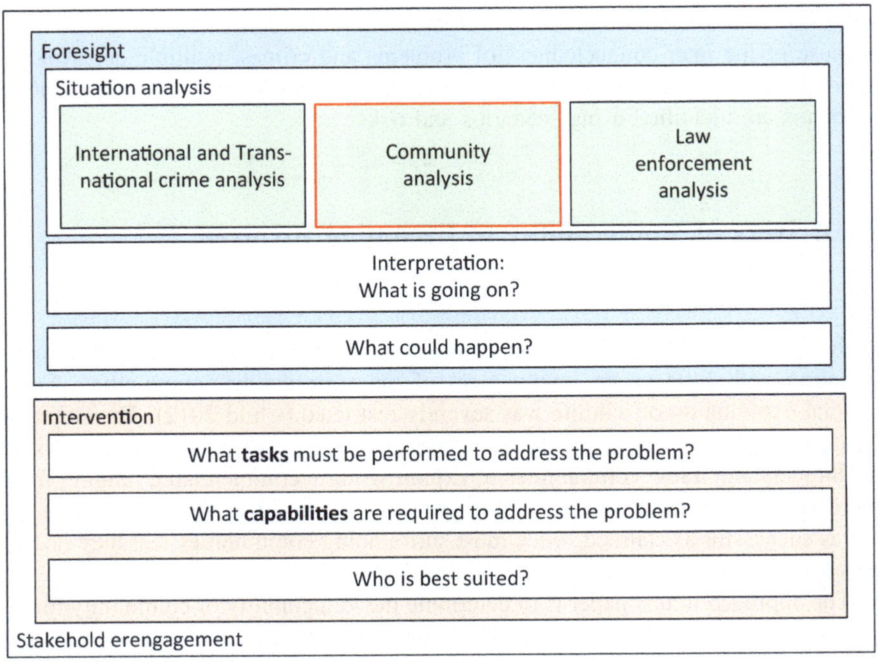

Fig. 2 The process for a Whole-of-Society approach

Interventions (a cross-department/organization strategy), which will be discussed shortly. Both of these require stakeholder engagement with government departments, local universities, NGOs and the private sector.

The threefold objective of foresight is to assess the situation, to create a common understanding of what is going on and to determine what could happen. Three areas are targeted as part of the situation analysis: International demand and Transnational Organized Crime networks, communities around game parks and reserves which are the focus of this paper and assessment of law enforcement capabilities. This work is conducted in collaboration with a network of experts. Situation analysis includes an integrated analysis of relevant constraints including legal constraints, inter- and intra-departmental policies and processes, as well as specific organizational design that potentially impact operational solutions and prevent effective and efficient operations. However, at a strategic level, legislation cannot be seen as a real constraint.

Getting stakeholders to talk is not enough—there must be a common purpose and trust to facilitate sharing of information, underpinned by requisite capabilities. Individually, organisations do not have the same purpose. Government organisations, for example, have different mandates. The common purpose is the combined purpose of a set of organisations in the context of their environment (Ackoff 1974). Common purpose or a shared future is created by building on a shared interpretation and understanding of the problem.

Our experience interacting with departments that do not have an environmental mandate is that wildlife crime is dismissed as somebody else's problem or "just" wildlife crime. This arises because of organizational and functional silos. However, because of the inter-connectedness of problems and crimes, multiple departments are required to address wildlife crime sustainably. In the context of security, future problems are identified using scenarios and risks.

3 Drivers of Vulnerability to Becoming Involved in Wildlife Crime

In the 1900s and 1930s, the approach was a strict governmental approach to conservation with little or no involvement of the surrounding communities. Commercial exploitation of wildlife was severely restricted (Child 2012). The situation changed in the 1960s to the 1980s when there was a drive to include private landholders and a few communities to exploit wildlife commercially, ranging from tourism to hunting and selling wildlife products (Child 2012). However, these are not as successful as claimed, since most surrounding communities feel they do not benefit from this arrangement (Anonymous 2014).

The approach in this paper is to determine the vulnerability of community members to being coerced into poaching based on socio-economic circumstances, using Statistics SA's Census 2011 data per sub-place as indicated in the red block in Fig. 3.

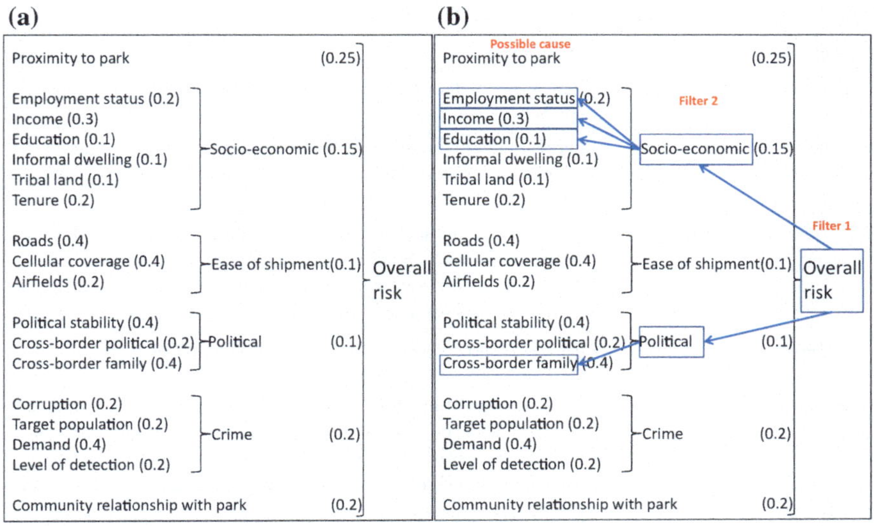

Fig. 3 Grouping of risks (**a**) and identifying risks (**b**)

The indicators listed in Table 1 have been identified based on Williams and Godson (2002), Warchol (2007), Abbott (2008), Duffy et al. (2013) and Bezuidenhout (2014).

Using the above, high vulnerability areas were determined. Then, working backwards to each indicator helped to establish possible interventions using the whole-of-society approach. This approach addresses the socio-economic environment at community level and not at an individual level. A poacher can be arrested and successfully convicted, but that will not address the community situation, since the community can still be vulnerable to coercion. The community needs to be economically uplifted in order to reduce the vulnerability to coercion. The next section discusses the methodology that was followed to determine these vulnerable communities.

4 Methodology

The main data source for the identified indicators was Census 2011 from Statistics SA. Each indicator was linked to various risk levels, ranging from very low risk (1) to very high risk (5). The indicators and associated risk levels will be discussed in this section. For all the indicators expressed as a percentage of persons or households per sub-place, the risk levels are given as very low risk (1) for the range 0 to 20%; low risk (2) for 20 to 40%; medium risk (3) for 40 to 60%; high risk (4) for 60 to 80%; and very high risk (5) for 80 to 100%.

For tenure, the risks have been reversed, meaning if the ownership is between 80 and 100% it is a very low risk (1) and if the ownership is between 0 and 20%, the

Table 1 Indicators of vulnerability based on the whole-of-society approach

Indicator	Description
Employment status	Percentage persons unemployed and discouraged work seekers per sub-place
Income	Percentage persons with no income and monthly income up to R1600 p.m. per sub-place. This is the most recent indicated poverty line by Stats SA (2015)
Proximity to game parks	Expressed as number of game parks within 20 km of each park. Game parks with rhinoceroses were given a higher risk. A 20 km buffer around a park was taken since this distance can be travelled comfortably on foot in one day or night
No education	Measured for heads of households and expressed as a percentage per sub-place
Informal dwelling	Informal dwelling, from shack to tent, expressed as percentage of households per sub-place. The assumption is that persons with more permanent structures might be more reluctant to participate in criminal activities due to potential asset forfeiture
Tribal land	Expressed as percentage households per sub-place. Persons on tribal land could be more vulnerable
Tenure	Ownership of land expressed as percentage households per sub-place. Land owners could be more reluctant owing to asset forfeiture
Road kilometer per square km per sub-place	The more roads, the easier the access to the parks as well as to ship poached rhino horn and other illegal products
Cell phone reception	Easier to organise shipment to the receivers of poached materials
Airfields	High value poached materials such rhino horn could be smuggled out of country from small airfields
Political stability	Unstable political situation more conducive environment to poaching than a stable political landscape. This includes the political situation in neighbouring countries
Level of corruption	If corruption is generally high, then higher probability for poaching
Community relationship with park	If the relationship between the community and the park is strained, then the community might tolerate poaching
Proximity of target population	Some crime syndicates may coerce target populations resident in certain areas to be involved. They could be recipients of poached materials. Example, Chinese nationals operating businesses might be coerced by crime syndicates to participate in poaching activities. Region of Birth from Census data used as a proxy
Demand at target population	The demand for poached materials at receiving populations such as China and Vietnam
Level of detection	The better the law enforcement activities, the higher the level of deterrence to poaching. This includes the targeting of middle men and kingpins

(continued)

Table 1 (continued)

Indicator	Description
Cross-border relationships: political	How good are the relationships between two neighbouring countries with respect to law enforcement?
Cross-border relationships: communities	Some communities have ties to communities across the border. These encourage illegal crossings to visit relations, and are easily exploited by criminal syndicates
Proximity of target population	Target population from Census 2011 data is percentage of persons born in Africa outside South Africa and SADC, Asia and unspecified nation of birth. Some sub-places have a high number of unspecified nations of birth and the assumption is that these persons are reluctant to give their nation of birth owing to possible stereotyping with regards to criminality, or that they might even be involved in crime

risk is very high (5). The following indicators, owing to lack of more detailed information during the study, were given a medium risk (3): political stability; level of corruption; community relationship with the park; and level of detection. South Africa has good relationships with its neighbours, hence a low level of risk (2) has been given for cross-border relationships: political. With regards to cross-border relationships: community, i.e., those areas with cross-border communities, were given a risk level of 3 and those further away from the border as risk level 1.

Note these are currently heuristic-based levels and further research into the refinement thereof is required. Several of these indicators were grouped into higher classes. The values in brackets are the weights assigned to each indicator in each group. These groups were in turn weighted to give the overall risk. Again, these weights were assigned heuristically and further research is required with regards to the assignment of weights. The groupings are shown in Fig. 3a.

The groups are socio-economic; ease-of-shipment; political and crime. The sub-place, which is a geographical region created by Stats SA to display Census 2011 data, was used as the map unit for calculating and displaying risk levels surrounding the game parks in South Africa. The aim was to create an overall risk map, identify areas of high risk, and then to use the model to determine which of the indicators needs to be addressed first to reduce the risk by backtracking from overall risk to the indicators on the left hand side of Fig. 3b. This process will be discussed in the next section.

5 Results and Discussions

As mentioned previously, the aim was to determine which indicators should be addressed first to reduce the vulnerability to being coerced into wildlife crimes. The overall risk map indicates vulnerable areas; however, the underlying reasons are not presented in this map. This map indicates overall risks and it requires further drill-down into a specific area to determine which indicators caused the high risk. Proximity to game parks or game reserves that have rhinoceroses, especially if communities are located close to several of these, are risk areas but they will not necessarily be vulnerable to be coerced into wildlife crimes if other risk factors are not present. Thus, it is important to determine other factors that will increase the likelihood of being coerced into wildlife crime. This is done by working backwards from the overall risk map to *each* of the indicators as illustrated in Fig. 3b.

Figure 4 shows the overall risk map for South Africa. The area indicated by the arrow will be used as an example to illustrate the drill-down concept. These are communities close to both the Kruger National Park and Bosbokrand Nature Reserve. The Kruger National Park has rhinoceroses, and faces the brunt of rhinoceros poaching in South Africa, while Bosbokrand Nature Reserve does not have rhinoceroses. From Fig. 3a, b, two indicators that were not grouped are proximity to game reserves (PTG) and the community relationship with the park (CRP). The reason is that these were viewed as the two crucial indicators whether a community could be vulnerable to coercion into poaching (Duffy et al. 2013 and Anonymous 2014). For the purpose of this discussion, it will be assumed that these two indicators are valid in the area of interest and will be addressed accordingly. A question remains: which other indicators in the area of interest could lead to vulnerability to coercion into wildlife crime?

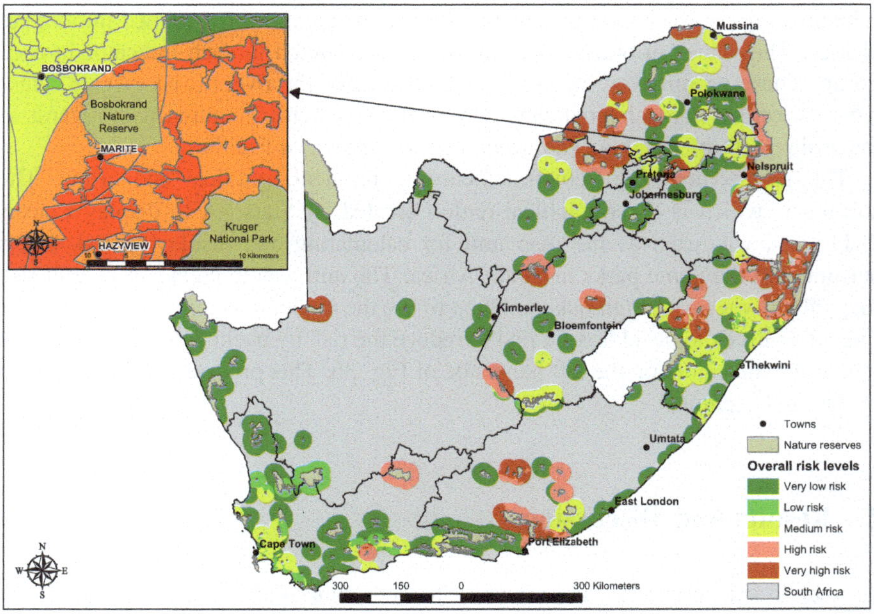

Fig. 4 Overall risk for coercion into poaching. Insert shows the area of interest

5.1 *The Four Groups*

From the four groups listed in Fig. 3a, b, the analysis of the data shows that, in most communities in this area, the two highest risks are socio-economic risk and crime. However, the communities at the bottom (south) of the map have crime and ease-of-shipment as the two highest risk factors. The reason is that there are a few airfields and dense road networks in these areas. For the purpose of the paper, the socio-economic and crime groups will be further analyzed, since these risks are high in the area of interest.

5.2 *The Socio-Economic Indicators*

The socio-economic indicator gives an indication of the socio-economic conditions around the parks. In most instances, the socio-economic conditions around the parks are linked to poverty and unemployment, thus making the community vulnerable to coercion. The socio-economic risk levels for the area of interest are shown in Fig. 5.

The green communities have low levels of unemployment and informal dwellings, few tribal land areas and high levels of land ownership (tenure), education and income. The yellow medium risk areas have slightly higher levels of unemployment

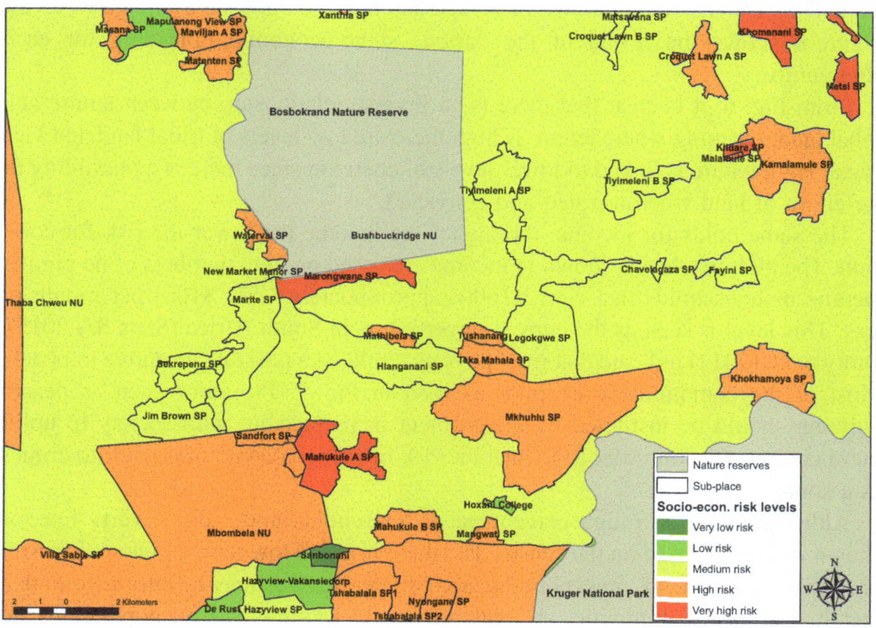

Fig. 5 Socio-economic risk levels

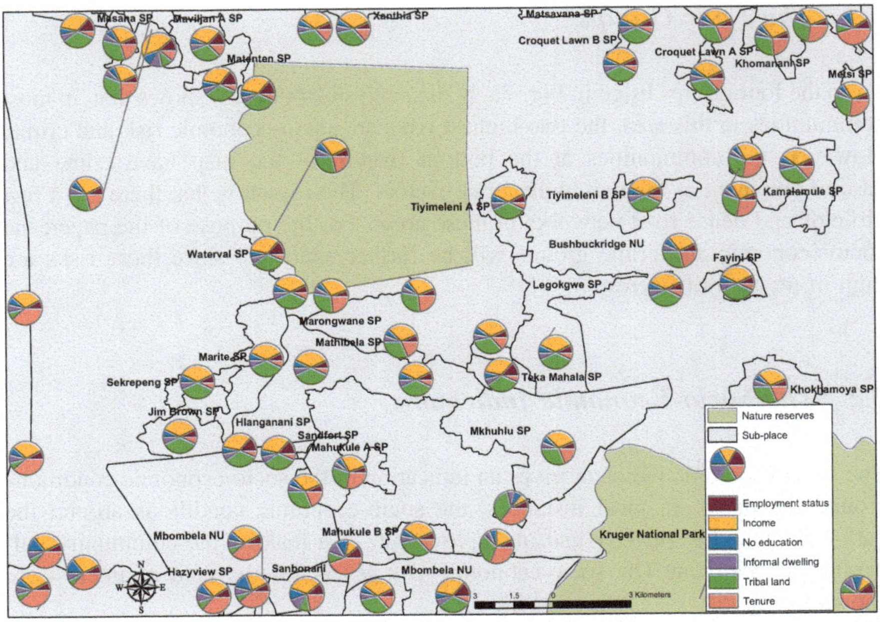

Fig. 6 The six socio-economic indicators

and informal dwellings and lower levels of education, income and land ownership. The high risk areas, shown in orange and red, are opposite to the green areas. Figure 6 shows the levels of the various socio-economic indicators for each community.

From Fig. 6, it is clear that there is an inverse relationship between tenure and tribal land, meaning where tenure is high there are low levels of tribal land and vice versa. High tenure is linked to lower levels of coercion since there is a possibility of forfeiture of land when arrested and convicted.

The same holds for income: the higher the income the lower the risk for coercion. The high risk levels shown in income are based on high numbers of no regular income or households that earn R1600 (approximately USD $100) per month or less. This level is seen as the current poverty line in South Africa (Stats SA 2015). Duffy et al. (2013) indicate that rural poverty could be a possible incentive to poach. Most of the communities are poor as seen in Fig. 7. Programs such as peace-building should be instituted by government in partnership with society to uplift these communities in order to reduce the risk of being coerced into wildlife crimes as a source of income.

Although there is a high percentage of households under the poverty line, as shown in Fig. 7, based on the census data there is also a low level of unemployment or of discouraged work seekers in the area of interest. This may be attributed to that fact that, in rural communities in South Africa, the economic active populations are moving towards the urban areas for better employment prospects or they are

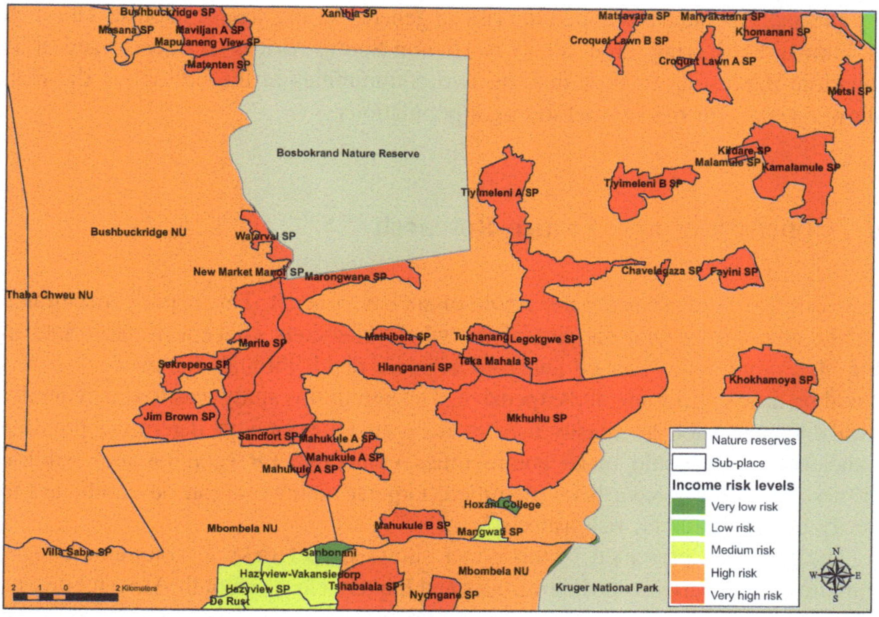

Fig. 7 Income risks based on high levels of poverty and no income

receiving government grants. This is illustrated by the small proportion related to "employment status" (i.e., being employed) in Fig. 6. There are also a large proportion of "No education" and "informal dwellings" in these areas. Most of the structures on tribal land are formal structures. The next section discusses the crime-related indicators.

5.3 *Crime Risk Indicators*

Section 4 lists the indicators which were given single values based on certain assumptions of which three are indicators in the crime group, namely corruption, demand for rhinoceros horn in the destination countries and level of detection. Table 1 also lists the proximity of a target population which is grouped in the crime group on Fig. 3. The target population, in this context, is persons who originate from the countries that have high demands for rhinoceros horn and reside in South Africa.

Some African nationals from outside of South Africa and the SADC region are often involved in organised crime and more easily work with persons from demand countries and so areas with a high number of the target population represent high risks of wildlife crime. Obviously, most foreign nationals are law abiding citizens, and therefore care should be taken when profiling this risk factor.

The analysis indicates a medium risk in general for the area of interest. This area is known for having high numbers of non-South African residents, and it is therefore interesting to note that the two communities at the bottom of the map show a very high risk related to target populations.

6 Conclusions and Future Research

This is a first iteration using the whole-of-society concept. This paper concentrated on the community analysis part of the whole-of-society concept as illustrated in Fig. 3. Communities can be analysed in more depth through fieldwork, but this could be time-consuming and expensive. The aim of this study was to use generally available data such as census data to do a first iteration to determine possible indicators which could make communities vulnerable for coercion into wildlife crimes. Once these have been identified, targeted fieldwork can be conducted to verify these, and other, risk aspects.

The risk map presented in this paper shows higher risks in communities surrounding game parks with rhinoceroses and that are close to South African borders, since the poaching of rhinoceros horns are currently very lucrative (Abbot 2008; Bezuidenhout 2014 and Duffy et al. 2013). At present the community relationships with parks are strained in general (Child 2012 and Duffy et al. 2013) and it is important to improve the relationship between the communities and the parks.

With regards to the socio-economic indicators, some communities show a higher risk than others as indicated in Fig. 6. According to the census data there is wide-spread poverty in the communities and high levels of unemployment, and it is recommended to determine the reasons for this and then implement programs to improve the income and employment levels of the community members.

Education shows a similar trend as employment. "No education" measured in this study concentrated only on heads of households without any education. Based on this, it is recommended to determine general levels of education in the communities, improve the education levels where necessary, and engage with the communities accordingly to sensitise them with regards to protecting wildlife and to improve their relationships with the parks.

Crime indicators that could be based on data from the census are mostly related to the number of persons from outside South Africa living in communities. This is important since Williams and Godson (2002) indicated that there are some instances where transnational organised criminals have the same ethnicity as communities that immigrated to a specific country. There are two groups included in the analysis, namely those from Africa but outside the SADC region and those from Asia. Care should be taken not to assume that these persons are all involved in illegal activities, but to use this risk indicator so that law enforcement agencies can be aware of areas with higher concentrations foreign nationals and the possible associated risk of them being involved in illicit activities.

The study shows great potential to identifying the underlying issues such as poverty, unemployment, tenure, park-community relationships, etc. that makes a community vulnerable to coercion into wildlife crime by crime syndicates. An article by Hübschle (2016) underlines the complexity of communities involved in poaching and further research is required to establish the validity and weights of current indicators, and to determine other indicators and the weights that should be allocated to these indicators. An alternative methodology which would be more aligned to the studies by Hübschle (2016) and others would be to establish a positive index such as the Fortunate Index (Schmitz et al. 2015; based on the Gross Happiness Index from Bhutan) which could guide the whole-of-society and/or peace building initiatives towards improving both the communities and conservation initiatives.

Acknowledgements We are grateful to the CSIR for the opportunity to publish this paper. Research was partly funded by INR. The authors would like to thank the reviewers and the editor for their comments and suggestions to improve the paper. The authors would like to express their gratitude to be able to present the paper at the ICC 2017.

References

Abbott, B. (2008). *The economics of endangered species poaching*. Resource Economics & Policy Analysis Research Group, University of Victoria, Victoria, Canada.

Ackoff, R. (1974). *Redesigning the future: A systems approach to societal problems*. s.l.: New York: Wiley.

Anonymous. (2014). *Wildlife trafficking exemplifies broader African issues*. Confidential report to the CSIR.

Bezuidenhout, C. (2014). *Rhino poaching*. Confidential report to CSIR delivered on 24 January 2014.

Child, B. (2012). The sustainable use approach could save South Africa's rhinos. *South African Journal of Science*. 108(7/8), Art. #1338, 4 pages. http://dx.doi.org/10.4102/sajs.v108i7/8.1338 Downloaded on 26 February 2016.

de Spiegeleire, S. (2011). Ten trends in capability planning for defence and security. *The RUSI Journal, 56*(5), 20–28.

Duffy, R., Emslie, R. H., & Knight, M. H. (2013). Rhino poaching: How do we respond? *Evidence on Demand.* http://dx.doi.org/10.12774/eod_hd087.oct2013.duffy.

Hoogervorst, J. (2009). *Enterprise governance and enterprise engineering*. s.l.: s.n.

Hübschle, A. M. (2016). The social economy of rhino poaching: Of economic freedom fighters, professional hunters and marginalized local people. *Current Sociology*, 9–21. doi:10.1177/0011392116673210.

International Federation of Red Cross and Red Crescent Societies (2000). *Improving Coordination*, s.l.: s.n.

Lindberg, H., & Sundelius, B. (2012). *Whole-of-society disaster resilience: The Swedish way*. Downloaded from https://www.msb.se/Upload/Nyheter_press/McGraw-Hill%20Homeland%20Security%20Handbook,%20Helena%20Lindberg%20and%20Bengt%20Sundelius.pdf on 26 February 2016.

Schmitz, P., Eksteen, S., & de Bruyn, S. (2015). Towards a new way of measuring service delivery in Gauteng: Calculating a Fortunate Index. *South African Journal of Geomatics, 4*(3), 317–325.

Stats, S. A. (2015). *Census 2011: Income dynamics and poverty status of households in South Africa*. Statistics South Africa, Private Bag X44, Pretoria, 0001, South Africa.

Warchol, G. L. (2007). The transnational illegal wildlife trade. *Criminal Justice Studies: A Critical Journal of Crime, Law and Society, 17*(1), 57–73.

Williams, P., & Godson, R. (2002). Anticipating organized and transnational crime. *Crime, Law & Social Change, 37,* 311–355.

Mapping Urban Landscapes Along Streets Using Google Street View

Xiaojiang Li, Carlo Ratti and Ian Seiferling

Abstract City streets are a focal point of human activity in urban centers. Citizens interact with the urban environment through its streetscape and it is imperative to, not only map city streetscapes, but quantify those interactions in terms of human well-being. Researchers now have access to fully digitized representation of streetscapes through Google Street View (GSV), which captures the profile view of streetscapes and, thus, shares equivalent viewing angles with those of the citizen. These two facets—a wealth of streetscape photographs at city-scale and a shared perspective with the end user—underscore the potential of these data in street-level urban landscape mapping. In this study, we introduce two examples that demonstrate GSV as a high-quality data source for mapping street greenery and openness. First, the modified green view index, which estimates the visibility of street greenery, was applied to static GSV images in order to map the spatial distribution of street greenery. Second, GSV panoramas were used to quantify and map the openness of street canyons by applying a geometrical transformation and image classification to the panoramas. The results of these two novel applications of street-level photographic data illustrate its utility for quantifying and mapping key urban environmental features at the same viewpoint in which we, as citizens, experience the urban landscape.

Keywords Google street view · Urban landscape · Street greenery · Street openness

X. Li (✉) · C. Ratti · I. Seiferling
MIT Senseable City Lab, Cambridge, MA, USA
e-mail: xiaojian@mit.edu

X. Li
Department of Geography, University of Connecticut, Storrs, CT 06269, USA

M.P. Peterson (ed.), *Advances in Cartography and GIScience*, Lecture Notes in Geoinformation and Cartography, DOI 10.1007/978-3-319-57336-6_24

1 Introduction

The street is the basic unit of the city and a focal point of human activity, acting as the foundation for transportation and information exchange. Moreover, because cities are complex systems that integrate the physical with the social spaces (Salesses et al. 2013), streetscapes play an important role influencing social interactions. Taken together, city streets become one of the most critical urban landscape features effecting, or reflecting, people's lifestyles and physical, mental, and social well-being (Miller and Tolle 2016; Li et al. 2016). It follows that a thorough quantification and understanding of the physical streetscape (i.e., features and dynamics) would offer great utility to those investigating the urban environment, its physical social interactions, and implications on human well-being.

A variety of methods and data sources to model and measure the urban physical environment exist and are widely exploited in urban studies. For example, high spatial resolution remote sensing has been used for the study of street greenery, building height estimation, and other urban feature extractions. However, high spatial resolution remotely sensed data are not always available. In addition, the profile view of streetscapes, that people experience and see from the ground, is different from the overhead view captured in remotely sensed imagery (Yang et al. 2009; Li et al. 2015). Those differences can be overcome with manual inventories and field surveys, however in situ data collection are laborious and time-consuming (i.e., low-throughput) and are prone to sampling errors, especially when non-expert or volunteer-based (Nowak et al. 2014).

Google Street View (GSV) was launched in 2007 and differentiates itself from traditional mapping software by directly capturing the ground-level visual appearance. By stitching pictures taken in different angles together, GSV creates what feels like a seamless tour of city streets and it can give people the feeling of "being there" (Li et al. 2015). Given GSV's ground-level sampling of digital photographs, one can argue that its data is a directly suitable representation of cityscapes, perhaps more so than remotely sensed images. In addition, the GSV is produced based on industrial criteria and collected for many cities all over the world. It would be an important and promising data source for urban landscape mapping along the streets.

GSV images have been widely studied in the computer vision community. These studies include commercial entities identification (Zamir et al. 2011), 3D city modeling (Torii et al. 2009; Mičušík and Košecká 2009). Recently, GSV images were also used for urban studies, such as public open space auditing (Edwards et al. 2013; Taylor et al. 2011), neighborhood environmental auditing (Charreire et al. 2014; Rundle et al. 2011; Odgers et al. 2012; Griew et al. 2013), and quantifying public safety perception (Naik et al. 2014). While these examples highlight the diverse potential of GSV data, there are few studies using street-level photographs for cartographic purposes. Two novel applications of GSV image data are introduced here that quantify environmental features and map them in the urban physical landscape;

those being street greenery and openness of the street canyons. In doing so, we show how to make GSV data an incredibly promising tool for urban environmental studies.

2 Google Street View (GSV) Data Collection

GSV is a free online service featured in Google Maps and Google Earth that provides panoramic views from positions along city streets throughout much of the world (Fig. 1). Google provides several types of application programming interfaces (APIs) for accessing the GSV data as static images and 360° panoramas.

2.1 *Collecting Static GSV Images*

Through the Google Static Image API, users can request a given static GSV image with a HTTP URL by specifying the location coordinates (Google 2014). In addition, the static GSV images in different horizontal directions and vertical angles can be requested by specifying the parameters of *heading* and *pitch*, respectively. Figure 2 shows one requested static GSV image and the corresponding URL.

2.2 *Collecting GSV Panoramas*

In addition to static images, GSV panoramas can also be downloaded from the Google Server through the API. Figure 3 shows 26 × 13 tiles of a GSV panorama together with corresponding URLs. In these URLs, the *panoid* represents the

Fig. 1 A snapshot of Google Street View

Coordinate: http://maps.googleapis.com/maps/api/streetview?size=400x400&location=41.935,-87.80524300000002&fov=60&heading=180&pitch=0&sensor=false

Fig. 2 A GSV image and the corresponding Uniform Resource Locator (URL)

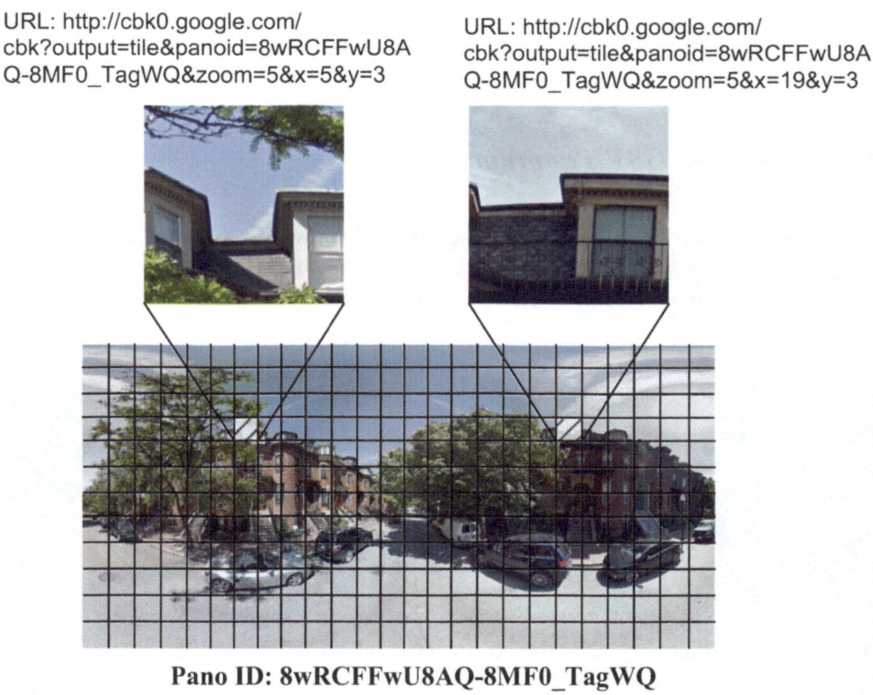

Fig. 3 The tiles of a GSV panorama and their corresponding URLs

unique panorama ID, x and y represent the column and row number of the tile, respectively. A complete panorama includes 26 × 13 tiles, therefore, the x ranges from 0 to 25 and y ranges from 0 to 12.

In the new version of Google Street View Image API (2016), the GSV image metadata can be accessed. The following is an example of accessing the metadata of a GSV panorama.

URL: https://maps.googleapis.com/maps/api/streetview/metadata?size=400x400&location=42.354489,-71.059150&fov=90&heading=235&pitch=10&key=YOUR_API_KEY

Metadata of the a GSV panorama


```
{
"copyright" : "© 2017 Google",
"date" : "2016-10",
"location" : {
"lat" : 42.3545003,
"lng" : -71.0591524
},
"pano_id" : 8a9h88JXDZ8k6FC2mPbDkA",
"status" : "OK"
}
```


The returned metadata includes the acquisition date, coordinate of the panorama, and the panorama ID. There is a discrepancy between the input coordinate in the URL and the returned coordinate in the returned metadata. This is because the GSV image API will return a nearby panorama and the coordinate of the panorama if there is no panorama at the place the user specified.

In this study, GSV panoramas were collected in two steps. The first step is to get the metadata of GSV panoramas using coordinates as inputs. The second step is to download GSV panorama tiles from Google servers and mosaic the tiles to complete GSV panoramas.

3 Urban Landscape Quantification and Mapping

This section introduces two applications using GSV to map the distribution of street greenery and the openness of street canyons.

3.1 Mapping the Visibility of Street Greenery

As an important component of urban greenery, street greenery has long played a critical role in the urban landscapes. Other than the instrumental functions such as urban heat island mitigation, storm-water reduction, airborne pollutant absorption, the street greenery also provides sensory functions to people. The existence of street greenery increases people's rating of streetscapes (Camacho-Cervantes et al. 2014; Balram and Dragićević 2005). Street greenery makes streets more walkable and beautiful (Li et al. 2015; Bain et al. 2012). In addition, the street greenery can also mitigate the visual intrusion of traffic, which would contribute to the walkability of streets. Assessment of the sensory functions of street greenery must be guaranteed for urban landscape management.

In situ manual inventories and remote sensing have been widely used to assess and map street greenery. The in situ survey is very time- and cost-consuming although it can measure detailed information about street greenery. Because of its virtues such as repeatability, synoptic view, and larger area coverage, remote sensing has become an important data source for assessment of urban greenery. With the emergence of new sensors such as LiDAR, that are now available at high, but sometimes feasible costs, it is possible to measure the canopy cover, canopy height, and even the volume of street greenery using high spatial resolution remotely sensed data with an acceptable margin of error. However, what people on the ground see or feel, which is influenced by the profile view of street greenery, would be very different compared with the overhead remotely sensed data (Li et al. 2015; Yang et al. 2009). Google Street View (GSV), providing street-level and profile views of urban landscapes, is a very promising data source for assessing the sensory function of street greenery. Considering the direct connection between the visibility of street greenery and human perceived greenery (Li et al. 2015), in this study, we mapped the visibility of street greenery in residential neighborhoods of Boston, Massachusetts, USA using street-level images.

In order to represent the street greenery in the residential area, three thousand sample sites were generated along the residential streets of the study area. In this study, we applied the modified green view index proposed by Li et al. (2015) to represent the distribution of street greenery.

The modified green view index is the average percentage of green vegetation in GSV images for 6 horizontal directions and 3 vertical directions at each sample site (Li et al. 2015) (Fig. 4). It was calculated using Formula (1),

$$Green\ View = \frac{\sum_{i=1}^{6}\sum_{j=1}^{3} Area_{g\_ij}}{\sum_{i=1}^{6}\sum_{j=1}^{3} Area_{t\_ij}} \times 100\,\% \qquad (1)$$

where, $Area_{g\_ij}$ is the number of green pixels in a GSV image, for each camera direction and vertical angle for each sample site, and $Area_{t\_ij}$ is the total number of pixels in each of the 18 GSV images.

The method has been shown to correlate well with the percent vegetation cover in GSV images as manually delineated by a human eye (Li et al. 2015). As should be expected, those sites with large street tree canopies and more street greenery tend to have larger modified green view index values than those with few trees or green cover (Fig. 5).

Figure 6 shows the spatial distribution of the modified green view index at the site level and the census tract level in Boston, Massachusetts. It is clear that the

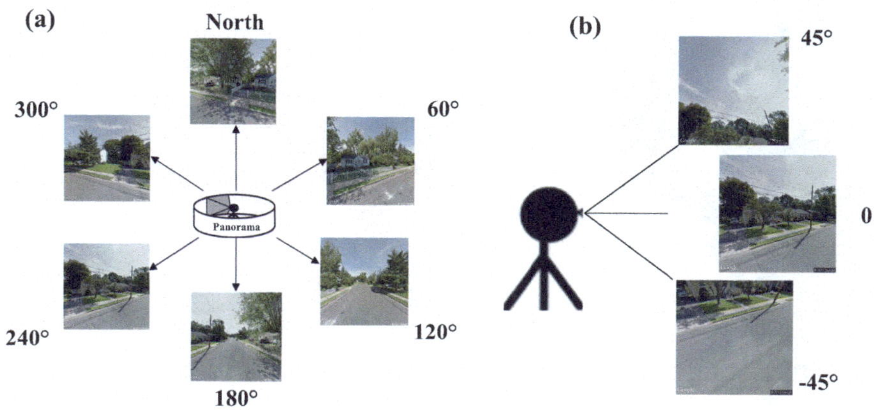

Fig. 4 GSV images of six *horizontal* directions at a sample site in the study area (**a**), and GSV images of three *vertical* view angles at a sample site (**b**)

Fig. 5 The modified green view index values in different sites of Boston

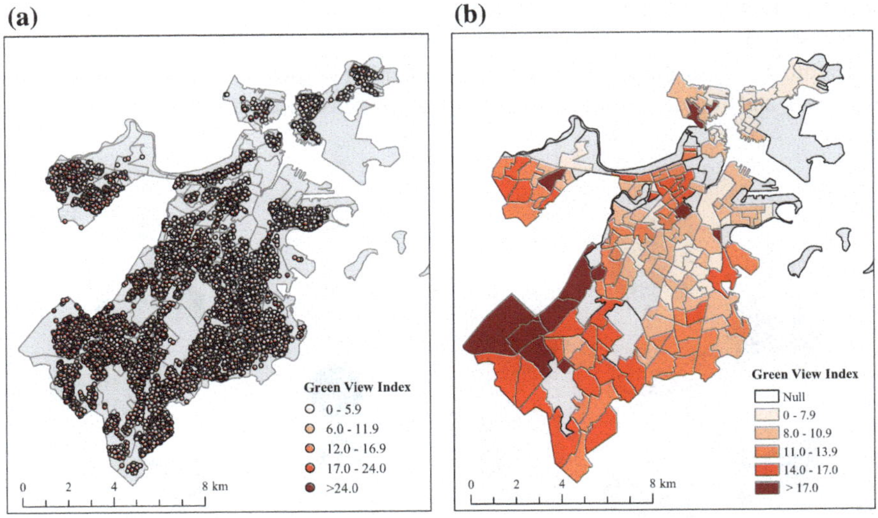

Fig. 6 The spatial distribution of the modified *green* view index in Boston, Massachusetts, (**a**) site level map, (**b**) the distribution of the *green* view index map at the census tract level

southwestern part of Boston has denser and more expansive street greenery than the northeastern part. The downtown areas of the city are characterized by low amounts of street greenery relative to the suburban areas.

3.2 Mapping the Openness of Street Canyons

The sky view factor (SVF) is a geometric quantification of the openness or the degree of sky visibility within street canyons (Oke 1981; Chapman and Thornes 2004; Ratti et al. 2002) and has a wide array of applications as an indicator of urban form, structure, and proxy of localized environmental conditions. The SVF has been applied in studies of forestry, urban climate, urbanization, air pollution, and urban heat island effects (Holmer et al. 2001; Grimmond et al. 2001; Debbage 2013; Hämmerle et al. 2011; Carrasco-Hernandez et al. 2015; Unger 2008; Eeftens et al. 2013; Chen et al. 2012; Svensson 2004; Lin et al. 2012).

The photographic method is one of the most widely used methods to estimate the SVF. However, taking fisheye images is usually required in fieldworks and the image classification is usually conducted manually. Both of these limit the photographic method to very small-scale SVF estimations. The digital surface model (DSM) simulation method is another way to estimate the SVF. In the simulation method, SVF values can be estimated based on the simulation of the light radiation in the DSM. However, the DSM is not always available and it is difficult to simulate the obstruction of tree canopies. In addition, the accuracy of SVF estimation result is influenced heavily by the spatial resolution of the DSM. Due to these limitations and discrepancies between methods, there still exists no widely used or standard method to compute the SVF at the city or regional scales.

In this study, we propose to use the publicly accessible GSV panoramas to estimate the SVF at regional scale based on geometrical transform and image classification. We collected 11,392 GSV panoramas for SVF estimation in Boston, Massachusetts. First, we created fisheye images by projecting those collected GSV panoramas from cylindrical projection to azimuthal projection. Figure 7 shows the geometric model for the transformation of cylindrical projection to azimuthal projection. The W_c and H_c are the width and height of the cylindrical panorama, so, the radius of the fisheye image should be, $r_0 = W_c/2\pi$, and the width and height of the fisheye image are W_c/π. Therefore, the center pixel of the result fisheye image (C_x, C_y) is,

$$C_x = C_y = \frac{W_c}{2\pi} \quad (2)$$

For pixel (x_f, y_f) on the result fisheye image, the corresponding pixel on the cylindrical panorama should be (x_c, y_c),

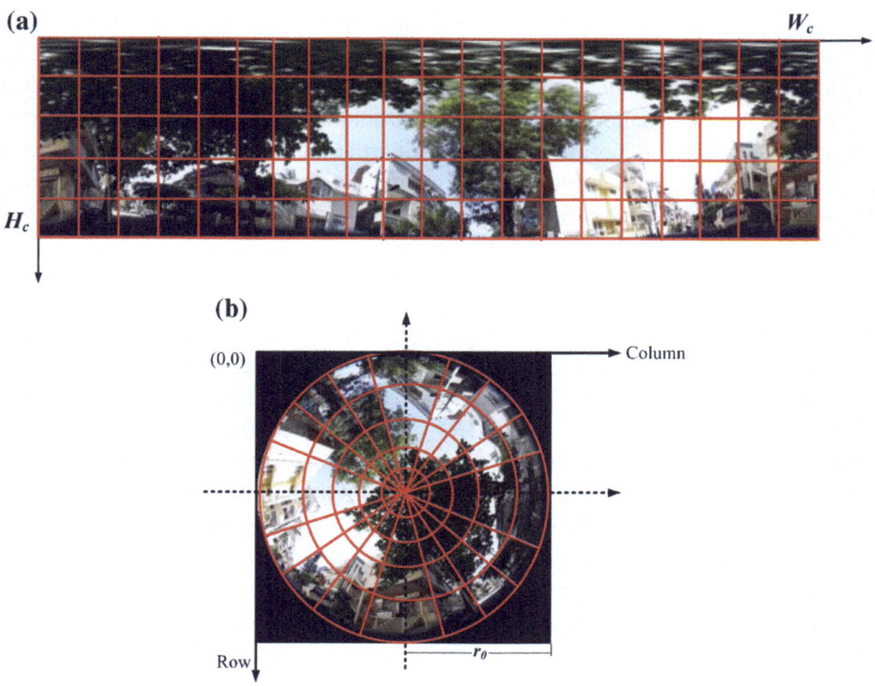

Fig. 7 Transform of cylindrical projection (**a**) to azimuthal projection (fisheye image) (**b**)

$$x_c = \frac{\theta}{2\pi} W_c$$
$$y_c = \frac{r}{r_0} H_c \tag{3}$$

where, θ and r are,

$$\theta = \begin{cases} \frac{3\pi}{2} - \arctan(\frac{y_f - C_y}{x_f - C_x}), x_f < C_x \\ \frac{\pi}{2} - \arctan(\frac{y_f - C_y}{x_f - C_x}), x_f > C_x \end{cases} \tag{4}$$

$$r = \sqrt{(x_f - C_x)^2 + (y_f - C_y)^2} \tag{5}$$

Based on the geometric model between the fisheye image and the cylindrical projection GSV panorama, we further applied affine transform to generate simulated fisheye images based on the cylindrical GSV panoramas. Figure 8 shows a generated fisheye image (Fig. 8a) based on a GSV panorama (Fig. 8b).

The sky extraction is a requisite step for SVF calculation in the photographic method. In this study, an unsupervised object-based image classification method was used to extract the sky pixels from the simulated fisheye images automatically. The *meanshift* algorithm was first used to segment the fisheye images (Comaniciu and Meer 2002). The image segmentation algorithm clusters nearby pixels that have similar spectral information into objects that would enhance the difference between the sky pixels and non-sky pixels. Considering the fact that sky pixels are usually brighter than non-sky pixels, we used the overall *brightness* of pixels to differentiate the sky pixels and non-sky pixels. The *brightness* image was calculated as,

$$Brightness = (red + green + blue)/3 \tag{6}$$

where, *red*, *green*, and *blue* are the pixel values in red, green, and blue bands, respectively. Those pixels have higher *Brightness* values are sky pixels. We further used Otsu's method (Otsu 1975) to find the optimum threshold to separate the sky pixels from the non-sky pixels based on the calculated *Brightness* images. The Otsu's method chooses a global threshold automatically by minimizing the within-class variance and maximizing the between-class variance. Those pixels have their *Brightness* values higher than the threshold are sky pixels. All other pixels other than sky were treated as obstruction pixels. Figure 8c shows the classified sky pixels in a fisheye image.

Based on the sky classification results, we applied the classical photographic method to calculate the SVF. The SVF is calculated as formula (7) (Johnson and Watson, 1984),

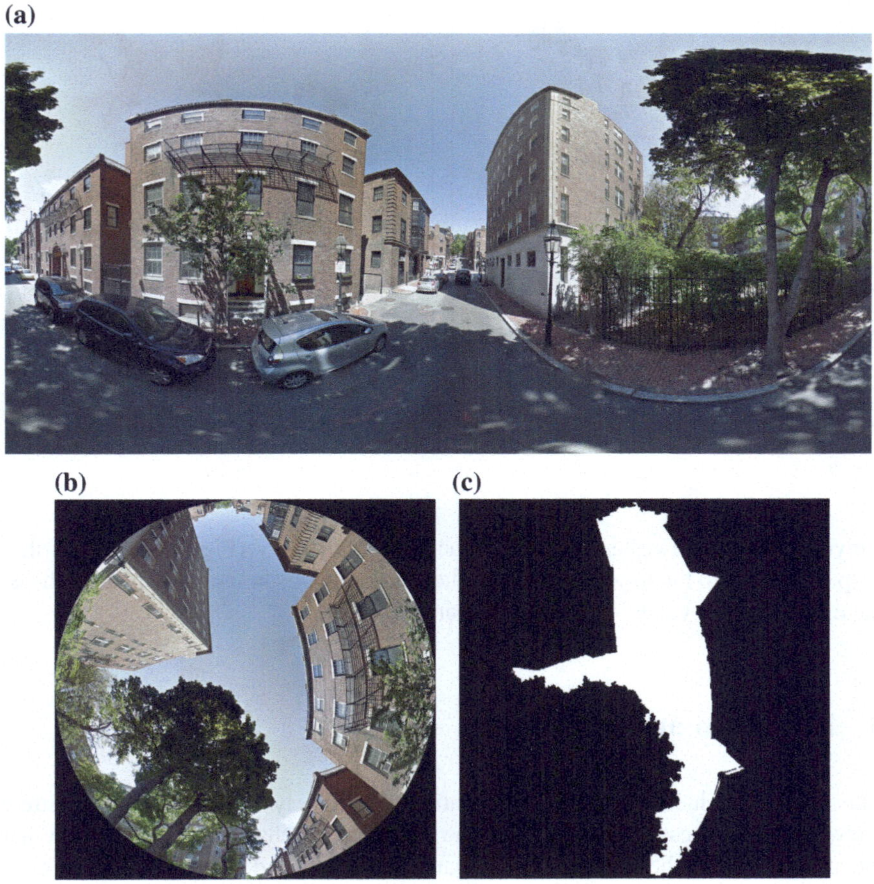

Fig. 8 GSV panorama for the SVF estimation, (**a**) a GSV panorama, (**b**) a generated fisheye image based on geometrical transform, (**c**) sky pixels classification result

$$SVF_P = \frac{1}{2\pi}\sin(\frac{\pi}{2n})\sum_{i=1}^{n}\sin(\frac{\pi(2i-1)}{2n})\alpha_i \quad (7)$$

where, i is the ring index, n is the number of rings (here is set to 37), and α_i is the angular width in ith ring.

By applying the above process to all retrieved GSV panoramas along the city streets, we mapped the distribution of estimated SVF values in Boston at both the point level and the census tract level (Fig. 9). It can be seen clearly that the downtown area and the southwestern part of the study area have much lower SVF values compared with other parts of the city. The SVF value is determined by the obstructions of building block and tree canopies. In the downtown area, the obstructions of high-rise buildings lead to low SVF values or relatively closed street

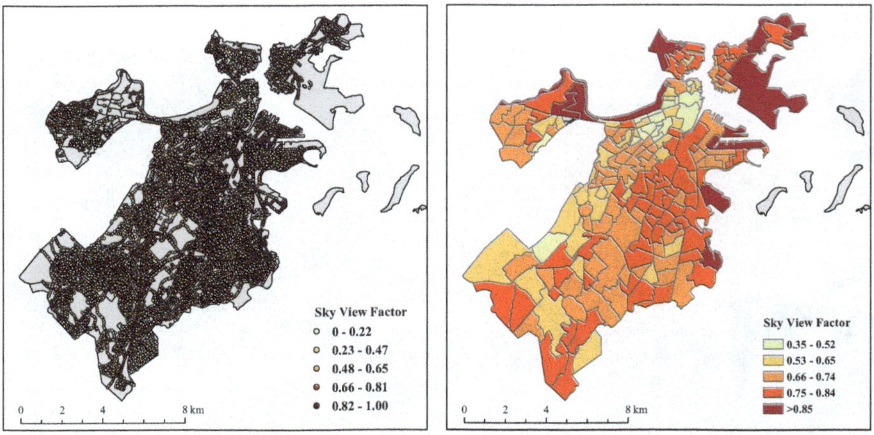

Fig. 9 The spatial distribution of the SVF values in Boston, Massachusetts

canyons. The southwestern part has much street greenery (Fig. 6), which further help contribute to the low SVF values. The street canyons in the north and southeast have higher SVF values than other regions.

4 Discussion and Conclusions

This study introduced two novel applications of publicly accessible Google Street View (GSV) images for urban streetscape mapping and characterization. Given that the street is one of the basic networks that forms cities, and one which represents a confluence of people, information and interactions, these streetscape mapping tools can contribute substantially to our understanding of the interplays between urban citizens and the built urban environment; a foundational objective al urban design and planning.

Street greenery is a critical part of the urban landscape and one that performs critical ecosystem functions and contributes directly to the health and well-being of urban citizens. There have been several metrics to measure and map urban greenness, however, few of them reflect the sensory benefits of urban greenery, not to mention achieving high spatial resolution at city-wide extents. Quantifying streetscape greenery at a ground-level perspective that is shared with the end-user (i.e., citizens) very likely captures scene information directly related to human perception and the social, economic, mental, and physical benefits that have been shown to relate to natural vegetation cover in cities. To exploit this shared perspective and take advantage of these spatially extensive data, we illustrated how one image-derived metric—the modified green view index—can provide a new perspective to study the distribution of the street greenery. Future urban planning practices would benefit from the new index to make the cities more livable.

The openness of street canyons is an important geometrical parameter for the study of urban microclimate, air pollution migration, and human perception of the environment. Currently, a generalized and widely applied method to compute SVF in urban landscapes does not exist. Here, we demonstrated an application of GSV panoramas to, rapidly and fairly simply, estimate the street openness of Boston, Massachusetts. We have shown how a geometric transformation and simple image analysis can be used to automate a workflow to describe SVF across cities, wherein the only input is the publicly accessible GSV data. As such, the GSV-based method is suitable for large-scale SVF estimation can help researchers, urban planners and managers better understand the influence of urban form on the urban microclimate, urban air pollution migration, and human perception of urban environment.

This study has presented two example projects for the streetscape mapping in terms of the visibility of street greenery and the openness of the street canyons. The GSV data should be utilized to generate further explorations and cartographies of the urban environment. Crowdsourcing of image data and machine learning techniques will make it possible to extract additional geospatial information in cities and, both, more complex or specified metrics (e.g., proxy measures of environmental conditions like temperature, urban typologies or object mapping). For example, the *Place Pulse* project of the MIT Media lab is a seminal project to study human perception of the environment based on GSV images and crowdsourcing data collected from volunteer participants.

Although the GSV is a very promising data source for urban studies, there remain some limitations in its use for cartographic purposes. First, the sampling and resampling time inconsistency of the GSV images produce some major challenges. As it stands, GSV image coverage varies both between cities and within cities with respect to the collection dates; images within a given city may be captured at different times, in some cases, representing differing seasonal patterns (e.g., growing season vs. non-growing season months) or simply different years. Thus, at present the user must ensure a temporal standardization across their dataset if the information they are extracting can be influenced by the image capture date (e.g., seasonality of vegetation growth). In addition, the street-view is a fake 3D (Boér et al. 2013) and the GSV images were not collected for the scientific purposes. Therefore, those GSV images or panoramas would have some geometrical distortions that would potentially influence the accuracy of the mapping results. In future applications of these data, supplementary, alternative data sources should be used to validate the accuracy of the results extracted from the GSV imagery. Finally, the impact of the elevation changes was not considered in our analysis. In future studies, a local terrain model should be considered for those areas with large elevation changes.

Considering the fact that Google has now catalogued a majority of the world's cities through GSV, this publicly accessible data will play a growing role as a geospatial data source to map and measure our urban landscapes. The GSV and other open image libraries may take on an even more significant role in areas and municipalities where geospatial data availability is an issue due to financial or accessibility constraints. In those places, GSV data has the ability to act as an important surrogate. What's more, as Google continues to update this service and library of images, it will become possible, and in fact likely very fruitful, to study and quantify the temporal changes of streetscapes in future. Recently, Google has launched a new project to using GSV cars to monitor the air pollutants. This is but one indication that the GSV collection platform will play an increasingly important role as sensor system and data source for urban environmental studies.

References

Bain, L., Gray, B., & Rodgers, D. (2012). *Living streets: Strategies for crafting public space*. New York: Wiley.

Balram, S., & Dragićević, S. (2005). Attitudes toward urban green spaces: Integrating questionnaire survey and collaborative GIS techniques to improve attitude measurements. *Landscape and Urban Planning, 71*(2), 147–162.

Boér, A., Çöltekin, A., & Clarke, K. C. (2013). An evaluation of web-based geovisualizations for different levels of abstraction and realism—What do users predict? In *International Cartographic Conference* (pp. 209–220).

Camacho-Cervantes, M., Schondube, J. E., Castillo, A., & MacGregor-Fors, I. (2014). How do people perceive urban trees? Assessing likes and dislikes in relation to the trees of a city. *Urban Ecosystem*, 1–13.

Carrasco-Hernandez, R., Smedley, A. R., & Webb, A. R. (2015). Using urban canyon geometries obtained from Google Street View for atmospheric studies: Potential applications in the calculation of street level total shortwave irradiances. *Energy and Buildings, 86,* 340–348.

Chapman, L., & Thornes, J. E. (2004). Real-time sky-view factor calculation and approximation. *Journal of Atmospheric and Oceanic Technology, 21*(5), 730–741.

Charreire, H., Mackenbach, J. D., Ouasti, M., Lakerveld, J., Compernolle, S., Ben-Rebah, M., et al. (2014). Using remote sensing to define environmental characteristics related to physical activity and dietary behaviours: A systematic review (the SPOTLIGHT project). *Health Place, 25,* 1–9.

Chen, L., Ng, E., An, X., Ren, C., Lee, M., Wang, U., et al. (2012). Sky view factor analysis of street canyons and its implications for daytime intra-urban air temperature differentials in high-rise, high-density urban areas of Hong Kong: A GIS-based simulation approach. *International Journal of Climatology, 32*(1), 121–136.

Comaniciu, D., & Meer, P. (2002). Mean shift: A robust approach toward feature space analysis. *IEEE Transactions on Pattern Analysis and Machine Intelligence, 24*(5), 603–619.

Debbage, N. (2013). Sky-view factor estimation: A case study of Athens. *Georgia. The Geographical Bulletin, 54*(1), 49.

Edwards, N., Hooper, P., Trapp, G. S., Bull, F., Boruff, B., & Giles-Corti, B. (2013). Development of a Public Open Space Desktop Auditing Tool (POSDAT): A remote sensing approach. *Applied Geography, 38,* 22–30.

Eeftens, M., Beekhuizen, J., Beelen, R., Wang, M., Vermeulen, R., Brunekreef, B., et al. (2013). Quantifying urban street configuration for improvements in air pollution models. *Atmospheric Environment, 72,* 1–9.

Google, (2014). Google Street View Image API. Retrieved October 2014, from https://developers.google.com/maps/documentation/streetview/.

Griew, P., Hillsdon, M., Foster, C., Coombes, E., Jones, A., & Wilkinson, P. (2013). Developing and testing a street audit tool using Google Street View to measure environmental supportiveness for physical activity. *International Journal of Behavioral Nutrition and Physical Activity, 10*(1), 103.

Grimmond, C. S. B., Potter, S. K., Zutter, H. N., & Souch, C. (2001). Rapid methods to estimate sky-view factors applied to urban areas. *International Journal of Climatology, 21*(7), 903–913.

Hämmerle, M., Gál, T., Unger, J., & Matzarakis, A. (2011). Comparison of models calculating the sky view factor used for urban climate investigations. *Theoretical and Applied Climatology, 105*(3–4), 521–527.

Holmer, B., Postgård, U., & Eriksson, M. (2001). Sky view factors in forest canopies calculated with IDRISI. *Theoretical and Applied Climatology, 68*(1–2), 33–40.

Johnson, G. T., & Watson, I. D. (1984). The determination of view-factors in urban canyons. *Journal of Climate and Applied Meteorology, 23*(2), 329–335.

Li, X., Zhang, C., Li, W., & Kuzovkina, Y. A. (2016). Environmental inequities in terms of different types of urban greenery in Hartford, Connecticut. *Urban Forestry & Urban Greening, 18*, 163–172.

Li, X., Zhang, C., Li, W., Ricard, R., Meng, Q., & Zhang, W. (2015). Assessing street-level urban greenery using Google Street View and a modified green view index. *Urban Forestry & Urban Greening, 14*(3), 675–685.

Lin, T. P., Tsai, K. T., Hwang, R. L., & Matzarakis, A. (2012). Quantification of the effect of thermal indices and sky view factor on park attendance. *Landscape and Urban Planning, 107*(2), 137–146.

Mičušík, B., & Košecká, J. (2009). Piecewise planar city 3D modeling from street view panoramic sequences. In *IEEE Conference on Computer Vision and Pattern Recognition, 2009. CVPR 2009* (pp. 2906–2912). IEEE.

Miller, H. J., & Tolle, K. (2016). Big data for healthy cities: Using location-aware technologies, open data and 3D urban models to design healthier built environments. *Built Environment, 42*(3), 441–456.

Naik, N., Philipoom, J., Raskar, R., & Hidalgo, C. (2014). Streetscore-predicting the perceived safety of one million streetscapes. In *Proceedings of the IEEE Conference on Computer Vision and Pattern Recognition Workshops* (pp. 779–785).

Nowak, D. J., Hirabayashi, S., Bodine, A., & Greenfield, E. (2014). Tree and forest effects on air quality and human health in the United States. *Environmental Pollution, 193,* 119–129.

Odgers, C. L., Caspi, A., Bates, C. J., Sampson, R. J., & Moffitt, T. E. (2012). Systematic social observation of children's neighborhoods using Google Street View: A reliable and cost-effective method. *Journal of Child Psychology and Psychiatry, 53*(10), 1009–1017.

Oke, T. R. (1981). Canyon geometry and the nocturnal urban heat island: Comparison of scale model and field observations. *Journal of Climatology, 1*(3), 237–254.

Otsu, N. (1975). A threshold selection method from gray-level histograms. *Automatica, 11*(285–296), 23–27.

Ratti, C., Di Sabatino, S., Britter, R., Brown, M., Caton, F., & Burian, S. (2002). Analysis of 3-D urban databases with respect to pollution dispersion for a number of European and American cities. *Water, Air, & Soil Pollution: Focus, 2*(5–6), 459–469.

Rundle, A. G., Bader, M. D., Richards, C. A., Neckerman, K. M., & Teitler, J. O. (2011). Using Google Street View to audit neighborhood environments. *American Journal of Preventive Medicine, 40*(1), 94–100.

Salesses, P., Schechtner, K., & Hidalgo, C. A. (2013). The collaborative image of the city: Mapping the inequality of urban perception. *PLoS ONE, 8*(7), e68400.

Svensson, M. K. (2004). Sky view factor analysis–implications for urban air temperature differences. *Meteorological Applications, 11*(03), 201–211.

Taylor, B. T., Fernando, P., Bauman, A. E., Williamson, A., Craig, J. C., & Redman, S. (2011). Measuring the quality of public open space using Google Earth. *American Journal of Preventive Medicine, 40*(2), 105–112.

Torii, A., Havlena, M., & Pajdla, T. (2009). From Google Street View to 3D city models. *IEEE 12th International Conference on Computer Vision Workshops* (pp. 2188–2195). Japan: Kyoto.

Unger, J. (2008). Connection between urban heat island and sky view factor approximated by a software tool on a 3D urban database. *International Journal of Environment and Pollution, 36* (1–3), 59–80.

Yang, J., Zhao, L., Mcbride, J., & Gong, P. (2009). Can you see green? Assessing the visibility of urban forests in cities. *Landscape and Urban Planning, 91*(2), 97–104.

Zamir, A. R., Darino, A., Patrick, R., & Shah, M. (2011). Street View challenge: Identification of commercial entities in Street View imagery. *Tenth International Conference on Machine Learning and Applications* (pp. 380–383). Hawaii, December: Honolulu.

Part VI
Numerical Analysis

Cross-Scale Analysis of Sub-pixel Variations in Digital Elevation Models

Mehran Ghandehari, Barbara P. Buttenfield and Carson J.Q. Farmer

Abstract Terrain is modeled on a grid of pixels, assuming that elevation values are constant within any single pixel of a Digital Elevation Model (DEM) (a 'rigid pixel' paradigm). This paradigm can generate imprecise measurements because it does not account for the slope and curvature of the terrain within each pixel, leading to precious information being lost. In order to improve the precision of interpolated points, this paper relaxes the rigid pixel assumption, allowing possible sub-pixel variations (a 'surface-adjusted' paradigm) to be used to interpolate elevation of arbitrary points given a regular grid. Tests based on interpolating elevation values for 5000 georeferenced random points from a DEM are presented, using the rigid pixel paradigm and different interpolation methods (e.g., Weighted Average, Linear, Bi-linear, Bi-quadratic, Bi-cubic, and best fitting polynomials) within different contiguity configurations (i.e., incorporating first and second order neighbors). The paper examines the sensitivity of surface adjustment to a progression of spatial resolutions (10, 30, 100, and 1000 m DEMs), evaluating sub-pixel variations that can be directly measured from 3 m resolution LIDAR data.

Keywords Digital elevation models · Rigid pixel paradigm · Sub-pixel estimation · Surface-adjusted analysis

---

M. Ghandehari (✉) · B.P. Buttenfield · C.J.Q. Farmer
Department of Geography, University of Colorado, Boulder, CO 80309, USA
e-mail: mehran.ghandehari@colorado.edu

B.P. Buttenfield
e-mail: babs@colorado.edu

C.J.Q. Farmer
e-mail: carson.farmer@colorado.edu

M.P. Peterson (ed.), *Advances in Cartography and GIScience*, Lecture Notes in Geoinformation and Cartography, DOI 10.1007/978-3-319-57336-6_25

1 Introduction

Terrain models guide scientists and planners in multiple ways and have fundamental impacts on society, safety, and resource management. For example, digital terrain data is used by natural hazards scientists to model flooding inundation and debris flows (Griffin et al. 2015) and by hydrologists to model water flow direction and accumulation (Stanislawski et al. 2015). Ecologists use terrain models to delineate habitats and nesting territory or to predict the presence of certain species (Czarnecka and Chabudziński 2014). Avalanche risk is measured using solar insolation metrics based on slope and aspect (Jaboyedoff et al. 2012). Civil engineers plan construction of road and railroad networks on the basis of elevation and slope in conjunction with other factors (Zhao et al. 2005). This research contests a prevailing paradigm for terrain-based analyses, which forms the foundation for nearly all environmental models.

A terrain surface is commonly modeled as a regular grid of elevation values (a DEM). Modeling assumes that each pixel is a horizontal planar surface with a constant elevation. We refer to this modeling assumption as a *'rigid pixel'* paradigm (Fig. 1a). The slope and curvature of the terrain is ignored in this model, resulting in imprecise estimates of elevation, which in turn can generate errors in distance, area and volume. In addition, the assumption of constant elevation within each pixel implies an abrupt height change (i.e., discontinuity) at pixel boundaries.

In contrast, a *'surface-adjusted'* paradigm (Buttenfield et al. 2016) relaxes the assumption of the rigid pixel paradigm by acknowledging that sub-pixel variations may occur in elevation, slope and curvature (Fig. 1b). Other researchers utilize partial surface-adjustment methods that account for elevation, slope and aspect of the terrain surface. The *surface-adjusted* paradigm incorporates elevation, slope, aspect, and curvature of DEM pixels into computations.

This paper demonstrates interpolation methods as one possible strategy for the incorporation of sub-pixel variations. Interpolation methods can reconstruct the three-dimensional (3D) surface of each pixel using contextual information from adjacent pixels. For example, a bilinear polynomial estimates a surface at each data point using the four closest pixel centroids. Thus, a series of contiguous bilinear

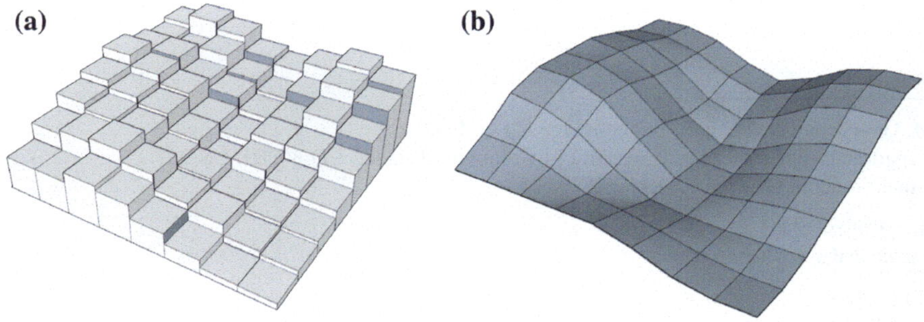

Fig. 1 (**a**) The rigid pixel paradigm and (**b**) the surface-adjusted paradigm

surfaces generate the whole terrain surface. Higher-order polynomials can be employed, but due to computational complexity, they are not normally used (though these higher-order polynomials are also evaluated in this paper). Initial results indicate that for any *individual* pixel, the improvement in measurements can be relatively small, however, the *additive* effect across the study area can become quite significant.

In the last decade, Light Detection and Ranging (LIDAR) has provided new opportunities for studying Earth surface processes, providing more precise measurements (Tarolli 2014). Although it is a fair assumption that elevation values are constant within any single pixel in LIDAR, this data provides some other challenges; LIDAR data needs to be filtered as a preprocessing step to remove artefacts and extract bare earth points. The filtering process reduces the accuracy of the original data and might eliminate or distort significant features (Passalacqua et al. 2015). LIDAR data has become more available but processing of this data is still a challenge due to data storage costs and long processing times. Furthermore, most GIS projects call for integration of various data layers. Ancillary data in the form of vegetation, precipitation, soil data, etc., is not generally available at fine resolutions, requiring filtering to harmonize LIDAR data with other data sources. Finally, and most importantly, the current availability of LIDAR data is far from comprehensive in developed nations, and non-existent in many rural and undeveloped regions.

The paper presents methods for cross-scale analysis of sub-pixel variations in DEMs. Different interpolation methods (Weighted Average, Linear, Bi-linear, Bi-quadratic, Bi-cubic, and best fitting polynomials) are compared, along with different contiguity configurations that incorporate first and second order neighbors. Methods and configurations are tested to interpolate the elevation of 5000 georeferenced random points from a DEM. This research examines the sensitivity of surface adjustments to a progression of spatial resolutions (10, 30, 100, and 1000 m DEMs), using sub-pixel variations that can be directly measured from a 3 m resolution LIDAR data benchmark. The analysis compares Root Mean Square Error (RMSE) between interpolated data and LIDAR to assess differences among the various methods and resolutions.

Discrepancies (error magnitudes) are expected to vary with DEM resolution, terrain roughness and landscape conditions (Buttenfield et al. 2016). With advances in computer processing speed and increased data availability, the scientific community must address limitations of existing modeling methods to provide decision-makers the best possible information about critical environmental issues including biophysical and anthropogenic factors and interactions.

2 Dataset and Study Area

The study area was chosen based on the availability of 3 m resolution LIDAR data to be used for validation. While other resolution terrain data are available for the continental US, coverage at 3 m resolution is much more limited. The study area in

North Carolina is centered on 35.798° N and 81.473° W. Its location at the southeast end of the Appalachian mountains provides elevations ranging from 209 to 1602 m. Its location where the Blue Ridge Mountains run down towards the coastal plains is a humid, hilly landscape, with annual precipitation averaging 51 in. (129.5 cm). The study area provides a mix of uninhabited land with smaller rural settlements.

DEM data will be tested at 10, 30, 100, and 1000 m resolutions, and compared against a 3 m LIDAR benchmark. The 3, 10, and 30 m resolutions are part of the USGS National Elevation Dataset (NED) and were downloaded from the Geospatial Data Gateway (https://gdg.sc.egov.usda.gov/). These are independently compiled DEMs generated according to the methodology outlined in Gesch (2007). The accuracy of NED varies spatially due to the diversity of data sources. The overall absolute vertical accuracy of this dataset has an RMSE of 2.44 m (Gesch et al. 2014). The source for 100 and 1000 m resolutions is the Shuttle Radar Topography Mission (SRTM) dataset (http://dds.cr.usgs.gov/srtm/version2_1/). The absolute vertical accuracy of this dataset has an RMSE of 4.01 m (Gesch et al. 2012). The DEMs are projected in the NAD 1983 UTM Zone 17 N coordinate system.

3 Methods

Tests will interpolate the elevation of 5000 georeferenced random points using different methods and contiguity configurations, across a progression of DEM resolutions. The tested methods all incorporate elevation, but differ in contextual information about surrounding pixels to gain a progression of surface adjustment methods. The methods are compared and validated against a 3 m LIDAR data benchmark (Fig. 2).

Figure 3 diagrams the research workflow that involves a three-factor research methodology. The first factor (along the x-axis) refers to tests of four spatial resolutions (10, 30, 100 and 1000 m). The second factor (along the y-axis) refers to tests for eight interpolation methods. The third factor (along the z-axis) indicates configurations of neighboring pixels to be tested. A range of contiguity configurations increases the number of pixels incorporated into elevation estimation. The second and third factors are not independent, since the interpolation method in some cases dictates the number of neighboring pixels. That is to say, tests for the polynomial fitting and Weighted Average interpolators utilize all configurations except the single pixel (1 neighbor). The other five interpolators use 1, 3, 4, 9, and 16 neighboring pixels, respectively. For the Weighted Average and polynomial fitting interpolation methods, all of the contiguity configurations are tested to find the optimum number of neighboring pixels.

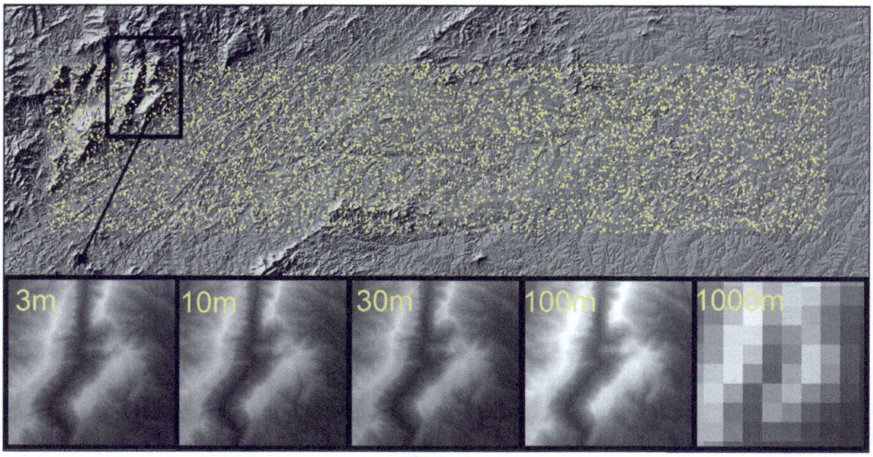

Fig. 2 3 m LIDAR DEM with four other resolutions. Dots show 5000 points sampled randomly within the study area

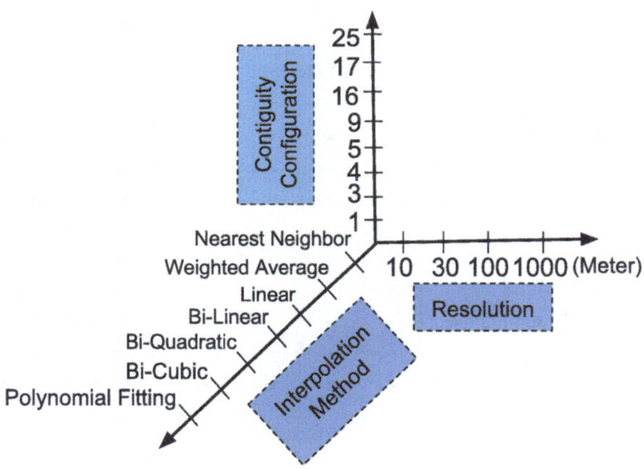

Fig. 3 The research methodology tests estimates of DEM elevations spanning four common spatial resolutions (*x-axis*), seven interpolation methods (*y-axis*) and eight contiguity configurations that incorporate information from a progression of neighboring pixels (*z-axis*). Nearest Neighbor, Linear, Bi-linear, Bi-quadratic, Bi-cubic interpolation methods are exact interpolation methods and utilize 1, 3, 4, 9, 16 neighboring pixels, respectively. The Weighted Average and polynomial fitting interpolation methods are tested in all but one contiguity configuration

3.1 Interpolation Methods

Given a regular elevation grid within the defined neighborhood, different interpolation techniques will generate differing elevation estimates. The methods compared in this paper include Nearest Neighbor, Weighted Average and exact polynomial surfaces (Linear, Bi-linear, Bi-quadratic, and Bi-cubic), all of which are exact interpolators, and least-squares polynomial fitting, an inexact interpolator.

• ***Nearest Neighbor (rigid pixel paradigm)***: This method is based upon the rigid pixel paradigm. Each point represents a small area, a "region of influence" (i.e., the pixel) surrounded by a horizontal planar surface. This interpolator is utilized in this paper as a "straw horse," a control method by which to compare the various surface adjusted interpolators. Estimation proceeds by assigning the elevation of the centroid for whatever pixel the sampled point lies within (Fig. 4a). Although this

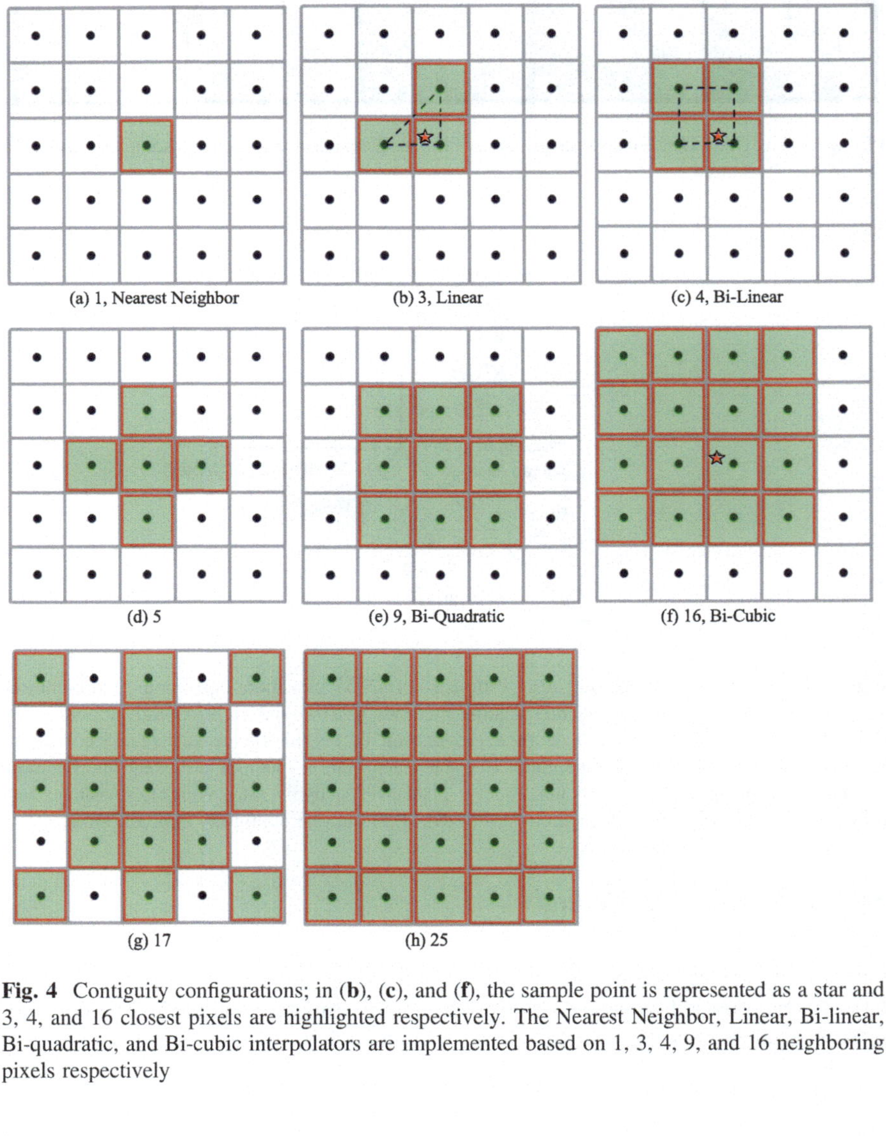

Fig. 4 Contiguity configurations; in (**b**), (**c**), and (**f**), the sample point is represented as a star and 3, 4, and 16 closest pixels are highlighted respectively. The Nearest Neighbor, Linear, Bi-linear, Bi-quadratic, and Bi-cubic interpolators are implemented based on 1, 3, 4, 9, and 16 neighboring pixels respectively

approach is quite simple and easy to implement, it does not incorporate terrain derivatives and thus there is no surface adjustment.

• ***Weighted Average***: This is a deterministic interpolator that computes an average elevation based on neighboring pixels. Distance to neighboring pixel centroids is commonly used as a weighting function, with shorter distances carrying larger weights. Weighted distance assumes that points that are closer together are more similar than those that are farther apart. Different distance functions can be used. The simplest and most common is the "Inverse Distance Weighting" (IDW) method, commonly used for continuous data such as terrain:

$$z(x,y) = \frac{\sum_{i=1}^{n} w_i z_i}{\sum_{i=1}^{n} w_i}; \quad w_i = \frac{1}{d_i^p}$$

where w_i and z_i are the weight and elevation of each source point, respectively. Distance (d) is given an exponent (p); and a larger exponent leads to progressively smaller weights at larger distances.

• ***Exact Polynomial Surfaces***: This is a set of deterministic methods whose estimates are based on a polynomial equation with varying amounts of terms and powers. By definition, the exact polynomial will generate a precise estimate of any original point. Polynomial equations with more terms create a surface with more freedom to undulate (Li et al. 2004). Four different exact polynomial interpolators are used in this research:

Linear: A triangle is generated from a three-term polynomial. Linear interpolation needs the three closest points to solve the equation (Fig. 4b). The mathematical function creates a plane:

$$z(x,y) = a_0 + a_1 x + a_2 y$$

Bi-linear: This is also a first order polynomial based upon four known points (Fig. 4c).

$$z(x,y) = a_0 + a_1 x + a_2 y + a_3 xy$$

Bi-quadratic: This second order polynomial uses nine terms, allowing for more freedom of undulation. Nine closest neighbors are selected to solve the Bi-quadratic equations (Fig. 4e).

$$z(x,y) = a_0 + a_1 x + a_2 y + a_3 xy + a_4 x^2 + a_5 y^2 + a_6 x^2 y^2 + a_7 x^2 y + a_8 xy^2$$

Bi-cubic: This third order polynomial increases the surrounding pixels to sixteen closest neighbors (Fig. 4f). This method calculates a surface with additional local maxima and minima.

$$z(x,y) = a_0 + a_1\mathrm{x} + a_2\mathrm{y} + a_3\mathrm{x\ y} + a_4x^2 + a_5y^2 + a_6x^2y^2 + a_7x^2y + a_8xy^2 + a_9x^3 + a_{10}y^3 + a_{11}x^3y^3 + a_{12}x^3y^2 + a_{13}x^2y^3 + a_{14}x^3\mathrm{y} + a_{15}xy^3$$

• ***Best Fitting Polynomials***: Terrain is a complex surface and using high-order polynomials in an exact fitting approach can lead to fluctuations in the resulting elevation estimate. Least-Squares Polynomial Best Fitting is a common way to resolve fluctuations by minimizing the sum of squared errors. This is an inexact interpolator, and is not necessarily a better approximation of the terrain surface. A best-fitting surface can be linear or a smoothed curved surface. Best fitting polynomials of order 1, 2, 3, and 4 are used as follows:

$$Order1: z(x,y) = a_0 + a_1\mathrm{x} + a_2\mathrm{y}$$

$$Order2: z(x,y) = a_0 + a_1\mathrm{x} + a_2\mathrm{y} + a_3\mathrm{x\ y} + a_4x^2 + a_5y^2$$

$$Oder3: z(x,y) = a_0 + a_1\mathrm{x} + a_2\mathrm{y} + a_3\mathrm{x\ y} + a_4x^2 + a_5y^2 + a_6x^2y + a_7xy^2 + a_8x^3 + a_9y^3$$

$$Order4: z(x,y) = a_0 + a_1\mathrm{x} + a_2\mathrm{y} + a_3\mathrm{x\ y} + a_4x^2 + a_5y^2 + a_6x^2y + a_7xy^2 + a_8x^3 + a_9y^3 + a_{10}x^4 + a_{11}y^4 + a_{12}x^2y^2 + a_{13}x^3\mathrm{y} + a_{14}xy^3$$

Note that the first order polynomial shows an equation identical to the Linear exact polynomial described above. The difference is that the Linear exact method is tested based on the three-neighbor contiguity configuration, but the first order best fitting method is tested based on the four and higher neighbor contiguity configurations using the least square equation.

3.2 Contiguity Configuration

Different contiguity configurations can be investigated based upon the number of neighbors (here 3, 4, 5, 9, 16, 17, 25). In this research, first and second order neighboring pixels are combined in different ways to configure the contiguity structure of sub-pixel variations (Fig. 4). A 1-pixel contiguity configuration is used as a control to compare the various surface-adjusted methods with the rigid pixel "Nearest Neighbor" method. Some interpolation methods are an exact interpolation technique and can only be tested on one contiguity configuration. For example, the Bi-linear interpolation method is based upon four neighbors. On the other hand, the Weighted Average can be tested on multiple contiguity configurations. The Nearest Neighbor, Linear, Bi-linear, Bi-quadratic, and Bi-cubic interpolators use 1, 3, 4, 9, and 16 neighboring pixels, respectively. For the Weighted Average, all the contiguity configurations except the single pixel are tested to find the optimum number of neighboring pixels. For the best fitting polynomial, the single pixel and the 3-pixel configurations are left out to insure inclusion of sufficient pixels in the least squares equations.

3.3 Workflow and Processing

The workflow involves several steps. Each resolution DEM is input one at a time. The 3 m benchmark DEM is loaded and a random sample of 5000 points is generated, creating a shapefile containing coordinate positions in UTM meters, and elevations in meters extracted from the 3 m benchmark DEM. At this point, the LIDAR DEM is unloaded, and the shapefile is stored on disk. For each of the test DEMs, a nested loop processes each point for all DEMs and for all interpolation methods, then moves on to the next sampled point. Estimated elevations are added as new attributes in the original shapefile. Within the same nested loop, residuals are calculated for each elevation (each sampled point). Once all sampled elevations have been estimated for all DEMs at all resolutions, the residuals are summarized with RMSE standard deviation of residuals, 95% confidence interval and elevation extrema.

The overall sequence of operations is as follows:

The overall sequence of operations is as follows:

Cross-scale analysis of sub-pixel variation in DEMs using various interpolation methods

---

(1) Load 4 Input DEMs (DEM10, DEM30, DEM100, DEM1000)
(2) Load benchmark (DEM3)
(3) Generate sample of 5000 random points in the study area
(4) For i in 5,000 random points:
    For j in 4 Input DEMs:
        For k in 7 interpolation methods:
            - Interpolate the elevation of point i based on the DEM resolution j using interpolation method k ($Elev_{ijk}$)
            - Compute $Residual_{ijk} = Elev_{ijk} - Elev_i$ of benchmark
(5) Accuracy Assessment (RMSE, Standard Deviation, etc.)

3.4 Accuracy Assessment

Accuracy assessment is conducted on the residuals. Prior to accuracy assessment, the residuals are tested for normality. There are several ways to examine the accuracy of interpolated sample points. We compare the interpolated elevation of sample point to 3 m resolution LIDAR data. The residuals are calculated by subtracting the lidar value from the interpolated value. Next, we summarize standard deviation (STD) of residuals, Root Mean Square Error (RMSE), Mean Absolute Error (MAE), and 95% confidence interval surrounding the RMSE values.

$$RMSE = \sqrt{\frac{\sum_{i=1}^{n} e_i^2}{n}}; \quad MAE = \frac{\sum_{i=1}^{n} |e_i|}{n}$$

4 Results and Discussion

4.1 Analysis of Residuals

Table 1 represents basic descriptive parameters of the residuals (mean, median, minimum, maximum, standard deviation, and 95th percentile–lower and upper bound) at different resolutions. These parameters are based on all sample points and for all of interpolation methods. The mean value is slightly positive (10, 100, and 1000 m) or negative (30 m) indicating overestimation and underestimation, respectively. Furthermore, the high values of minima and maxima are due to the occurrence of both positive and negative outliers. The standard deviation values are small, relative to the mean and data range.

Two standard normality tests are run on the residuals—the Kolmogorov–Smirnov test and the Anderson–Darling test. The results of these tests verify that residuals are not normally distributed. Based on the American Society for Photogrammetry and Remote Sensing (ASPRS) (Flood 2004) and the National Digital Elevation Program (NDEP) (NDEP 2004) guidelines, the 95th percentile can be used for reporting the vertical accuracy of elevation date when the error distribution is not normal.

Outliers are an indication to inspect the spatial pattern of residuals more closely. Local Moran's I (Anselin 1995) was calculated for the residuals in order to visualize clusters whether residuals with similar values (either high values or low values are clustered) and outliers. The inverse distance squared method was used and the neighborhood delineated that ensured every point has at least one neighbor. Residuals display spatial heterogeneity due to the varied character of terrain. The circles in Fig. 5 highlight the extreme residuals and illustrate that many are situated

Table 1 Summary statistics of residuals for different resolutions

Resolution	Mean	Median	Min	Max	Standard Deviation	95th percentile-lower bound	95th percentile-higher bound
10	0.01	0.0	−10.91	10.63	1.07	−1.53	1.60
30	−0.02	−0.12	−35.53	41.26	3.94	−5.60	6.28
100	5.11	3.71	−40.90	66.37	7.95	−4.75	19.52
1000	4.06	4.09	−399.81	381.55	41.89	−35.52	47.39

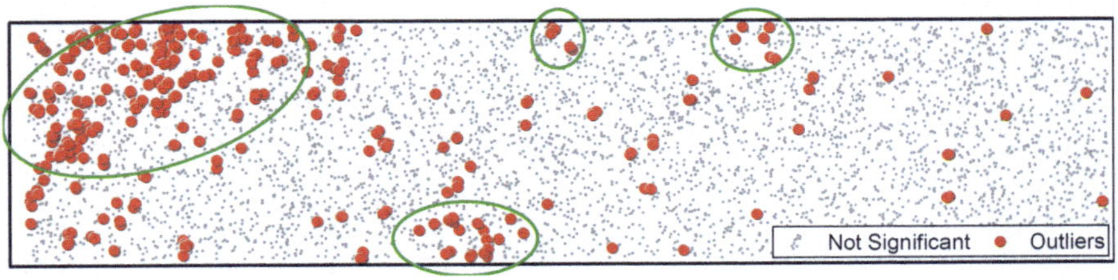

Fig. 5 Outliers of residuals in 10 m DEM using the Bi-linear interpolation

Table 2 The correlation and p-values in 95% confidence interval between residuals and elevation, slope, aspect, curvature, local relief, and roughness (30 m DEM is used here). P-values are so small that even at 4 or 5 decimal places they do not differ meaningfully from zero

Variable	Elevation	Slope	Aspect	Curvature	Local relief	Roughness
Correlation	0.38	0.62	−0.06	0.33	0.59	0.53
P-values	0.0000	0.0000	1.6e-05	0.0000	0.0000	0.0000

in rough terrain (in the northwest corner). These imply that terrain type is an important factor needing further investigation in future research.

Table 2 reports the correlation between the residuals and elevation, slope, aspect, curvature, local relief (the range of elevation in a 10 by 10 focal window), and roughness (the standard deviation of elevation in a 5 by 5 window) based on the 30 m DEM. All of the correlation coefficients are highly significant, though the relationship with aspect is weaker than with other terrain derivatives. The high significance levels could be a consequence of the large number of sampled pixels. Slope has the highest correlation with the residual, indicating the accuracy of DEM in terrain with high slopes is lower than with flatter slopes. In addition, there is a strong relationship between residuals and terrain roughness and local relief; and this also reflects the slope relationship.

4.2 Optimal Configuration for Weighted Average Interpolator

Table 3 shows the RMSEs resulting from the Weighted Average method for all contiguity configurations, and across all resolutions. The range of RMSEs in each resolution is quite small. The 16 neighboring pixel configuration is chosen as the optimum contiguity configuration as it shows the lowest RMSE in most resolutions for this interpolator, and is chosen for the further analysis in Sect. 4.4.

Table 3 RMSEs for the weighted average method using different number of neighbors in different resolutions; boldface indicates the lowest RMSE in each resolution

Contiguity Configuration	10 m	30 m	100 m	1000 m
WA3	1.09	4.01	**9.46**	42.57
WA4	**1.08**	3.97	**9.46**	42.24
WA5	1.10	4.02	9.51	42.53
WA9	**1.08**	3.95	9.50	41.56
WA16	**1.08**	**3.88**	9.53	**40.93**
WA17	**1.08**	3.91	9.54	41.15
WA25	**1.08**	3.89	9.61	40.80
Range of RMSEs (m)	*0.02*	*0.14*	*0.15*	*1.64*

4.3 Optimal Configuration for Best Fitting Polynomials

Table 4 shows the RMSEs for the best fitting polynomial method. Different orders and number of neighbors are tested. In the Table, polynomial order 1 shows the lowest RMSE. Reading down the columns for a given order, there is a general trend where RMSE increases with increasing numbers of neighboring pixels. At 1000 m resolution, the opposite occurs: polynomial order 1 with contiguity configuration 4 has the highest RMSE. Again, there is not an optimum method in all of the resolutions. Polynomial order 1 with 5 neighbors was selected for further analysis, having the lowest RMSE.

Table 4 RMSEs for the best fitting method using various polynomial orders and different number of neighbors in different resolutions. Boldface indicates the lowest RMSE in each resolution

Polynomial order	Contiguity Configuration	9 m	30 m	100 m	1000 m
1	4	**1.07**	3.94	**9.45**	42.09
	5	**1.07**	3.87	9.47	40.85
	9	1.08	**3.82**	9.65	**39.24**
	16	1.13	3.90	10.12	39.45
	25	1.22	4.14	10.76	40.75
2	9	1.08	**3.82**	9.65	39.26
	16	1.13	3.90	10.12	39.48
	17	1.18	4.02	10.44	40.17
	25	1.22	4.14	10.76	40.77
3	9	1.08	**3.82**	9.67	39.51
	16	1.13	3.90	10.12	39.50
	17	1.18	4.01	10.44	40.20
	25	1.22	4.14	10.76	40.80
4	16	1.14	3.91	10.18	39.58
	17	1.19	4.02	10.51	40.71
	25	1.22	4.14	10.76	40.82

4.4 Comparing Surface-Adjusted Elevations with the Rigid Pixel Paradigm

The comparison of interpolated elevations with the rigid pixel paradigm is accomplished by comparing results discussed so far with the Nearest Neighbor (NN) method. Recall that the NN method uses only a single pixel for estimating elevation. Table 5 shows the method comparison. Methods are ordered across the table columns by the number of pixels in each configuration. Only the optimal configurations selected for Weighted Average (WA) and for Best Fitting (BF) are included here, for comparison.

Error magnitudes vary with DEM resolution and with interpolation method. Accuracy assessment parameters show a general trend of increase in the residuals at coarser resolutions. The Nearest Neighbor method shows highest magnitude errors for all resolutions, except the 100 m in which the Bi-cubic method has the highest amount of error. Clearly, these results are mixed. Best Fitting Linear methods show the best performance (lowest RMSE) at 10 m resolution, with Linear, Bi-linear and Best Fitting Linear showing lowest residuals at 30 and 100 m. The Bi-quadratic RMSE is lowest at the coarsest 1000 m resolution. Readers should note very similar values for all metrics at each resolution indicating that this analysis should be repeated for other terrain surfaces before a definitive conclusion can be reached.

Table 5 Accuracy assessment parameters for different interpolation methods in different resolutions; Columns show interpolation methods (left to right): Nearest Neighbor, linear using 3 neighbors, Bi-linear using 4 neighbors, best fitting polynomial of order 1 using 5 neighbors, Bi-quadratic using 9 neighbors, Bi-cubic using 16 neighbors, weighted average using 16 neighbors. Boldface indicates the lowest RMSE in each resolution. STD = standard deviation of residuals

		NN	Li3	BiLi4	BF1_5	BiQ9	BiC16	WA16
10 m	*RMSE*	1.19	1.08	**1.07**	**1.07**	1.08	1.13	1.08
	RMSE (95%)	0.75	0.66	**0.65**	0.67	0.68	0.73	0.67
	MAE	0.72	0.65	0.65	0.65	0.66	0.71	0.66
	STD	1.19	1.08	1.07	1.07	1.08	1.13	1.09
30 m	*RMSE*	4.34	3.95	3.94	**3.87**	3.82	3.90	3.88
	RMSE (95%)	2.75	2.58	2.58	**2.53**	2.55	2.64	2.57
	MAE	2.70	2.53	2.53	2.50	2.50	2.61	2.51
	STD	4.34	3.95	3.95	3.87	3.82	3.90	3.88
100 m	*RMSE*	9.85	**9.45**	**9.45**	9.47	9.65	10.12	9.53
	RMSE (95%)	8.25	**7.97**	**7.97**	7.99	8.06	8.37	8.01
	MAE	6.87	6.61	6.61	6.63	6.75	7.08	6.68
	STD	8.42	7.94	7.95	7.99	8.19	8.72	8.037
1000 m	*RMSE*	45.08	42.20	42.09	40.85	**39.26**	39.50	40.93
	RMSE (95%)	23.77	22.22	22.34	22.04	**21.51**	21.84	21.87
	MAE	24.69	23.52	23.50	22.97	22.36	22.42	22.95
	STD	44.92	41.85	42.00	40.65	39.05	39.28	40.72

5 Summary

This work advances understanding of how spatial error propagates through resolutions, a cross-scale terrain analytics component that is at present not well understood. The current paradigm for estimating terrain elevations assumes that the DEM terrain surface is uniform within each pixel, ignoring slope and curvature. This research employs realistic surface geometries of terrain using different interpolation methods and the information from adjacent pixels, searching for a more accurate and precise interpolation of points. Sub-pixel variations in DEMs through different resolutions are investigated to develop a foundation of surface-adjusted computations on DEMs, resulting in more realistic analysis in terrain models.

Findings for this DEM, and for the interpolators tested here, indicates that different interpolators generate lowest RMSE values at various resolutions, with some type of linear polynomial performing well at finer and intermediate resolutions. RMSE values are so close for this analysis that it is difficult to conclude one method is preferable to the other. Ideally, one would prefer an interpolation method that carries the lowest computational intensity, such as the Weighted Average method. In this experiment, however, the Weighted Average RMSEs do not outperform the other methods of estimation.

The next stage of research will investigate how the results vary with geographic conditions (different types of terrain) and how terrain differences result in elevation estimations that affect modeling applications. Finally, the assumption of non-rigid pixels carries an additional computational load, as it must incorporate not only elevation and pixel size, but also slope and curvature, in effect, adjusting for the changing terrain surface at sub-pixel resolutions. The longer-term goal of this research addresses the balance between the increased computations needed to measure surface-adjusted elevation against the improvement in precision.

Acknowledgements This research is supported in part by the Grand Challenge Initiative "Earth Lab" (http://www.colorado.edu/grandchallenges/) funded by the University of Colorado. We acknowledge the USGS Center for Excellence in Geospatial Information Science (CEGIS) for analytic advice.

References

Anselin, L. (1995). Local indicators of spatial association—LISA. *Geographical Analysis, 27*(2), 93–115.

Buttenfield, B. P., Ghandehari, M., Leyk, S., Stanislawski, L. V., Brantley, M., & Qiang, Y. (2016). Measuring distance "As the Horse Runs": cross-scale comparison of terrain-based metrics. In *Proceedings of the 9th International Conference on Geographic Information Science, GIScience 2016*, Montreal, Canada.

Czarnecka, B., & Chabudziński, Ł. (2014). Assessment of flora diversity in a minor river valley using ecological indicator values, geographical information systems and digital elevation models. *Open Life Sciences, 9*(2), 220–231.

Flood, M. (2004). ASPRS guidelines: Vertical accuracy reporting for LiDAR data. Ownership by ASPRS Lidar Committee (PAD) Version.

Gesch, D. B. (2007). Chapter 4—The national elevation dataset. In D. Maune (Ed.), *Digital elevation model technologies and applications: The DEM users manual* (2nd ed., pp. 99–118). Bethesda, Maryland: American Society for Photogrammetry and Remote Sensing.

Gesch, D., Oimoen, M., Zhang, Z., Meyer, D., & Danielson, J. (2012). Validation of the ASTER global digital elevation model version 2 over the conterminous United States. *International Archives Photogrammetry Remote Sensing and Spatial Information Science,* pp. 281–286.

Gesch, D. B., Oimoen, M. J., & Evans, G. A. (2014). Accuracy assessment of the US geological survey national elevation dataset, and comparison with other large-area elevation datasets: SRTM and ASTER: *US Geological Survey Open-File Report* 2014–1008, 10 p.

Griffin, J., Latief, H., Kongko, W., Harig, S., Horspool, N., & Hanung, R., et al. (2015). An evaluation of onshore digital elevation models for modeling tsunami inundation zones. *Frontiers in Earth Science*, *3*(32).

Jaboyedoff, M., Choffet, M., Derron, M. -H., Horton, P., Loye, A., & Longchamp, C., et al. (2012). Preliminary slope mass movement susceptibility mapping using DEM and LiDAR DEM. In *Terrigenous mass movements* (pp. 109–170), Springer.

Li, Z., Zhu, C., & Gold, C. (2004). Digital terrain modeling: principles and methodology, CRC press.

NDEP. (2004). *Guidelines for Digital Elevation Data*, Version 1.0. National Digital Elevation Program.

Passalacqua, P., Belmont, P., Staley, D. M., Simley, J. D., Arrowsmith, J. R., Bode, C. A., et al. (2015). Analyzing high-resolution topography for advancing the understanding of mass and energy transfer through landscapes: A review. *Earth-Science Reviews, 148,* 174–193.

Stanislawski, L. V., Falgout, J., & Buttenfield, B. P. (2015). Automated extraction of natural drainage density patterns for the conterminous united states through high-performance computing. *The Cartographic Journal, 52*(2), 185–192.

Tarolli, P. (2014). High-resolution topography for understanding earth surface processes: opportunities and challenges. *Geomorphology, 216,* 295–312.

Zhao, J., Wang, Q., & Wang, X. (2005). Real-time 3D road animation based on digital terrain model. *Computer Engineering*, *12*, 071.

Extraction of Ridge Lines from Grid DEMs with the Steepest Ascent Method Based on Constrained Direction

Wenping Jiang, Daping Xi, Xiaolong Deng, Lina Huang and Shen Ying

Abstract The extraction of ridge lines from digital elevation models is fundamental for generalizing digital elevation models (DEM), analyzing digital valleys, remote sensing analysis, and topographic indices analysis. In this paper, the authors propose a method to extract ridge lines from a gridded DEM called the Steepest Ascent Method Based on Constrained Direction (SAMBCD). In the SAMBCD method, based on the overland flow simulation method and the steepest ascent method, the authors define control points and constrain the direction of connecting points in the process of organizing major ridge lines and minor ridge lines, respectively, among the discrete points of the catchment boundary. Specifically, with SAMBCD, the points on hill peaks and saddle points are first sorted and used as control points to constrain the connecting direction. Second, the major ridge lines are organized by connecting the relevant discrete points of the catchment boundary, following the constrained direction. Finally, the remaining discrete points of the catchment boundary are connected to form the minor ridge lines with the same constrained connecting direction. Many tests are performed to extract ridge lines in different regions. The results show that the method of SAMBCD is effective for decreasing the fracture, cross and rotation of extracted ridge lines. Moreover,

W. Jiang · L. Huang · S. Ying
School of Resource and Environmental Sciences, Wuhan University,
Wuhan, Hubei, China
e-mail: 1759228429@qq.com

L. Huang
e-mail: 151030490@qq.com

S. Ying
e-mail: 42071077@qq.com

D. Xi (✉)
Faculty of Information Engineering, China University of Geosciences,
Wuhan, Hubei, China
e-mail: 46619441@qq.com

X. Deng
CMB Network Technology Company Limited, Shenzhen, Guangdong, China
e-mail: xiaol_deng@163.com

M.P. Peterson (ed.), *Advances in Cartography and GIScience*, Lecture Notes in Geoinformation and Cartography, DOI 10.1007/978-3-319-57336-6_26

SAMBCD has better performance in the continuity and integrity of ridge lines than the two basic methods mentioned above.

Keywords Grid DEM · Ridge lines · Constrained connecting direction · Overland flow simulation · Steepest ascent method

1 Introduction

Topographic characteristic lines include the skeleton lines of geomorphologic shape and the control lines for describing and processing digital terrain, mainly ridge lines and valley lines. The automatic extraction of topographic characteristic lines plays an increasingly important role in performing hydrographic analysis, generating a DEM of high quality, allowing cartographic generalization, analyzing soil erosion, processing remote-sensing digital images and analyzing digital terrain, etc. (O'Callaghan and Mark 1984).

There are many mature applications in such commercial software as ESRI ArcGIS. But, in some research such as generalizing contours automatically and generalizing DEMs, it is necessary to analyze the automatic extraction method of topographic characteristic lines. Because of the wide availability of DEMs and continuous improvement in data accuracy, methods for the automatic extraction of topographic characteristic lines have been widely researched and applied (O'Callaghan and Mark 1984; Matsuura and Aniya 2012).

In recent decades, many algorithms have been proposed to detect ridge lines based on surface geomorphologic analysis but all of these algorithms have their respective shortcomings, such as judgment of feature points, connection of ridge, consecutiveness, and hierarchy (Chang and Sinha 2007; Yokoyama et al. 2007; Koka et al. 2011, 2012).

The method of overland flow simulation based on grid DEM has been widely researched since it was proposed in 1984 (Martz and Garbrecht 1995; Barnes et al. 2014; Zhang and Huang 2009; Jana et al. 2007; Garbrecht and Martz 1997; Lee et al. 1992). Many studies have been proposed to improve depression filling (Jenson and Domingue 1988; Martz and Garbrecht 1998; Planchon and Darboux 2002; Wang and Liu 2006; Zhu et al. 2006). However, the flat area resulting from depression-filling makes flow direction determination difficult (Bai et al. 2015; Barnes et al. 2014; Martz and Garbrecht 1998; Zhu et al. 2004; Saraiva and Paz 2014; Li et al. 2013). Moreover, depression filling is difficult to implement and has unacceptable processing times, especially a mass of depression in the detection region (Bai et al. 2015; Barnes et al. 2014).

Shinji Koka proposed the steepest ascent method, in which the steepest ascent lines are obtained by selecting the maximum inclined direction from eight neighbors (Koka et al. 2011). The major ridge lines that are extracted by this method may fracture at the peak point of the hill and are not continuous. Moreover, these ridge lines may intersect with valley lines because this method can't avoid the effect of valley points when linking ridge lines.

Thus, we propose an improved Steepest Ascent Method Based on Constrained Direction (SAMBCD) to extract ridge lines from the DEM. In this method, we improve the steepest ascent method by defining control points and constraining the direction of connecting points during organizing major ridge lines and minor ridge lines among the discrete points on the catchment boundary.

In this paper, the SAMBCD method can connect ridge lines in the watershed perfectly and is easy to implement and has a low processing time; furthermore, this algorithm has great performance in the consecutiveness and hierarchy of the detected ridge lines.

2 Related Works

2.1 The Steepest Ascent Method

The steepest ascent method proposed by Koka et al. (2011) is used to extract ridge lines. The basic idea of this algorithm is, at first, to determine the point with minimal elevation among points not yet dealt with in the experimental region. Then, the point will be set as starting point and added into ridge points array. During the cyclical process of connecting ridge points into ridge lines, once the next ridge connecting point is found, it will be regarded as new center point to participate in comparing elevations with its 8 neighboring points. After the point with maximal elevation is found, and this point has not been connected into ridge lines and has a greater elevation than the center point, it will be added into ridge points array as next connecting point. Otherwise, tracing and connecting will end (see Fig. 1).

2.2 The Method of Overland Flow Simulation

The overland flow simulation method was introduced by O'Callaghan and Mark (1984). In this method, at first, the matrix of elevation will be reversed, and the

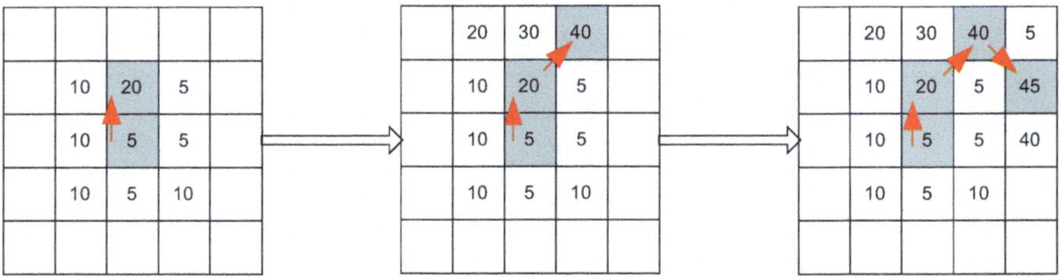

Fig. 1 Process of the steepest ascent method (After Koka et al. 2011)

drainage direction matrix and drainage accumulation matrix will be generated after filling depressions and treating flat areas. The points of the drainage accumulation matrix are discrete ridge points with no drainage inputs. These points have an input drainage direction and no output drainage direction, and their drainage accumulations all are 0. These ridge points can be connected into ridge lines by some methods.

3 The Steepest Ascent Method Based on Constrained Direction (SAMBCD)

3.1 The Algorithm of SAMBCD

In the method of overland flow simulation, when extracting topographic characteristic lines based on the DEM, the matrix of drainage accumulation in the experimental area will be generated simultaneously. In this matrix, those points with a drainage accumulation value of 0 are called discrete points on the watershed. The SAMBCD algorithm is used to extract ridge lines from these discrete points on the watershed (see Fig. 2).

In connecting ridge lines, hill points and saddle points, which are control points of connecting direction, will be sorted out from the discrete points on the watershed and added into the control points array. With the constraint of direction, these hill points and saddle points will be connected with discrete points on the watershed to

Fig. 2 Process of SAMBCD

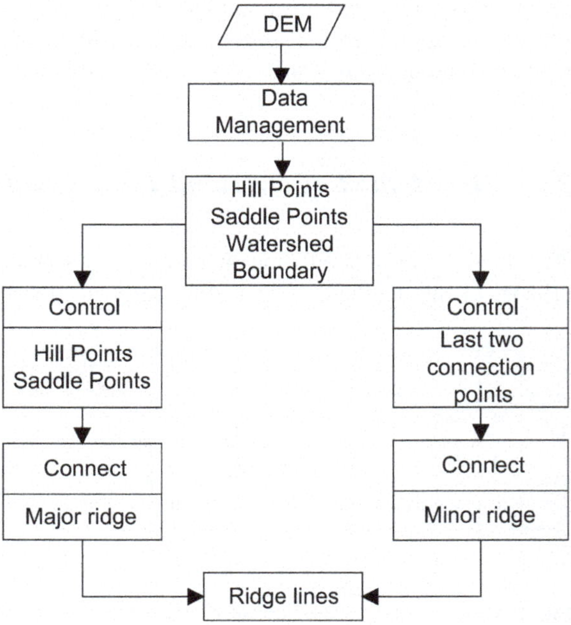

form main ridge lines. Then, the connected points of the major ridge lines will be set as a connected flag, and neighboring points of points connected will also be set as the detected flag. The remaining points will be connected into minor ridge lines with the modified method of steepest ascent, in which the connecting direction is constrained by the last two points of the ridge lines being connected. The short minor ridge lines will be deleted if their length is less than the threshold.

3.2 The Implementation of SAMBCD

3.2.1 The Major Ridge Lines

The process of generating major ridge lines is divided into 4 steps as follows:

Step 1. **Generating the queue of hill points**

We deal first with traverse grid cells whose value is equal to 0 in the drainage accumulation matrix and determine hill points and saddle points in discrete points on the watershed. When dealing with one of the discrete points, we create a 3 by 3 moving window centered on this point. If the elevation of the center point is larger than that of all eight neighboring points, we record this point as a hill point. As shown in Fig. 3, point c2 and point c5 are hill points. After analyzing the two orthogonal directions through this point we find that the terrain profile in one direction is concave and another is convex, then this center point is a saddle point (Lee et al. 1992). The point c3 is a saddle point in Fig. 3.

We then form the queue of hill points. We will create a 2D coordinate system whose origin locates on a current hill point. The coordinate system is divided into eight parts, which are the 1st quadrant, the 2nd quadrant, the 3rd quadrant, and the 4th quadrant, positive x semi-axis, positive y semi-axis, negative x semi-axis and negative y semi-axis as shown in Fig. 4. Every part corresponds to one of eight hill points respectively. Every other hill point will be examined in the part that it lies and added into the corresponding hill points array. All of the points in each hill points array in every part will be sorted by the distance between every hill point and origin point of the coordinate system in ascending order to form hill points queue.

Fig. 3 Hill and saddle points

	1	2	3	4	5	6
a						
b	70	65	45	60	75	55
c	60	80	50	65	95	70
d	45	50	30	65	65	75
e						

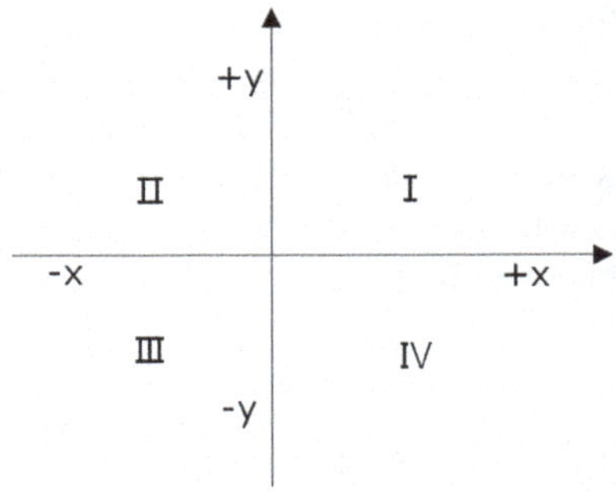

Fig. 4 Coordinate system of hill point

Step 2. **Establishing coordinate system for connecting ridge points;**

In this step, we will connect respectively the points of the hill point's queue in eight parts with the queue in each part being traversed in turn. The hill points to be connected in the each queue is called Ending Hill Points (EHP), and the hill point on the origin point is called the Starting Hill Points (SHP). The last hill point connected is called as Connecting Ridge Points (CRP). While connecting ridge lines, each point is successively popped from the hill point's array and judged whether it has been connected to SHP. If yes, the next point is popped from the queue and the traverse continues. Otherwise, an empty array is created for storing ridge points and this starting point will be added into the new array. The last point of ridge points array will be set as CRP. If this CRP is exactly EHP, the connecting of the current ridge line ends. Otherwise, the next step is to establish another 2D coordinate system whose origin is this CRP and the plane is divided into 16 parts, as shown in Fig. 5.

Step 3. **Establishing a priority queue for connecting;**

There are many discrete points on the watershed to be potentially connected in the eight neighborhoods of the CRP, as shown in Fig. 6. In this algorithm, these surrounding discrete points have the highest priority when being added into a ridge line, where px represent the priority level of point x. The priorities level of discrete

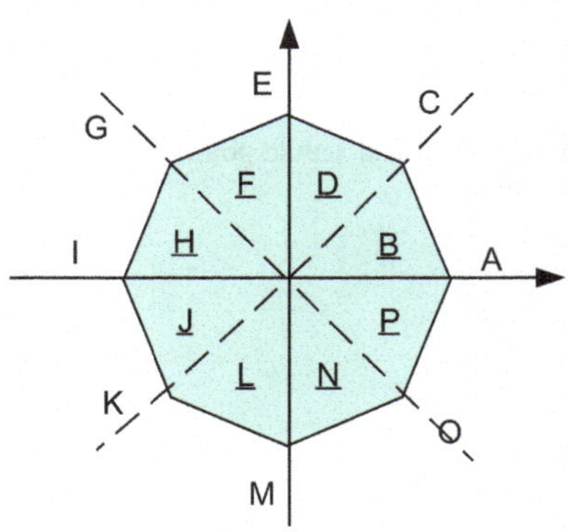

Fig. 5 Coordinate system of connection point

Fig. 6 Eight neighborhoods

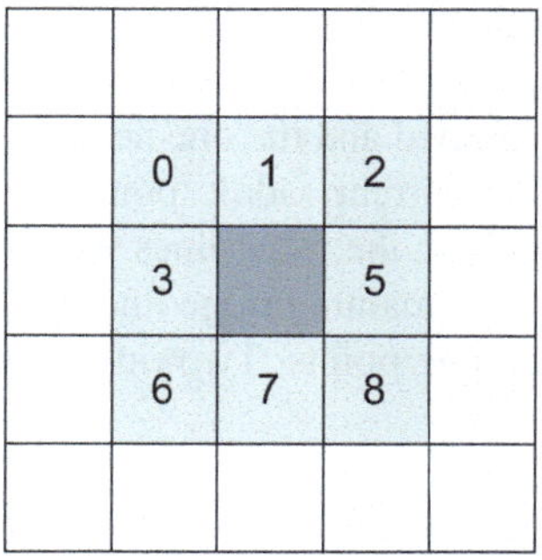

points in the 8 neighboring regions of the current CRP are determined by the position of EHP in the coordinate system of the CRP. If the EHP is located in the A region of the coordinate system above, then points 2, 5, and 8, as shown in Fig. 6, are the candidate connection points, of which the priority order is p2 = p8 and p5 > p2. The possible positions and priorities of an EHP are shown in Table 1.

If the priorities of two candidate connection points are the same, the point with larger elevation will be added into the ridge lines. To avoid intersections of ridge lines, the points connected will be neglected.

Step 4. **Connecting saddle points;**

After the three steps above, most of saddle points have been added into the ridge lines. The remaining saddle points will be examined one-by-one. As shown in Fig. 5, for every saddle point, a 2D coordinate system will be established with the

Table 1 Priorities of candidate connection points

Location of EP in the coordinate system	Candidate connecting points	Priority of candidate connecting points
A	2, 5, 8	p2 = p8, p5 > p2
B	1, 2, 5, 8	p2 = p5, p1 = p8, p2 > p8
C	1, 2, 5	p1 = p5, p2 > p5
D	0, 1, 2, 5	p2 = p1, p0 = p5, p2 > p5
E	0, 1, 2	p0 = p2, p1 > p2
F	0, 1, 2, 3	p1 = p0, p3 = p2, p1 > p2
G	0, 1, 3	p3 = p1, p0 > p1
H	0, 1, 3, 6	p3 = p0, p6 = p1, p0 > p1
I	0, 3, 6	p6 = p3, p0 > p3
J	0, 3, 6, 7	p7 = p0, p6 = p3, p3 > p7
K	3, 6, 7	p7 = p3, p6 > p3
L	3, 6, 7, 8	p7 = p6, p8 = p3, p6 > p8
M	6, 7, 8	p8 = p6, p7 > p8
N	5, 6, 7, 8	p5 = p6, p8 = p7, p7 > p5
O	5, 7, 8	p5 = p7, p8 > p7
P	2, 5, 7, 8	p5 = p8, p2 = p7, p5 > p2

point as origin. The hill point closest to the origin will be calculated in each of 16 parts. Among these hill points, two points, which are not in the same part, are selected and the one nearer to the origin is selected as the current saddle point. With the current saddle point as starting point and two different hill points as ending points, the ridge lines will be connected in two directions. With the same method, the remaining ridge lines will be handled under the control of starting point and two ending points. Two ridge lines with same starting points will be merged at the end.

3.2.2 The Minor Ridge Lines

The major ridge lines have been constructed based on the SAMBCD following the steps above. The next step is to connect the remaining discrete points on the watershed into minor ridge lines. Before connecting, it is necessary to examine all the discrete points one-by-one. If a point has participated in connecting major ridge lines, it will be set as the connected flag (square in Fig. 7). In addition, all points within the eight neighboring areas that are not used will be set as detection flags (pentagon symbol as shown in Fig. 7). The remaining discrete points unconnected are set as circle symbols as shown in Fig. 7 where the triangle symbol represents hill points.

Following this, we look for the points whose elevation is the minimum among the discrete points on the watershed and has not been connected and detected, and then start dealing with the ridge line. When forming a ridge line, the last point from the point array of ridge lines connected will be selected. If the number of the point array is more than 1, the point prior to the last point will be removed.

Fig. 7 Set symbols of points

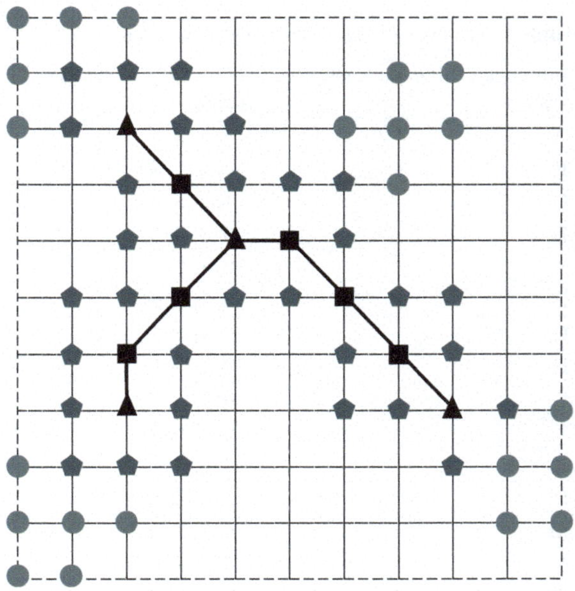

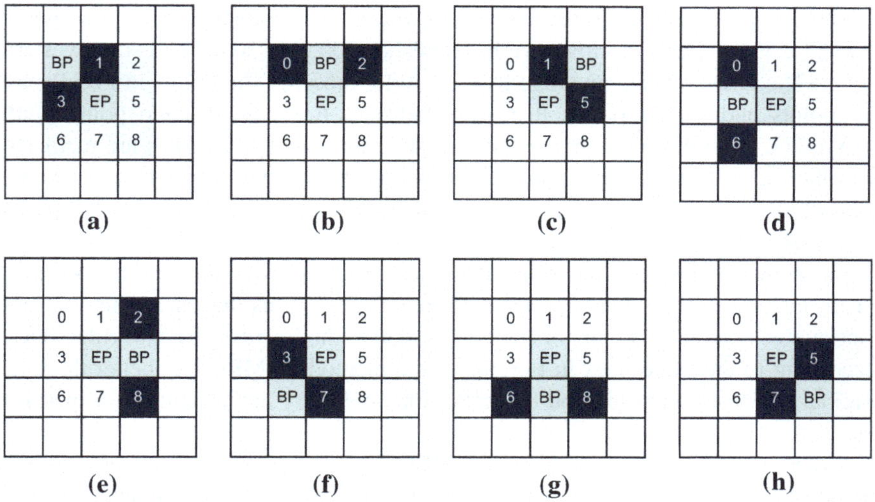

Fig. 8 Constrained connecting direction

According to the position of a connection point in the last point's eight neighboring areas, the ridge line direction will be constrained to avoid rotating and crossing of ridge lines. In Fig. 8, the connecting directions of the eight positions are shown, where EP is the last connecting point of the ridge line and BP is the previous point of EP. In addition, the darker points are points excluded, and the lighter points are the candidate connecting points. The points on black blocks may lead to rotating and crossing of ridge lines.

Next, we will examine the discrete points in the EP's eight neighboring areas in the watershed rather than in exclusive blocks, and seek the point whose elevation is the greatest in all the candidate points and greater than EP's elevation. If this point has not been connected and detected, it will be directly added into the point array of ridge line and will be subsequently traced. Otherwise, the point will be added into the point array and the tracing will end. As a result, the current ridge line will be connected and new routine will start. If the point has been detected, other points in the point's eight neighboring areas will be examined to see whether it has been connected. If another point exists, add and record the ridge point with maximum elevation into connected points array one by one and stop connecting. If not, stop connecting the ridge line.

4 Comparison and Analysis

In this paper, the ridge lines from the watershed are detected based on SAMBCD. The experimental data are grid DEM (90 m × 90 m) from the SRTM data. Experimental areas with different types were used in this test.

The experimental area covers DEM data whose coordinate of upper left corner is (100.85°, 30.18°) and the lower right corner is (100.99°, 30.04°) in southwestern China. The DEM elevations range from 2640 to 4665 m. The points in Fig. 9a are hill points and saddle points that were detected from the discrete points with SAMBCD. With these points as control points, the major ridge lines are extracted with the method of SAMBCD as shown in Fig. 9b, and the minor ridge lines are connected. After deleting the short minor ridge lines, the ridge lines extracted based on SAMBCD are shown in Fig. 9c.

As shown in Fig. 9d, the ridge lines are extracted with the original steepest ascent method in the same experimental area. It is obvious that these lines are not continuous, especially the major ridge lines, due to the fracture of major ridge lines. According to the ridge lines in Fig. 9c, d, the method SAMBCD is better than the original steepest ascent method regarding the continuity and stratification of ridge lines.

Moreover, this paper compares SAMBCD with the original steepest ascent method in other two experimental regions. Figure 10a, b are ridge maps that were detected in the first experimental region whose elevation range is from 200 to 800 m.

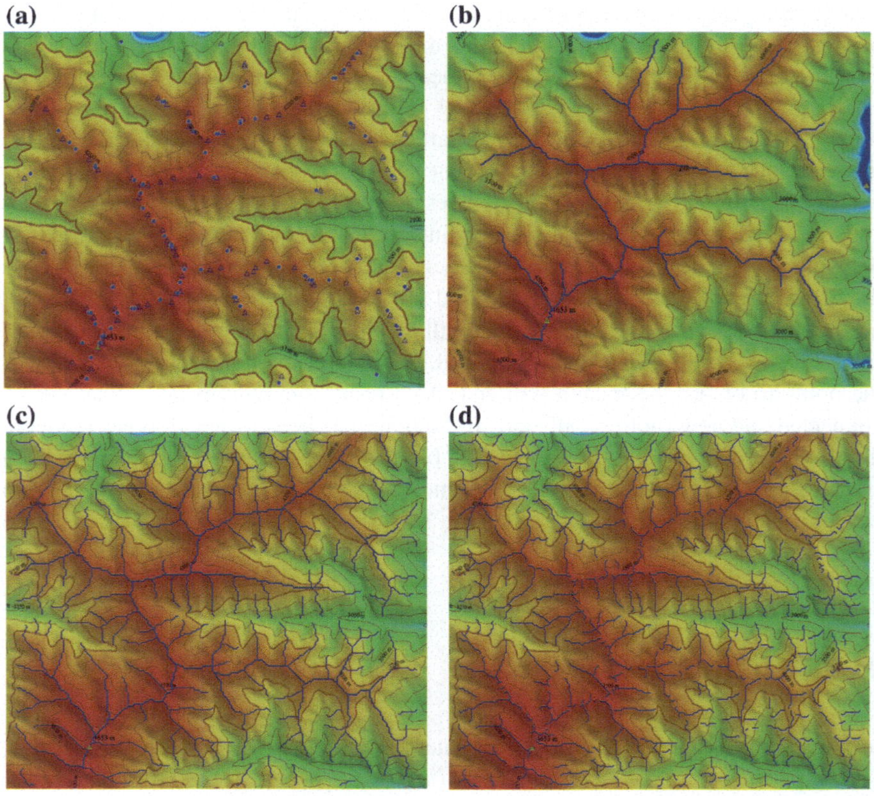

Fig. 9 (**a**) Hill and saddle points. (**b**) The major ridge lines. (**c**) Result with SAMBCD method. (**d**) Result with steepest ascent method

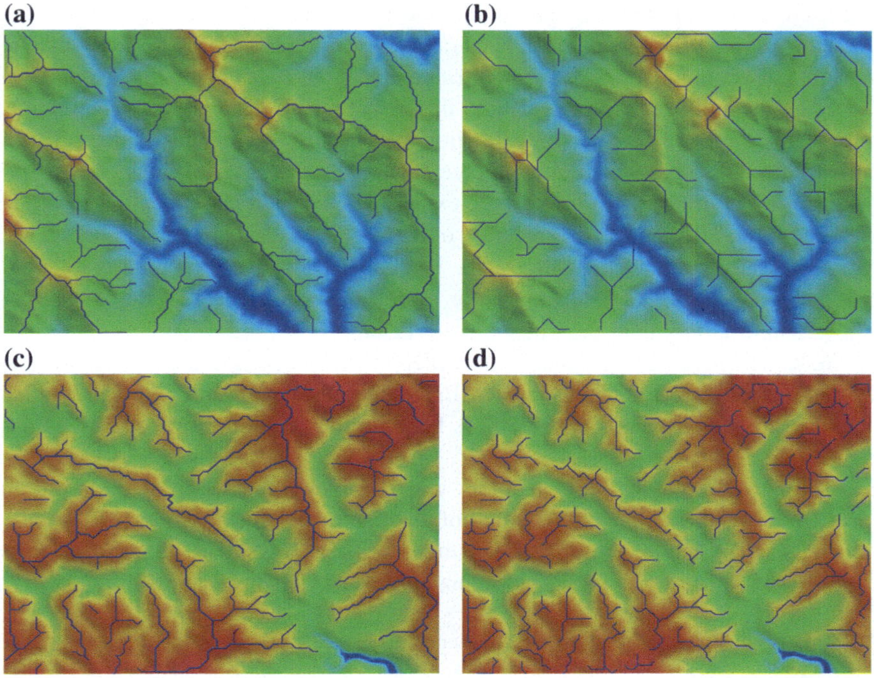

Fig. 10 Experimental results and comparison

Figure 10c, d are ridge maps that were detected in the second experimental region whose elevation range is from 500 m to 1000 m. In Fig. 10a, c are produced with SAMBCD; (b) and (d) are generated with steepest ascent method. According to the experimental results with different geomorphologic types, the method SAMBCD has a better performance in ridge detection than the original steepest ascent method.

5 Conclusion

For decades, many researchers have studied many methods for extracting terrain characteristic lines. For generalization of contour or DEM, terrain characteristic lines, which are extracted by geometric calculation from DEM, are the geomorphic skeleton of the research area. They are abstract and a generalization of ground truth, but are also real and objective.

In this paper, the authors propose the steepest ascent method based on automatic constrained connecting direction for extracting ridge lines from a watershed. With the constraint of connecting direction, the major and the minor ridge lines are extracted based on the discrete points of watershed. The general connecting direction of major ridge lines is constrained by the SHP and the EHP and the real-time direction of connecting major ridge lines is constrained by the last point of the ridge line and the EHP. In addition, this test connects discrete watershed points into ridge lines with connecting direction constrained by saddle points and hill points not connected, which is a supplement for the major ridge lines. The major ridge lines and minor ridge lines in the experimental region can be extracted based on SAMBCD. The connecting direction of SAMBCD is also constrained by the last two points of the connecting ridge line. The aim of SAMBCD method is to avoid crossing ridge lines. After all the minor ridge lines are connected, the short ones whose length is less than the threshold are deleted.

The experiment in three areas with different geomorphologic types shows that the ridge lines extracted based on SAMBCD are more continuous and well-bedded than the original steepest ascent method. In addition, the intersection of valley lines is also avoided.

As for the future work, the authors will consider more specific conditions to further improve the connecting method of the major ridge lines. Moreover, not only hill points and saddle points but also more feature points will be considered as control points for improving and optimizing the SAMBCD method.

Acknowledgements This research was financially supported by the National Natural Science Foundation of China (No. 41371428 and No. 41201473) and the National Fund for Talent Training in Basic Science of China (No. J1103409).

References

Bai, R., Li, T., Huang, Y., Li, J., & Wang, G. (2015). An efficient and comprehensive method for drainage network extraction from DEM with billions of pixels using a size-balanced binary search tree. *Geomorphology, 238,* 56–67.

Barnes, R., Lehman, C., & Mulla, D. (2014). An efficient assignment of drainage direction over flat surfaces in raster digital elevation models. *Computers & Geosciences, 62,* 128–135.

Chang, Y.-C., & Sinha, G. (2007). A visual basic program for ridge axis picking on DEM data using the profile-recognition and polygon-breaking algorithm. *Computers & Geosciences, 33* (2), 229–237.

Garbrecht, J., & Martz, L. W. (1997). The assignment of drainage direction over flat surfaces in raster digital elevation models. *Journal of Hydrology, 193*(1–4), 204–213. doi:http://dx.doi.org/10.1016/S0022-1694(96)03138-1.

Jana, R., Reshmidevi, T., Arun, P., & Eldho, T. (2007). An enhanced technique in construction of the discrete drainage network from low-resolution spatial database. *Computers & Geosciences, 33*(6), 717–727.

Jenson, S., & Domingue, J. (1988). Extracting topographic structure from digital elevation data for geographic information system analysis. *Photogrammetric engineering and remote sensing, 54* (11), 1593–1600.

Koka, S., Anada, K., Nomaki, K., Sugita, K., Tsuchida, K., & Yaku, T. (2011). Ridge detection with the steepest ascent method. *Procedia Computer Science, 4*(0), 216–221. doi:http://dx.doi.org/10.1016/j.procs.2011.04.023.

Koka, S., Anada, K., Nakayama, Y., Sugita, K., Yaku, T., & Yokoyama, R. (2012). A comparison of ridge detection methods for DEM data. In *2012 13th ACIS International Conference on, Software Engineering, Artificial Intelligence, Networking and Parallel & Distributed Computing (SNPD), 8–10 Aug. 2012* (pp. 513–517). doi:10.1109/SNPD.2012.46.

Lee, J., Snyder, P. K., & Fisher, P. F. (1992). Modeling the effect of data errors on feature extraction from digital elevation models. *Photogrammetric Engineering and Remote Sensing, 58*(10), 1461–1467.

Li, R., Tang, Z., Li, X., & Winter, J. (2013). Drainage structure datasets and effects on LiDAR-derived surface flow modeling. *ISPRS International Journal of Geo-Information, 2*(4), 1136–1152.

Martz, L. W., & Garbrecht, J. (1995). Automated recognition of valley lines and drainage networks from grid digital elevation models: A review and a new method—Comment. *Journal of Hydrology, 167*(1), 393–396.

Martz, L. W., & Garbrecht, J. (1998). The treatment of flat areas and depressions in automated drainage analysis of raster digital elevation models. *Hydrological Processes, 12*(6), 843–855.

Matsuura, T., & Aniya, M. (2012). Automated segmentation of hillslope profiles across ridges and valleys using a digital elevation model. *Geomorphology, 177,* 167–177.

O'Callaghan, J. F., & Mark, D. M. (1984). The extraction of drainage networks from digital elevation data. *Computer Vision, Graphics, and Image Processing*, *28*(3), 323–344. doi:http://dx.doi.org/10.1016/S0734-189X(84)80011-0.

Planchon, O., & Darboux, F. (2002). A fast, simple and versatile algorithm to fill the depressions of digital elevation models. *CATENA, 46*(2), 159–176.

Saraiva, A. G. S., & Paz, A. R. (2014). Multi-step change of scale approach for deriving coarse-resolution flow directions. *Computers & Geosciences, 68,* 53–63.

Wang, L., & Liu, H. (2006). An efficient method for identifying and filling surface depressions in digital elevation models for hydrologic analysis and modelling. *International Journal of Geographical Information Science, 20*(2), 193–213.

Yokoyama, R., Kureha, A., Motohashi, T., Ogasawara, H., Yaku, T., & Yoshino, D. (2007) Geographical concept recognition with the octgrid method for learning geography and geology. In *Seventh IEEE International Conference on Advanced Learning Technologies, ICALT 2007, 18–20 July 2007* (pp. 470-471). doi:10.1109/ICALT.2007.150.

Zhang, H., & Huang, G. (2009). Building channel networks for flat regions in digital elevation models. *Hydrological Processes, 23*(20), 2879–2887.

Zhu, Q., Tian, Y., & Zhao, J. (2006). An efficient depression processing algorithm for hydrologic analysis. *Computers & Geosciences, 32*(5), 615–623.

Zhu, Q., Zhao, J., Zhong, Z., & Sui, H. (2004). The extraction of topographic patterns based on regular grid dems. *Acta Geodaetica Et Cartographic Sinica, 33,* 77–82.

Using the A* Algorithm to Find Optimal Sequences for Area Aggregation

Dongliang Peng, Alexander Wolff and Jan-Henrik Haunert

Abstract Given two land-cover maps of different scales, we wish to find a sequence of small incremental changes that gradually transforms one map into the other. We assume that the two input maps consist of polygons that constitute a planar subdivision and belong to different land-cover classes. Every polygon in the small-scale map is the union of a set of polygons in the large-scale map. In each step of the sequence that we compute, the smallest area in the current map is merged with one of its neighbors. We do not select that neighbor according to a prescribed rule but define the whole sequence of pairwise merges at once, based on global optimization. An important requirement for such a method is a formalization of the problem in terms of optimization objectives and constraints, which we present together with a solution that is based on the so-called A* algorithm. This algorithm allows us to limit the exploration of the search space such that we can compute solutions of high quality in reasonable time. We tested the method with a dataset of the official German topographic database ATKIS and discuss our results.

Keywords Continuous map generalization · Land cover · Type change · Compactness · Graph · Optimization

1 Introduction

A *land-cover map* is a planar subdivision in which each area belongs to a land-cover class or *type*. Suppose that there are two land-cover maps of different scales that cover the same spatial region. We consider the problem of finding a sequence of small incremental changes that gradually transforms the large-scale map (the *start map*) into the small-scale map (the *goal map*). We use this sequence to generate and

D. Peng (✉) · A. Wolff
Chair of Computer Science I, University of Würzburg, Würzburg, Germany
e-mail: dongliang.peng@uni-wuerzburg.de

J.-H. Haunert
Institute of Geodesy and Geoinformation, University of Bonn, Bonn, Germany
e-mail: haunert@igg.uni-bonn.de

M.P. Peterson (ed.), *Advances in Cartography and GIScience*, Lecture Notes in Geoinformation and Cartography, DOI 10.1007/978-3-319-57336-6_27

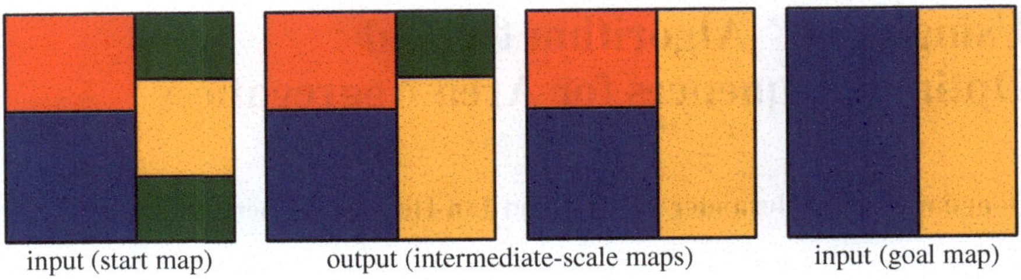

Fig. 1 The input and a possible output for an instance of our problem

show land-cover maps of intermediate scales (see Fig. 1). In this way, we can avoid large and sudden changes that distracts a user when zooming in or out.

With the same motivation, a strategy of hierarchical schemes has been proposed. This strategy generalizes a more-detailed representation to obtain a less-detailed representation based on small incremental changes, e.g., the Generalized Area Partitioning tree (for short, GAP-tree), which can be constructed if only the large-scale map is given (van Oosterom and Schenkelaars 1995) or if both the large-scale map and the small-scale map are given (Haunert et al. 2009).

Typically, the next change in such a sequence is determined locally, in a greedy fashion. If we insist on finding a sequence that is, according to some global measure, optimal, the problem becomes complicated.

We assume that there exist many-to-one relationships between the areas of the start map and the areas of the goal map. This assumption is based on the fact that there are many algorithms (e.g., Haunert and Wolff 2010; van Smaalen 2003; Cheng and Li 2006) that aggregate some land-cover areas into one single land-cover area. The input and the generalized results of these methods can be used as our input. We term the areas of the goal map *regions*. That is, every region is the union of a set of areas on the start map.

The type of a region may differ from the types of its components. For example, a small water area together with multiple adjacent forest areas may constitute a large forest region in the smaller scale. We assume, however, that every region contains at least one area of the same type. Our assumptions hold if the goal map has been produced with an automatic method for area aggregation, for example, with the method of Haunert and Wolff (2010). That method produces a land-cover map at a single output scale, given a land-cover map at a larger scale. It is based on global optimization and minimizes type changes as well as a cost for non-compact shapes while satisfying constraints on the sizes of the output regions. Although the method of Haunert and Wolff leads to results of high quality, it does not yield land-cover maps at intermediate scales.

As stated by Chimani et al. (2014), "a sequence of well-generalized maps is not necessarily the same as a good sequence of maps." To generate a good sequence of maps, we must consider the sequence of maps as a whole instead of considering each intermediate-scale map separately. On the other hand, to simplify the problem, we first consider each region of the goal map (with its components on the start map) independent of the other regions. Once we have found a sequence of maps for each region, we integrate all the sequences into an overall sequence that transforms the start map into the goal map. Our aggregation sequence may be stored in data structures such as the GAP-face tree (van Oosterom 2005) or the map cube model (Timpf 1998) to support on-the-fly display.

Contribution. We give general ideas and basic concepts of our method, and analyze the size of our problem (see Sect. 2). We present a new global optimization approach to solve this problem. Our approach is based on the A* algorithm (see Sect. 3) which handles the combinatorial explosion that we will discuss in Sect. 2. Then, we do a case study using a dataset of the German topographic database ATKIS to assess the effectiveness and efficiency of our method (see Sect. 4). Finally, we conclude the paper (in Sect. 5).

We do not deal with the simplification of lines in this paper, since this can be handled separately from the aggregation of areas, for example, by using the method of Douglas and Peucker (1973) or Saalfeld (1999). Those methods can be used to set up the Binary Line Generalized tree (van Oosterom and Schenkelaars 1995) (for short, BLG-tree), which is a hierarchical data structure that defines a gradual line simplification process.

2 Preliminaries

To allow us to describe our method more easily, we assume that the goal map has only one region. On the start map, the same region consists of n land-cover areas (components). In other words, the union of the n land-cover areas is the only region on the goal map. We show how to compute an optimal aggregation sequence for a region. For a goal map with more than one region, we simply "interleaf" aggregation sequences of different regions (see Sect. 3.6).

To find a sequence of small changes that transforms the start map into the goal map, we require that every change involves only two areas of the current map. More precisely, in each step the smallest area a is merged with one of its neighbors (adjacent areas) b such that a and b are replaced by their union. The union must take over the type of either a or b. If the union uses the type of a, we say that area b is *aggregated into* area a, and vice versa. How to aggregate exactly is decided by optimizing a global cost function. This requirement ensures that the size of the smallest area in the map increases in each step and thus the sequence reflects a gradual reduction of the map's scale. From another perspective, we consider the smallest area as the least important, instead of involving more rules for importance, to avoid making our problem too complicated. Even though the requirement reduces the number of possible solutions, there is still enough room for optimization since we leave open with which of its neighbors the smallest area is aggregated. We term a sequence of changes that adheres to our requirement an *aggregation sequence*.

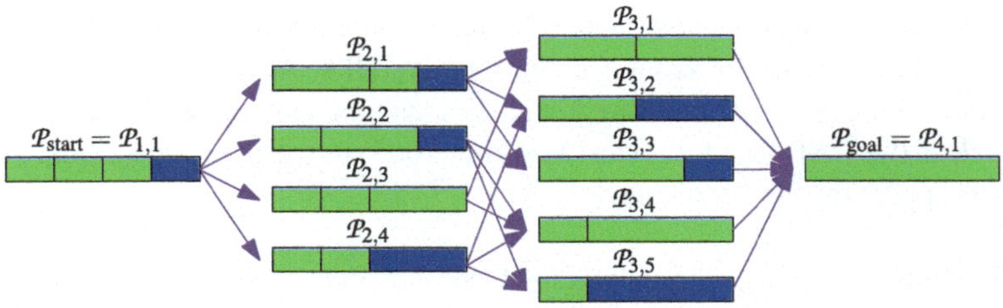

Fig. 2 The subdivision graph G. The nodes of the graph are the subdivisions. There is an arc from node $\mathcal{P}_{t,i}$ to node $\mathcal{P}_{t+1,j}$ if $\mathcal{P}_{t+1,j}$ is the result of aggregating the smallest area p of $\mathcal{P}_{t,i}$ with a neighbor of p

Model. We consider a directed graph G, which we call the *subdivision graph* (see Fig. 2). The node set V of G contains a node for every possible map (or *subdivision*, including the start map, all possible intermediate-scale maps, and the goal map). The arc set E of G contains an arc uv between any two maps $u, v \in V$ such that v can be reached from u with a single aggregation operation involving the smallest area. Then, any directed path in G from the start map to the goal map defines a possible aggregation sequence. Our idea is to compute an optimal aggregation sequence by computing a minimum-weight path from start to goal. This idea obviously requires that the arc weights are set such that a minimum-weight start–goal path does actually correspond to an aggregation sequence of maximum cartographic quality. Moreover, putting the idea to practice is far from trivial since the graph G can be huge as we will argue below. Thus, exploring all of its nodes must be avoided. To limit the search space when computing an optimal path, we use the A* algorithm (Hart et al. 1968).

Notation. Each land-cover area is represented by a polygon with a type. A *patch* is a set of polygons whose union is connected. The patch also has the property type.

Recall that there are n land-cover areas on the start map. Hence, the desired aggregation sequence consists of $n - 1$ steps. There are n subdivisions on a path from the start map to the goal map. We use $t \in \mathcal{T} = \{1, 2, \ldots, n\}$ to denote the *time* of the path. When $t = 1$, the subdivision consists of n patches, and there is only 1 patch when $t = n$. The subdivision graph consists of layers $\mathcal{L}_1, \ldots, \mathcal{L}_n$, where layer $\mathcal{L}_t = \{\mathcal{P}_{t,1}, \ldots, \mathcal{P}_{t,n_t}\}$ contains every possible subdivision $\mathcal{P}_{t,i}$ with $n - t + 1$ patches (see Fig. 2).

Exponential lower bound. We now analyze the size of G. Consider an instance that consists of a start map $\mathcal{P}_{1,1}$ with $n = 2k$ unit squares in a row and a goal map $\mathcal{P}_{n,1}$ that is simply the union of the n squares, that is, an $(n \times 1)$-rectangle. One of the intermediate subdivisions in layer $\mathcal{L}_{k+1}$—let's call it $\mathcal{P}_{k+1,1}$—consists of a row of k (2×1)-rectangles. A lower bound for the number of nodes of G is the number of ways the (2×1)-rectangles in $\mathcal{P}_{k+1,1}$ can be subdivided into unit squares: each way corresponds to a unique node in a layer between the start map and $\mathcal{P}_{k+1,1}$. Clearly, there are exactly $2^k = 2^{n/2}$ ways.

3 Using the A⋆ Algorithm

3.1 Formalizing Area Aggregation as a Pathfinding Problem

We model our aggregation problem as finding a shortest path in the subdivision graph G, which is weighted. We define the weight of an arc to be the cost of the corresponding aggregation step. We try to find a shortest path from subdivision $\mathcal{P}_{\text{start}}$, the start map, to $\mathcal{P}_{\text{goal}}$, the goal map. There are various algorithms that compute shortest paths in graphs, most notably Dijkstra's algorithm. As the graph G—our search space—can be of exponential size (see Sect. 2), we avoid computing the whole graph explicitly.

Once we have found a shortest path, we can generate an optimal aggregation sequence. Note that we only know subdivisions $\mathcal{P}_{\text{start}}$ and $\mathcal{P}_{\text{goal}}$ at the beginning. To save time and memory, we generate a subdivision $\mathcal{P}_{t,i}$ only when we want to visit it.

We decided to use the A⋆ algorithm to find a shortest path from the start node $\mathcal{P}_{\text{start}}$ to the goal node $\mathcal{P}_{\text{goal}}$ because A⋆ uses a clever best-first strategy. For a node $\mathcal{P}_{t,i}$, A⋆ considers the exact cost of a shortest path from $\mathcal{P}_{\text{start}}$ to $\mathcal{P}_{t,i}$ and estimates the cost to get from $\mathcal{P}_{t,i}$ to $\mathcal{P}_{\text{goal}}$. The A⋆ algorithm can be seen as a refinement of Dijkstra's algorithm where nodes are explored earlier if they are estimated to be close to the goal.

More formally, we define $g(\mathcal{P}_{t,i})$ to be the exact cost of a shortest path from $\mathcal{P}_{\text{start}}$ to $\mathcal{P}_{t,i}$, and we define $h(\mathcal{P}_{t,i})$ to be the estimated cost to get from $\mathcal{P}_{t,i}$ to $\mathcal{P}_{\text{goal}}$. Then, the total cost at node $\mathcal{P}_{t,i}$ is

$$F(\mathcal{P}_{t,i}) = g(\mathcal{P}_{t,i}) + h(\mathcal{P}_{t,i}). \tag{1}$$

If $h(\mathcal{P}_{t,i})$ is always bounded from above by the exact cost of a shortest path from $\mathcal{P}_{t,i}$ to $\mathcal{P}_{\text{goal}}$, A⋆ is guaranteed to find a shortest path from $\mathcal{P}_{\text{start}}$ to $\mathcal{P}_{\text{goal}}$, that is, an optimal aggregation sequence. Using the estimation, A⋆ is able to reduce the search space. The better the estimation, the more A⋆ can reduce the search space. In the following, we show how we define $g(\mathcal{P}_{t,i})$ and $h(\mathcal{P}_{t,i})$.

3.2 Cost Functions

We consider two aspects of cost for our aggregation problem. First, we want to minimize the total type change, over all aggregation steps. Second, for the patches of a subdivision, we consider the one with the minimum compactness. We want to maximize the sum of the minimum compactnesses, over all intermediate subdivisions.

We define the cost of type change as follows. Suppose that we are at the step of generalizing from subdivision $\mathcal{P}_{s-1,i}$ to subdivision $\mathcal{P}_{s,j}$. In this step, patch p is aggregated into patch q. We denote the types of the two patches by $T(p)$ and $T(q)$. We define the cost of type change of this step by

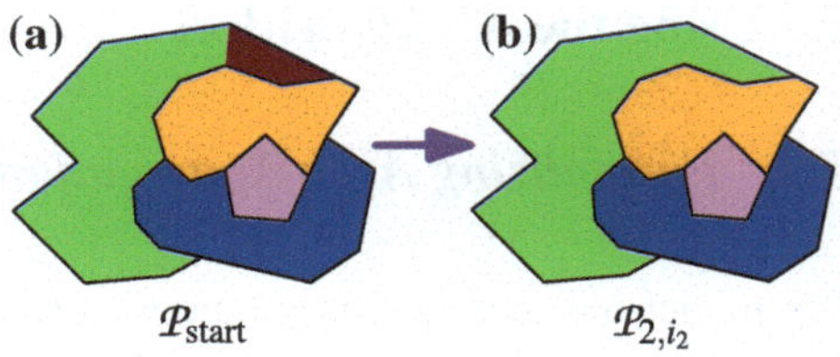

Fig. 3 An aggregation step, where the *brown* patch is aggregated into the *green* patch

$$f_{\mathrm{type}}(\mathcal{P}_{s-1,i}, \mathcal{P}_{s,j}) = \frac{A_p}{A_R} \cdot \frac{Cost(T(p), T(q))}{T_{\max}}, \tag{2}$$

where A_p is the area of patch p, and A_R is the area of region R. The constant $T_{\max}$, the maximum cost over all type changes, is used to unify the cost of type change and is known from the input. The input specifies, for each pair (T_1, T_2) of types, the cost $Cost(T_1, T_2)$ of changing type T_1 to type T_2.

For an example, refer to Fig. 3a, where the brown and the green patch play the roles of patches p and q in Eq. 2. The shaded area in Fig. 4a shows the contribution of type change to the cost.

For path $\Pi = (\mathcal{P}_{1,i_1}, \mathcal{P}_{2,i_2}, \ldots, \mathcal{P}_{t,i_t})$, we define the exact cost of type change over all steps by

$$g_{\mathrm{type}}(\Pi) = \sum_{s=2}^{t} f_{\mathrm{type}}(\mathcal{P}_{s-1,i_{s-1}}, \mathcal{P}_{s,i_s}) \tag{3}$$

Recall that we also want to maximize the sum of the minimum compactnesses over all intermediate subdivisions. However, the A* algorithm tries to find a solution that minimizes a cost. To adapt our requirement to the A* algorithm, we modify our request to minimize the sum of a *shape* cost

$$f_{\mathrm{shape}}(\mathcal{P}_{s,k}) = \frac{1 - \min_{p \in \mathcal{P}_{s,k}} c_p}{n-2}, \tag{4}$$

where $c_p = 2\sqrt{\pi A_p}/L_p$ is the compactness of patch p (Frolov 1975), and L_p is the perimeter of p. We use $n-2$ as the denominator to balance between the costs of type change and shape. Take Fig. 4b for example, the compactness of the shaded

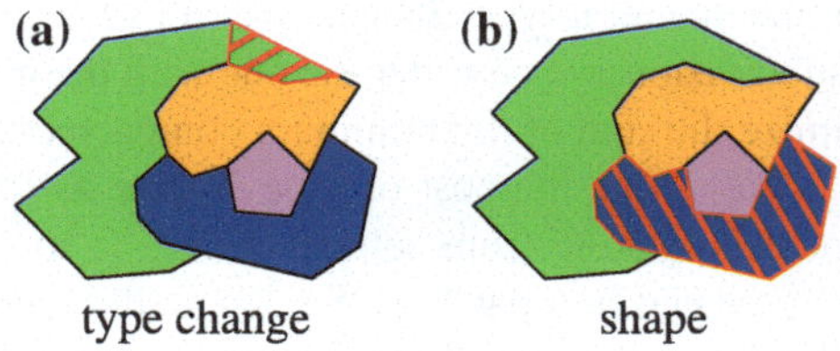

Fig. 4 The shaded parts show the contributions to the costs

blue patch will be used to compute f_{shape} because the blue patch is the least compact one among the four patches.

For path $\Pi = (\mathcal{P}_{1,i_1}, \mathcal{P}_{2,i_2}, \ldots, \mathcal{P}_{t,i_t})$, we define the exact cost of shape over all intermediate subdivisions by

$$g_{\text{shape}}(\Pi) = \sum_{s=2}^{t-1} f_{\text{shape}}(\mathcal{P}_{s,i_s}). \tag{5}$$

Combining type change and shape, we define the total exact cost of path Π by

$$g(\Pi) = (1 - \lambda)g_{\text{type}}(\Pi) + \lambda g_{\text{shape}}(\Pi), \tag{6}$$

where $\lambda \in [0, 1]$ is a parameter to balance f_{type} and f_{shape}. In most cases, we use $\lambda = 0.5$ in our experiments. We want to find a path from $\mathcal{P}_{\text{start}}$ to $\mathcal{P}_{t,i}$ that minimizes, among all such paths, $g(\Pi)$. Slightly abusing notation, we denote the length (or cost) of this shortest path from $\mathcal{P}_{\text{start}}$ to $\mathcal{P}_{t,i}$ by $g(\mathcal{P}_{t,i})$.

To narrow down the search space, we set up estimation functions for type change (Sect. 3.3) and shape (Sect. 3.4). These functions are meant to direct A⋆ towards the goal.

3.3 Estimating the Cost of Type Change

As illustrated in Sect. 3.1, if the estimated cost is always a lower bound of the exact cost of a shortest path from $\mathcal{P}_{t,i}$ to $\mathcal{P}_{\text{goal}}$, A⋆ is guaranteed to find a shortest path. To find a lower bound of the cost of type change, we simply assume that every patch will be aggregated into a patch with type T_{goal}, where T_{goal} is the type of the only patch on the goal map. As long as the cost of type change is a metric, this aggregation strategy indeed returns a lower bound. For subdivision $\mathcal{P}_{t,i}$, let $(\mathcal{P}_{t,i_t}, \mathcal{P}_{t+1,i_{t+1}}, \ldots, \mathcal{P}_{n,i_n})$, $\mathcal{P}_{t,i_t} = \mathcal{P}_{t,i}$ and $\mathcal{P}_{n,i_n} = \mathcal{P}_{\text{goal}}$, be the path that always changes the type of the smallest patch to T_{goal}. Then, the estimated cost of type change is

$$h_{\text{type}}(\mathcal{P}_{t,i}) = \sum_{s=t}^{n-1} f_{\text{type}}(\mathcal{P}_{s,i_s}, \mathcal{P}_{s+1,i_{s+1}}). \tag{7}$$

For an example, after the aggregation step of Fig. 3, we compute h_{type} according to the aggregation sequence of Fig. 5. Note that the step from $\mathcal{P}_{2,i_2}$ to $\mathcal{P}_{3,i_3}$ in Fig. 5 is impossible in reality because the violet patch cannot be assigned color green if none of its neighbors is green. However, this assignment is allowed for estimation because we may find a shortest path as long as the estimated cost is no more than the exact cost of a shortest path.

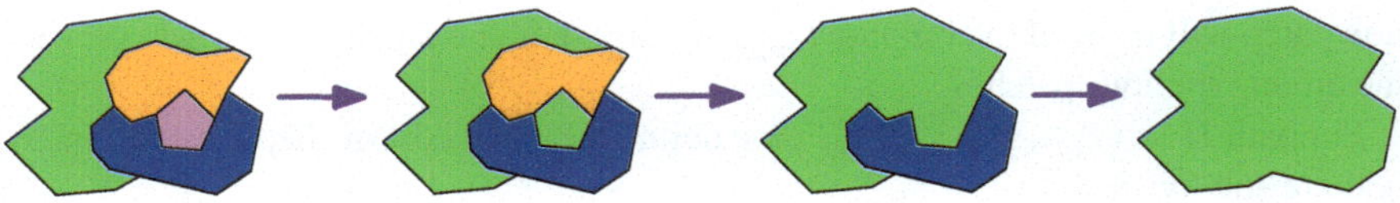

Fig. 5 An "aggregation sequence" for computing h_{type}, based on the aggregation result of Fig. 3b

3.4 Estimating the Cost of Shape

We find a lower bound of $f_{\text{shape}}(\mathcal{P}_{s,k})$, in Eq. 4, based on the edge number of the boundaries. We use n_{ext} to denote the edge number of the region's exterior boundaries. As the exterior boundaries will not be changed by aggregation, n_{ext} is a constant. We use $B(\mathcal{P}_{s,k})$ to denote the set of interior boundaries of subdivision $\mathcal{P}_{s,k}$. Note that a boundary between two patches is not necessarily connected; for an example, see the orange boundary in Fig. 7a. The number of interior edges is

$$e_{\text{int}}(\mathcal{P}_{s,k}) = \sum_{b \in B(\mathcal{P}_{s,k})} \|b\|, \tag{8}$$

where $\|b\|$ is the edge number of boundary b. Considering that each interior edge is used by two patches, the total number of edges over all patches is

$$e_{\text{sum}}(\mathcal{P}_{s,k}) = e_{\text{ext}} + 2e_{\text{int}}(\mathcal{P}_{s,k}). \tag{9}$$

To compute a lower bound of $f_{\text{shape}}(\mathcal{P}_{s,k})$ is to find an upper bound of $\min_{p \in \mathcal{P}_{s,k}} c_p$. As the regular polygon is the most compact one among all polygons with the same number of edges, we construct regular polygons using the edges we have, ignoring their lengths. A regular polygon with more edges is more compact than one with fewer edges. We now suppose that we know an upper bound of edge number $e_{\text{sum}}(\mathcal{P}_{s,k})$. We denote this upper bound by $E(\mathcal{P}_{s,k})$. We will show how to compute $E(\mathcal{P}_{s,k})$ later. Now, we define the upper bound of $\min_{p \in \mathcal{P}_{s,k}} c_p$ based on $E(\mathcal{P}_{s,k})$.

At time s, there are $n - s + 1$ patches. As each patch has 3 edges at least, $E(\mathcal{P}_{s,k}) \geq 3(n - s + 1)$. If we are allowed to freely construct the patches, we should evenly distribute the edges and construct $n - s + 1$ regular patches in order to maximize the minimum compactness. In this case, each patch has

$$\rho_{s,i} = \lfloor E(\mathcal{P}_{s,k})/(n - s + 1) \rfloor$$

edges at least. The smallest compactness of such a set of regular patches is

$$C_{\min}(\mathcal{P}_{t,i}) = \sqrt{\frac{\pi}{\rho_{s,i}} \Big/ \tan \frac{\pi}{\rho_{s,i}}}.$$

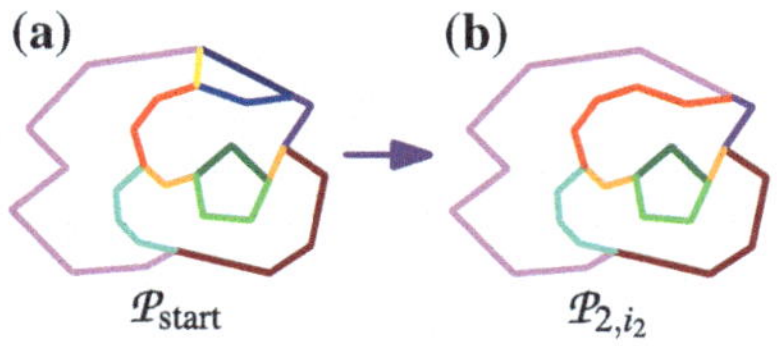

Fig. 6 An example to show the change of the remained boundaries after removing a boundary. (**a**) Original boundaries. (**b**) The boundaries after removal. This example corresponds to the aggregation step in Fig. 3. Note that the *pink* boundary and the navy boundary in (**a**) join together after we remove the *yellow* boundary. So do the *red* boundary and the *blue* boundary

This value describes the largest minimum-compact situation, and thus is an upper bound for $\min_{p \in \mathcal{P}_{s,k}} c_p$. The lower bound of $f_{\text{shape}}(\mathcal{P}_{s,k})$ based on the edge number can be computed, according to Eq. 4, by

$$l(\mathcal{P}_{s,k}) = \frac{1 - C_{\min}(\mathcal{P}_{s,k})}{n - 2}.$$

In our case, the patches can never be circles, thus $C_{\min}(\mathcal{P}_{s,k}) < 1$ and $l(\mathcal{P}_{s,k}) > 0$. This property is important when we overestimate (see Sect. 3.5).

One remaining task is to compute upper bound $E(\mathcal{P}_{s,k})$. Removing a boundary may result in joining several boundaries to one (see Fig. 6 for an example). At each step of the aggregation, we remove one interior boundary. To make sure that we have an upper bound, we always remove the original interior boundary, from the set of interior boundaries $B(\mathcal{P}_{t,i})$, with the fewest edges (see Fig. 7 for an example). Then at each step we compute $E(\mathcal{P}_{s,k})$ according to Eqs. 8 and 9. Note that we do this only for estimation. (When computing the real compactness, we consider the boundary length and the area of a patch; see Sect. 3.2.)

Let $(\mathcal{P}_{t,j_t}, \mathcal{P}_{t+1,j_{t+1}}, \ldots, \mathcal{P}_{n,j_n})$, $\mathcal{P}_{t,j_t} = \mathcal{P}_{t,i}$ and $\mathcal{P}_{n,j_n} = \mathcal{P}_{\text{goal}}$, be the path from $\mathcal{P}_{t,i}$ to $\mathcal{P}_{\text{goal}}$ where we repeatedly remove the interior boundary with the fewest edges. We define the estimated cost of shape by

$$h_{\text{shape}}(\mathcal{P}_{t,i}) = \sum_{s=t}^{n-1} l(\mathcal{P}_{s,j_s}), \tag{10}$$

where we define $l(\mathcal{P}_{1,i_1}) = 0$.

We define the total estimated cost to be

$$h(\mathcal{P}_{t,i}) = (1 - \lambda) h_{\text{type}}(\mathcal{P}_{t,i}) + \lambda h_{\text{shape}}(\mathcal{P}_{t,i}),$$

where λ is the same as in Eq. 6. Recall that we compute the total cost at node $\mathcal{P}_{t,i}$ by Eq. 1.

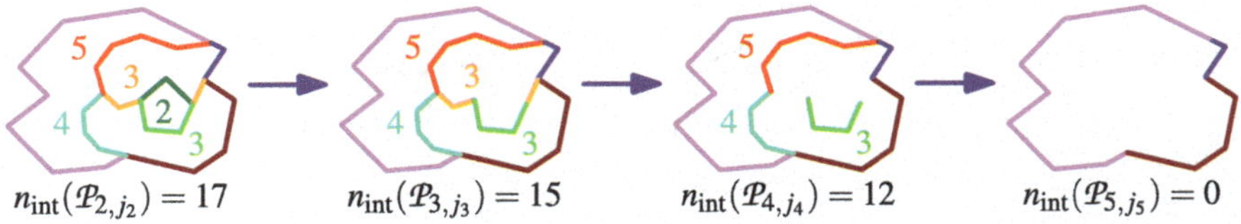

Fig. 7 An "aggregation" sequence for computing h_{shape} (see Eq. 10) based on the number of edges. At each step we remove the boundary with the fewest edges. At the last step, i.e., from $\mathcal{P}_{4,j_4}$ to $\mathcal{P}_{5,j_5}$, we simply remove all the interior boundaries because we know the exact shape from the input. The numbers represent the edge numbers of the interior boundaries. This example corresponds to the aggregation step in Fig. 3

3.5 Overestimation

Since the number of subdivisions can be exponential, it can happen that we run out of main memory before we find an optimal solution. To solve this problem, we overestimate the cost. Overestimation is a popular technique in the A* algorithm. For example, Pohl (1973) used a strategy of dynamic weighting to overestimate. We propose another strategy in order to find a solution faster. We first try finding a solution by the A* algorithm. If we cannot find one after we have visited a predefined number M of nodes of G, then we overestimate the cost by a factor of 2 and try again. We repeat this until we have a feasible solution. We revise Eq. 1 that defines the total cost at a node $\mathcal{P}_{t,i}$ to

$$F(\mathcal{P}_{t,i}) = g(\mathcal{P}_{t,i}) + K \cdot h(\mathcal{P}_{t,i}), \tag{11}$$

where $K = 2^k$ is the overestimation factor, and $k \geq 0$ is the number of repetitions. Whenever overestimating ($k \geq 1$), we cannot guarantee optimality anymore. We need to make sure that $h(\mathcal{P}_{t,i}) > 0$, otherwise the overestimation is of no effect.

3.6 Integrating Aggregation Sequences of Different Regions

We compute an aggregation sequence for each region. Then, we integrate the sequences of different regions with respect to the order of the smallest patches (see Fig. 8 for an example). This integration is similar to the merge step in the sorting algorithm Mergesort (see Cormen et al. 2009, pp. 29–37).

4 Case Study

We have implemented our methods based on C# (Microsoft Visual Studio 2012) and ArcObjects SDK 10.2. We ran our case study under 64-bit Windows 7 on a 3.3 GHz dual core CPU with 8 GB RAM. We measured processing time by the built-in C# class *Stopwatch*. We tested our method on a dataset from the German topographic

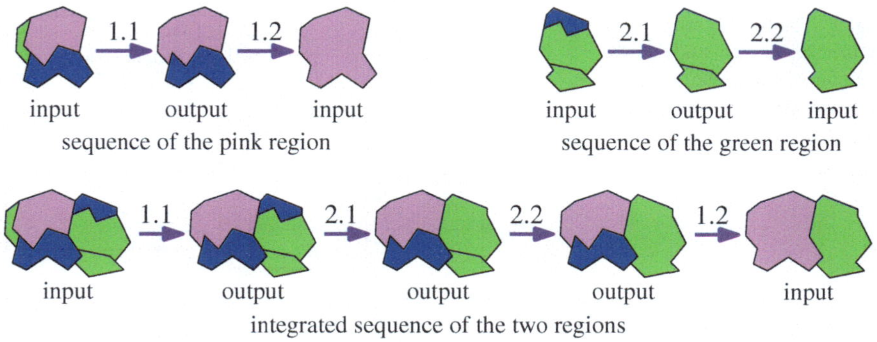

Fig. 8 Integrating two aggregation sequences of different regions: the resulting sequence contains the given sequences as subsequences and always takes the subdivision with smallest patch next

database ATKIS DLM 50. The dataset represents the place "Buchholz in der Nordheide" at scale 1:50,000. Our start map is the map after the collapse of areas by Haunert (2009, pp. 61–66), which has 5,537 polygons. Our goal map was generalized from the start map by Haunert and Wolff (2010) setting the scale at 1:250,000 (see Fig. 10). The goal map has 734 polygons, which means that there are $N = 734$ regions. We used a tree-based method introduced by Schwering (2008) to define the cost of type change; the cost is the distance between two types in the type hierarchy (see Fig. 9). For example, the distance between type fence and type street is 4. In this tree, the maximum distance is 6. Therefore, $T_{\max} = 6$ in our case study for Eq. 2.

We overestimated whenever we could not find a solution after having visited M nodes (see Sect. 3.5). We tried $M = 200{,}000$ and $M = 400{,}000$. The results are shown in Table 1. When $M = 200{,}000$, we have found optimal aggregation sequences for 692 of the 734 regions (94.3%); see column #over in Table 1, where we overestimated for the other 42 regions.

Recall Eq. 11, for region with ID i we define k_i as the least repetitions that we do to find a feasible solution, and we define the least feasible overestimation factor $K_i = 2^{k_i}$. We define the total repetitions as $k_{\text{sum}} = \sum_{i=1}^{N} k_i$, where N is the number of regions. After we increased M to 400,000, we found optimal aggregation sequences for only three more regions, but the total repetitions k_{sum} decreased quite a bit, from 113 to 96. The numbers of regions that needed a certain overestimation factor are shown in Fig. 11. Besides, we visited more arcs and nodes, used more time, but got (slightly) less cost when $M = 400{,}000$. Although the number of regions that needed overestimation is relatively small, A⋆ spent most of the running time on these few regions: 5.7 and 5.3% of the regions cost 88.7and 92.2% of the total running time, respectively (see Table 1).

The details of some regions are presented in Table 2. According to the entries with overestimation factor $K_i = 1$, we often have $R_{\text{type}} = 1$ and a large R_{shape}. Recall that when $K = 1$, $R_{\text{type}} \geq 1$ and $R_{\text{shape}} \geq 1$; see Sect. 3.1. $R_{\text{type}} = 1$ means that we have the best estimation for the cost of type change. A large R_{shape} means that our estimation for the cost of shape is poor.

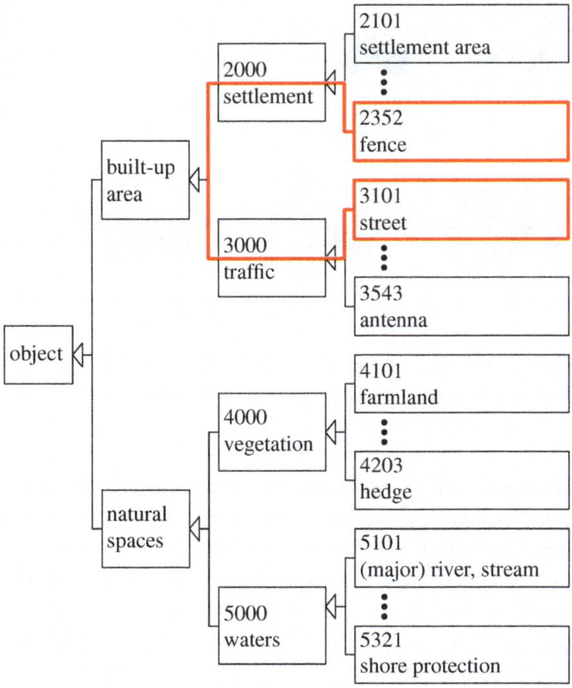

Fig. 9 The type distance defined in our case study. For example, the distance between type fence and type street is 4

Table 1 A comparison of the A* algorithm for two different settings that depend on the maximum number of nodes, M, that we allowed the A* algorithm to explore. #over is the number and percentage of regions (out of 734) that needed overestimation. Variable k_{sum} is the total repetitions. Symbols #nodes and #arcs are the total numbers of nodes and arcs that the A* algorithm visited. (For instances where we needed overestimation, only the final attempt is counted.) We used g_{t}, g_{s}, and g to denote the sums of $g_{\mathrm{type}}(\mathcal{P}_{\mathrm{goal}})$, $g_{\mathrm{shape}}(\mathcal{P}_{\mathrm{goal}})$, and $g(\mathcal{P}_{\mathrm{goal}})$, respectively, over all instances (see Eqs. 3, 5, and 6). The percentage in the time column is the fraction of the total runtime spent on solving the instances that needed overestimation

M	#over	k_{sum}	#nodes	#arcs	g_{t}	g_{s}	g	Time (h)
200,000	42 (5.7%)	113	$2.4 \cdot 10^6$	$4.9 \cdot 10^6$	51.5	259.1	155.3	0.6 (88.7%)
400,000	39 (5.3%)	96	$4.3 \cdot 10^6$	$8.9 \cdot 10^6$	51.4	259.0	155.2	1.1 (92.2%)

Interestingly enough, region 543 has neither the most patches nor the most adjacencies, but it needs the largest overestimation factor, which is 512 (see the first entry in Table 2). The reason is that all the patches of this region are with the same type (see region 543 with 20 patches in Fig. 12). The type change has no contribution to the estimation, while the estimation of shape is poor to help find an aggregation sequence.

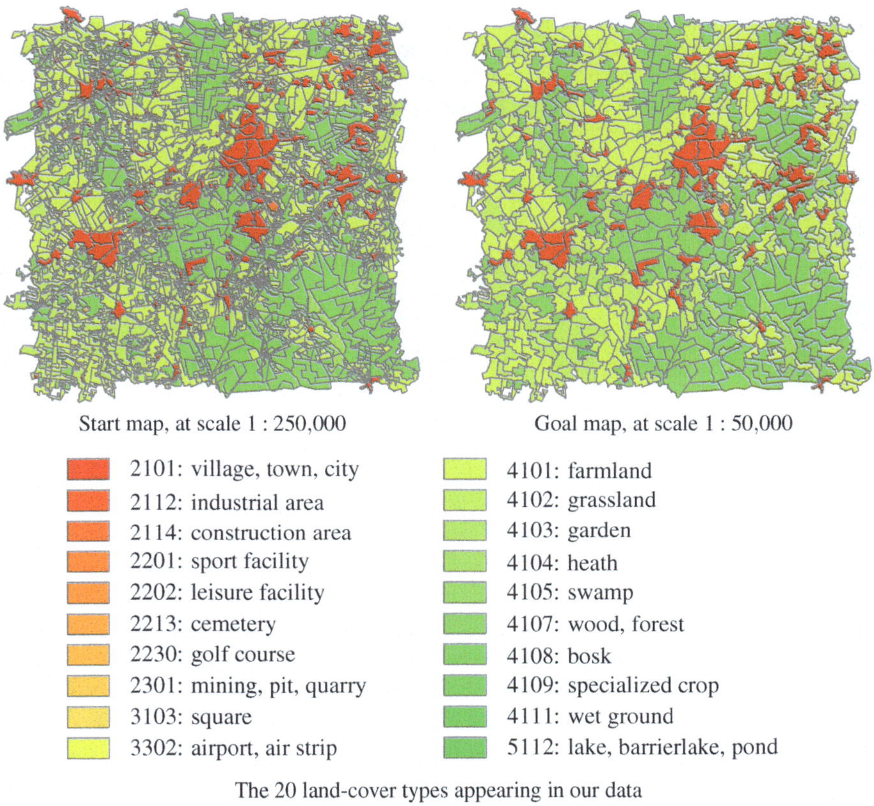

Fig. 10 The data of our case study

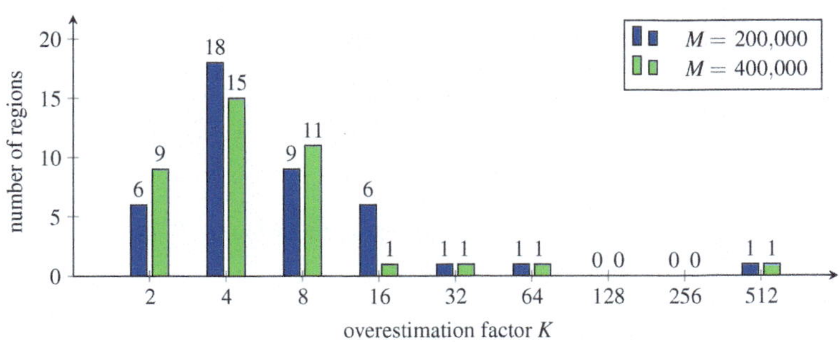

Fig. 11 The numbers of regions where our approach was forced to use the given overestimation factors in order to find a solution without exploring more than $M \in \{200{,}000, 400{,}000\}$ nodes of the subdivision graph

Table 2 The costs in detail of some regions, where $M = 200{,}000$. The parameters n and m are the numbers of patches and their adjacencies on the start map, respectively. Parameter K_i is the overestimation factor defined in this section. We evaluate the quality of our estimations for type change and shape by listing the numbers $R_{\text{type}} = g_{\text{type}}(\mathcal{P}_{\text{goal}})/h_{\text{type}}(\mathcal{P}_{\text{start}})$ and $R_{\text{shape}} = g_{\text{shape}}(\mathcal{P}_{\text{goal}})/h_{\text{shape}}(\mathcal{P}_{\text{start}})$. Note that if $h_{\text{type}}(\mathcal{P}_{\text{start}}) = 0$, we define $R_{\text{type}} = 1$

ID	n	m	K_i	R_{type}	R_{shape}	Time (s)
543	20	40	512	1.000	0.516	175.7
94	32	74	64	0.016	4.577	163.6
190	22	45	32	0.034	21.535	84.9
301	30	61	16	0.066	21.895	78.5
⋮	⋮	⋮	⋮	⋮	⋮	⋮
569	15	32	2	0.514	69.011	16.7
358	21	32	1	1.000	1020.133	1.3
97	20	35	1	1.000	179.132	0.9
257	20	33	1	1.000	463.178	5.9
⋮	⋮	⋮	⋮	⋮	⋮	⋮

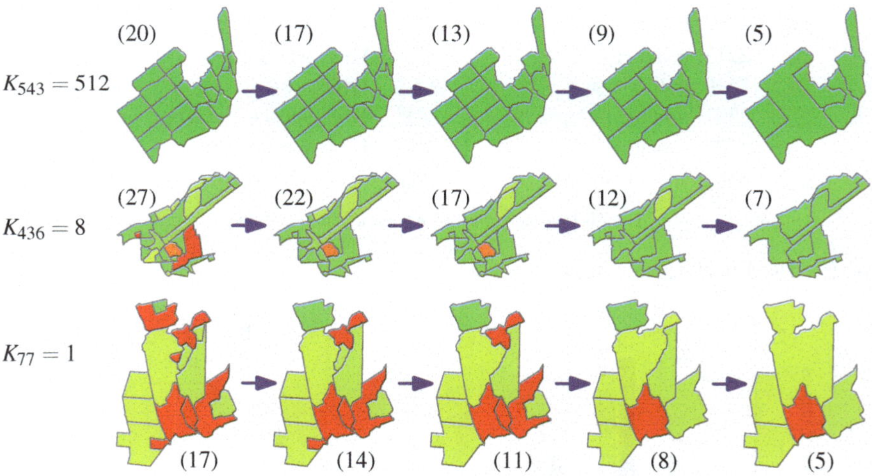

Fig. 12 Some intermediate aggregation results of regions 543, 436 and 77. The numbers in parentheses are the numbers of remaining patches

We show the intermediate aggregation results of some regions in Fig. 12. Although we had to overestimate for regions 543 and 436, the results seem fine (see the first row and the second row of Fig. 12). In Fig. 12, the third row shows an optimal aggregation sequence for region 77. However, we noticed some inappropriate aggregations (see Fig. 13 for an example). This inappropriate aggregation could have been caused by the overestimation or by our definition of the shape cost.

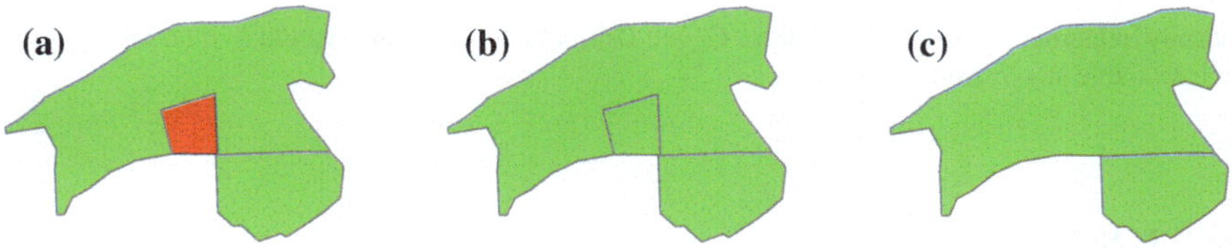

Fig. 13 An inappropriate result and the result we prefer. Figures (**a**) and (**b**) are extracted from region 436 when there are 27 and 22 patches (see Fig. 12). Our method aggregates the *red* patch of figure (**a**) into the *lower-right* patch (see Fig. (**b**)). We prefer aggregating the *red* patch into the *upper green* patch (see Fig. (**c**))

It turned out that the objective function g seems rather robust against changes of λ. For region 77, we varied λ in the range $\{0.1, 0.5, 0.9\}$, but the resulting aggregation sequences stayed almost the same. When we used even larger values for λ (e.g., 0.99), the (good) influence of type change was mitigated and, as a result, we needed to overestimate (with factor $K = 8$).

5 Conclusions

In this paper, we proposed a method for finding an optimal aggregation sequence between two maps of land-cover areas based on the A*algorithm. This method reduces the search space of attaining an optimal result. We have a good estimation for the cost of type change, which helps a lot to reduce the search space. In contrast, our estimation for the cost of shape is poor. If we investigate the shapes of the patches more carefully, we may improve our estimation for the cost of shape. Our current method did not manage to find optimal aggregation sequences for some of the regions in our datasets. For these regions, we overestimate in order to find feasible aggregation sequences. Note that a larger overestimation factor does not necessarily mean a poorer aggregation sequence. We compared our A*-based method with a method based on integer linear programming, and found that, within a fixed time frame, the A*-based method managed to solve more instances to optimality than the integer linear program.

To keep our problem concise, we did not incorporate many cartographic rules. An example is that a settlement is important and should be remained longer if the settlement is surrounded by farmlands. We did not consider the case like aggregating farmland with a hedge results in a land-cover area with type vegetation. These issues may be considered in our future work. Our method provides a way for continuous generalization of land-cover areas. We proposed our cost functions (i.e., type change and shape) based on cartographic requirements. However, to what extent our output remains faithful to real-world geography still needs to be evaluated.

Acknowledgements We thank Thomas C. van Dijk, Joachim Spoerhase, and Sabine Storandt for their valuable suggestions.

References

Cheng, T., & Li, Z. (2006). Toward quantitative measures for the semantic quality of polygon generalization. *Cartographica, 41*(2), 487–499.

Chimani, M., van Dijk, T. C., & Haunert, J.-H. (2014). How to eat a graph: Computing selection sequences for the continuous generalization of road networks. In *Proceedings of 22nd ACM SIGSPATIAL International Conference on Advances in Geographic Information Systems (ACMGIS'14)* (pp. 243–252).

Cormen, T. H., Leiserson, C. E., Rivest, R. L., & Stein, C. (2009). *Introduction to algorithms* (3rd ed.). The MIT Press.

Douglas, D. H., & Peucker, T. K. (1973). Algorithms for the reduction of the number of points required to represent a digitized line or its caricature. *Cartographica, 10*(2), 112–122.

Frolov, Y. S. (1975). Measuring the shape of geographical phenomena: A history of the issue. *Soviet Geography, 16*(10), 676–687.

Hart, P. E., Nilsson, N. J., & Raphael, B. (1968). A formal basis for the heuristic determination of minimum cost paths. *IEEE Transactions on Systems, Science, and Cybernetics, 4*(2), 100–107.

Haunert, J.-H. (2009). Aggregation in map generalization by combinatorial optimization. *Dissertation. Leibniz Universität Hannover, Germany.*

Haunert, J.-H., & Wolff, A. (2010). Area aggregation in map generalisation by mixed-integer programming. *International Journal of Geographical Information Science, 24*(12), 1871–1897.

Haunert, J.-H., Dilo, A., & van Oosterom, P. (2009). Constrained set-up of the tGAP structure for progressive vector data transfer. *Computers and Geosciences, 35*(11), 2191–2203.

Pohl, I. (1973). The avoidance of (relative) catastrophe, heuristic competence, genuine dynamic weighting and computational issues in heuristic problem solving. In *Proceedings of the 3rd International Joint Conference on Artificial Intelligence IJCAI'73* (pp. 12–17). Stanford, USA: Morgan Kaufmann Publishers Inc.

Saalfeld, A. (1999). Topologically consistent line simplification with the Douglas-Peucker algorithm. *Cartography and Geographic Information Science, 26*(1), 7–18.

Schwering, A. (2008). Approaches to semantic similarity measurement for geo-spatial data: A survey. *Transactions in GIS, 12*(1), 5–29.

Timpf, S. (1998). Hierarchical structures in map series. *Dissertation. Technical University Vienna, Austria.*

van Oosterom, P. (2005). Variable-scale topological datastructures suitable for progressive data transfer: The GAP-face tree and GAP-edge forest. *Cartography and Geographic Information Science, 32*(4), 331–346.

van Oosterom, P., & Schenkelaars, V. (1995). The development of an interactive multi-scale GIS. *International Journal of Geographical Information Systems, 9*(5), 489–507.

van Smaalen, J. (2003). Automated aggregation of geographic objects. *Dissertation. Wageningen University, The Netherlands.*

Quantitative Expressions of Spatial Similarity in Multi-scale Map Spaces

Haowen Yan, Liming Zhang, Zhonghui Wang, Weifang Yang, Tao Liu and Liang Zhou

Abstract Spatial similarity plays an important role in automated map generalization, spatial computations, spatial reasoning, spatial retrieval and the construction of multi-scale spatial databases. Nevertheless, the features of spatial similarity relations have never been investigated systematically, yet. To fill the gap, this paper focuses on quantitative expressions of the features of spatial similarity relations in map spaces. It firstly summarizes and analyzes the features of similarity in other fields, such as computer science, psychology and geography; then it proposes a number of features of spatial similarity relations in map spaces, i.e., equality, minimality, auto-similarity, symmetry (reflectivity), non-transitivity, triangle inequality, and scale dependency. Mathematical expressions of the features are presented and explained. Finally, potential research issues in spatial similarity relations are reviewed.

Keywords Spatial similarity relations · Features · Spatial reasoning · Map spaces

---

H. Yan (✉) · L. Zhang · Z. Wang
Department of GIS, Lanzhou Jiaotong University, Lanzhou 730070, China
e-mail: haowen2010@gmail.com

L. Zhang
e-mail: zhanglm8@gmail.com

Z. Wang
e-mail: 1449041349@qq.com

W. Yang · T. Liu · L. Zhou
Gansu Provincial Engineering Laboratory for National Geographic State Monitoring, Lanzhou 730070, China
e-mail: yangwf@163.com

T. Liu
e-mail: 25951804@qq.com

L. Zhou
e-mail: zhougeo@126.com

M.P. Peterson (ed.), *Advances in Cartography and GIScience*, Lecture Notes in Geoinformation and Cartography, DOI 10.1007/978-3-319-57336-6_28

1 Origination and Significance

Multi-scale vector databases are a fundamental component of spatial data infrastructures. They provide spatial positioning for various services in politics, economy, military, environment, traffic, transportation, and telecommunication, etc., and play an important role in the construction of digital nations/regions/cities (Yan 2010).

Automated map generalization, a technique for solving spatial conflicts and congestions in the process of generating smaller scale maps from larger scale ones using various appropriate algorithms and operators (e.g., selection, displacement, simplification, etc.) under defined conditions (e.g., map scale, map purpose, etc.), has been viewed as a most feasible way for constructing multi-scale vector map databases (Ruas 1998; Weibel and Jones 1998). In essence, map generalization is a type of similarity transformation. Take Fig. 1 as an example: the islands on the map at scale 1:250K are generated from the map at scale 1:100K. Although the original map has been simplified in the process of scale change (islands have been smoothed, combined or deleted), the two maps of the same area keep their basic similarity.

Currently, fully automated map generalization has not been achieved. A major reason is that quantitative expressions of spatial similarity relations among the same objects on multi-scale maps have not been solved. This hampers the full automation of map generalization because: (1) if quantitative spatial similarity relations are not obtained, a map generalization system/software does not know to what extent an original map should be generalized to produce a resulting map that has the same look as the original; and (2) the system/software also does not know when to terminate the map generalization procedure because its quantitative parameters that depend on spatial similarity relations are usually unknown. For example, the Douglas-Peucker Algorithm (Douglas and Peucker 1973) cannot stop line simplification automatically by itself because its quantitative parameter, "distance tolerance," cannot be obtained before carrying out the map generalization procedure. It needs the input and expertise of the cartographer.

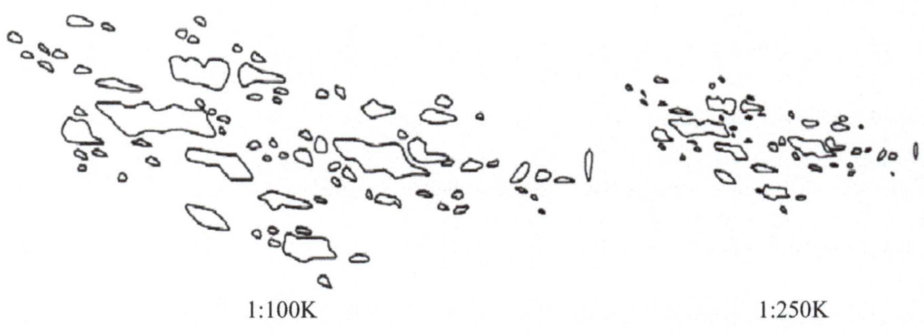

Fig. 1 Map generalization is a type of spatial similarity transformation

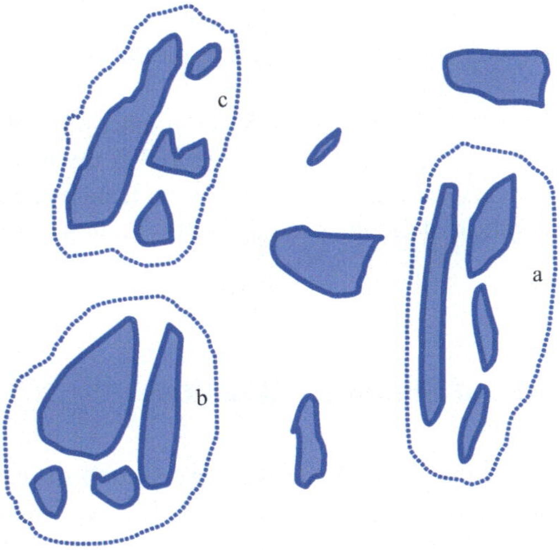

(a) Horizontal similarity relations

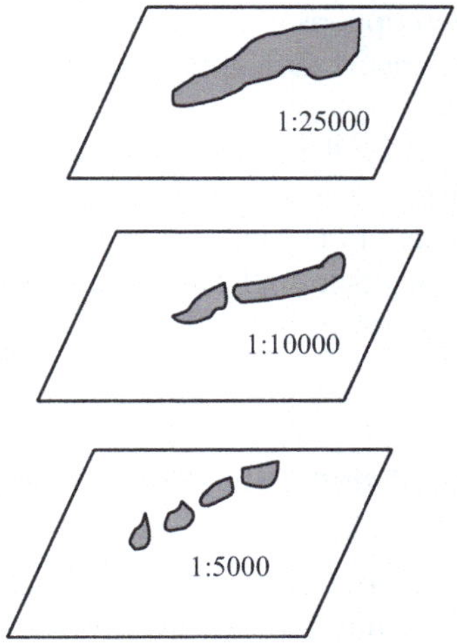

(b) Perpendicular similarity relations

Fig. 2 Two types of similarity relations on maps

Spatial similarity relation has been of interest for many years in cartography (Yan 2010) and geographic information science (Rodríguez and Egenhofer 2003, 2004). It is not only an important component of the theory of spatial relations which, in addition, includes distance (Hong 1994), topology (Egenhofer and Franzosa 1991; Du et al. 2008) and direction (Peuquet 1986; Goyal 2000; Yan et al. 2006), but is also an element in spatial retrieval, spatial inference (Markman 1997; Goldstone 2004) and spatial cognition (Li and Fonseca 2006).

Spatial similarity on maps exists among/between either different objects at the same scale or the same objects at different scales. Here, the former is named horizontal spatial similarity, and the latter perpendicular spatial similarity. For example, Fig. 2a shows the horizontal spatial similarity relations among three groups of objects A, B and C, that are all at the same scale while Fig. 2b shows the perpendicular spatial similarity relations among a group of land patches at scales 1:5000, 1:10000 and 1:25000. This paper ignores the horizontal spatial similarity relation, because it has been addressed in academic circles and is popularly used. On the contrary, little research work has been done on perpendicular spatial similarity relation, even though it is of significance in cartography, specifically in automated map generalization or multi-scale representation of vector map databases (Wang 1993; Ruas 2001).

The definitions, classification systems and calculation models for spatial similarity in multi-scale map spaces have been previously explored (Yan 2015; Yan et al. 2016) while the features of spatial similarity relations in multi-scale map spaces have not. This study focuses on this issue and aims at giving mathematical descriptions of the features so that they can be used in the calculation of the degree of spatial similarity, spatial data computation, spatial reasoning, spatial retrieval, etc.

After the introduction, Sect. 2 presents the features of similarity in other relevant fields such as computer science, psychology and geography. Section 3 is the main body of the paper which proposes the mathematical expressions of the features and gives their applications/examples in multi-scale map spaces. Finally, a number of concluding remarks are given (Sect. 4).

2 Literature Review: Features of Similarity Relations

Similarity is seemingly a simple concept that is used almost every second in our daily life. For example, people can recognize familiar people by their faces when they meet. When judging the similarity of faces, someone may say that two human faces are similar if they have a common skin color, while someone else would require the identity of the geometric structure of facial features like the eyes, the nose, the mouth, etc. Similarity relation also plays a crucial role in many fields in social and natural sciences (Bronstein et al. 2009). The following presents the features of similarity relations in a number of fields for the purpose of finding the features of spatial similarity relations in multi-scale map spaces.

2.1 Similarity Relations in Computer Sciences

In computer science, similarity metrics is the most closely related area of research related to similarity (Belohlavek 2000). Similarity metrics (also called string metrics) are a class of metrics that are used for measuring similarity (closeness) and dissimilarity (distance) between two character strings for approximate matching or comparison in fuzzy string searching. Cilibrasi and Vitanyi (2006) presented three features of similarity applicable for processing text strings in computer science: equality, symmetry and triangle inequality.

Let Ω be a nonempty set and R^+ be the set of non-negative real numbers. A distance function for describing the dissimilarity between two text strings is $D\colon \Omega \times \Omega \rightarrow R^+$. Based on this function, three features of similarity relations between text strings can be obtained.

(1) Equality: $D(x, y) = 0$, *iff* $x = y$;
(2) Symmetry: $D(x, y) = D(y, x)$; and
(3) Triangle inequality: $D(x, y) \leq D(x, z) + D(z, y)$.

The value $D(x, y)$ is the distance between $x, y \in \Omega$.

Here, equality means a string has no difference with itself; symmetry means the difference between strings x and y is equal to that between y and x; and triangle inequality means the difference between strings x and y is less than the sum of the difference between x and z and between y and z.

2.2 Similarity Relations in Psychology

Spatial similarity relation roots itself in psychology, thus a glimpse of the achievements of similarity in psychology might be of help for understanding spatial similarity relations in multi-scale map spaces. Similarity in psychology refers to the psychological nearness or proximity of two mental representations. Four models/approaches for assessing the proximity of two mental representations have been developed in past research, i.e., mental distance approaches, featural approaches, transformational approaches and social psychological approaches. Each of these is based on particular set of assumptions.

Mental distance approaches lay their foundation on an assumption that mental representations can be expressed as some concepts in a kind of mental space (Shepard 1962). Usually, the concepts are represented using points in space. Then, the similarity between two concepts is a function that can be used to calculate the distance between two points (i.e., two concepts) in the space. If the distance between a pair of points is shorter than that of another pair of points, the concepts represented by the former two points are said to be nearer to each other than that of the latter two points.

To overcome the shortcomings in mental distance approaches, featural approaches (Tversky 1977) were proposed. A typical shortcoming in the mental distance approach is that they assume that spaces are symmetric (because the distance between any two points is the same regardless of which point we start from to calculate the distance). However, psychological similarity is not always symmetric. In many cases, people actually can only state similarity in one direction. For example, it feels more natural to say "John Smith looks very like his father Alex Smith" than to say "Alex Smith looks very like his son John Smith". The featural approach assesses similarity between two objects by comparing a list of features that describe the properties of the object. The more commonalties they share, the more similar they are.

Developed to evaluate similarity independently of the type of mental representation, the basic idea of transformational approaches (Hahn et al. 2003) is as follows: it assumes that any mental representation can be transformed into another one by a number of steps. So, it is possible to define some necessary steps to complete this transformation. The more the number of steps in the transformation, the less similar the two representations are. However, Larkey and Markman (2005) found some evidence that contradicts this idea. Their work has shown that the number of steps to transform the colors and shapes of geometric objects does not predict people's similarity judgments for those objects.

In social psychology, researchers use similarity to describe the closeness or nearness of attitudes, personality, values, interests and culture between people. It is interesting that research has revealed that interpersonal attraction is based on similarity between people, and many forms of similarity have been shown to increase attraction. For example, similarities in opinions, interpersonal styles, communication skills, demographics, and values have all been shown in experiments to increase how much people like each other. Several explanations have been offered to explain how similarity increases interpersonal attraction. First, people with similar interests tend to put themselves into similar types of settings. For example, two people interested in literature are likely to run into each other in the library and form a relationship. Another explanation is that we notice similar people, expect them to like us, and initiate relationships. Also, having relationships with similar people helps to validate the values we hold. People tend to make negative assumptions about those who disagree with them on fundamental issues, leading to feelings of repulsion.

Four features of similarity have been discussed and analyzed in psychology (Tversky 1977; Goldstone 1994), i.e., symmetry, asymmetry (or directionality), minimality and triangle inequality.

(1) Symmetry: Is based on two assumptions. The first one is that the similarity from A to B equals to the similarity from B to A; the second one is that judgments of similarity and difference are complementary (the more similarity, the less difference, and vice versa). Mathematically, this can be expressed as $D(A,B) = D(B,A)$.
(2) Asymmetry and directionality: The contrast model proposed by Tversky (1977) has proved that "feature commonalities tend to increase perceived similarity more than feature differences can diminish it." In addition, the structure alignment model has shown that similarity judgments focus on matching relations between items, while difference judgments focus on the mismatching attributes (Medin et al. 1993; Markman 1997). Therefore, when A is more similar to T than B is, it is still possible that A is also more different from T than B.
(3) Minimality: $D(A,B) \geq D(A,A)$ (Tversky 1977). This should be obvious, because it is impossible that the dissimilarity between identical objects is greater than that between different objects.
(4) Triangle inequality: $D(A,B) + D(B,C) \geq D(A,C)$ (Tversky 1977).

Where, $D(A,B)$ is the distance/dissimilarity function, similar to the one used above for the features in computer science.

2.3 *Similarity Relations in Geography*

The concepts of reflexivity (or equality), symmetry, non-transitivity, self-similarity, and scale-dependency in geography have been proposed by Yan (2010).

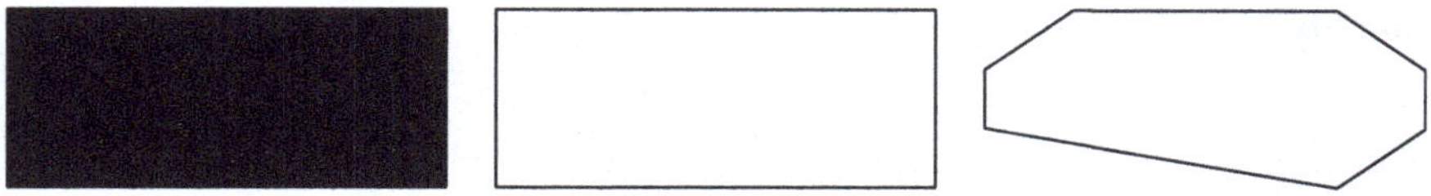

(a) settlement (b) agricultural land (c) agricultural land

Fig. 3 An example of the non-transitivity of spatial similarity relations

(1) Reflexivity: any object has similarity relations with itself.
(2) Symmetry: if object A has similarity relations with object B, object B have the same similarity relations with object A.
(3) Non-transitivity: We cannot conclude that object A has similarity relations with object C, even if object A has similarity relations with object B, and object B has similarity relations with object A.

 For example, in Fig. 3, if we take {shape, land type} as the properties for detecting similarity relations, the property set of (a), (b), and (c) are $C_a = \{\text{rectangle, settlement}\}$, $C_b = \{\text{rectangle, vegetable land}\}$, and $C_c = \{\text{irregular polygon, vegetable land}\}$. $C_a \cap C_b = \{\text{rectangle}\}$ denotes that the objects in (a) and (b) have similarity relations; $C_b \cap C_c = \{\text{vegetable land}\}$ denotes that the objects in (b) and (c) have similarity relations; but the conclusion that the objects in (a) and (c) have similarity relations cannot be made, for $C_a \cap C_c = \Phi$.
(4) Self-similarity on maps at multiple scales: Geographic objects can be symbolized using different patterns and symbols on maps at different scales. The objects on maps at different scales must have spatial similarity.
(5) Scale-dependence of self-similarity degree at multiple scales: The degree of spatial similarity of objects on maps at different scales depends on scale change. The greater the scale change from the original map to a generalized map, the less the degree of similarity between two maps.

3 Features and Their Mathematical Expressions

This section addresses the features of spatial similarity in multi-scale map spaces, aiming at explaining the applicability of the features in the geographic space and describing the features using mathematical expressions. Supposing that A, B, C and T are objects on maps, $Sim(A, B) \in [0, 1]$ represents the spatial similarity relation/degree between A and B, and $Dif(A, B) = 1 - Sim(A, B)$.

Equality

$$\forall(\mathrm{A}), Sim(\mathrm{A}, \mathrm{A}) = 1 \tag{1}$$

This seems self-evident: any object on the map is totally similar to itself, regardless of the scale and space.

Minimality

$$\forall(\mathrm{A}, \mathrm{B}), Sim(A, \mathrm{A}) \geq Sim(\mathrm{A}, \mathrm{B}) \tag{2}$$

The degree of similarity between two identical objects on the map is greater than that between two different ones.

Auto-similarity

$$\forall(A, B), Sim(A, B) = Sim(A, A) \Leftrightarrow A = B \tag{3}$$

This is an inference from Formula (1) and (2).

Symmetry (reflexivity)

$$\forall(A, B), Sim(A, B) = Sim(B, A) \tag{4}$$

The spatial similarity degree between *A* and *B* should be equal to that between *B* and *A*.

Non-transitivity

$$\forall(A, B, C), (Sim(A, B) > 0 \wedge Sim(B, C) > 0), \quad \exists Sim(A, C) > 0 \tag{5}$$

That *A* is similar to *B* and *B* is similar to *C* does not guarantee that *A* is similar to *C*.

This feature is popular in geographic and map spaces. For example, in Fig. 3, *A* is a settlement, *B* and *C* are agricultural lands. If we take {shape, land type} as the properties for detecting similarity, the property set of (a), (b), and (c) are C_a = {rectangle, settlement}, C_b = {rectangle, agricultural land}, and Cc = {irregular polygon, agricultural land}. $C_a \cap C_b$ = {rectangle} denotes that the objects in (a) and (b) have similarity relations. $C_b \cap C_c$ = {agricultural land} denotes that the objects in (b) and (c) have similarity relations, but the conclusion that the objects in (a) and (c) have similarity relations cannot be made because $C_a \cap C_c = \Phi$. In other words, we cannot say that *A* is similar to *C* even though $Sim(A, B) > 0$ and $Sim(B, C) > 0$.

Triangle Inequality

$$\forall(A, B, C), \mathrm{Sim}(\mathrm{A}, \mathrm{B}) + \mathrm{Sim}(\mathrm{B}, \mathrm{C}) \geq \mathrm{Sim}(\mathrm{A}, \mathrm{C}) \tag{6}$$

The degree of similarity between *A* and *B* plus that of *B* and *C* is greater than that of *A* and *C*.

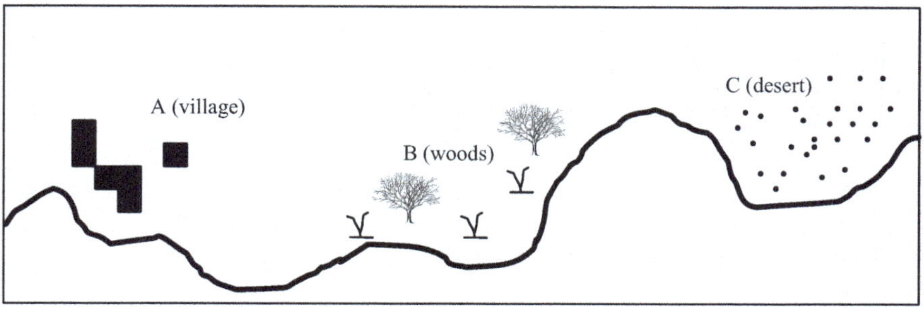

Fig. 4 Triangular inequality of spatial similarity relations

Suppose that there are three objects alongside of a river bank, a village, a patch of woods, and a desert (Fig. 4). Their property set contains three elements: history, size, owner, and shape. Possible values for these elements are as follows:

History: ancient, modern, current, unknown
Size: large, small, and
Owner: public, private, unknown
Shape: regular, irregular

The properties of the three objects are:

A: $\{modern, large, public, irregular\}$;
B: $\{current, large, private, irregular\}$; and
C: $\{ancient, small, public, irregular\}$.

To simplify the calculation of the degree of spatial similarity, suppose that the weight of each property in similarity evaluation is the same and the degree of similarity is proportional to the number of the same properties shared by the two objects. Using existing approach (Yan and Li 2014), the similarity degree can be calculated by

$$Sim(A,B) = 0.25 \times 2 = 0.5$$
$$Sim(B,C) = 0.25 \times 1 = 0.25$$
$$Sim(A,C) = 0.25 \times 2 = 0.5$$
$$\therefore Sim(A,B) + Sim(B,C) > Sim(A,C)$$

Scale dependency

Scale dependency may be explained in this way: if object A on the map at scale S is gradually generalized to get objects $A_1, A_2, \ldots A_n (n > 1)$ on maps at scales $S_1, S_2, \cdots\cdots S_n$ and $S_1 > S_2 > \ldots\ldots S_n$. If objects $A_1, A_2, \ldots\ldots A_n$ are compared with A, respectively, Formula (7) should be correct.

Fig. 5 Scale dependency of spatial similarity relations in multi-scale map spaces

$$Sim(A, A_1) \geq Sim(A, A_2) \geq \cdots \geq Sim(A, A_n) \tag{7}$$

To express this in a simple way: the more an object is simplified (generalized), the less similar it is compared with the original object. Figure 5 is shows this fundamental characteristic. Scale dependency of spatial similarity relations has been proved and quantitative relations between the degree of spatial similarity and change in map scale in multi-scale map spaces has been demonstrated (Yan 2015; Yan et al. 2016).

4 Conclusions

This paper proposes a number of features of spatial similarity in map spaces, i.e., equality, minimality, auto-similarity, symmetry (reflectivity), non-transitivity, triangle inequality, and scale dependency. Mathematical formulae were presented to quantitatively describe the features which can be used in spatial computation, reasoning and retrieval. However, the following two points need to be made regarding spatial similarity: (1) because similarity judgments root in human's psychological activities, the validity and adaptability of some features are difficult to prove mathematically; these can be tested by psychological experiments and/or social surveys. (2) People of different cultures, nations, ages, education backgrounds and gender usually have different cognitions towards the features of spatial similarity, and therefore, if the environment is different, some of the mathematical expressions of the features proposed in the paper would need to be tested.

Acknowledgements The work described in this paper is funded by the Natural Science Foundation Committee, China (No. 41371435; No. 71563025).

References

Belohlavek, R. (2000). Similarity relations in concept lattices. *Journal of Logic and Computation, 10*(6), 823–845.

Bronstein, A. M., Bronstein, M. M., Bruckstein, A. M., & Kimmel, R. (2009). Partial similarity of objects, or how to compare a centaur to a horse. *International Journal of Computer Vision, 84* (2), 163–183.

Cilibrasi, R., Vitanyi, P. (2006). Similarity of objects and the meaning of words. In *Proceedings of the 3rd Annual Conference on Theory and Applications of Models of Computation (TAMC'06)* (Vol. 3959, pp. 21–45). Lecture Notes in Computer Science.

Douglas, D. H., & Peucker, T. K. (1973). Algorithms for the reduction of the number of points required to represent a digitised line or its caricature. *Canadian Cartographer, 10*(2), 112–122.

Du, S. H., Qin, Q. M., Wang, Q., & Ma, H. J. (2008). Reasoning about topological relations between regions with broad boundaries. *International Journal of Approximate Reasoning, 47* (2), 219–232.

Egenhofer, M., & Franzosa, R. (1991). Point-set topological spatial relations. *International Journal of Geographical Information Systems, 5*(2), 161–174.

Goldstone, R. L. (1994). Similarity, interactive activation and mapping. *Journal of Experimental Psychology. Learning, Memory, and Cognition, 20*(1), 3–28.

Goldstone, R. L. (2004). Similarity. In R. A. Wilson & F. C. Keil (Eds.), *MIT encyclopedia of the cognitive sciences*. MIT Press.

Goyal, R. K. (2000). *Similarity assessment for cardinal directions between extended spatial objects*. PHD thesis, The University of Maine, USA.

Hahn, U., Chater, N., & Richardson, L. B. (2003). Similarity as transformation. *Cognition, 87*(1), 1–32.

Hong, J. (1994). *Qualitative distance and direction reasoning in geographic space*. PHD thesis, University of Maine, USA.

Larkey, L. B., & Markman, A. B. (2005). Processes of similarity judgment. *Cognitive Science, 29*, 1061–1076.

Li, B., & Fonseca, F. T. (2006). TDD—a comprehensive model for qualitative spatial similarity assessment. *Spatial Cognition and Computation, 6*(1), 31–62.

Markman, A. B. (1997). Constraints on analogical inference. *Cognitive Science, 21*(4), 373–418.

Medin, D. L., Goldstone, R. L., & Gentner, D. (1993). Respects for similarity. *Psychological Review, 100*(2), 254–278.

Peuquet, D. (1986). The use of spatial relationships to aid spatial database retrieval. In *Proceedings of the 2nd International Symposium on Spatial Data Handling (SDH)*, Zurich, Switzerland.

Rodríguez, A., & Egenhofer, M. (2003). Determining semantic similarity among entity classes from different ontologies. *IEEE Transactions on Knowledge and Data Engineering, 15*(2), 442–456.

Rodríguez, A., & Egenhofer, M. (2004). Comparing geospatial entity classes: An asymmetric and context-dependent similarity measure. *International Journal of Geographical Information Science, 18*(3), 229–256.

Ruas, A. (1998). A method for building displacement in automated map generalization. *International Journal of Geographical Information Science, 12*(8), 789–803.

Ruas, A. (2001). Automating the generalization of geographical data. In *Proceedings of the 20th International Cartographic Conference*, Beijing, China, 1943–1953.

Shepard, R. N. (1962). The analysis of proximities: Multidimensional scaling with an unknown distance function. *International Psychometrika, 27*(2), 125–140.

Tversky, A. (1977). Features of similarity. *Psychological Review, 84*(4), 327–352.

Wang, J. Y. (1993). *Principles of general map generalization*. Beijing: Surveying and Mapping Press. (in Chinese).

Weibel, R., & Jones, C. B. (1998). Computational perspectives on map generalization. *GeoInformatica, 2*(4), 307–314.

Yan, H. W. (2010). Fundamental theories of spatial similarity relations in multi-scale map spaces. *Chinese Geographical Sciences, 20*(1), 18–22.

Yan, H. W. (2015). Quantitative relations of spatial similarity degree and map scale change of individual linear objects in multi-scale map spaces. *GeoCarto International, 30*(4), 472–482.

Yan, H. W., Chu, Y. D., Li, Z. L., & Guo, R. Z. (2006). A quantitative direction description model based on direction groups. *Geoinformatica, 10*(2), 177–196.

Yan, H. W., Li, J. (2014). *Spatial similarity relations in multi-scale map spaces*. Springer International Publishing Switzerland.

Yan, H. W., Shen, Y. Z., & Li, J. (2016). Approach to calculating spatial similarity degrees of the same river basin networks on multi-scale maps. *GeoCarto International, 31*(7), 765–782.

Balanced Allocation of Multi-criteria Geographic Areas by a Genetic Algorithm

Shahin Sharifi Noorian and Christian E. Murphy

Abstract The balanced partitioning of geographic space into regions is a common problem. This Territory Design Problem (TDP) of assigning smaller areas to larger regions with equal potential is a task mainly done manually. Therefore, the result becomes subjective and provides only a roughly approximated balanced result. This work presents an automated allocation of independent areas to regions using the Genetic Algorithm (GA), which finds an optimally balanced configuration of regions based on multiple criteria. Thereby, spatial constraints are fully respected as (1) all areas remain contiguous within a region and (2) the automated allocation facilitates a compact region shape. The developed algorithm was tested on a case study in the field of sales territory planning. The target of sales territory planning is the optimal distribution of balanced and fair sales areas based on market potentials. Results of our case study demonstrate the effectiveness of our proposed technique to find an optimal structure of sales territories in a reasonable time. The distance that salesmen need to travel is 16% lower than the existing sales territory configuration. This means that the regions are more balanced and more compact. Due to the independent nature of the GA, this method demonstrates a high flexibility to the optimization problem. It can be easily altered to any objective in territory planning as well as to familiar multi-criteria spatial allocation problems in other disciplines.

Keywords Territory design problem · Combinatorial optimization · Graph partitioning · Genetic algorithm

---

S. Sharifi Noorian (✉) · C.E. Murphy
Chair of Cartography, Technische Universität München, Arcisstraße 21, 80333 Munich, Germany
e-mail: shahin.sharifi@tum.de

C.E. Murphy
e-mail: christian.murphy@tum.de

M.P. Peterson (ed.), *Advances in Cartography and GIScience*, Lecture Notes in Geoinformation and Cartography, DOI 10.1007/978-3-319-57336-6_29

1 Introduction

Spatial aggregation is a procedure for grouping discrete geographical units (such as zip-code areas, municipality area, company trading areas and electoral districts) into larger geographic clusters called territories in such a way that those clusters are eligible according to the planning criteria and satisfy the conditions (Drexl and Haase 1999).

Territory Design is the aggregation of basic geographical units into larger geographic regions. Territory Design Problems (TDPs) are motivated by a broad range of applications such as administrative segmentation, design of school territories, social facilities, emergency services, sales and service territory design. Businesses with direct sales departments need sales and service territories, as well-planned decisions enable an efficient market penetration and lead to decreased costs and improved customer service. For instance, a beverage manufacturer will deploy a sales force in order to promote its products for retailers and supermarkets. This planning will be effective when some kind of balance is maintained between the workload of sales representatives, which involves the distance that representatives need to travel, and the number of customers that they have to visit.

When some kind of balancing among the territories is desired, it is very difficult to generate an acceptable solution by manual adjustment and, in some cases, it is almost impossible. Besides the balancing of territories in terms of workload or the number of customers, contiguity and compactness should be considered as other core criteria for the TDP. Therefore, an automatic algorithm is required to handle all criteria simultaneously and resolve the complexity of the TDP.

In recent years, a number of TDP heuristic and non-heuristic approaches have appeared in the literature. For some examples, Drexl and Haase (1999) proposed a non-linear mixed-integer programming in order to solve the sales force deployment problem. Kalcsics et al. (2005) utilized computational geometry techniques for solving districting problems, which are suitable for interactive use. Some others suggested heuristic approaches for TDPs such as GRASP (Ríos-Mercado and Fernández 2009), Simulated Annealing (D'Amico and Fernández 2002), and Tabu Search (Blais et al. 2003).

These approaches are mostly applied to relatively small to medium-sized problems in terms of space. In addition, these approaches usually do not take the geographical nature of the TDP into consideration and thus, totally neglect geographic obstacles such as mountain chains, river, etc. On the other hand, territory design is an iterative process which is based on planning parameters and related data. Therefore, the proposed approach must be flexible enough to deal with different kinds of objectives and constraints.

This motivated us to present a generic objective-independent approach that supports core criteria for all TDPs such as compactness, contiguity and balancedness in addition to a customized objective function and multiple problem-specific constraints like organizational limitation, locked basic areas (the basic areas that have to be excluded from planning), etc.

In this paper, we apply the genetic algorithm to the TDP. Considering the fast growing web technologies and immense utilization of software as a services (SaaS) (Gil 2016), we also used techniques such as parallelization of algorithms, simplification of search space, etc., to improve the performance of our method. This prepares a level of interactivity for planning and enables us to publish our method as a web service for a broad range of problems.

2 Territory Design by a Multi-objective Genetic Algorithm

TDP involves combinatorial optimization in which a set of feasible solutions are available. The goal is to find the best solution in a finite collection of solutions. Each feasible solution usually has a concise demonstration. But, the number of solutions is huge. Therefore, an exhaustive search such as checking all objects one-by-one and selecting the best one is not practical. The traveling salesman problem, the vehicle routing problem, and the knapsack problem are some other well-known problems which are classified as combinatorial optimization (Schrijver 2003).

2.1 Usability of Graphs in Order to Simulate the Problem

TDP has multiple objectives as well as several constraints which have to be taken into consideration. Moreover, the scale of the problem changes for every use case. A comprehensive model is needed for representing the problem. Due to the flexibility and scalability of graphs, we model the problem in the form of a graph. The Graph structure allows us to model all kinds of scenarios which are based on relationships. According to the definition of spatial aggregation and the geographical essence of territory design, using graphs for problem modelling is the best choice (Płuciennik and Płuciennik-Psota 2014). Generally in our graph model, each node represents an entity (person, location, product and etc.), and each relationship between nodes represents how two entities are associated.

2.2 Mapping the Territory Design Problem into a Graph Model

The goal of territory design specifies how basic units should be grouped into clusters (Ríos-Mercado and Ferández 2009). Based on the domain of a problem, a certain weight will be assigned to each basic unit that is calculated based on its attributes. As a simple example, the government wants to define a certain number of census territories in the level of zip-codes. Each territory is supposed to have an equal population. Therefore, the number of people in each basic unit indicates its weight. In the graph model, each basic unit (centroid) is represented as a node which can contain one or several attributes. Each node also has a label that specifies the cluster to which it belongs. The model structure is flexible enough to represent all types of spatial relationships between two basic units such as neighbourhood, higher administrative division, distance, etc.

As the territory design is a multi-constraint problem, our graph model is also capable to support multiple constraints (see Fig. 1). Feasible solutions of the problem are defined based on the constraints. For instance, contiguity is a common criteria for all TDPs. It means that only those territories that are not divided into separated parts are acceptable. One can easily check the contiguity constraint by considering the connectivity of the graph. In Fig. 2a, the cluster (blue arrays) is not totally connected and the contiguity constraint is violated. Figure 2b shows a connected generated cluster. This solution is valid.

It is also possible to define certain constraints based on the problem. For example, in some cases, it is necessary to reserve a group of basic areas exclusively for a certain territory. From the graph modelling point of view, the corresponding nodes will be labelled constantly in the initial graph and therefore are excluded from the process of optimization. Figure 3 illustrates how constraints are applied into the graph model.

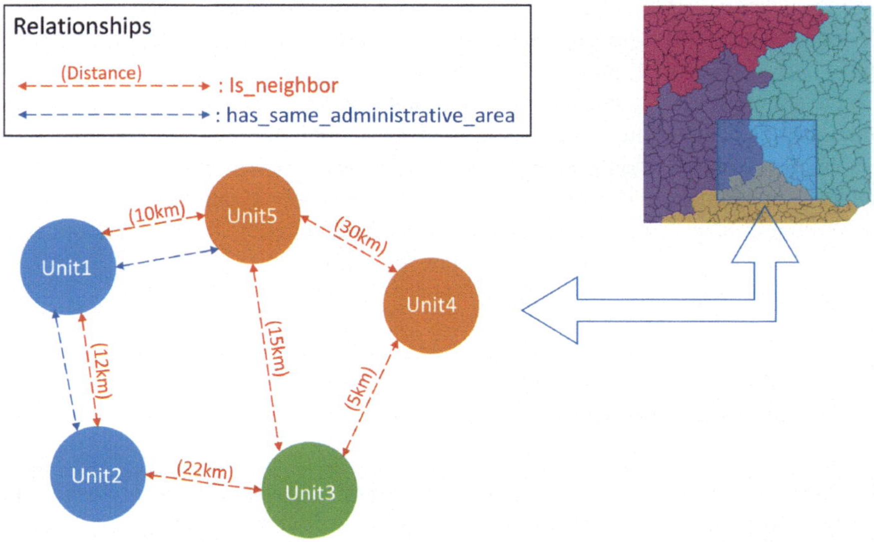

Fig. 1 Proposed scheme for mapping a real-world problem into the graph model

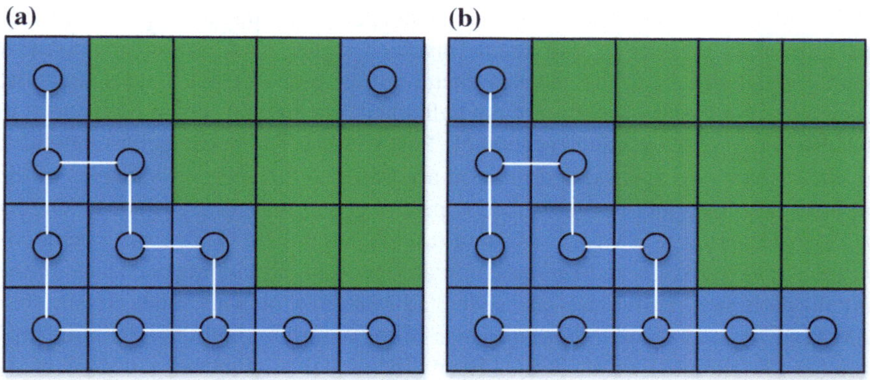

Fig. 2 (**a**) The contiguity constraint for the territory is violated; (**b**) the territory is connected

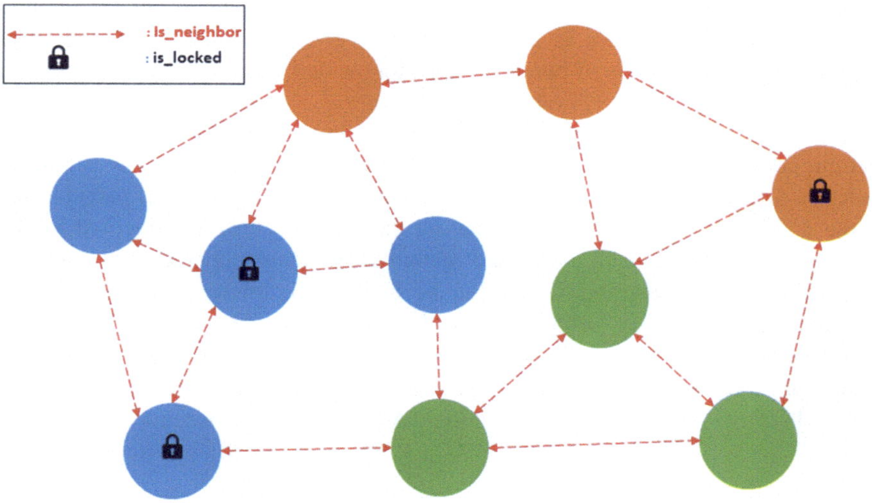

Fig. 3 Graph modelling of constraints: locked nodes are not modified by the algorithm

2.3 The Core of the GA

The GA is a heuristic search technique which was inspired from the principal of natural evolution to solve optimization problems. The GA is a subclass of evolutionary algorithms which attempts to mimic some of the processes that occur in natural selection based on "Survival of the fittest" (Sivanandam and Deepa 2007). The GA looks for the best solution among a set of feasible solutions. The set of feasible solutions is called Search Space. It explores the search space and evaluates the fitness of candidate solutions in order to find the best one. The GA has two distinguished elements which are *individuals* and *population*.

An individual is a single candidate solution to the optimization problem while the population is a set of individuals involved in the optimization process. GA applies variation only to the genetic structure of the problem, not the real structure. Therefore, observable properties of the individual (phenotype) should be encoded to the abstract genetic structure (genotype) (Fogel 1995). The genotype is subdivided to some small parts (genes) and mapped according to the decision variables of the problem. The mapping between phenotype and genotype is one of the challenging parts of designing a GA for a real-world application. The GA´s main operations *initialization*, *selection* and *reproduction* are based on a random procedure. In the initialization step, individuals are generated randomly or with some certain consideration for improving the performance of the algorithm and the quality of final solution (Kazimipour et al. 2014). Every individual is assigned a degree of its excellence by means of a *fitness function*. The fitness function is an evaluator that not only shows how good the solution is, but also measures how close our potential solution is to the target. The selection process involves a method for selecting two parent genotypes based on their fitness value. Those selected genotypes are given to the reproduction procedure as input in order to generate a new population. The reproduction step consists of two main operators which are *crossover* and *mutation* (Fogel 1995).

Crossover is a recombination method in which two parent genotypes are cut at a certain point and the halves are connected to each other to make a new candidate solution (offspring) which contains the characteristics of the parents. If parents have some good features, we can expect that all the good features are inherited by the generated offspring. Mutation, as the second reproduction operator, is a way of generating a new offspring which can be a little bit different from its parents (Sivanandam and Deepa 2007). Mutation does not happen frequently in nature, but it is a very important operation in the artificial GA, because the mutation operator helps the algorithm to rapidly explore the search space (Fonseca and Fleming 1995). In Fig. 4, an abstract scheme of the GA is outlined in a pseudo-code fashion.

Algorithm 1 Genetic Algorithm

```
initialize a population with random individuals;
while (termination condition is satisfied) do
    evaluate each individual;
    select fitter individuals and copy into a temporary set;
    while (size of new generation = size of population) do
        select a pair of individuals from temporary pool;
        recombine parent individuals to make an offspring;
        mutate the generated offspring;
        copy the offspring into the next generation;
    end while
end while
return the fittest individual(solution)
```

Fig. 4 Abstract scheme of the GA

Self-adaptive GA:

Local optima is one of the common issues in all GAs. It means that the GA can easily converge to an incorrect solution. Thus, the GA should be capable to escape from a local optimum and to converge to the global optimum. In order to tackle this issue, we propose a self-adaptive GA in which the probability of crossover and mutation, Pc and Pm, are varied depending on the fitness values of the solutions. Fitness assignment of SEA realizes the twin goals of maintaining diversity in the population and guiding the population to the global optimum (Cao et al. 2007). In other words, the mutation probability should be increased when the diversity of candidate solutions is decreasing and should be decreased when the diversity is increasing.

Parallelization of the GA:

GAs operate on a population in which each individual represents a potential solution to the problem. This feature of the GA makes it extremely amenable to parallelization which can significantly improve the efficiency vof the algorithm (Luque and Alba 2011). The GA which is designed in this research project, is capable to be executed concurrently so that the population is divided into smaller populations and each subpopulation is assigned to one processor of the parallel processing system.

3 Case Study: Sales Territory Planning

Sales Territory Planning is considered the problem of grouping smaller area units, so-called sales coverage units (SCUs), into larger geographic clusters called territories in such a way that those clusters are eligible according to the planning criteria and satisfy the conditions. Each territory must have a responsible sales representative. To be responsible means that the sales representative has to provide service for all customers located in the corresponding sales territory (Zoltners and Sinha 1983). For example, a sales representative who is working for a company that provides materials concerning dental surgery will be responsible for all dentists practicing in her sales territory. She is only allowed to sell products in SCUs that belong to her sales territory (Haase and Müller 2014). Reasons for aligning existing or designing new territories can be changes in the number of sales representatives or balancing their workload.

Nowadays, marketers and decision-makers face three different scenarios in sales territory planning. These scenarios are the following (Zoltners et al. 2009):

1. We have neither the location of sales representatives nor the existing sales territory structure. In this case, the number of needed territories should be determined by decision-makers. Then, the algorithm consists of two parts, respectively:

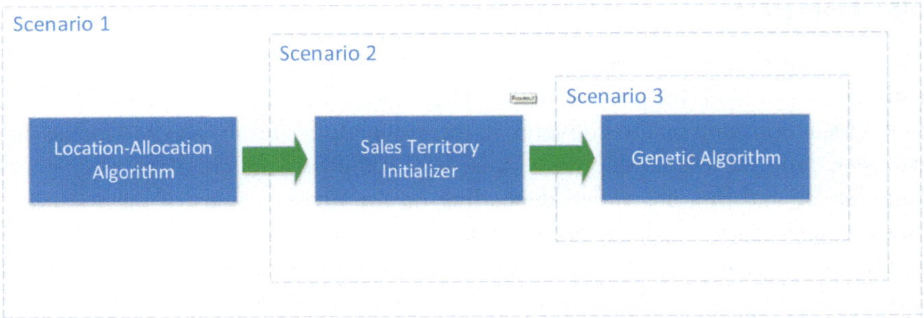

Fig. 5 A generic workflow for sales territory planning

- Finding the best location for each sales representative. The number of sales representatives are equal to the number of territories.
- Performing a heuristic search algorithm for finding an acceptable sales territory structure.

This scenario is commonly called Green Field Planning.

2. We have a number of sales representatives with fixed locations but we do not have any existing territory structure.
3. We have either fixed sales representatives or an existing sales territory structure. But, the territories should be optimized due to the new strategy of the sales organization.

For each scenario, a respective strategy is chosen. But, we are going to design a generic approach for all scenarios. So, we should have a workflow which supports all scenarios. Figure 5 presents how our model handles all scenarios:

For the case study, we considered a real situation in which an agricultural machinery manufacturer is selling a certain product (tractor) in Germany. The sales team consists of 27 representatives who must cover the entire country. Each salesman must be responsible for one territory only. The territory planning is carried out on the level of municipalities (administrative districts), which consist of 11,212 basic areas as SCUs. For this case study, we considered the number of farms inside each coverage unit.

4 Results

To assess the correctness and effectiveness of our method, we performed some experiments for both evaluating the balance of workload among sales representatives and evaluating the travel time improvement.

5 Tuning of the GA Parameters

The Genetic Algorithm is very sensitive to parameters such as population size (N), elitism percentage (E), cross-over probability (Pc) and mutation probability (Pm). The elitism rate identifies the percentage of individuals that should be transferred to the next generation without any changes (Sivanandam and Deepa 2007). After several preliminary experiments, we have tuned the parameters as follows:

- N: 30
- E: 26%
- Pc: [0.6 – 0.95]
- Pm: [0.005 – 0.05]

Since the algorithm is self-adaptive, a range of probabilities are defined for Pc and Pm in which the appropriate probability is chosen depending on the average fitness of the population.

6 Location-Allocation and Initializing the Territories

As we explained in Sect. 3, we have three scenarios. Since the first scenario (Green Field Planning) includes other scenarios, we evaluate here only the Green Field Planning case. We defined three steps for solving the Green Field Planning problem. Firstly, we found the best positions for locating salesmen. In order to create the initial form of territories, we clustered the administrative districts based on the nearest salesman. Then, the location of salesmen and the initial form of territories are used as input for the GA. Since the residence of salesmen is an essential input for sales territory planning, finding the best location is crucial.

In order to avoid unnecessary computational effort, we have selected a subset of municipalities that have more convenience for locating the sales team such as accessibility, and amenities, etc. For choosing the best candidate municipalities, we considered the capital municipalities of higher-level administrative districts (German *Landkreis*). Figure 6 illustrates the candidate municipalities and the connections between them.

After simplification of the dataset, the size of basic units decreases from 11,212 to 402 administrative districts. Similar to the graph model in Sect. 2, each candidate district is shown as a node. In addition, we have considered the highway network for showing the connections between nodes in our graph model. Thus, the highway which connects two districts is shown as an edge which is labelled by distance. We assigned a certain weight to each node by summing up the total travel time from the node to all other nodes. Then, the mapping of the location-allocation graph model

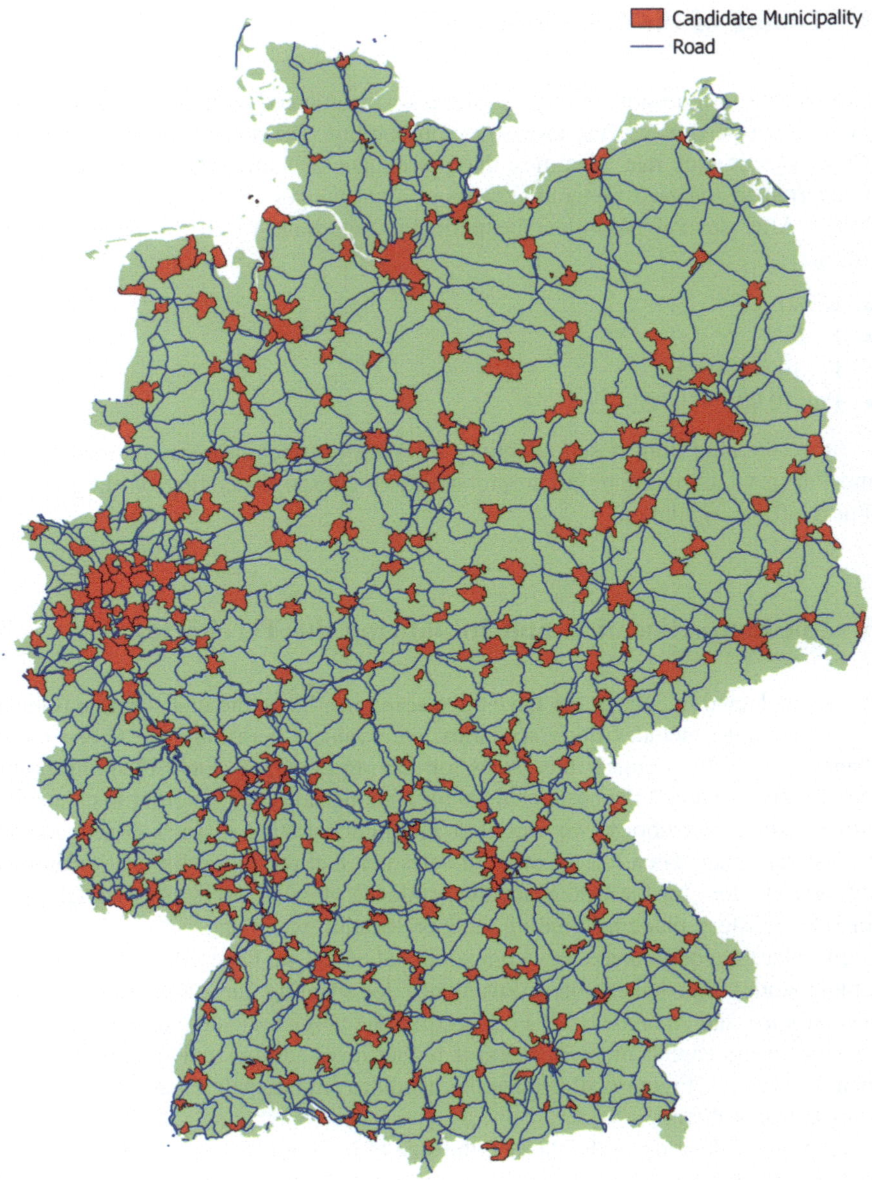

Fig. 6 Location possibilities of the sales teams in the candidate municipalities of Germany

into the genotype is done (see Sect. 2). The objective is to minimize the total travel time (DT) by which those municipalities are chosen that are more accessible. The fitness value for each solution in the GA is calculated as:

$$minimize\, D_T = \sum_{i=1}^{L} \sum_{j=1}^{M} d_{ij} \quad , i \neq j \tag{1}$$

The variables in Eq. (1) are as follows:

- *M: number of candidate municipalities*
- *L: number of needed locations*
- *d_ij: the travel time from municipality i to municipality j*

The best found locations in this step are just used as a starting point. The ultimate locations for salesmen will be determined at the final stage based on the optimal form of territories. Figure 10a, b show the allocated locations for salesmen before and after balancing the workload.

6.1 Evaluating the Balance of Workload

Since balancing the workload among sales representatives (territories) is one of the main objectives in the sales territory planning, we want to prove the efficiency of our algorithm for this objective. Figure 7 shows the unbalanced distribution of workload after generating the initial structure of territories. Since each territory only belongs to one salesman, the ID of salesmen in Fig. 8 represents the belonging territory.

To balance the workload, we consider the standard deviation (σ) as the fitness value of each solution which is to be minimized. The smaller the standard deviation, the more equality will be gained. We formulated the objective function as:

Fig. 7 Distribution of workload deviations from the mean before optimization

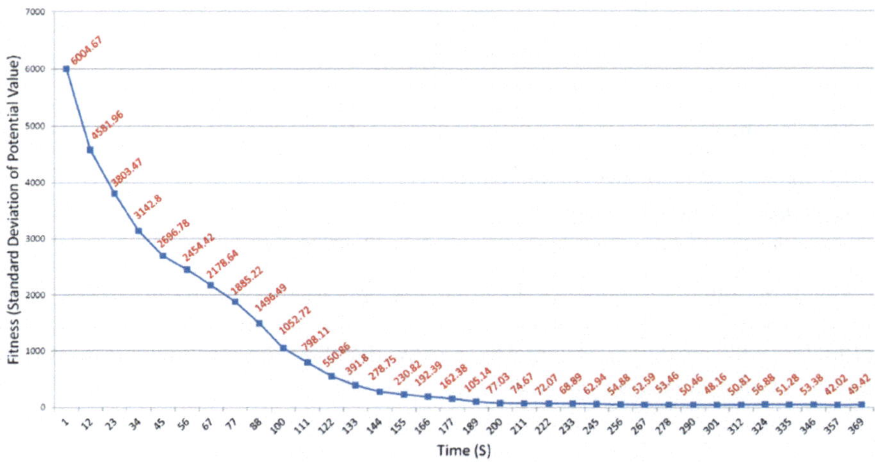

Fig. 8 Fitness progress of the GA

$$minimize\ \sigma = \sum_{i=0}^{T} |R_i - \bar{r}| \tag{2}$$

where,

$$\bar{r} = \frac{1}{T}\sum_{j=1}^{n} r_j, R_i = \sum_{j \in T_i} r_j \quad and\ |R_i - \bar{r}| \le \delta\bar{r}$$

The variables in Eq. (2) are as follows:

- *n: number of basic areas (SCUs)*
- *T: number of territories*
- *T_i: the ith territory*
- *r_j: the workload contained in 〖SCU〗_j*
- *R_i: the total workload in territory i*
- *δ: the maximum allowable percentage (0 ≤ δ ≤ 1)*

The GA can be terminated in different ways such as elapsed time, generation count, stagnation, target fitness and intentional abort. In this experiment, we have chosen the stagnation strategy which halts the algorithm if no improvement in the average fitness of population is observed within a predetermined number of generations or during an interval of time. The processing of our GA has finished after 369 s, which is a promising computational time for the scale of our problem. Figure 8 represents the progress of our algorithm in which the standard deviation progressively decreases over time until improvement becomes negligible.

As shown in Fig. 8, the standard deviation has decreased from ~6000 to ~50. Figure 9 also illustrates that the workload is almost completely balanced between salesmen.

After balancing the potential between territories, the location of sales representatives are updated according to the new structure of territories. Thus, we run our location-allocation algorithm to find the best location for the sales persons based on the current structure of territories. Figure 10 illustrates the structure of territories as well as the locations of salesmen before and after optimization.

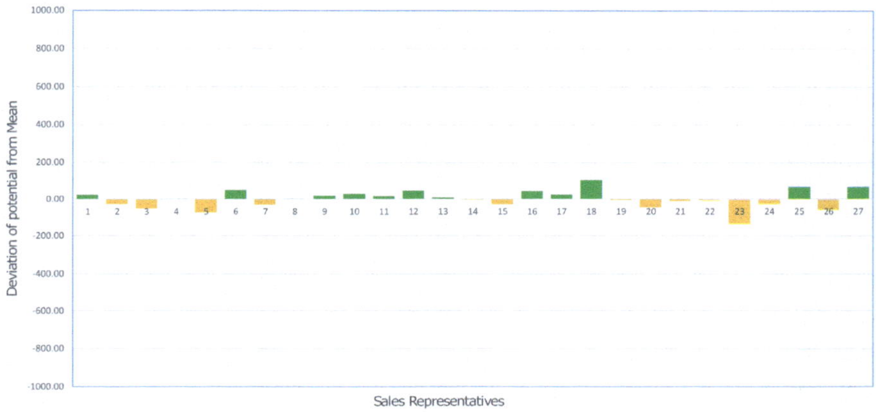

Fig. 9 Distribution of workload deviations from the mean after optimization

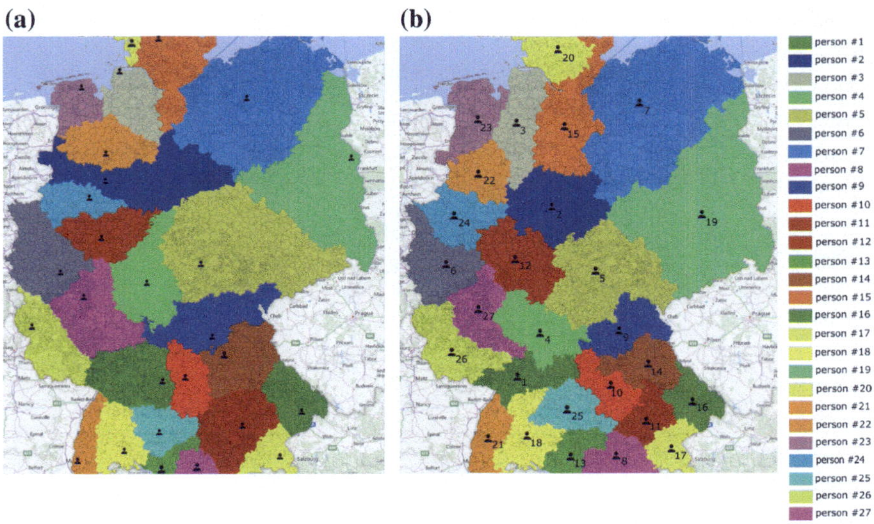

Fig. 10 The shape of sales areas before (**a**) and after (**b**) optimization

6.2 Evaluating the Travel Time Improvement

As the travel cost is one of the important parameters in sales territory planning, we have defined the minimizing of travel time and distance as an objective in our algorithm—besides the balancing of the potential. In our case study, a low travel cost is not the first priority. So we defined a lower weight for minimizing the travel cost than balancing the potential among salesmen.

The calculation of travel cost depends on several parameters and can vary among different sales organization. In our algorithm, we assumed that the salesmen only visit the potential customers of one district in each journey. In addition, we did not

Table 1 Travel time optimization by salesman and in total

Salesman	Travel time before optimization (Hour)	Travel time after optimization (Hour)
Person #1	45.8	112.6
Person #2	2177.9	2767.00
Person #3	2423.4	1775.7
Person #4	359.5	126.8
Person #5	479.6	169.5
Person #6	732.0	525.6
Person #7	212.6	204.2
Person #8	748.8	533.1
Person #9	440.7	256.6
Person #10	1093.2	1086.8
Person #11	137.7	217.8
Person #12	639.6	466.8
Person #13	157.7	140.9
Person #14	1402.2	590.7
Person #15	738.2	631.6
Person #16	124.7	398.6
Person #17	1513.6	464.5
Person #18	467.2	257.2
Person #19	203.07	144.3
Person #20	97.5	148.2
Person #21	24.4	195.2
Person #22	415.04	711.2
Person #23	78.9	83.6
Person #24	201.6	314.6
Person #25	108.4	137.6
Person #26	542.3	481.0
Person #27	164.2	248.2
Total travel time	15730.975	13191.324

take empty districts into account. The total travel cost for each salesman is the sum of all travel costs from the living location to the area (excluding empty ones) belonging to the salesman. Table 1 shows, the total travel time after the optimization process has decreased by around 2,539 h. This is a travel time saving of 16%.

6.3 Evaluating the Contiguity and Compactness

In addition to the balanced workload and the minimum travel cost, we also have two other objectives, compactness and contiguity. This is guaranteed by using a clustering method for initializing the first population of candidate solutions. In order to keep the size of the search space and the performance of the algorithm on a feasible level, we first create a temporary structure of territories which are compact and totally connected, and then we start balancing the potential among salesmen from the borders of territories. In other words, we randomly select some districts on the border of territories and assign them to the neighbour territories. If this operation decreases the standard deviation value, we transfer the new structure of territories as a good solution into the next generation. Moreover, we define operations to detect and reject the invalid solutions to keep the compactness and contiguity of the temporary structure.

7 Conclusions

This work presents a decision-support system for the Territory Design Problem (TDP), the balanced partitioning of geographic areas into regions. This can substitute the subjective, time consuming and mostly less balancing manual allocation. The decision support system has been developed by applying a Genetic Algorithm (GA) in combination with a graph model, and can automatically allocate large numbers of areas to regions, through simultaneous, multi-criteria potential balancing.

Sales territory planning was selected as the case study, as this is one of the most important issues for a commercial organization. An intelligent decision support system for sales territory planning is highly needed. One major benefit of a decision support system is the ability to plan the outcomes of various scenarios before making a real investment. The results of the case study reveal highly balanced regions with contiguous and very compact region shapes. According to our results, locating the sales team based on the distribution of potential, extremely reduces the total calculation time of Green Field Planning. The decision support system proves to be an effective approach for optimizing territories on a large scale.

Due to the independent nature of the GA and the graph model, our method is easily configured to any other area allocating objective. By changing the attributes of nodes and edges in the graph model, applying constraints in the graph model and changing the objective function based on the problem, our method can be used to solve other TDPs.

Acknowledgements This research was supported by the WIGeoGIS GmbH. We wish to thank WIGeoGIS for kindly providing the case study data. We specially thank Michael Steigemann for insightful discussions in the conception of this work.

References

Blais, M., Lapierre, S. D., & Laporte, G. (2003). Solving a home-care districting problem in an urban setting. *The Journal of the Operational Research Society, 54,* 1141–1147.

Cao, R., Li, G., & Wu, Y. (2007). A self-adaptive evolutionary algorithm for multi-objective optimization. In D.-S. Huang, L. Heutte, & M. Loog (Eds.), *Advanced intelligent computing theories and applications with aspects of artificial intelligence* (pp. 553–564). Berlin, Heidelberg: Springer.

D'amico, S. J., Wang, S.–J., & Batta, R., et al. (2002). A simulated annealing approach to police district design. *Computers & Operations Research, 29,* 667–684.

Drexl, A., & Haase, K. (1999). Fast approximation methods for sales force deployment. *Management Science, 45,* 1307–1323.

Fogel, D. B. (1995). Phenotypes, genotypes, and operators in evolutionary computation. In *Proceedings of 1995 IEEE International Conference on Evolutionary Computation* (Vol. 1).

Fonseca, C. M., & Fleming, P. J. (1995). An overview of evolutionary algorithms in multiobjective optimization. *Evolutionary Computation, 3,* 1–16.

Gil, P. "What is 'SaaS' (Software as a Service)?". Retrieved June 13, 2016 from http://netforbeginners.about.com/od/s/f/what_is_SaaS_software_as_a_service.htm.

Haase, K., & Müller, S. (2014). Upper and lower bounds for the sales force deployment problem with explicit contiguity constraints. *European Journal of Operational Research, 237,* 677–689.

Kalcsics, J., Nickel, S., & Schröder, M. (2005). Towards a unified territorial design approach—Applications, algorithms and GIS integration. *Reports of the Fraunhofer Institute for Industrial Mathematics, 71,* 1–56.

Kazimipour, B., Li, X., & Qin, A. K. (2014). A review of population initialization techniques for evolutionary algorithms. In *2014 IEEE Congress on Evolutionary Computation (CEC)* (pp. 2585–2592).

Luque, G., & Alba, E. (2011). *Parallel genetic algorithms: Theory and real world applications* (Vol. 367). Springer.

Płuciennik, T., & Płuciennik-Psota, E. (2014). Using graph database in spatial data generation. In D. A. Gruca, T. Czachórski, & S. Kozielski (Eds.), *Man-machine interactions* 3 (pp. 643–650). Springer International Publishing.

Ríos-Mercado, R. Z., & Fernández, E. (2009). A reactive GRASP for a commercial territory design problem with multiple balancing requirements. *Computers & Operations Research, 36,* 755–776.

Schrijver, A. (2003). *Combinatorial optimization: Polyhedra and efficiency*. Berlin, Heidelberg: Springer.

Sivanandam, S. N., & Deepa, S. N. (2007). *Introduction to Genetic Algorithms*. Berlin Heidelberg: Springer.

Zoltners, A. A., Sinha, P., & Lorimer, S. E. (2009). *Building a winning sales force: Powerful strategies for driving high performance*. New York: AMACOM A division of American Management Association.

Zoltners, A. A., & Sinha, P. (1983). Sales territory alignment: A review and model. *Management Science, 29,* 1237–1256.

Rethinking the Buffering Approach for Assessing OpenStreetMap Positional Accuracy

Qi Zhou

Abstract OpenStreetMap (OSM) is a free source of spatial data based on crowd-sourcing. Although OSM data are widely used in applications such as the generation of 3D models, routing and navigation, the quality issue is still one of the significant concerns when using these data. Extensive studies have focused on assessing the quality, especially the positional accuracy, of OSM data. One method for assessing accuracy is the buffering approach where a buffer is created around a validated road network using a predefined buffer radius. The percentage of OSM road lengths that lie within this buffer is then calculated. While existing studies have used the buffering approach, the method itself has not been evaluated either theoretically and experimentally. It is found that the percentage of OSM road length calculated based on the buffering approach may be imprecise if the validated road network and the OSM road network are not matched one-to-one. Therefore, this study suggests that it is necessary to first match the OSM road network with the validated road network before using the buffering approach.

Keywords OpenStreetMap · Quality assessment · Positional accuracy · Buffering approach

1 Introduction

In the past few years, free sources of spatial data based on crowd-sourcing have appeared online, termed Volunteered Geographic Information (VGI) (Goodchild 2007). The OpenStreetMap project (OSM) may be the most representative of these. OSM is a well-known source of freely available, volunteer collected worldwide geodata. OSM data are now accessible, editable and useable by anyone. A range of applications have been proposed based on OSM data, such as automatically

Q. Zhou (✉)
School of Information Engineering, China University of Geosciences, Wuhan, Hubei, China
e-mail: sevenlaand@hotmail.com

M.P. Peterson (ed.), *Advances in Cartography and GIScience*, Lecture Notes in Geoinformation and Cartography, DOI 10.1007/978-3-319-57336-6_30

generating 3D models (Over et al. 2010; Goetz 2013), integration with authoritative data (Du et al. 2012), mapping land-use patterns (Arsanjani et al. 2013), mapping population at a fine resolution (Bakillah et al. 2014), identifying urban sprawl (Jiang and Liu 2012), and routing/navigation (Zielstra and Hochmair 2012; Bakillah et al. 2013).

The use of OSM data has many benefits. First, it is free of use, whereas most data or maps produced by a national mapping agency either are not open to the public or have limited use. Second, OSM data are updated on a daily basis (Zielstra and Zipf 2010) and thus are up-to-date. Third, OSM data provide global coverage, even for the less developed parts of the world (http://download.geofabrik.de/).

Despite these benefits, some general concerns exist about OSM data. The most significant concern is: "How good is the quality of OSM data?" The majority of OSM data is collected by non-specialists and amateur geographers (Goodchild 2008; Haklay 2008). A number of investigations have been carried out evaluating the quality of OSM data.

Among the OSM feature sets, the road network has received more attention than the others. The accuracy of buildings, coastlines and vegetation has not been of major concern. Among the different elements of road network quality, completeness and positional accuracy have been the primary focus (Haklay 2008; Koukoletsos et al. 2012; Zielstra et al. 2013; Fan et al. 2014; Hochmari et al. 2015). Evaluating how well the coordinate values of a road in OSM relate to the reality on the ground (Haklay 2008; Fan et al. 2014) is especially essential for location and navigation. This study also examines the positional accuracy of OSM roads.

Positional accuracy has been evaluated by comparing an OSM road network with a reference road network, which may come from either official or commercial sources. There are two typical approaches to compare the OSM road network and official road network. The first approach is based on calculating the distance between corresponding road intersections in two different road networks. For instance, Girres and Touya (2010) compared an OSM road network in Hendaye (a region of France) and a reference one produced by the French National Mapping Agency. They manually selected 207 pairs of corresponding road intersections, and calculated the average Euclidean distance between these road intersections as the positional accuracy.

Helbich et al. (2012) compared an OSM road network in Germany and a commercial dataset from Tele Atlas. They first extracted 121 corresponding road intersections and then applied a bi-dimensional regression analysis to calculate the positional accuracy of these intersections. This approach is based on linear objects, i.e., roads, rather than road intersections.

The most common approach in examining positional accuracy involves a buffer (Goodchild and Hunter 1997). As stated by Goodchild and Hunter (1997), the buffering approach "compares a tested source to a reference source by computing the percentage of the length of the tested source lying within a given distance of the reference source." One of the preliminary comparison analyses was carried out by Haklay (2008). He used the buffering approach to assess the positional accuracy of the OSM road network in Great Britain by comparing it with the dataset produced by the Ordnance Survey, Great Britain's national mapping agency.

Compared with other approaches, buffering has several advantages. First, it is easy to implement. Almost all of GIS software (e.g., ArcGIS and QGIS) have an automated tool for buffer analysis. Second, there is no need to match the OSM and reference road networks. Third, as a road network consists of both intersections and segments, the analysis based on roads, rather than road intersections, seems to better estimate the positional accuracy of a road network.

Despite these benefits, to our knowledge, very few studies have focused on evaluating the effectiveness of using the buffering approach to assess the positional accuracy of an OSM road network. This study, therefore, aims to evaluate the buffering approach. We found that, if the OSM road network and the reference road network are not matched one to one, the percentage of OSM roads that lie within a buffer calculated using the buffering approach may be imprecise. More importantly, use of this percentage as a measure to compare the positional accuracy of different regions or road types may be problematic. We further suggest that it is better to first determine the corresponding roads from the different road networks, and then apply the buffering approach to assess the positional accuracy of these corresponding roads.

This study is structured as follows: Sect. 2 contains a theoretical analysis of the limitations of the buffering approach with schematic road networks; Sect. 3 then designs experiments to evaluate the buffering approach, and Sect. 4 shows the experimental results and analyses; Sect. 5 discusses the reasons for needing road matching and also the limitations of using an automated road matching approach; Sect. 6 contains our conclusions.

2 Theoretical Analysis of the Buffering Approach

This section introduces the principles of the buffering approach, and also theoretically analyzes the limitations of this approach in assessing the positional accuracy of an OSM road network.

2.1 Principles of the Buffering Approach

The buffering approach was originally used to assess the positional accuracy of coastlines (Goodchild and Hunter 1997), and it has also been applied to OSM road networks (Haklay 2008). For the assessment of the positional accuracy of an OSM road network, a reference road network is needed, either from governmental or commercial sources. The buffering approach simply compares the OSM and official

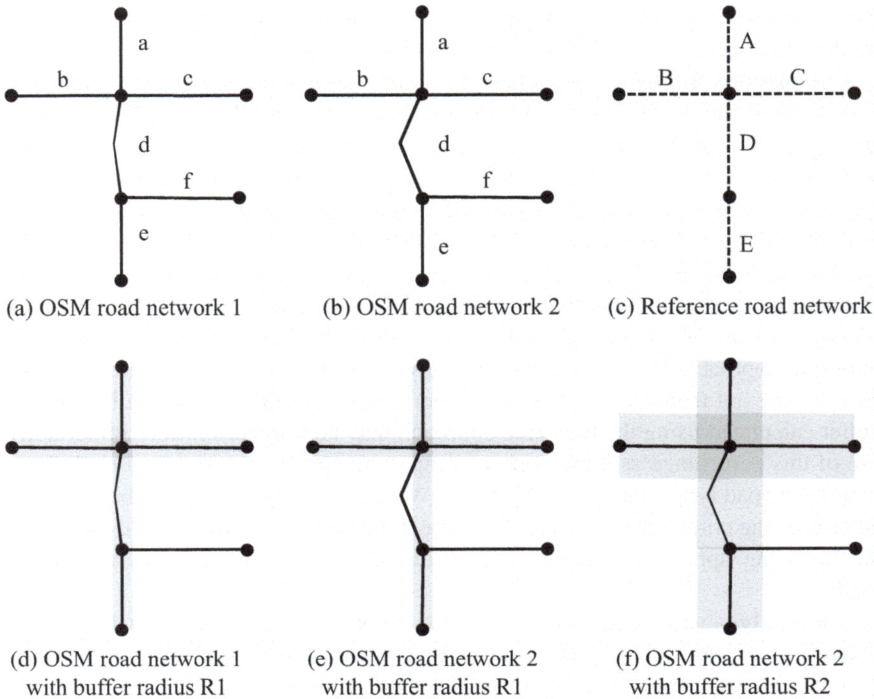

Fig. 1 Principles of the buffering approach in evaluating the positional accuracy of a road network

datasets through two steps: (a) create a buffer of width x around the reference road network; and (b) calculate the percentage p(x) of the OSM road length that lies within the buffer according to Eq. 1. As an example, Fig. 1 shows two schematic OSM road networks (Fig. 1a, b) and a reference road network (Fig. 1c). Figure 1d–f depict the buffers created around the reference road network with different buffer radii. It can be seen in Fig. 1. That with the same buffer radius (Fig. 1d, e), the larger the percentage p(x), the better the geometric quality of an OSM road network. Second, the larger the buffer radius, the more the OSM roads that lie within the buffer (Fig. 1e, f).

$$p(x) = \frac{Length(x)}{Total\ length} \tag{1}$$

where, p(x) denotes the percentage of the OSM road length that lies within a buffer of width x; Length(x) denotes the length of the OSM roads that lie within a buffer of width x; Total length is the total length of the OSM roads for testing.

2.2 *Limitations of the Buffering Approach to Assessing the Positional Accuracy*

Using the buffering approach to assess the positional accuracy may be imprecise if the OSM road network and the reference road network are not initially fully matched. Taking Fig. 1 as an example, the schematic OSM road network has six (6) road segments (i.e., a, b, c, d, e and f) but the reference road network only has five (5) road segments (i.e., A, B, C, D and E). This means that not every road segment (e.g., f) in the OSM road network has a corresponding one in the reference road network. A calculation of the percentage p(x) based on the inconsistent road networks may produce two types of errors:

- The first is that the OSM road length that lies within the buffer (i.e., Length(x)) becomes longer than the actual length because the OSM road segment f in Fig. 1 is partly included; and
- The total length of the actual OSM roads for comparison (i.e., Total length) also becomes longer because the unmatched road segment f in Fig. 1 is included.

The above flaws can be solved by first matching the road networks from the OSM and reference datasets. Then, the percentage p(x) is calculated based on the matched OSM roads only, rather than all of the OSM roads (Eq. 2)

$$p(x) = \frac{\text{Matched Length}(x)}{\text{Matched length}} \tag{2}$$

where, Matched Length(x) denotes the length of the matched OSM roads that lie within buffer of width x; Matched length denotes the length of the matched OSM roads for testing.

The process of matching the different road networks is called road matching. However, most of the existing studies used the buffering approach to assess the positional accuracy of an OSM road network without considering road matching. So, in the next sections, a series of experiments are described that investigate whether the process of road matching is necessary.

3 Design of Experiments for Evaluating the Buffering Approach

3.1 *Experimental Data*

The experimental data involve two digital road networks in the same region (i.e., Hawke's Bay, a region of New Zealand on the east coast of the North Island). Figure 2a shows the digital road network obtained from the OSM dataset (http://download.geofabrik.de/) and Fig. 2b shows the digital road network obtained from

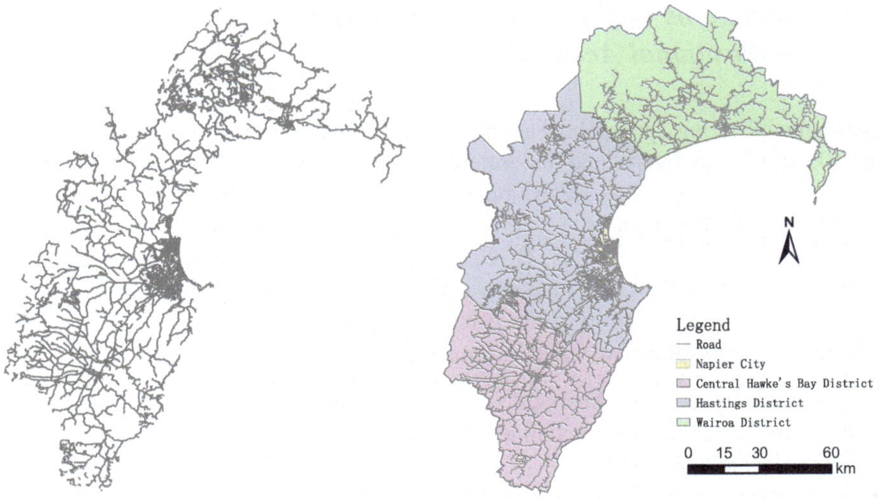

(a) OSM road network (b) Reference road network

Fig. 2 Study area

the Land Information of New Zealand (LINZ), which is used as the reference dataset. The map scale of the dataset downloaded from LINZ is 1:50,000, which is the largest scale of all of the digital datasets freely available for this study area (http://www.linz.govt.nz/land/maps/linz-topographic-maps/topo50-maps).

3.2 Approaches, Steps and Implementation for Assessing the Positional Accuracy

In the experiments, two approaches were used to assess the positional accuracy of the tested OSM road network in Fig. 2a. The first approach is the buffering approach, widely used in previous studies. The second approach adds a process of road matching before using the buffering approach. The two approaches are compared in order to investigate whether the buffering approach without road matching is still effective or whether the process of road matching is necessary.

Two software packages, ArcGIS 10.0 and Road Matcher 1.4, are involved in implementing the above two approaches. ArcGIS was used for the buffering approach, and the freeware Road Matcher 1.4 (Vivid Solutions) was used for road matching. Theoretically, it is possible to manually select the corresponding roads from the OSM road network and the reference road network (Girres and Touya 2010; Helbich et al. 2012). But, the tested OSM road network has a total length of 6356 km, which makes manual road matching tedious. Therefore, an automated process of road matching in Road Matcher was employed.

3.3 *Evaluation of the Buffering Approach*

The buffering approach was evaluated in both quantitative analysis and visual inspection.

3.3.1 Quantitative Assessment

The percentage p(x) (see Eq. 1) has been widely used for quantitatively assessing the positional accuracy of an OSM road network. However, most of the existing studies only set a single threshold (e.g., 20 m) for the buffer radius (Haklay 2008; Arsanjani et al. 2013; Sithole and Zlatanova 2012; Wang et al. 2013; Zhou et al. 2014). The set threshold seems to be arbitrary and subjective. In this study, a number of different buffer radii were set to investigate how the percentage p(x) varies with the buffer radius. The detailed steps are listed as follows:

(1) Create buffers with different radii, i.e., 5, 10, 25, 50, 100 and 200 m;
(2) Calculate the percentage p(x) for each buffer. The percentages p(x) for approach I and II were calculated according to Eq. 1 and Eq. 2, respectively;
(3) Plot the relationship between the buffer radius and corresponding percentage p(x).

The above steps were also applied to the four administrative regions (i.e., Wairoa, Hastings, Napier and Central Hawke's Bay) of New Zealand (Fig. 1b) and various road types (e.g., primary, secondary, track and trunk) for testing in order to investigate whether the buffering approach is able to compare the positional accuracy of different administrative regions or road types (Wang et al. 2013; Zhou et al. 2014).

3.3.2 Visual Inspection

The percentage p(x) is useful for analyzing the positional accuracy of an entire OSM road network but it cannot distinguish in detail which OSM roads have relatively higher positional accuracy than others. So, the distribution of the tested OSM road network in terms of positional accuracy was also investigated. The detailed steps include:

(1) Divide the tested OSM roads according to the different buffer ranges (e.g., 0–5, 5–10, 10–25 and >25 m). As an example, the buffer range 5–10 m means the OSM roads or parts of an OSM road that lie within the buffer area of a 10 m radius but lie outside the buffer area of a 5 m radius.
(2) Visualize the OSM roads that lie within various buffer ranges with different colors because the different buffer ranges somehow represent the positional accuracy of the different OSM roads. For instance, the OSM roads that lie within the buffer range of 5–10 m may have a higher positional accuracy than those that lie within the buffer range greater than 25 m.

4 Results and Analyses

4.1 Quantitative Analysis

Figure 3 plots the relationship between the buffer radius (varying from 5 to 200 m) and the corresponding percentage p(x), in consideration of the two different approaches.

Figure 3 shows that:

(1) The distributions of the two different approaches are similar. More precisely, the percentage p(x) increases rapidly along with an increase in the buffer radius from 5 to 25 m. This is because the larger the buffer radius, the more of the OSM roads that lie within a buffer (Fig. 1e, f). But, when the buffer radius is larger than 25 m, the percentage p(x) increases slowly or tends to be stable. This indicates that most of the OSM roads lie within a buffer with a 25 m radius, which is close to the user accuracy (i.e., ±22 m) of the reference road network.

(2) However, the distributions of the different approaches also show differences. For the buffering approach with road matching (dotted line), the percentage p (x) is close to 100% when the buffer radius was set at 25 m, and it reaches 100% when the buffer radius was set at 50 m. So the maximum displacement in position is between 25 and 50 m. For the buffering approach without road matching (solid line), the percentage p(x) increases slowly even when the buffer radius was set at 200 m. This is because of the first type of flaw. On the other hand, for the same buffer radius (e.g., 25 m), the percentage p(x)

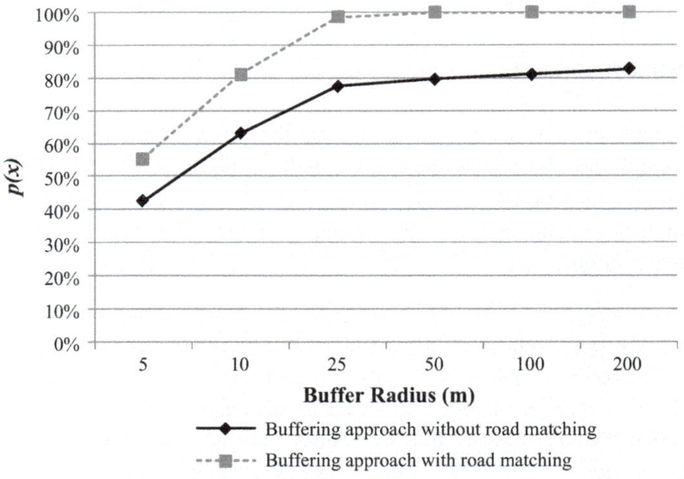

Fig. 3 Plot of the relationship between the buffer radius and the corresponding percentage p(x). The percentage p(x) calculated using the buffering approach is represented by the *solid line*, and the percentage calculated using road matching is represented by the *dotted line*

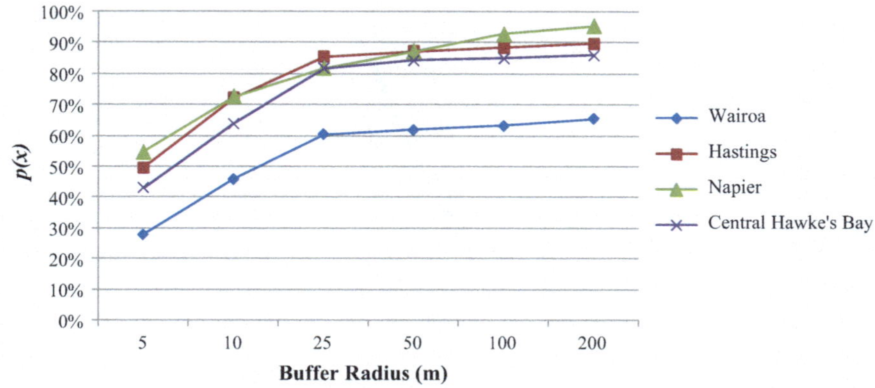

(a) Buffering approach without road matching

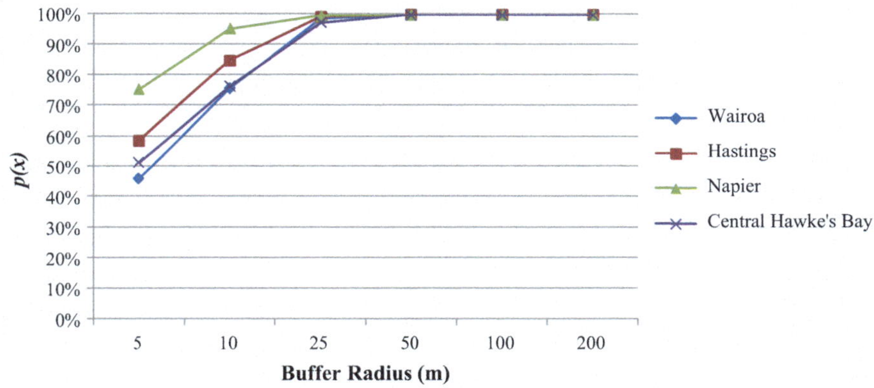

(b) Buffering approach with road matching

Fig. 4 Plot of the relationship between the buffer radius and corresponding percentage p(x) considering the different administrative regions

(e.g., 78%) for the buffering approach without road matching is much lower than that (e.g., 99%) for the approach with road matching. This is because of the second type of flaw.

Figures 4 and 5 plot the relationship between the buffer radius and corresponding percentage p(x), in consideration of the different administrative regions and the different road types, respectively.

Figures 4 and 5 show that:

(1) Most of the distributions of the two approaches are quite different, either from the different administrative regions or the different road types. As an example, by using the buffering approach without road matching, the percentage p(x) for the road type-path increases slowly, and it is no more than 20% even when the

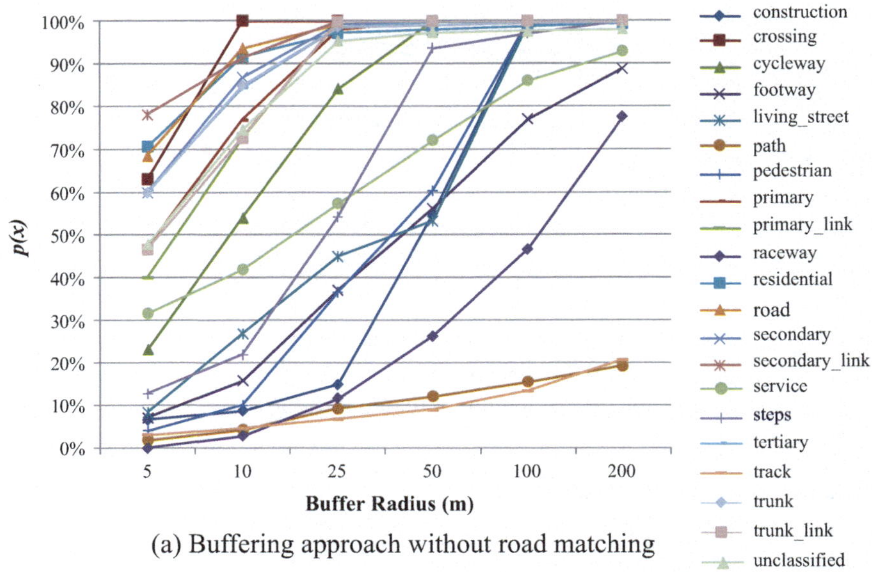

(a) Buffering approach without road matching

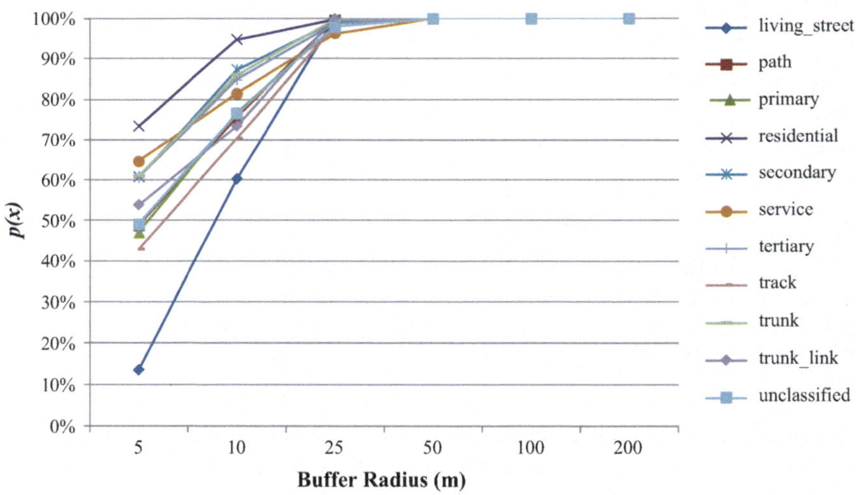

(b) Buffering approach with road matching

Fig. 5 Plot of the relationship between the buffer radius and corresponding percentage p(x) in consideration of the different road types

buffer radius was set at 200 m. But, this is not the case using the buffering approach with road matching. That is, the percentage p(x) for the road type-path increases rapidly along with an increase in the buffer radius from

5 to 25 m, and it is close to 100% when the buffer radius was set at 25 m. The reason is that a number of paths in the OSM road network did not match with a corresponding one in the reference road network, so the percentage p(x) for the buffering approach is much lower because of the second type of flaw. Furthermore, use of the percentage p(x) calculated based on the buffering approach without road matching to compare the positional accuracy of different administrative regions or road types may draw an incorrect conclusion.

(2) In addition, there are 21 road types in the tested OSM road network, but the number decreases to 11 in the matched OSM road network. After looking into the details of the "disappeared" road types in Google Maps, it was found that most of these OSM roads did not match with a corresponding one in the reference road network. Without excluding these OSM roads, the percentage p (x) calculated using the buffering approach may be imprecise.

4.2 *Visual Inspection*

Figure 6 shows the distributions of the tested OSM road network (a) and the matched OSM road network (b) in terms of the positional accuracy. More precisely, the OSM roads that lie within the four buffer ranges, i.e., 0–5, 5–10, 10–25 and >25 m, are represented in four different colors, i.e., dark blue, light blue, light red and dark red, respectively.

Figure 6 shows that:

(1) By using the buffering approach without road matching, the distribution shows that a number of roads (e.g., see dark red in regions A, B, C, D, E and F in Fig. 6) lie outside the buffer with a radius of 25 m. There may be two reasons for this. First, these OSM roads cannot match with a corresponding one in the reference road network; second, the positional accuracy of these roads is relatively low. Nevertheless, it is hard to distinguish these roads for the above reasons.

(2) By using the buffering approach with road matching, most of the unmatched roads were excluded. It is clear to see that the positional accuracy of some roads in region G is relatively low (>25 m). It was further found that the OSM roads in the urban areas (see rectangle 2 in Fig. 6b) seem to have relatively higher positional accuracy than those in the rural areas (see rectangle 1 in Fig. 6b). This finding, being consistent with previous studies (Hagenauer and Helbich 2012), has been explained by the fact that there is a denser population and also more contributors in the urban areas, which results in a better quality of the corresponding OSM data.

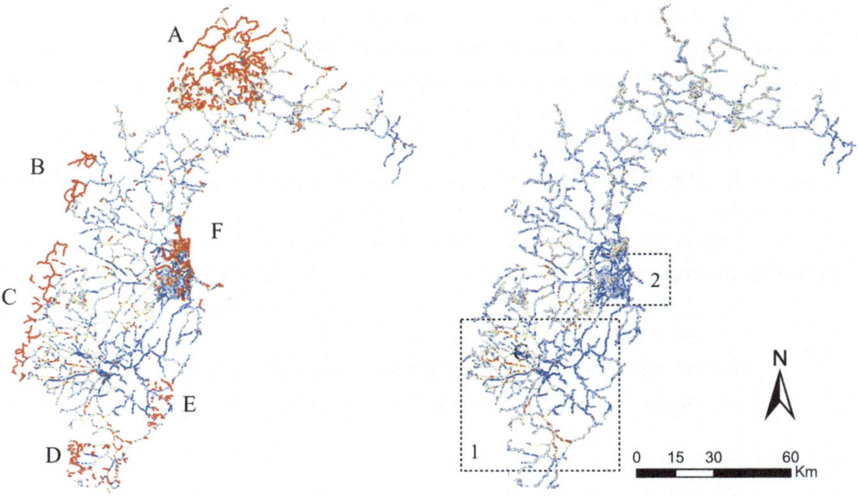

(a) Buffering approach without road matching

(b) Buffering approach with road matching

Fig. 6 Distributions of the tested OSM road network (**a**) and the matched OSM road network (**b**) in terms of positional accuracy

5 Reasons for Inconsistency Between the OSM Road Network and Reference Road Network

(1) The two road network datasets being compared were usually produced at different times. This is because the OSM dataset is updated on a daily basis, but the reference one may only be updated every year or even less frequently.
(2) The reference road network dataset is produced by either a national mapping agency or a commercial company, which has detailed cartography specifications or rules to represent a road network at a specific scale (e.g., how a long or wide road should be retained). But, most of the OSM dataset is produced by volunteers or non-experts, without considering such cartography specifications or rules.
(3) There are more road types defined in an OSM dataset than those in a reference road network. For instance, in an OSM dataset, there may be road types such as crossings, steps, raceways and subways (Fig. 5a). These are rarely recorded in a reference road network.

6 Conclusions

This study reports a theoretical and experimental analysis of the widely used buffering approach for assessing the positional accuracy of an OSM road network. It was found that the buffering approach has limitations when used to assess positional accuracy if the tested OSM road network and the validated road network are not matched one to one. That is, the percentage of the OSM road length that lies within a buffer calculated using the buffering approach is not precise. A further analysis based on imprecise percentages may draw an incorrect conclusion in terms of the positional accuracy of different regions or road types, and the distribution of a tested OSM road network. Therefore, we further suggest that it is necessary to first match the tested OSM road network with the validated road network before using the buffering approach.

Although an automated approach to matching road networks from the different sources can decrease manual work and identify 93–97% or even more of the corresponding roads, a fully automated approach is still not available. So, in future work, a comparison of various automated approaches to road matching for the OSM dataset is worth discussing, and better approaches with fewer matching errors need to be proposed.

Acknowledgements The project was supported by National Natural Science Foundation of China (No. 41301523).

References

Arsanjani, J. J., Helbich, M., Bakillah, M., Hagenauer, J., & Zipf, A. (2013). Toward mapping land-use patterns from volunteered geographic information. *International Journal of Geographical Information Science, 27*(12), 2264–2278.

Bakillah, M., Lauer, J., Liang, S., Zipf, A., Arsanjani, J. J., Mobasheri, A., et al. (2013). Exploiting big VGI to improve routing and navigation services. In H. Karimi (Ed.), *Big data techniques and technologies in geoinformatics*. Boca Raton: CRC Press.

Bakillah, M., Liang, S., Mobasheri, A., Arsanjani, J. J., & Zipf, A. (2014). Fine-resolution population mapping using OpenStreetMap points-of-interest. *International Journal of Geographical Information Science, 28*(9), 1940–1963.

Du, H., Anand, S., Alechina, N., Morley, J., Hart, G., Leibovici, D., et al. (2012). Geospatial information integration for authoritative and crowd sourced road vector data. *Transactions in GIS, 16*(4), 455–476.

Fan, H. C., Zipf, A., Fu, Q., & Neis, P. (2014). Quality assessment for building footprints data on OpenStreetMap. *International Journal of Geographical Information Science, 28*(4), 700–719.

Girres, J. F., & Touya, G. (2010). Elements of quality assessment of French OpenStreetMap data. *Transactions in GIS, 14*(4), 435–459.

Goetz, M. (2013). Towards generating highly detailed 3D CityGML models from OpenStreetMap. *International Journal of Geographical Information Science, 27*(5), 845–865.

Goodchild, M. F. (2007). Citizens as sensors: The world of volunteered geography. *GeoJournal, 69*(4), 211–221.

Goodchild, M. F. (2008). Assertion and authority: The science of user-generated geographic content. In: *Proceedings of the Colloquium for Andrew U. Frank's 60th Birthday*. Vienna University of Technology, Department of Geoinformation and Cartography.

Goodchild, M. F., & Hunter, G. J. (1997). A simple positional accuracy measure for linear features. *International Journal of Geographical Information Science, 11*(3), 299–306.

Hagenauer, J., & Helbich, M. (2012). Mining urban land-use patterns from volunteered geographic information by means of genetic algorithms and artificial neural networks. *International Journal of Geographical Information Science, 26*(6), 963–982.

Haklay, M. (2008). How good is volunteered geographical information? A comparative study of OpenStreetMap and ordnance survey datasets. *Environment and Planning B: Planning and Design, 4*(37), 682–703.

Helbich, M., Amelunxen, C., Neis, P., & Zipf, A. (2012). Comparative spatial analysis of positional accuracy of OpenStreetMap and proprietary geodata. In J. Strobl, T. Blaschke, & G. Griesebner (Eds.), *Angewandte Geoinformatik 2012*. Berlin: Herbert Wichmann Verlag.

Hochmari, H. H., Zielstra, D., & Neis, P. (2015). Assessing the completeness of bicycle trail and lane features in OpenStreetMap for the United States. *Transactions in GIS, 19*(1), 63–81.

Koukoletsos, T., Haklay, M., & Ellul, C. (2012). Assessing data completeness of VGI through an automated matching procedure for linear data. *Transactions in GIS, 16*(4), 477–498.

Jiang, B., & Liu, X. (2012). Scaling of geographic space from the perspective of city and field blocks and using volunteered geographic information. *International Journal of Geographical Information Science, 26*(2), 215–229.

Over, M., Schilling, A., Neubauer, S., & Zipf, A. (2010). Generating web-based 3D city models from OpenStreetMap: The current situation in Germany. *Computers, Environment and Urban Systems, 34*(2010), 496–507.

Sithole, L. S., & Zlatanova, S. (2012). Assessment of the homogeneity of volunteered geographic information in South Africa. In *International archives of the photogrammetry, remote sensing and spatial information sciences* (Vol. XXXIX-B4). *XXII ISPRS Congress*, August 25–September 01, 2012, Melbourne, Australia.

Wang, M., Li, Q. Q., Hu, Q. W., & Zhou, M. (2013). Quality analysis of open street map data. In: *International archives of the photogrammery, remote sensing and spatial information sciences* (Vol. XL-2/W1). *2013, 8th International Symposium on Spatial Data Quality*, May 30–June 1, 2013, Hong Kong.

Zielstra, D., & Hochmair, H. H. (2012). Using free and proprietary data to compare shortest-path lengths for effective pedestrian routing in street networks. *Transportation Research Record, 2299*, 41–47.

Zielstra, D., Hochmair, H. H., & Neis, P. (2013). Assessing the effect of data imports on the completeness of OpenStreetMap—A United States case study. *Transactions in GIS, 17*(3), 315–334.

Zielstra, D., & Zipf, A. (2010). A comparative study of proprietary geodata and volunteered geographic information for Germany. In *13th AGILE International Conference on Geographic Information Science*, May 11–14, 2010, Guimarães, Portugal.

Zhou, P. P., Huang, W., & Jang, J. (2014). Validation analysis of OpenStreetMap data in some areas of China. In *The international archives of the photogrammetry, remote sensing and spatial information sciences* (Vol. XL-4). *ISPRS Technical Commission IV Symposium*, May 14–16, 2014, Suzhou, China.

Data Classification for Highlighting Polygons with Local Extreme Values in Choropleth Maps

Jochen Schiewe

Abstract Following the general demand for task-orientation in map design, one specific task will be examined here: the preservation and highlighting of local extreme values in choropleth maps. Extreme value polygons are ones that show a larger (local maximum) or smaller (local minimum) attribute value compared to all directly neighboring polygons. For a visual identification in a classified choropleth map, such a polygon must belong to a class other than the surrounding polygons. However, data classification methods that are commonly used in the process of generating choropleth maps are data-driven, i.e., the intervals are determined solely on the basis of the present frequency distribution of the original values. With such a division along the number line, the spatial context of the underlying data is completely neglected and with that the desired categorization for local extreme values is not guaranteed. As a consequence, a new method (called PLEX) is presented for this purpose. The application and the effectiveness of this method will be demonstrated using real-world examples.

Keywords Choropleth mapping · Data classification · Task-oriented

1 Introduction

The choropleth method is the most frequently used for thematic mapping (Slocum et al. 2009: 251). The representation of the attribute data in these maps can be either unclassified or classified. While the first option preserves the actual values and their relationships in an exact manner, the second option—much more frequently used—creates more contrast and has the advantage of being able to match a shading in the map to one in the legend.

For data-classification purposes, different methods are available in standard GIS and cartography software packages, e.g., the division into intervals of equal size

J. Schiewe (✉)
HafenCity University Hamburg, Hamburg, Germany
e-mail: jochen.schiewe@hcu-hamburg.de

M.P. Peterson (ed.), *Advances in Cartography and GIScience*, Lecture Notes in Geoinformation and Cartography, DOI 10.1007/978-3-319-57336-6_31

(*equidistant*), the building of classes with the same number of observations (*quantiles*), the setting of interval limits at gaps in the value range (*natural breaks* or *Jenks optima,* respectively) or the use of arithmetic, geometric or harmonic progressions. It is well known that the choice of the method—together with the variation of the number of classes, the representation of border classes (open vs. closed) and the consideration of outliers—has a significant influence on the class membership (Monmonier 1991). This influence then extends to affect visual perception, interpretation and finally the decision-making by the user.

The above mentioned classification methods are data-driven, i.e., the intervals are determined solely on the basis of the present frequency distribution of the original values. The *spatial* context of the underlying data, which is relevant for many applications, is completely neglected when using such divisions. Information about spatial relationships or patterns—and also potential desired "messages" of a map—may be lost. In practice (and in a best case), a subjective selection out of several classification options is performed in order to produce a suitable classification in the form of a map.

Consequently, a task-oriented approach should be considered. Although it is not always possible to assign a specific task or "message" to each choropleth map in advance, there are enough applicable situations in common practice to justify this approach. In this paper, we will concentrate on one specific task, namely the preservation and highlighting of local extreme values in choropleth maps. Local extreme value polygons are ones that show a larger (local maximum) or smaller (local minimum) attribute value compared to all directly neighboring polygons. For a visual identification in a classified choropleth map, such a polygon must belong to a class other than the surrounding polygons.

After discussing previous work on a task-oriented approach in general and data classification in particular (Sect. 2), Sect. 3 describes the necessary methodological development for this specific task (called *PLEX—Polygon Local EXtreme values*). Section 4 will demonstrate the application and the effectiveness of this method using real-world examples. Finally, Sect. 5 will summarize the work done so far and point out aspects for future developments.

2 Previous Work

2.1 Task-Oriented Approach

In textbooks, only rather coarse categorizations of (choropleth) map interpretation tasks can be found (e.g., by Hake et al. 2002: orientation, education, exploration, etc.). Also, the concept of the *Map Use Cube* by MacEachren and Kraak (1997), which considers the parameters information content, user expertise and interactivity of the map, allows only a very broad categorization. The approach of Peuquet (1994) uses the simple spatio-temporal questions "what", "when" and "where," for which usually two aspects are given and the missing third is sought. The task taxonomy proposed by Andrienko and Andrienko (2006) is much more detailed: it divides into elementary tasks ("low level"), which require a separate (independent) view of individual data elements, and synoptic tasks ("high level"), which need a simultaneous view of the entire data set (or sub-sets thereof). In Schiewe (2016) the taxonomy of Andrienko and Andrienko has been implemented in the task-oriented use of choropleth maps, identifying the most common spatio-temporal questions that are addressed in this context.

In contrast to elementary tasks, the synoptic ones like pattern recognition, pattern comparison or search for similarities pose more challenges with regard to an optimized cartographic design. Not only is the choice of a suitable color scheme, but in particular the prior data processing (especially, data classification) of great importance for an effective and efficient use. As already mentioned, the conventional methods for data classification function in a data-driven manner, so that the preservation (or possibly even the amplification) of a required spatial property is not taken into account.

Of course, the idea of a task-oriented design of (choropleth) maps is not entirely new in cartography. However, what is missing is a systematic implementation that—in opposition to the "one-size-fits-all idea"—takes into account an adaptive data processing (here: data classification) and visualization.

2.2 *Data Classification*

In the literature, the subject of data classification for cartographic purposes is comprehensively covered; for example, the overview presented by Cromley and Cromley (1996). The clear focus here is on data-driven methods. A number of empirical studies deal with the comparison of such methods to answer-map use tasks (e.g., Goldsberry and Battersby 2009; Brewer and Pickle 2002; Mersey 1990). Interactive tools have been developed in order to find the "optimal" configuration for a given application, like the use of linked views between a data histogram and choropleth map (Andrienko et al. 2001). In view of this research work, it is not surprising that only data-driven data-classification methods are implemented in current GIS or cartography software.

On the other hand, relatively little has been published on the problem of neglecting the spatial context or, alternatively, using a task-oriented approach; Armstrong et al. (2003) provide an overview on this. They also propose to carry out several data classifications for each application that together can lead to the development of genetic algorithms for the generation of multi-criteria class intervals.

Schiewe (2016) presents a method called *Edge Preserving Index* (*EPI*) to preserve or even amplify significant value differences between neighboring polygons. It is demonstrated numerically as well as visually that the desired effect can be achieved.

3 PLEX Method

3.1 Definition of Local Extreme Values

In the context of choropleth mapping, we define local extreme value polygons as ones that show a larger (local maximum) or smaller (local minimum) attribute value compared to all directly neighboring polygons. In an ideal case, those polygons can be easily determined based on a topological data structure that consists of neighborhood information.

Depending on the given application, it might be useful to reduce the total number of extreme value polygons for visualization purposes, i.e., to avoid an overload of these elements in the map. Furthermore, it might make sense to ignore those polygons that show only a very small value difference to their neighbors. A necessary threshold can be derived from the acquisition accuracy or other statistical or practical considerations.

As it might not be possible to preserve all desired local extreme polygons, the goal is to keep at least the most significant ones. For this purpose, the following significance measures are used:

- In the case of depicting just local *minimum* polygons: the difference between the polygon under observation (with attribute value x) and the maximum value of all n (with i = 1, …, n) neighboring polygons (having all smaller values than x): $\Delta x_{MIN} = x - MAX(x_i)$. The larger this difference is, the more significant the polygon.
- In the case of depicting just local *maximum* polygons: the difference between the polygon under observation (with attribute value x) and the minimum value of all n (with i = 1, …, n) neighboring polygons (having all larger values than x): $\Delta x_{MAX} = MIN(x_i) - x$. Again, the larger this difference is, the more significant the polygon.
- In the case of depicting *all* local extreme polygons: $\Delta x = \Delta x_{MIN}$ *or* $\Delta x = \Delta x_{MAX}$.

Obviously, other definitions for the significance can be applied: for instance, by taking a larger neighborhood than the direct neighbors into account, or by considering all neighboring values via the standard deviation. However, for the following partitioning step (Sect. 3.2), just the ordering of local extreme values according to their importance (i.e., significance values) is needed; the methodology for obtaining this order may be adapted to the context.

3.2 Partitioning

For preserving or highlighting a local extreme polygon, it is necessary to always place at least one class boundary between it and all its neighbors during the process

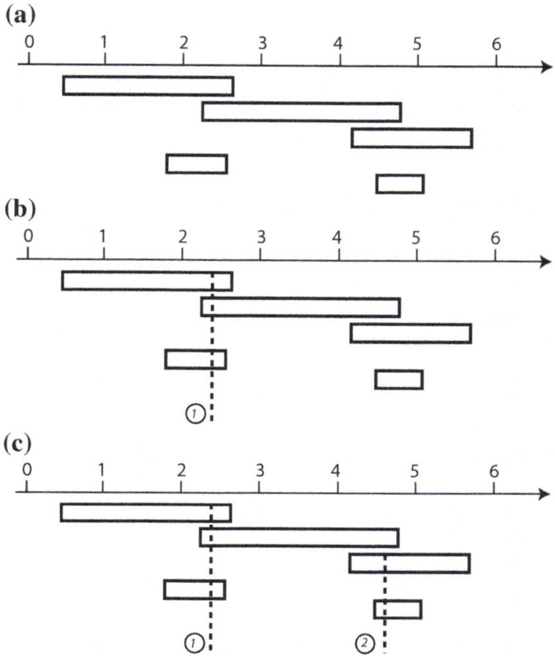

Fig. 1 Principle of graphical partitioning (see text for explanation of steps (**a**) to (**c**))

of data classification. Obviously, with more than one boundary in between, the visual effect of highlighting or sharpening is increased. As a second-order condition, it is required that the total number of classes be kept as small as possible.

For this purpose the above-defined significance values Δx (or, depending on the application, Δx_{MIN} or Δx_{MAX}, resp.) are sorted from large to small. In a graphical approach, these intervals are drawn on a number line as boxes (see Fig. 1a), considering the actual value range:

- for local minimum values: $[x; MIN(x_i)]$
- for local maximum values: $[MAX(x_i); x]$

In contrast to the plane-sweep approach that generally follows the direction of the number line, we have to consider the significance of the intervals. Hence, we start with the maximum of Δx and draw a class boundary line that crosses as many intervals below as possible (see Fig. 1b). After that the next largest interval is taken that has not yet been crossed by the previous line. Again, another class boundary line is drawn that not only crosses as many intervals as possible below but also above this interval (see Fig. 1c). The consideration of intervals above leads to a further division of upper intervals—producing even more intermediate classes and with that a better (visual) separability. Since there are multiple options to place a crossing line having the same number of intersections, one can either chose the average value between the left and right border of the possible intersection interval on the x-axis or try to find a "round number" in between.

With each crossing line an additional class is generated (the number of classes is equal to the number of crossing lines plus one). With that it is possible to stop if a certain pre-defined maximum number of classes is reached. Obviously, it is also possible to introduce a stop criterion that requires that all significant local extreme values as defined in the beginning of the procedure have been recognized.

4 Example

4.1 Data Sets

These two data sets were used in the following:

- *States*: Election turnout (in %) for state elections between 2009 and 2013 in the 16 states of Germany;
- *Counties*: Average age (in years) of inhabitants in the 439 counties of Germany in 2011.

Table 1 summarizes the statistical characteristics of these data sets. Data set *States* shows a rather uniform distribution with random spatial occurrence of values. Data set *Counties* is rather homogeneous, but in contrast to *States* it is based on many more values and shows more clusters.

Figure 2 localizes the desired local extreme values in both data sets. According to Sect. 3.1 the definition of local extreme values is as follows: For *States* all (i.e., six) detected polygons are used. For *Counties* a maximum number of 10 polygons for local minimum and 10 for local maximum values are allowed, with the additional requirement that the value difference to the numerically closest neighbor shall exceed 0.9 years (equal to 0.5 times the RMSE). Using these rules, 16 extreme value polygons are defined for this data set.

4.2 Usage of Conventional Methods

For both data sets, three conventional data-classification methods—equidistant (equal interval) partition, quantiles, and the Jenks algorithm—have been applied (using *ArcGIS* software), varying the number of classes from 3 to 6. Table 2 shows

Table 1 Statistical description of used data sets (*N* number of polygons)

Data set	N	Mean	Min	Max	RMSE	Skewness
States	16	59.8	51.2	73.2	5.97	0.48
Counties	439	43.6	39.1	48.6	1.71	0.39

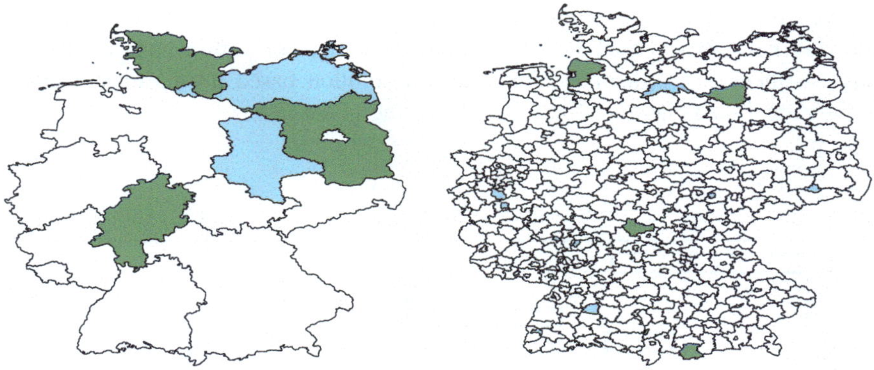

Fig. 2 Local extreme values in data sets *States* (*left* showing election turnout) and *Counties* (*right* showing average age)—local maximum polygons are displayed in *dark green*, local minimum polygons in *light blue* (not all polygons of interest may be visible due to the small size)

Table 2 Number of detected local extreme polygons (*in parentheses* total number of a priori defined polygons)

Method	# classes	Data set *States*			Data set *Counties*		
		Local MAX (3)	Local MIN (3)	All local extremes (6)	Local MAX (6)	Local MIN (10)	All local extremes (16)
Equidistant	3	1	2	3	0	7	7
	4	2	1	3	3	9	12
	5	3	1	4	4	7	11
	6	2	1	3	5	9	14
Quantile	3	1	1	2	4	4	8
	4	2	1	3	2	7	9
	5	2	1	3	5	8	13
	6	1	2	3	4	9	13
Jenks	3	1	1	2	2	4	6
	4	1	2	3	3	8	11
	5	1	2	3	4	10	14
	6	1	2	3	6	10	16

the number of local extreme polygons detected by these algorithms in contrast to the number of desired polygons (in total, 6 for *States*, 16 for *Counties*).

Generally, one can expect that with an increasing number of classes, a better separability and with that a better preservation of local extreme values can be achieved. However, the number of classes is limited due to practical reasons (e.g., due to matchable color intensity steps). As Table 2 shows, in contrast to this general expectation, there is no guaranteed improvement with an increasing number of classes. This is caused by the non-homogeneous distribution of intervals along the number line.

Even more important is the confirmation of the known fact that—except by allowing non-efficient trial-and-error approaches together with a huge number of classes—there is no guarantee for a perfect solution based on these (and other) conventional data-classification methods for the given task.

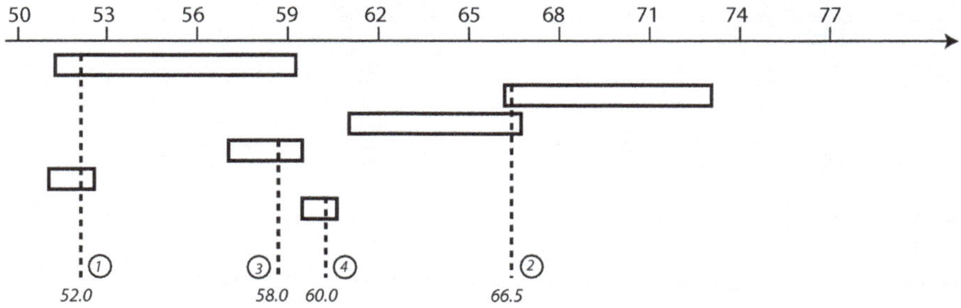

Fig. 3 Graphical partitioning for data set *States* leading to 4 intersection lines (numbers in *circles* describe the sequence of the partitioning process)

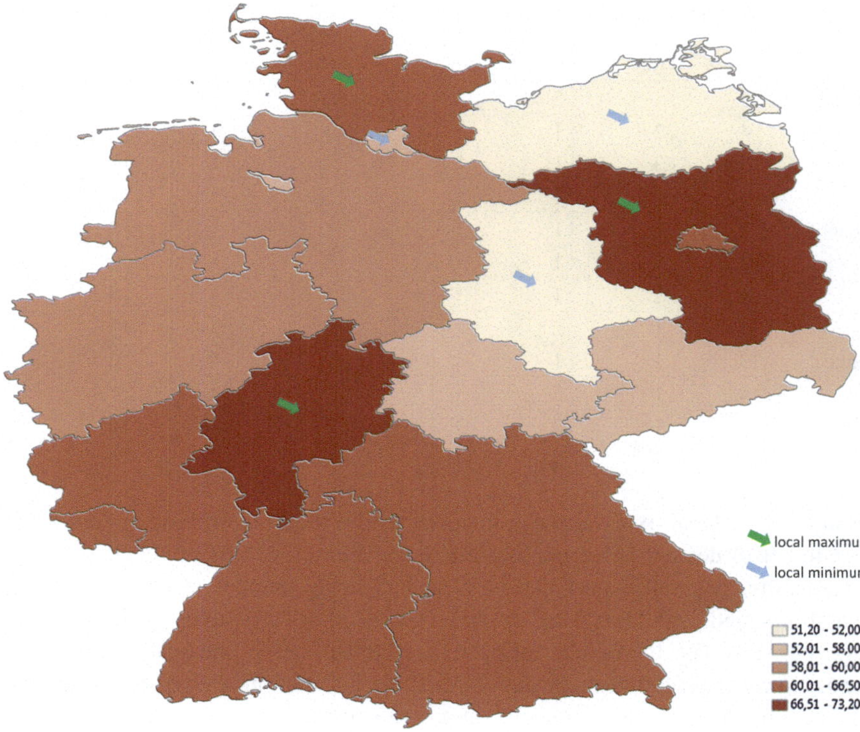

Fig. 4 Final result of PLEX method for data set *States* (showing election turnout) indicating all six required local extreme value polygons

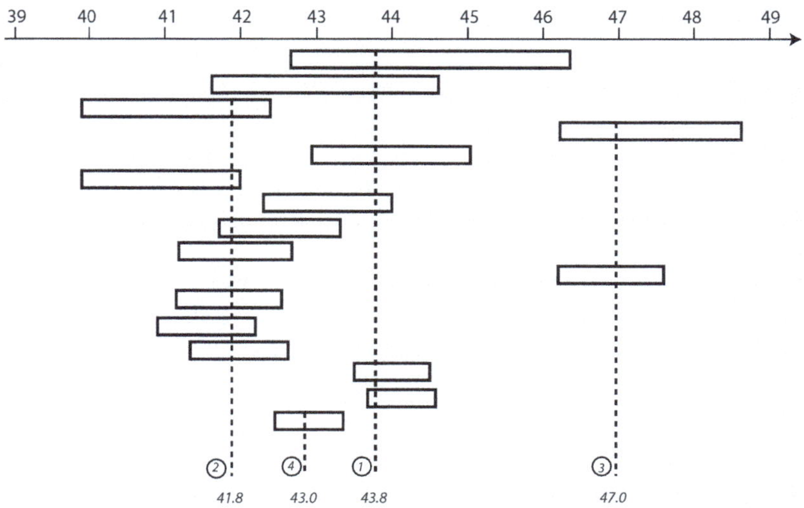

Fig. 5 Graphical partitioning for data set *Counties* leading to 4 intersection lines (numbers in *circles* describe the sequence of the partitioning process)

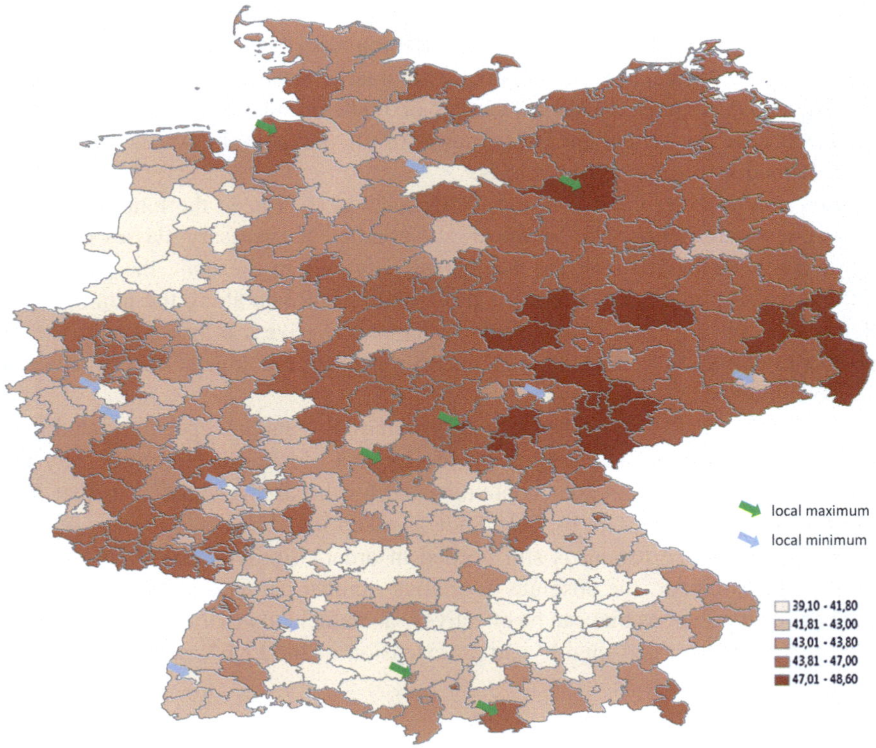

Fig. 6 Final result of PLEX method for data set *Counties* (showing average age) indicating all 16 required local extreme value polygons

4.3 Usage of PLEX Method

Figure 3 shows the graphical partitioning process for data set *States*. It can be seen that 4 intersections (i.e., 5 classes) are necessary in order to cover all desired 6 polygons (a 2-class solution would cover only 2 polygons, 3-class 4 polygons, 4-class 5 polygons). Figure 4 shows the final result (with 5 classes) that fulfills the condition of placing all extreme value polygons into different classes compared to all of their direct neighbors.

Figure 5 shows the graphical partitioning process for data set *Counties*. Again, 4 intersections (i.e., 5 classes) are necessary in order to cover all desired 16 polygons (a 2-class solution would cover 6 polygons, 3-class 13 polygons, 4-class 15 polygons). Figure 6 shows the final map result.

5 Conclusions and Outlook

This paper stresses the relevance of a task-oriented approach in the course of (choropleth) map design. One specific contribution is presented—the highlighting of polygons with local extreme values. For this purpose, a data classification has to ensure that there is always at least one class boundary between the extreme value polygon and all of its direct neighbors. In contrast to standard data-driven methods like equidistant (equal interval), quantiles or Jenks optimal, the PLEX method as described here guarantees the preservation of all polygons in advance that have been defined as local extreme objects.

As a second-order condition, the PLEX method shall keep the total number of classes as small as possible. However, the number of necessary classes can vary significantly—depending on the number of polygons in the map, the number of extreme value polygons and the actual distribution of data along the number line. Hence, the current work deals with the attempt to derive a priori an approximate number of classes from the histogram together with the topological data structure of the given data set.

Ostensibly, the PLEX method can be transferred to another specific spatial map task, the preservation or amplification of Hot or Cold Spots. Here, not only the neighborhood of one polygon (containing the local extreme value), but that of a group of polygons (the Hot or Cold Spot) is taken into account.

Finally, empirical testing is still needed in order to prove the usability of this method. In particular, the absolute or relative number, as well as the area size of local extreme polygons that can be effectively and efficiently extracted by map users is of central interest and might yield an extension of the PLEX method.

References

Andrienko, N., & Andrienko, G. (2006). *Exploratory analysis of spatial and temporal data: A systematic approach*. Berlin, Heidelberg, New York: Springer.

Andrienko, G., Andrienko, N., & Savinov, A. (2001). Choropleth maps: Classification revisited. In *Proceedings International Cartographic Conference*, Beijing (Vol. 2, pp. 1209–1219).

Armstrong, M. P., Xiao, N., & Bennett, D. A. (2003). Using genetic algorithms to create multicriteria class intervals for choropleth maps. *Annals of the Association of American Geographers, 91*(3), 595–623.

Brewer, C. A., & Pickle, L. (2002). Evaluation of methods for classifying epidemiological data on choropleth maps in series. *Annals of the Association of American Geographers, 92*(4), 662–681.

Cromley, E. K., & Cromley, R. G. (1996). An analysis of alternative classification scheme for medical atlas mapping. *European Journal of Cancer, 32A*(9), 1551–1559.

Goldsberry, K., & Battersby, S. (2009). Issues of change detection in animated choropleth maps. *Cartographica, 44*(3), 201–215.

Hake, G., Grünreich, D., & Meng, L. (2002). *Kartographie* (8th ed.). Verlag De Gruyter.

MacEachren, A. M., & Kraak, M.-J. (1997). Exploratory cartographic visualization: Advancing the agenda. *Computers & Geosciences, 23*(4), 335–343.

Mersey, J. E. (1990). Choropleth map design—A map user study. *Cartographica, 27*(3), 33–50.

Monmonier, M. (1991). *How to lie with maps*. University of Chicago Press.

Peuquet, D. J. (1994). It's about time: A conceptual framework for the representation of temporal dynamics in geographic information systems. *Annals of the Association of American Geographers, 84*(3), 441–461.

Schiewe, J. (2016). Preserving attribute value differences of neighboring regions in classified choropleth maps. *International Journal of Cartography, 2*(1), 6–19. doi:10.1080/23729333.2016.1184555.

Slocum, T. A., McMaster, R. B., Kessler, F. C., & Howard, H. H. (2009). *Thematic cartography and geovisualization* (3rd ed.). Prentice Hall.

Part VII
Routing

A Confidence-Based Approach for the Assessment of Accessibility of Pedestrian Network for Manual Wheelchair Users

Amin Gharebaghi, Mir-Abolfazl Mostafavi, Geoffrey Edwards, Patrick Fougeyrollas, Patrick Morales-Coayla, François Routhier, Jean Leblond and Luc Noreau

Abstract An obvious way to assist people with disabilities in their efforts to move around in urban areas is to offer information regarding accessible path segments that take into account their specific needs. Providing safe navigation in urban areas for people with disabilities can substantially enhance their opportunities for full social participation and the exercise of their basic rights. In order to offer these people information on accessible paths, we need to assess the accessibility of pedestrian networks. Accessibility is the result of interactions between the individual and the environment. These interactions need to be more carefully considered when we evaluate accessibility for persons with disabilities. This paper proposes a new

A. Gharebaghi (✉) · M.-A. Mostafavi · G. Edwards · P. Morales-Coayla
Center for Research in Geomatics, Laval University, Quebec, QC, Canada
e-mail: amin.gharebaghi.1@ulaval.ca

M.-A. Mostafavi
e-mail: mir-abolfazl.mostafavi@scg.ulaval.ca

G. Edwards
e-mail: geoffrey.edwards@scg.ulaval.ca

P. Morales-Coayla
e-mail: patrick.morales-coayla.1@ulaval.ca

A. Gharebaghi · M.-A. Mostafavi · G. Edwards · P. Fougeyrollas · F. Routhier · J. Leblond · L. Noreau
Center for Interdisciplinary Research in Rehabilitation and Social Integration, Laval University, Quebec, QC, Canada
e-mail: Patrick.Fougeyrollas@cirris.ulaval.ca

F. Routhier
e-mail: Francois.Routhier@rea.ulaval.ca

J. Leblond
e-mail: Jean.Leblond@cirris.ulaval.ca

L. Noreau
e-mail: Luc.Noreau@rea.ulaval.ca

M.P. Peterson (ed.), *Advances in Cartography and GIScience*, Lecture Notes in Geoinformation and Cartography, DOI 10.1007/978-3-319-57336-6_32

approach for assessing accessibility, based on user confidence while moving around in urban space. For validation purposes, the proposed approach is applied in the Saint-Roch area of Quebec City using confidence information from 127 manual wheelchair users. The results are presented and discussed and further research perspectives are proposed.

Keywords Accessibility · Pedestrian networks · Confidence · Wheelchair user · Modeling · Dynamic segmentation

1 Introduction

Ensuring full social participation of people with disabilities is a challenging issue. This is because the special needs of these people are often not taken into consideration in the development of cities, public places, new technologies and services. People with disabilities who cannot move about autonomously cannot carry out their daily activities such as going to work and school, shopping, or participating in community and family life. The needs for better participation in social activities by people with disabilities are usually referred to as "hidden demands" (Marston and Golledge 2002). These are activities that people with disabilities are unable to carry out despite their desire to do so. According to a Canadian Survey on Disability (CSD) in 2012, pain, motor disability[1] and lack of adapted services, are the most common causes for social exclusion among Canadian adults. Indeed, mobility problems constitute 7% of the aforementioned disability issues. Providing people with disabilities with assistive technologies to improve their mobility will help to increase their effective participation in society. To address this issue, the research reported here aimed to develop a new approach for evaluating the accessibility of pedestrian networks for people with motor disabilities. Since mobility is a result of interactions between humans and their environment (Gharebaghi and Mostafavi 2016), our approach emphasizes on human capabilities and their confidence levels across a range of environmental interactions during displacements.

Several studies have attempted to address this issue by modeling the interactions between humans and the environment. For example, Warren (1984) analyzed the dynamics of an human-environment system using Gibson's affordance theory (Gibson 1977). He showed that, for example, stair climbing is made possible by the interaction between leg length and stair height. Matthews et al. (2003) and Beale et al. (2006) developed a GIS-based system for modeling access for wheelchair users in urban areas. Kasemsuppakorn and Karimi (2009) developed a personalized routing model for wheelchair users focusing on user priorities and sidewalk

[1]Disabilities related to the movement and maintenance of body position (Fougeyrollas 1998).

properties. Jonietz et al. (2013) and Jonietz and Timpf (2013) proposed a framework for modeling spatial-suitability of pedestrian networks based on affordance theory[2]. Tajgardoon and Karimi (2015) proposed an approach based on a weighted linear model for different characteristics of sidewalk segments to evaluate the accessibility of sidewalks for persons with disabilities. Although all of these studies proposed valuable methods to model the human-environment interactions, (a) they did not fully consider the individual's capabilities for implementation of their approaches, (b) their validation are not generally supported by human involvement and finally (c) each of them identified and employed their own set of properties describing pedestrian networks. However, it does not appear that they fully took into account users' perception regarding pedestrian network's properties. As a result, the users' requirements are not fully met.

We propose a new approach that considers the perception and capabilities of manual wheelchair users to evaluate the accessibility of pedestrian network for persons who use a manual wheelchair in their daily activities. We account for both actual and perceived capabilities of the users. User confidence is considered as a criterion to measure the user's perceived capabilities. The accessibility assessment process is undertaken in seven steps:

1. Capturing the pedestrian network data;
2. Partitioning the pedestrian networks into segments;
3. Gathering the user profile information;
4. Linking segment properties with the corresponding user confidences;
5. Aggregating the confidence levels for each segment;
6. Evaluating the accessibility level of each segment based on the total confidence; and finally,
7. Visualizing the accessibility level of each segment on the pedestrian network map.

The remainder of this paper is organized as follows: In Sect. 2, we determine a specification for the database that will contain the information about the pedestrian network, identifying the most important environmental criteria in relation to the mobility of people with manual wheelchairs. In addition, we describe the segmentation process. Section 3 proposes the confidence-based framework to the assessment of the accessibility of the pedestrian networks for manual wheelchair users. Section 4 presents the results of the accessibility assessment for the Saint-Roch area of Quebec City for 127 individuals. Finally, Sect. 5 presents conclusions and future work.

[2]The affordances of the environment are what it offers the animal, what it provides or furnishes, whether for good or ill (Gibson 1977).

2 Pedestrian Network Database

One of the fundamental components of a navigation system is a geospatial database. A geospatial database provides the necessary information for performing navigation, mapping and routing functions. Nowadays, road network databases are designed to serve diverse navigation applications. However, these databases are not generally usable for navigation of persons with disabilities. Pedestrian network databases require information about the environment in much greater details such as obstacles on the sidewalks. In addition, to employ such databases for navigation of persons with different capabilities, these databases should be adapted to their needs. Therefore, a suitable database for their navigation should contain both information on user profiles and the pedestrian network itself and its characteristics in relation to the mobility task.

A pedestrian network is one of the most important environmental entities in outdoor mobility that persons with disabilities interact with in their daily activities. It contains the geometric and topological relations between pedestrian path segments (Karimi and Kasemsuppakorn 2013). Pedestrian networks are typically classified into sidewalks, crosswalks, footpaths, building entrances, trails, pedestrian bridges, and tunnels. Each of these classes consist several segments. Two points, starting from one point and ending at another, define a segment. Each segment has properties that can be either permanent or temporal. The spatial database for enabling the mobility of people with disability should take into account both permanent and temporal properties (events) using appropriate algorithms (see below for more details about these processes). To create these algorithms, first it is necessary to determine the most relevant entities of the pedestrian network that will affect the mobility of persons with motor disabilities. Following this, the segmentation of the network can be carried out based on the static and dynamic nature of those entities.

2.1 Determining the Most Important Environmental Criteria for Enabling the Mobility of Persons with Manual Wheelchairs

As mentioned above, the variation of pedestrian network's characteristics or the presence of different obstacles on the pedestrian network may affect the network segmentation process. Furthermore, this set of characteristics may vary from one context to another. For example, while the quality of the pavement is likely to be of central importance for persons with motor disability, it may not be a significant factor for persons with hearing impairment. Therefore, to evaluate the accessibility of a segment for persons with disabilities, the main properties that affect their daily activities should be determined. So far, a number of studies have investigated these properties. For example, Matthews et al. (2003) and Kasemsuppakorn and Karimi

(2009) used width, length, slope, sidewalk surface, steps, sidewalk conditions and sidewalk traffic. Although pertinent, there is no evidence that these properties were determined according a rigorous study involving the perception of people with disabilities or the experience of expert groups. To overcome the limitation of these works, we use properties that are identified in the Wheelchair Mobility Confidence Scale (WheelCon) (Rushton et al. 2011). This survey instrument is a questionnaire that is developed specifically for studying the mobility of wheelchair users. WheelCon is one of the most reliable approaches for evaluating capabilities of wheelchair users. The questionnaire includes 65 items, which are identified by a three round Delphi[3] survey among a panel of experts (43 experts), of which 30% were wheelchair users.

Here, we are specifically interested to the characteristics of the outdoors that affect the mobility of people with disabilities. For this reason, we have generated a short version of the main questionnaire. A statistical process was used to analyze and sort out the items in terms of relevancy, correlation, and adaptation. The WheelCon questionnaire includes 65 items. Among them, 17 are especially relevant to this study. Data from 127 participants indicated that those selected items can effectively predict the 65-item WheelCon total score (SPSS 23; linear regression; adjusted $R^2 = 0.923$; $p < 0.0001$). A principal component analysis on these 17 items identified two factors that obtained respectively 65% and 9% of the total variance. The sampling adequacy was high (Kaiser-Meyer-Olkin index = 0.93). In order to further reduce the number of items of an abridged version of the WheelCon specific to our study, we iteratively removed one item at a time in order to check the impacts of this removal on the adjusted R^2 and KMO index. In order to preserve the most detailed information, we followed the strategy to remove first the item with the smallest variance. Optimally, this process identified a set of 12 items that has a minimal impact on the adjusted R^2 (0.888; $p < 0.0001$) and the KMO index (0.92). These 12 items are similarly sensitive to two components that obtained respectively 67% and 10% of the total variance. Figure 1 outlines this process.

As a result of the explained process, 12 items among the 65 original items of WheelCon are selected. These items are used to guide the segmentation of the pedestrian networks. They are classified into six categories related to the segments that include slope, curb cut, surface quality, presence of snow on the segment, intersections with and without traffic lights, and sidewalk traffic. Among these properties, the slope, curb cut, and intersections are classified as permanent properties and surface quality, snow, and sidewalk traffic are classified as temporal properties. Different algorithms for the segmentation of the pedestrian network are proposed for each class of properties. These algorithms are described in the following section.

[3]The Delphi method is a structured communication technique or method, originally developed as a systematic, interactive forecasting method which relies on a panel of experts (https://en.wikipedia.org/wiki/Delphi_method).

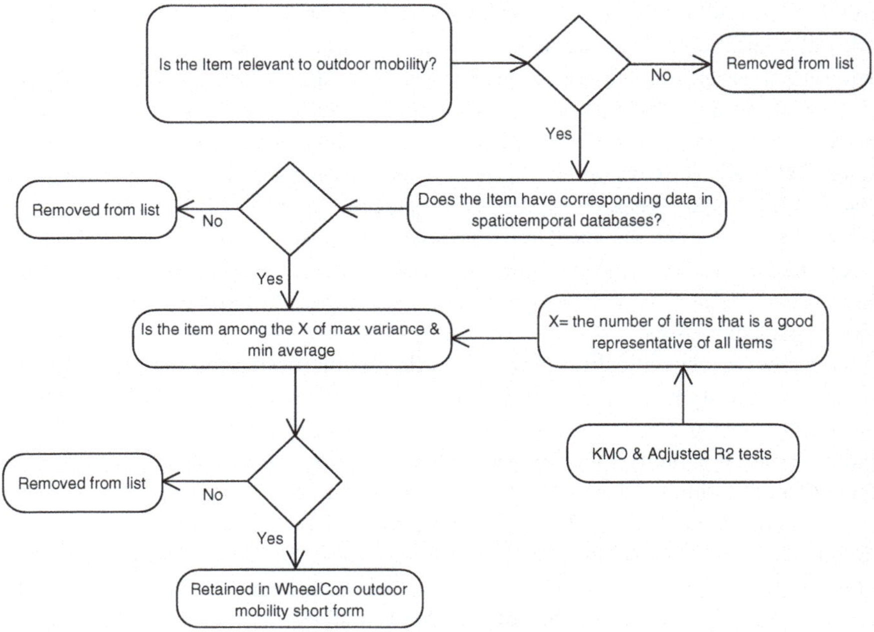

Fig. 1 Selection of the most pertinent items for the mobility of manual wheelchair users from WheelCon questionnaire

2.2 Pedestrian Network Segmentation

The segmentation of the pedestrian network is the first step for its accessibility assessment and mapping. The segmentation should be carried out based on both the relevant properties of the network as described earlier and the user requirements. This process includes four steps:

1. Extracting the centerlines of the network;
2. Ensuring their connectivity and topological consistency;
3. Carrying out the segmentation based on its permanent properties; and
4. Carrying out the segmentation based on the temporal properties.

Figure 2a shows the permanent segmentation in our research area considering slope changes of more than 5°, as well as the presence of curb cuts. The process is extended to generate new segments based on temporal properties. Here, we use a dynamic segmentation process (Weigang and Guiyan 2009), which is a method for identifying segments with changing attributes in time. Instead of splitting segments whenever there is a change in attribute values, here we apply a linear reference system to indicate the start and end of a temporal property without splitting the segment (Cadkin 2002). Figure 2b shows an example resulting from this process. Once the segmentation of pedestrian network is carried out the accessibility of each

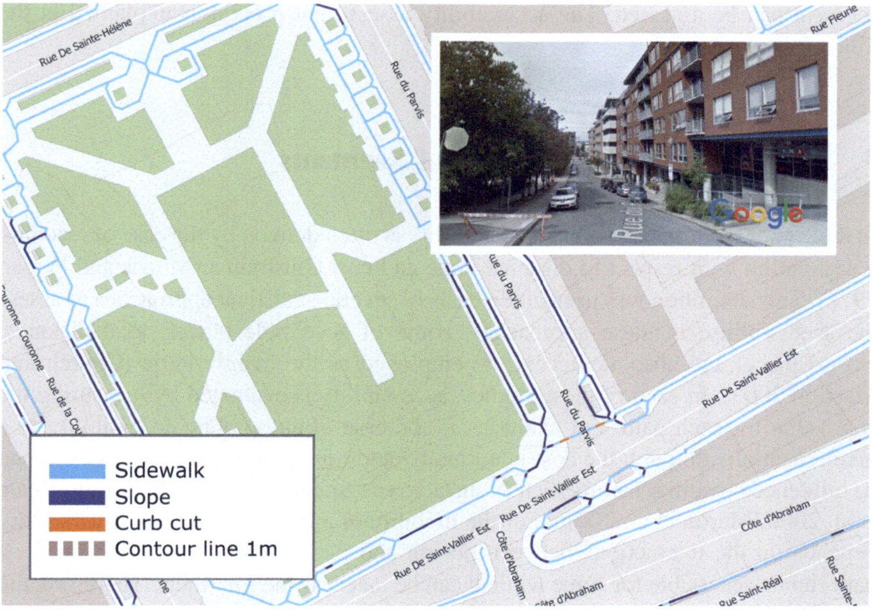

(a) Permanent segmentation

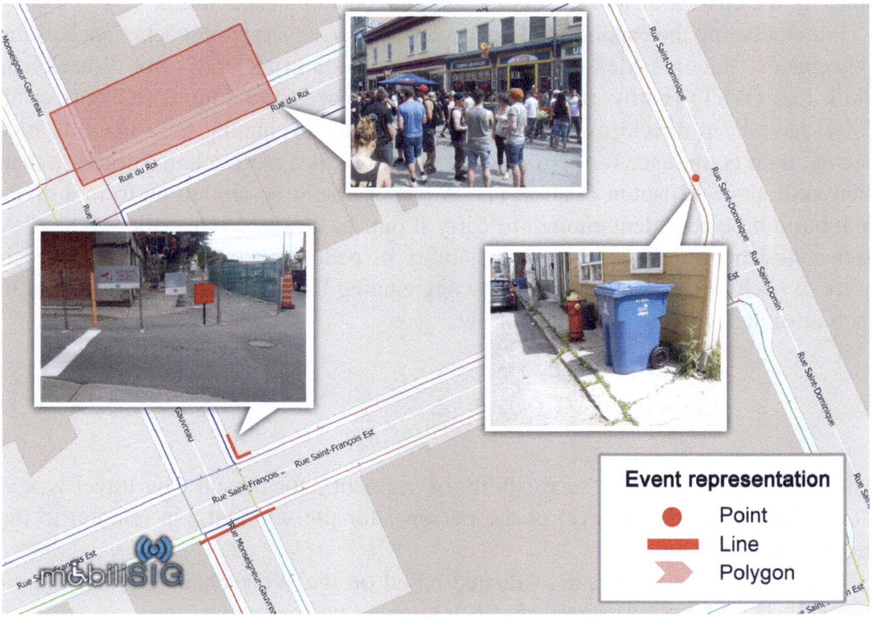

(b) Dynamic segmentation

Fig. 2 Segmentation of the pedestrian network

segment can be assessed. The accessibility assessment process is explained in the following section.

3 Evaluating the Accessibility of Segments

In the context of urban mobility, accessibility is often defined by the ease of reaching a destination with respect to distance, time and cost constraints (Morris and Wigan 1978). This definition was used at the basis of several statistical approaches to assess the accessibility. In these approaches, count of accessible places, total distance, closest place, and absolute access are employed as the main criteria (Church and Marston 2003). In each of these studies, accessibility is evaluated by measuring the cost value for each path. For example, in the case of automobile accessibility to a given location, although the cost is normally determined in terms of distance, it can be influenced by other factors such as surface type, speed limit and congestion (Beale et al. 2006; Martin et al. 2001). For the mobility of a person, however, accessibility depends on the interactions between person and the environment. For example, a path can be accessible for some while it can be inaccessible for others. Therefore, the accessibility of each segment of a pedestrian network should be assessed considering not only the environmental factors but also the person's capabilities.

For evaluating the capability of a wheelchair user, several approaches such as the Wheelchair Outcome Measure (WhOM) (Mortenson et al. 2007), the Wheelchair Skill Test (WST) (Kirby et al. 2002), and the Wheelchair Circuit (Kilkens et al. 2004) have been developed. On the other hand, for evaluating the capability of a person, their confidence for performing a given task is a more reliable criterion than their skill alone (Rushton et al. 2011). A person might be able to perform a given task but not be confident enough to carry it out. In this research work, we employ user confidence to evaluate the accessibility of pedestrian segments. The accessibility of each segment is calculated by aggregating user confidence with respect to each attribute of the segment as follow.

$$A_{ijl} = \sum_{p=1}^{7} Con_{ijp} \tag{1}$$

where, A_{ijl} is the index of accessibility of segment j for person i by travel type l; Con_{ijp} is the confidence level of the person i for the segment j in relation to the property p;

User confidence levels are evaluated based on the Wheelchair Mobility Confidence Scale (WheelCon) approach. Each segment has more than one property, so it will take more than one confidence value. These confidence values are aggregated for each segment and a total confidence value for a given segment is calculated. Hence, the accessibility assessment process is completed in three steps:

1. Setting the user profile information that includes the user confidence level via the 12 questions of the WheelCon short-form questionnaire,
2. Aggregating the confidence levels for each segment with respect to its properties, and
3. Evaluating the accessibility level of each segment based on the total confidence.

3.1 Aggregation of Confidence Levels

In order to assess the accessibility of each segment, the aggregation of confidence values related to properties of that segment is required. There are several methods that can be employed to aggregate these values such as weighted linear models and if-then rules approaches. Aggregation process is a very important step in the accessibility assessment approach. Weighted linear models might have some limitations in complex situations. For instance, in a scenario of mobility the user may (a) move down from a standard curb, (b) cross an intersection, and (c) pass over a hole on the sidewalk. If the confidence level of the user with respect to each of these properties is at medium level, the aggregated confidence level might not be medium for this user. A weighted linear model would result a medium confidence level for this scenario, which does not reflect the reality.

Here we use If-then rules approaches, namely fuzzy logic approach, which provide more realistic values for aggregated confidence level. Fuzzy logic is a widely used if-then rule approach. Fuzzy logic is introduced by Zadeh et al. (1965) to model the vagueness that is associated with human cognitive processes. In recent years this approach has been widely used in many different applications including routing and transportation planning (Kasemsuppakorn and Karimi 2009) and (Karimanzira et al. 2006). To employ fuzzy logic, three steps must be followed: (1) build the rule set and define the membership functions (fuzzification), (2) make a fuzzy inference system (FIS) using if-then rules and (3) merge the outputs of the rules and ensure defuzzification of the results using a different set of membership functions to derive output variables (Mamdani and Assilian 1975). The objective of the Fuzzification step is to transform the numerical confidence values to qualitative values (linguistic variables) by defining a membership function. For example, the values between 0–20 correspond to a very Low confidence. Very Low (VL), Low (L), Medium (M), High (H), and Very High (VH) labels are defined for describing the confidence levels. Then the if-then rules are defined to aggregate the individual user confidences and, consequently, calculate the total confidence of the user for each segment. For example:

If (the confidence level of slope is low) & (the confidence level of poor surface is low) Then (the total confidence level is very low)

Table 1 indicates the defined rules. Rules are stated as if aggregation would be applied on two components, in case that we have more than two parameter the result of first aggregation would be aggregated with the next parameter and this

Table 1 If-Then rules

Rule no.	Confidence level		Aggregated confidence	Rule no.	Confidence level		Aggregated confidence
	p	q			p	q	
1	V.L	V.L	V.L	9	L	V.H	L
2	V.L	L	V.L	10	M	M	L
3	V.L	M	V.L	11	M	H	M
4	V.L	H	V.L	12	M	V.H	M
5	V.L	V.H	V.L	13	H	H	M
6	L	L	V.L	14	H	V.H	H
7	L	M	V.L	15	V.H	V.H	V.H
8	L	H	L				

process is continued till the total confidence is obtained. Since these rules directly affect the result of the process, they need to be validated. In our research, an expert who is also a wheelchair user carries out the validation step. However, we understand that further investigation is needed for a more rigorous validation of these rules by participation of experts and wheelchair users.

Once the aggregation of confidence levels is achieved and a unique confidence level is calculated for each segment, a numerical confidence value must be recovered. To address this issue, a defuzzification technique should be applied to produce exact numerical values from the fuzzy values, based on the defined membership functions and rules. The output values are utilized for determining the accessibility levels of pedestrian network segments in four categories of Not Accessible (NA), Low Accessible (LA), Accessible (A), and Very Accessible (VA).

4 Cases Study

To illustrate the validity and utility of the proposed approach, the whole process of accessibility of the Saint-Roch area in Quebec City is assessed. To fulfill this assessment each user determines his/her confidence level by choosing a corresponding number from the rating scale from 0 to 100. The accessibility assessment process is visualized for two users. First, it is carried out for a 61-years-old female wheelchair user called user#1. In a second case, we suppose a user with mean confidence called user#2. The mean confidence value is obtained using information from 127 users to assess the average accessibility of the study area. The data on users confidences is obtained in collaboration with another team who conducts a survey on manual wheelchair users skills. The mean confidence values for a given

Table 2 Confidence values of manual wheelchair users for each parameter

No.	User confidence items	User#1	User#2
1	Move up a curb (15 cm)	15	22
2	Move down a curb (15 cm)	15	41
3	Move across 3 m of flat, unpacked gravel	15	51
4	Move along a sidewalk with 5 cm of snow	0	51
5	Move through a pothole	10	54
6	Move along a flat dirt path or trail	25	59
7	Move up a dry steep slope (>5° incline)	30	68
8	Move along a paved sidewalk that is cracked	40	70
9	Cross a street with no traffic lights	10	73
10	Move through a crowd of people	10	73
11	Move across 3 m (10 ft) of flat, freshly mowed, dry grass	10	74
12	Cross a street before the traffic light changes	10	75
–	Mean confidence	17	52

Fig. 3 Accessibility map for user#1

segment in sidewalk are provided in Table 2. In this example the mean values for the user#1 and user#2 are 17 and 52, respectively.

The accessibility assessment process starts with the segmentation step based on the properties of each segment and related events observed in the environment. Then, the attributes of each segment for selected properties are stored in a spatio-temporal database. Next, the confidence value of the corresponding attributes for each user is imported into the database. The fuzzification, aggregation, and calculation of the total confidence for each segment are carried out for each user. Finally, the accessibility level for each segment in the study area is evaluated and visualized. Figure 3 shows the accessibility map for user#1 where "Not Accessible" segments are represented with red, "Low Accessible" segments are yellow, "Accessible" segments are green, and "Very Accessible" segments are dark green. As the accessibility map shows, there are a significant number of inaccessible segments for user#1. This process indicates that for this manual wheelchair user, 924 out of 2427 segments are not accessible. This represents 38% of the whole network. Most of the inaccessible segments are located in the intersections.

The next map shown in Fig. 4 is generated for the same area employing the mean confidence values for all 127 users. Since the average confidence level for all participants is significantly higher than the confidence level of user#1, the number of inaccessible segments is negligible. According to the statistical analysis, 62% of the St-Roch area's pedestrian network is very accessible, 25% is accessible, 12% is low accessible, and about 1% is not accessible. This map can be used by city

Fig. 4 Accessibility map for user#2

authorities as a valuable decision making tool to locate the inaccessible and low accessible segments and propose an accessibility improvement plan in the area.

5 Conclusion and Future Work

In this paper, we proposed a novel method to assess the level of accessibility of a pedestrian network for manual wheelchair users for their mobility. First, we presented a segmentation process for pedestrian network incorporating a range of segment properties and related events. The study also drew on the results of the WheelCon survey for assessing accessible paths based on users' perceptions and capabilities to identify a set of 12 characteristics. The key contribution of the present paper was to propose a confidence-based approach for the assessment of the accessibility of pedestrian network. To our knowledge, this is the first time that a confidence based approach is used to evaluate the accessibility in an urban area. To illustrate the utility of the proposed approach, we presented two scenarios using data from the Saint-Roch area in Quebec City. The results demonstrated different levels of accessibility for a specific user compared to an average manual wheelchair user. City authorities for further analysis and decision-making could use the resulting maps. Further investigation is needed for more rigorous validation of the if-then rules proposed here for aggregating confidence values in relation to different segment properties. We also plan to carry out further validation process with experts and wheelchair users in more realistic application scenarios.

Acknowledgements This research was supported by the MobiliSIG project, thanks to a researrch grant provided by the Canadian Institutes of Health Research and the Natural Science and Engineering Research Council of Canada. We are grateful to all members of the MobliSIG project for their valuable contributions to the discussions that led to this paper.

References

Beale, L., Field, K., Briggs, D., Picton, P., & Matthews, H. (2006). Mapping for wheelchair users: Route navigation in urban spaces. *The Cartographic Journal*, *43*(1), 68–81. https://doi.org/10.1179/000870406X93517.

Church, R., & Marston, J. (2003). Measuring accessibility for people with a disability. *Geographical Analysis*, 1. Retrieved January, from http://onlinelibrary.wiley.com/doi/10.1111/j.1538-4632.2003.tb01102.x/abstract.

Fougeyrollas, P. (1998). The Quebec classification: Disability creation process. *INDCP, International network on the disability creation process.* CSICIDH.

Gharebaghi, A., & Mostafavi, M. A. (2016). A new ontological perspective for integration of social and physical environments: Disability and rehabilitation context. *ISPRS Annals of Photogrammetry, Remote Sensing & Spatial Information Sciences*, 3(2).

Gibson, J. J. (1977). *The theory of affordances*. USA: Hilldale.

Cadkin, J. (2002). Understanding dynamic segmentation. *ArcUser October-December*. Retrieved December, from http://www.srnr.arizona.edu/rnr/rnr420/dynseg_2.pdf.

Jonietz, D., Schuster, W., & Timpf, S. (2013). Modelling the suitability of urban networks for pedestrians: An affordance-based framework. In *Geographic information science at the heart of Europe* (pp. 369–382). Springer.

Jonietz, D., & Timpf, S. (2013). An affordance-based simulation framework for assessing spatial suitability. In *Spatial information theory* (pp. 169–184). Springer.

Karimanzira, D., Otto, P., & Wernstedt, J. (2006). Application of machine learning methods to route planning and navigation for disabled people. In *MIC* (Vol. 6, pp. 366–371).

Karimi, H. A., & Kasemsuppakorn, P. (2013). Pedestrian network map generation approaches and recommendation. *International Journal of Geographical Information Science*, *27*(5), 947–962. https://doi.org/10.1080/13658816.2012.730148.

Kasemsuppakorn, P., & Karimi, H. A. (2009). Personalised routing for wheelchair navigation. *Journal of Location Based Services, 3*. https://doi.org/10.1080/17489720902837936.

Kilkens, O. J., Dallmeijer, A. J., De Witte, L. P., Van Der Woude, L. H., & Post, M. W. (2004). The wheelchair circuit: Construct validity and responsiveness of a test to assess manual wheelchair mobility in persons with spinal cord injury. *Archives of Physical Medicine and Rehabilitation, 85*(3), 424–431. https://doi.org/10.1016/j.apmr.2003.05.006.

Kirby, R. L., Swuste, J., Dupuis, D. J., MacLeod, D. A., & Monroe, R. (2002). The wheelchair skills test: A pilot study of a new outcome measure. *Archives of Physical Medicine and Rehabilitation, 83*(1), 10–18. https://doi.org/10.1053/apmr.2002.26823.

Mamdani, E., & Assilian, S. (1975). An experiment in linguistic synthesis with a fuzzy logic controller. *International Journal of Man-Machine Studies*, 1–13. Retrieved from http://www.sciencedirect.com/science/article/pii/S0020737375800022.

Marston, J. R., & Golledge, R. G. (2002). The hidden demand for participation in activities and travel by persons who are visually impaired. *Journal of Visual Impairment & Blindness, 97*(8), 475–488.

Martin, A., Owende, P., & Holden, N. (2001). Designation of timber extraction routes in a GIS using road maintenance cost data. *Forest Products …* Retrieved from http://cat.inist.fr/?aModele=afficheN&cpsidt=14062443.

Matthews, H., Beale, L., Picton, P., & Briggs, D. (2003). Modelling access with GIS in urban systems (MAGUS): Capturing the experiences of wheelchair users. *Area, 35*, 34–45. doi:10.1111/1475-4762.00108.

Morris, J. M., & Wigan, P. L. D., & Wigam, M. R. (1978). Accessibility indicators for transport planning. *Transportation Research* 13A, 91–109.

Mortenson, W. B., Miller, W. C., & Miller-Pogar, J. (2007). Measuring wheelchair intervention outcomes: Development of the wheelchair outcome measure. *Disability and Rehabilitation: Assistive Technology, 2*(5), 275–285.

Rushton, P. W., Miller, W. C., Lee Kirby, R., Eng, J. J., & Yip, J. (2011). Development and content validation of the Wheelchair Use Confidence Scale: A mixed-methods study. *Disability and Rehabilitation: Assistive Technology*, *6*(1), 57–66. https://doi.org/10.3109/17483107.2010.512970.

Tajgardoon, M., & Karimi, H. A. (2015). Simulating and visualizing sidewalk accessibility for wayfinding of people with disabilities. *International Journal of Cartography*, *1*(1), 79–93. https://doi.org/10.1080/23729333.2015.1055646.

Warren, W. H. (1984). Perceiving affordances: Visual guidance of stair climbing. *Journal of Experimental Psychology. Human Perception and Performance*, *10*(5), 683–703. Retrieved from http://www.ncbi.nlm.nih.gov/pubmed/6238127.

Weigang, Z., & Guiyan, J. (2009). A new extended arc-node data model of dynamic segmentation technology in urban GIS-T. In *WRI World Congress on Software Engineering*, 208–212. https://doi.org/10.1109/WCSE.2009.407.

Zadeh, L. A., Introduction, I., & Navy, U. S. (1965). *Fuzzy Sets, 353,* 338–353.

Accessibility in Pedestrian Routing

Megen Brittell, Christine Grummon, Amy Lobben, Masrudy Omri and Nicholas Perdue

Abstract Pedestrian route planning presents unique challenges compared to relatively well established car routing applications. The challenges are even more pronounced when also considering a diverse user group, which includes people with disabilities. Existing research has identified environmental features that are important to pedestrians with disabilities, but there are still a number of outstanding problems that have hindered the development of a widespread, usable solution. We developed the EUG Access web application (http://geog.uoregon.edu/geocog/eugaccess) to implement an approach to pedestrian routing that incorporates environmental accessibility and responds to the needs and preferences of a diverse user population. This paper reviews related literature and reports our experience through three phases of development: personalization, route calculation, and interface design. It concludes by drawing attention to issues that we propose are critical to successfully incorporating environmental accessibility into future pedestrian routing applications.

Keywords Human centered computing · Accessibility systems · Routing systems · Geographic information systems

M. Brittell · C. Grummon · A. Lobben (✉) · M. Omri
University of Oregon, Eugene, OR, USA
e-mail: lobben@uoregon.edu

M. Brittell
e-mail: megen@uoregon.edu

C. Grummon
e-mail: cgrummon@uoregon.edu

M. Omri
e-mail: masrudyomri@gmail.com

N. Perdue
Humboldt State University, Arcata, CA, USA
e-mail: nicholas.perdue@humboldt.edu

M.P. Peterson (ed.), *Advances in Cartography and GIScience*, Lecture Notes in Geoinformation and Cartography, DOI 10.1007/978-3-319-57336-6_33

1 Introduction

The ubiquity of map-based routing apps, such as Google Maps, has changed how pedestrians find their way around a city. Google Maps reimagined what maps could be. In Nielsen 2015, both Google Maps and Apple Maps ranked in the top 10 most used apps of the year (Nielsen 2015). Their ubiquity is owed to their usefulness. With great ease, most people can use their associated routing capabilities to quickly identify a route between two or more locations. And, the same group of people will usually be able to follow the route without encountering issues of environmental accessibility. The routes do not take into account issues of environmental accessibility. With the exception of being able to request a scenic or toll-free route, the selected routes are predominantly one-size-fits-all.

Some assistive technologies for pedestrian travel have been developed. Notable systems include Trekker Breeze by Humanware (http://store.humanware.com/hus/trekker-breeze-plus-handheld-talking-gps.html) and Sendero Group's (https://www.senderogroup.com/products/GPS/allgps.htm) GPS-enabled navigation systems. But, both of those systems use map data that were originally optimized for car navigation and route calculation is based on length, not accessibility. AXS Map (https://www.axsmap.com), on the other hand, includes a primary feature that allows users to rate accessibility of different locations, including businesses. This information about individual locations is crowdsourced and shared with other users in the AXS Map network. While useful, again, the routes are not systematically calculated based on barriers or facilitators to environmental access.

Travelers with disabilities have a need for routing applications that calculate routes based on accessibility of environmental features rather than on distance alone. Toward that goal, we developed a prototype routing application that optimizes accessibility in route calculation. The remainder of this paper outlines the conceptual and technological approach used to create EUG Access.

2 Related Work

In contrast to the datasets that support vehicles, pedestrian navigation systems require more detailed data sets and new techniques (Karimi and Kasemsuppakorn 2013; Vöklel et al. 2008; Zielstra and Hochmair 2012). While vehicles are restricted to travel along the established road network, pedestrians have more flexibility to travel through open areas and may travel in either direction along one way roads (as noted by Stark et al. 2007). In addition, pedestrian routing may benefit from considering subjective values, e.g., preferences and risk-taking propensity requires individual-level information (Adomavicius and Tuzhilin 2005). Interest in addressing pedestrian routing needs has led to multi-criteria routing algorithms (e.g., Li 2013) and personalized routing (e.g., McGinty and Smyth 2002).

Personalization is not technology—it's a strategy. The use of personalization systems for directing online interaction is very common (Yang and Padmanabhan 2005). The goal is to provide a person what they need without requiring them to ask for it directly—or even know that they need it. The product that results from the personalization systems approach provides an individual with "tailored products, services, information or information relating to products or services" (Mulvenna et al. 2000, p. 124).

Personalization becomes particularly important to meet the needs of a diverse population. Environmental features impact individual navigators in different ways (Adomavicius and Tuzhilin 2005). What may be a positive assistance feature for one person may pose a barrier for someone else. In one related example, almost half of ratings in a bike routing system lacked internal consistency (Priedhorsky et al. 2012). Allowing the end user to set parameter values to the routing algorithm provides one way to respond to individual needs and differences.

Multi-criteria routing contends with potentially competing interests, and can be sensitive to individual differences. One solution is to allow the end-user to complete a profile indicating their preferences (e.g., Priedhorsky et al. 2012). An alternative is to create user groups that use a sub-group average that is more representative of an individual who is a member of a specific sub-group than of the overall population average. For example, survey responses from blind pedestrians highlight a barrier posed by irregular sidewalks and complex [road] crossings (Zeng 2013) and interviews with wheelchair users highlight the barrier of curbs without curb cuts (Bromley et al. 2007; Meyers et al. 2002, cited in Kasemsuppakorn and Karimi 2009).

Computational methods to determine network paths have been implemented and are available in both commercial and open source software. As an example of commercial software, ESRI's Network Analyst provides functionality to encode properties for each line segment within the network. For example, given the gross weight of the vehicle, a simple comparison against bridge weight limits can be used as the criteria to penalize or entirely exclude a potential route. Within the open source realm, igraph (http://igraph.org) provides network analysis tools for R, Python, and C that can be parameterized by placing weights or restrictions on the graph vertices or edges.

To apply these routing algorithms, criteria for establishing weights are required. Researchers have used survey techniques to solicit lists of environmental features that are important to navigation. As an example, researchers have identified that multiple attributes of sidewalks, such as surface material, slope, and curbs (Lobben and Perdue 2016; Neis and Zielstra 2014), are important in route selection. Attribute data that includes such details about environmental features improves the suitability of a dataset for use in accessible routing applications.

While personalization techniques and the identification of critical environmental features begin to address some questions about how to determine values for the algorithm parameters (restrictions or weights), implementations are often plagued by a lack of data. Expanded data sets are required to feed into the multi-criteria routing systems, and lack of data is a commonly cited limitation in pedestrian routing applications (e.g., Karimi and Kasemsuppakorn 2013; Karimi et al. 2013;

Zeng 2013; Vöklel et al. 2008). The lack of data for use as input to multi-criteria systems for pedestrian routing is even more acute for populations with disabilities (Holone et al. 2008).

Responses to the lack of organized municipal or proprietary data sets include manual digitization of remote sensing images (Karimi and Kasemsuppakorn 2013) and contributed data from the public (Neis and Zielstra 2014). Despite its potential to amortize data collection costs across many people, the actual number of input data points may still be relatively small compared to the number of possible path segments (Priedhorsky et al. 2012). VGI for a specialized interest being even more susceptible to sparse data than topics of interest to the "average" end user.

In instances where data is available (or generated by the application developers) and routing algorithms have been parameterized to incorporate user needs and preferences, there remains the issue of presenting those routes to the user. Graphical map-based depictions are common, but are not accessible to people who are blind. Touch interaction on a mobile device with a small screen may not be accessible to people who have limited fine motor control. Interface design that incorporates multiple, possibly redundant input and output methods reaches a broad audience of users, and allows for personalization of the interface (in addition to personalization of the routing algorithm). The application described in this paper does not claim to demonstrate a perfect user interface solution, but serves as an example of interface design that targets a diverse audience.

3 Methods

EUG Access (http://geog.uoregon.edu/geocog/eugaccess) was developed as a web-based application to explore a complete system that implements accessible pedestrian routing. Application development was conducted in three parts: personalization, configuration of a network dataset for route calculation, and interface development.

3.1 Personalization

Personalization is conducted in several phases: data collection, building consumer profiles, matchmaking, and delivery and presentation (Adomavicius and Tuzhilin 2005).

In the EUG Access App, personalization functionality is provided with two options, a pre-set and a custom-set. These options are discussed in the Interface Development section below. In both cases, empirical data gathered through a survey, which was created, deployed, and analyzed in prior research (Lobben and Perdue 2016), were used to construct the consumer profiles. The survey was

Table 1 Example survey questions. 1 = Strongly disagree, 2 = Disagree, 3 = Somewhat disagree, 4 = Neutral, 5 = Somewhat agree, 6 = Agree, 7 = Strongly agree

Statement	Response						
I seek routes that do not have steep hills	1	2	3	4	5	6	7
I avoid traveling on narrow sidewalks	1	2	3	4	5	6	7
I avoid sidewalks along roads with consistent heavy traffic	1	2	3	4	5	6	7
I avoid traveling along routes that are busy during rush hour	1	2	3	4	5	6	7

designed to evaluate the impact that specific environmental features such as curb cuts, sidewalk slope, talking pedestrian crossing controls, and sidewalk texture have on travel choice. People with disabilities were asked to evaluate features along a five-point Likert scale and identify the extent to which features served as barriers or facilitators to environment access. Example questions are included in Table 1.

In the analysis of the surveys, Lobben and Perdue (2016) conducted both a factor analysis and a linear discriminant analysis. The results from the factor analysis identified four distinct groups with different perceptions of environmental features as either facilitators or barriers to route planning and navigation. Those groups include: people who are blind or low vision, people who use power wheel chairs, people who use manual motor assistive devices (i.e. manual wheel chairs or walkers), and people who are Deaf or hard of hearing. The preferences of those groups were then incorporated into pre-set personalization options in the EUG Access routing application.

The linear discriminant analysis from the prior research (Lobben and Perdue 2016) indicated user preference or avoidance (values ranging from positive one to negative one, respectively) for each environmental feature. Those LDA coefficients were used to parameterize restriction attributes in a network data set. The coefficients for each user group was classed into seven equal interval classes. The seven classes represented three degrees of avoidance, a neutral value, and three degrees of preference. The choice to use seven classes reflects the options available to parameterize restrictions in the ESRI Network Analyst tools. Restriction attributes identify the contribution that each feature makes in the calculation of routes. The restriction attributes are assigned a decreased cost reward for preferred features, and an increased cost penalty for features that people try to avoid.

Within the data set, each environmental feature was coded as a binary variable for each line segment. For example, a single line segment either had a steep slope or not. Thresholds used in EUG Access were subjectively determined by the researchers. The cost associated with traversing a given line segment was then based on whether or not the line segment exhibited each environmental feature, and, if so, the respective degree of preference (or avoidance) for that environmental feature.

In the EUG Access application, personalization options provide values for the restrictions and the routing algorithm calculates a least cost route based on the distance, but augmented with high-cost restrictions and low-cost rewards.

3.2 Routing Implementation

The GIS data set was obtained from the Lane Council of Governments, Oregon. While the data set was spatially robust and included useful metadata, it did not include all of the relevant environmental features. Notably missing were crosswalk location, types of pedestrian crossing controls (lighted, audio, stop sign), and curb cuts. As a result, missing data were field collected at each intersection, recording the location and type of features. Each feature was entered as a field within the GIS database as a binary variable. The fields were weighted based on their LDA coefficient value for each user group (blind or low vision, Deaf or hard of hearing, power mobility devices, manual mobility aids, and no disability).

The personalization process configured the degree to which route cost was rewarded (low cost) or penalized (high cost) by environmental attributes. For line segments that were avoided or given preference, the restriction usage was set with a high, medium, or low value for the degree of the preference or avoidance. Using ESRI's Network Analyst tools, line segments were avoided by applying a scalar multiplier to the cost of traversing the segment. Similarly, line segments were given preference by applying a fractional multiplier to the respective cost.

The network was built and published as a network service on ArcGIS for Server 10.3.1. The network service exposes a Representational State Transfer (REST) endpoint through an internet universal resource locator (URL), making it available to any (authorized) web-enabled device (desktop or mobile). Applications send locations between which the lowest cost route is calculated. The request includes a list of restrictions that should be used to determine the route, and the degree to which these restrictions should impact cost. Omitting a restriction from the list ignores the respective environmental variable during route calculation. In other words, this personalized accessible mobile routing allows a user to create a route along transportation networks that are most accessible to them.

3.3 Interface Design

Three screens make up the web-based interface to our network dataset. A route display (Fig. 1) provides input mechanisms through which to specify start and end points, a visual display of the street network, and both visual and text-based displays of the computed route. A pair of customization screens solicit input from the user at two levels: user group (pre-set configurations, Fig. 2) and individual (Fig. 3). Aria markup facilitates screen reader access to all text content and interaction widgets.

The route display is the main screen of the interface and provides functionality to specify the start and end points of a route and presents the results of the routing calculation. Users enter their origin and destination points by either clicking directly on the graphical map representation (touch screen or mouse), or by typing the

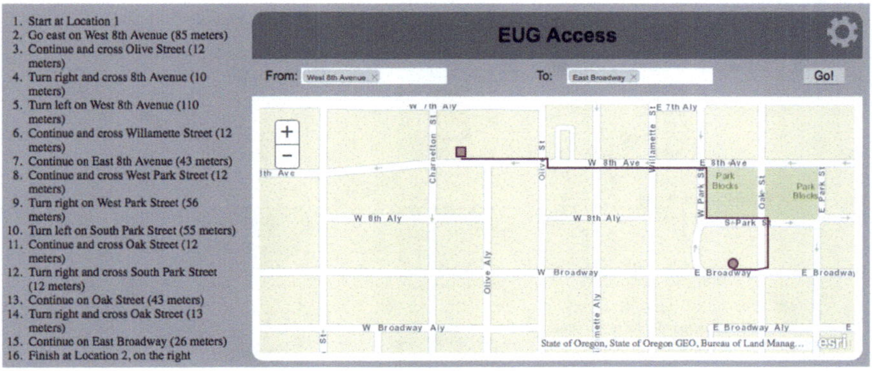

Fig. 1 EUG Access map interface accepts input through touch interaction with the map, and keyboard input to the text fields. The resulting routes are displayed graphically on the map and in a text list that is available to screen readers. In this example, a strong preference for auditory crossing signals and streets with low traffic produced a lowest cost route that is longer than the shortest path between the selected points

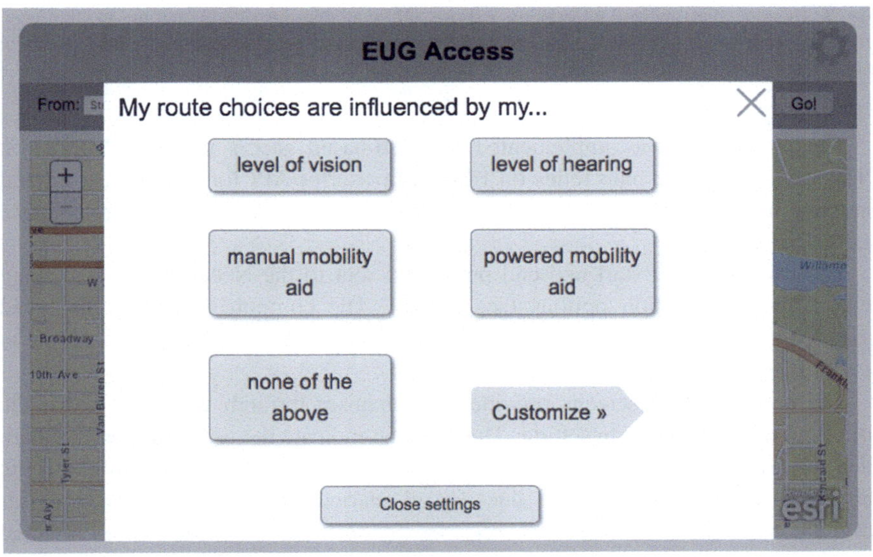

Fig. 2 Pre-set personalization options, by user group

addresses in text fields (keyboard or accessibility device). Click input from the map is snapped to the nearest line segment on the network grid, and street names associated with those lines are displayed in the text fields. For text entry, the EUG Access application performs a pseudo-geocoding operation to convert Eugene street names into locations; the approach could be directly generalized by implementing or connecting to a full geocoding service.

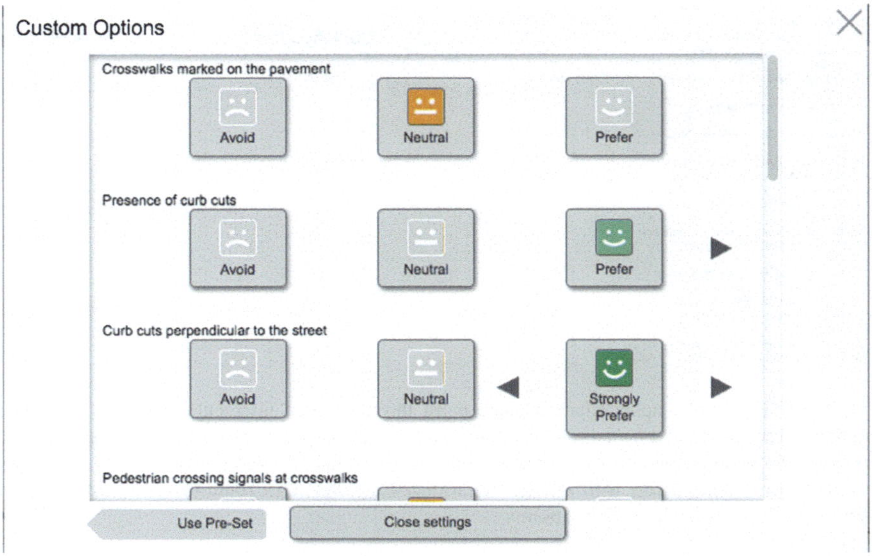

Fig. 3 Custom personalization options expose all available restriction setting to the user, soliciting their input through a series of input widgets: large graphic buttons (touch screen or mouse) and groups of radio buttons (screen reader)

The application was implemented for web-based access using HTML, CSS, PHP, Javascript. The map relies on ESRI's Javascript API for map rendering and interaction with the REST endpoint. The HTML includes ARIA markup to enhance accessibility.

The user specified start and end points are sent to the Network Service, along with any customization options (see below). The computed route is displayed graphically as a line feature on the map, and turn by turn directions are available to screen readers.

The customization screens provide a mechanism through which to match the user with a personalization setting. The user can indicate the user group with which their preferences may align, or they can indicate their preferences for specific environmental features. Pre-set values for all restrictions are available for five user groups). Selecting a user group configures pre-set values for each environmental feature restriction based on the survey results, and those settings can be further customized.

Individual environmental features are listed in the Custom form along with a seven point Likert-style rating system. The seven-level preference corresponds to the restriction usage parameters in Network Analyst. Preferred environmental features (coded as restrictions) receive a low-cost reward by multiplying the standard distance-based cost by a scalar value between zero and one. Environmental features to avoid are penalized by multiplying the standard distance-based cost by a value greater than one.

To minimize complexity of the interface, the preference scale is presented using three large size buttons (visual display), one for each of the three types of response: avoid, neutral, and prefer. However, all levels are still available to the user by pressing the "avoid" or "prefer" buttons multiple times. The full resolution of the seven-level restriction is indicated in both the visual and text-based displays with qualifiers that stepped through low, medium, and high levels.

Interface accessibility was addressed by emphasizing order in the design that reflected functionality, reducing complexity, providing keyboard access, and incorporating markup available to accessibility devices. Accessibility considerations that were implemented in EUG Access include:

- Linear ordering of elements to facilitate access using a screen reader or keyboard shortcuts
- Colorblind safe colors
- Limited number of elements in the graphical display (three buttons, compared to the total of seven options, saves screen space, simplifies initial experience)
- Keyboard access to screen elements
- Aria markup to annotate display elements (e.g., buttons and headings).

4 Challenges and Discussion

As we stepped through the process of implementing the EUG Access application, we encountered a number of non-trivial issues for which we are not aware of existing solutions. Our firsthand experience highlighted the following gaps in available solutions.

- Pedestrian networks, which are distinct from vehicle navigation, have not yet been fully defined.

As noted in the established literature, pedestrians are not cars. They are neither limited to the street network (e.g., potential shortcut across a parking lot or the grass in a park) nor its rules (e.g., one way streets). This difference creates a need for additional data sets. Some of these data may already be available, but distributed across multiple data sources. For example, how might a sidewalk data set be combined with alleyways, a portion of the street network. Still others may not yet exist.

Further, the way that pedestrians choose to move through space is also distinct from movement by drivers. The possibility of free movement across open spaces also creates a need for additional strategies within routing algorithms. And, algorithms that do reasonably well when there is a specific objective and established parameters to optimize, struggle when exploratory or undirected movements are expected. The use case of the "wanderer" (Perdue 2016) is missing from the car navigation applications. How, for example, can the purpose of the trip be incorporated into the suggested routing? An individual person may have different preferences when they are walking to the market for groceries, compared to when they are out for a stroll on a sunny afternoon.

- Data to support pedestrian routing for a diverse population of users is limited.

As approaches to pedestrian (and also more generally non-vehicle) navigation develop, new data sets will be created. In the case of pedestrian navigation, crowd sourced or volunteered geographic information (VGI) solutions represent on approach to collecting new data. This approach, however, makes an assumption of universality among a large number of users. As seen in the literature, however, routing preferences are highly variable between individuals (e.g., Adomavicius and Tuzhilin 2005; Priedhorsky et al. 2012). Effective accessible routing applications will rely on both new ways to use or combine existing data and the creation of additional data sets, for example, the EUG Access personalization interface, which allows users to select preferences for each environmental feature contained in the database to determine the feature role in the calculated route.

For the EUG Access implementation, the municipal data set included sidewalks along the street, but that data set was not readily available to the public. We manually added several footpaths. But, we excluded paths through park areas and alleyways. Paths through park areas are difficult to determine. The parks are effectively open space without prescriptive lanes of travel. The alleyways present potentially less clean or less well-lit paths, which could have safety implications. Excluding these options, however, restrict knowledge of alternative paths that may be available to the general population.

- Representation of the critical attributes of environmental features requires both values and relationships.

Addressing the needs of a diverse pedestrian population is more complicated that simply augmenting a dataset with additional fields. Despite a number of efforts to bolster datasets with municipal data and VGI, there is not yet a standard approach to representing data is a way that supports its efficient use. Within OpenStreetMap (http://wiki.openstreetmap.org/wiki/Disabilities), a set of tags have been established to capture information about environmental accessibility. But, tags related to accessibility and disability are optional and rely on application developers to determine how to incorporate them into their designs. Similarly, ESRI's Network Analyst provides a mechanism to set restrictions, but again leaves it up to the developer to choose features and establish relative weights.

Through our experience we also noted, unsurprisingly, the relationships between two pairs of environmental features. Curb cuts and pedestrian crossing controls are two examples of features in the pedestrian network that have inherent structure. The presence of curb cuts (and pedestrian crossing controls) represent a first level feature. Within that feature, there are critical characteristics: direction of curb cut and audible crossing signal, respectively. The survey data solicited preferences for

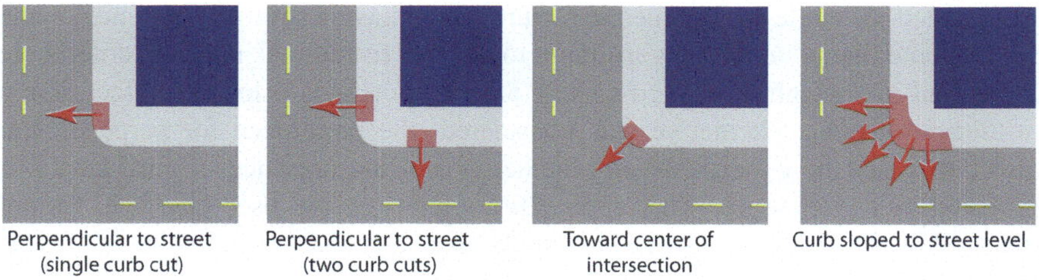

Fig. 4 Sidewalks in downtown Eugene, Oregon, are punctuated with multiple configurations of curb cuts. In our data set, curb cuts were annotated as being either perpendicular to the street (*left*) or at other angles (*right*)

each feature separately, but did not directly query the connection between these paired features.

Taking one example from EUG Access, we augmented our dataset with information about curb cuts. For each intersection in the street network we recorded the presence (binary: present or absent) and direction of curb cuts (binary: perpendicular to the street or at any other angle relative to the street). However, these two seemingly simple binary variables are deceptively complex. In the relatively small area of downtown Eugene, Oregon, where we collected data we found many possible combinations of curb cut arrangements (Fig. 4).

As we tried to translate these into restrictions (implemented in Network Analyst), two challenges arose. First, the assignment of the curb cut attribute to either a line segment or node of the network (sidewalks and crosswalks were represented as line segments that met as a single point on each street corner) did not adequately capture the experience of traversing the intersection. Assigning the curb cut attribute to the sidewalk line segment unnecessarily weighted (either positively or negatively) a route that turned the corner, bypassing the curb cut. Adding the attribute to the crosswalk line segment, however limited the influence of the restriction to those routes that actually involved a transition from the sidewalk to the street level. To add the curb cut attribute to the crosswalk line segment, however, required two pieces of data: one for each end of the crosswalk (curb cuts on opposite sides of the street were not always symmetric). Ultimately, we simplified the curb cut data by creating a single attribute that reflected the least preferred configuration, effectively rounding down. If either end of the crosswalk lacked a curb cut, the line segment was annotated as "no curb cut". Only if both ends of the crosswalk had curb cuts perpendicular to the street was the line segment assigned a "perpendicular curb cut" attribute.

A second challenge was encountered when establishing weights for the restrictions. The survey data provided a quantitative metric, but when the routing process was opaque and preferences tended to cancel each other out or to redundantly contribute to the computed costs, and overwhelming other preferences. By including both a moderate preference for the presence of curb cuts and a moderate preference for directional curb cuts, the influence of the curb cut would be exaggerated. A single environmental feature would have a stronger influence on the overall route than a very strong aversion to another feature. As multiple criteria are considered in a weighted sum cost, there is a real potential to lose the effect of any customization. One alternative was considered: the preference for perpendicular curb cuts could be conditional on whether or not the presence of curb cuts was indicated as a preference. This approach complicated the user interface, further obscured the route calculate process from the user, and had a minimal impact on the results (within our largely homogeneous data set).

- Binary criteria for whether or not a line segment should be characterized by a restriction substantially simplified the available data.

Although "accessibility" is often reduced to a binary metric (does a space meet a minimum objective threshold), accessibility is in practice much more nuanced. In our Network Analyst implementation, however, the simplification to a binary each line segment was assigned a binary attribute for each environmental feature (e.g., whether or not the slope was "steep").

- Interface to routing algorithm (accessible interfaces)—response: keyboard interaction and text-based route display
- The graphical depiction of the calculated route conveys more information than the text-based display.

The server responds to a route request with both a spatial feature to render on the graphical map and a list of turn-by-turn directions. Although both versions describe the least cost route, the graphical depiction also provides information about surrounding landmarks and overview of the route shape.

5 Conclusions

Incorporating environmental accessibility into pedestrian routing requires explicit consideration of accessibility at all stages of design from selection of data, which will be available in the system, to presentation of routes through a digital interface. There are a number of ways in which small changes can lead to substantial improvements to usefulness and usability of routing applications. Some of those changes involve implementation of known techniques, and some remain open questions for future research.

Existing literature has identified environmental features that are critical to navigation. Inclusion of details related to environmental accessibility (e.g., presence, direction, and grade of curb cuts) in data sets through the course of normal street maintenance and upgrades and as part of crowd sourced or VGI efforts will provide rich detail to support pedestrian routing. Augmenting established data sets with details of these environmental features will create a strong foundation.

New strategies for incorporating such detail into routing algorithms will be necessary to effectively serve a diverse population of pedestrians. Outstanding questions include the appropriate representation of accessibility-related environmental features within geospatial data sets and strategies to incorporate those environmental features into the calculation of qualitatively optimal routes.

The design of interfaces that deliver routing details can immediately be improved by incorporating existing tools, but also require new techniques to improve accessibility to geospatial information. In the context of web-based interfaces, standards, such as Accessible Rich Internet Applications, ARIA, (https://www.w3.org/TR/wai-aria), have been developed to increase usability. The ARIA standard, as an example, handles text data and interface navigation, but additional work is necessary to improve accessibility of geospatial map displays.

In addition to these general concepts, future work with the EUG Access application could incorporate additional landmark features and provide information about accessibility of places to which the routes guide people (e.g., BluePath (https://blue-path.org/) data, or wheelchair accessibility as demonstrated in WheelMap (http://wheelmap.org)). The implementation of a network dataset that provides accessibility-based weighting can also be automated to provide a measure of relative accessibility within the pedestrian network for various user groups.

Acknowledgements This work undertaken as part of the graduate student seminar Accessible Maps, and we acknowledge and are grateful for the contribution of work by our peers and colleagues who designed and collected survey data as part of a Research Methods course. We would like to also thank Jacob Bartruff for his assistance and guidance through the design and implementation of the network.

References

Adomavicius, G., & Tuzhilin, A. (2005). Personalization technologies: A process-oriented perspective. *Communications of the ACM*, *48*(10), ACM: USA, 83–90. doi:10.1145/1089107.1089109.

Bromley, R., Matthews, D., & Thomas, C. (2007). City centre accessibility for wheelchair users: The consumer perspective and the planning implications'. *Cities, 24,* 229–241.

Golledge, R. G. (2008, July). Behavioral geography and the theoretical/quantitative revolution. *Geographical Analysis, 40*(3), 239–257. doi:10.1111/j.1538-4632.2008.00724.x

Holone, H., Misund, G., Tolsby, H., & Kristoffersen, S. (2008). Aspects of personal navigation with collaborative user feedback. In *Proceedings of the 5th Nordic Conference on Human-computer Interaction: Building Bridges (Lund, Sweden, October 18–22, 2008) NordiCHI '08.* (pp. 182–191). New York, NY, USA: ACM. doi:10.1145/1463160.1463180.

Karimi, H. A., & Kasemsuppakorn, P. (2013). Pedestrian network map generation approaches and recommendation. *International Journal of Geographical Information Science, 27*(5), 947–962. doi:10.1080/13658816.2012.730148.

Karimi, H. A., Zhang, L., & Benner, J. G. (2013). Personalized accessibility maps (PAMs) for communities with special needs. In S. Liang, X. Wang, & C. Claramunt (Eds.), *Proceedings of the 12th International Conference on Web and Wireless Geographical Information Systems (Banff, Alberta, Canada, April 4–5, 2013) W2GIS '13* (pp. 199–213). Springer: Heidelberg. doi:10.1007/978-3-642-37087-8_15.

Kasemsuppakorn, P., & Karimi, H. A. (2009). Personalised routing for wheelchair navigation. *Journal of Location Based Services, 3*(1), 24–54.

Li, R. (2013). An adaptive compromise programming method for multi-objective path optimization. *Journal of Geographical Systems, 15*(2), 211–228. doi:10.1007/s10109-012-0172-1.

Lobben, A. K., & Perdue, A. K. (2016). *Personalized environmental spaces*. Manuscript in Preparation.

McGinty, L., & Smyth, B. (2002). Shared experience in personalized route planning. In S. M. Haller & G. Simmons (Eds.), *Proceedings of the 15th International Florida Artificial Intelligence Research Society Conference (Pensacola Beach, Florida, USA, May 14-16, 2002) FLAIRS '02* (pp. 111–115). USA: Association for the Advancement of Artificial Intelligence (AAAI) Press, Menlo Park.

Meyers, A. R., Anderson, J. J., Miller, D. R., Shipp, K., & Hoenig, H. (2002). Barriers, facilitators, and access for wheelchair users: Substantive and methodologic lessons from a pilot study of environmental effects. *Social Science Medicine, 55*(8), 1435–1446.

Mulvenna, M. D., Anand, S. S., & Büchner, A. G. (2000). Personalization on the net using web mining. *Communications of the ACM, 43*(8), 122–125. New York, NY, USA: ACM. doi:10.1145/345124.345165.

Neis, P., & Zielstra, D. (2014). Generation of a tailored routing network for disabled people based on collaboratively collected geodata. *Applied Geography*, 47, 70–77. doi:10.1016/j.apgeog.2013.12.004.

Nielsen, J. (2015). Tops of 2015: *Digital*. http://www.nielsen.com/us/en/insights/news/2015/tops-of-2015-digital.html.

Perdue, N. (2016). Cognitive agents and pedestrian redevelopment. *Ph. D. Thesis*. University of Oregon.

Priedhorsky, R., Pitchford, D., Sen, S., & Terveen, L. (2012). Recommending routes in the context of bicycling: Algorithms, evaluation, and the value of personalization. In *Proceedings of the ACM 2012 Conference on Computer Supported Cooperative Work, (Seattle, Washington, February 11–15, 2012) CSCW '12.*(pp. 979–988). New York, NY, USA: ACM. doi:10.1145/2145204.2145350.

Stark, A., Riebeck, M., & Kawalek, J. (2007). How to design an advanced pedestrian navigation system: Field trial results. In *4th IEEE Workshop on Intelligent Data Acquisition and Advanced Computing Systems: Technology and Applications (Dortmund, Germany, September 6–8, 2007) IDAACS 2007* (pp. 690–694). New York, NY, USA: IEEE. doi:10.1109/IDAACS.2007.4488511.

Völkel, T., Kühn, R., & Weber, G. (2008). Mobility impaired pedestrians are not cars: Requirements for the annotation of geographical data. In K. Miesenberger, J. Klaus, W. Zagler & A. Karshmer (Eds.) *Proceedings of the 11th International Conference on Computers Helping People with Special Needs (Linz, Austria, July 9–11, 2008) ICCHP 2008* (1085–1092). Springer: Heidelberg. doi:10.1007/978-3-540-70540-6_163.

Yang, Y., & Padmanabhan, B. (2005, May). Evaluation of online personalization systems: A survey of evaluation schemes and a knowledge-based approach. In *Journal of Electronic Commerce Research 6*(2), 112–122.

Zeng, L. (2013). Acquisition of spatial environmental information from tactile displays. Ph.D. Thesis, Technische Universität Dresden.

Zielstra, D., & Hochmair, H. H. (2012). Using free and proprietary data to compare shortest-path lengths for effective pedestrian routing in street networks. *Transportation Research Record: Journal of the Transportation Research Board, 2299,* 41–47.

Visualization of Traffic Bottlenecks: Combining Traffic Congestion with Complicated Crossings

Andreas Keler, Jukka M. Krisp and Linfang Ding

Abstract Daily mobility patterns in highly populated urban environments rely on a well-functioning effective road network. Nevertheless, traffic bottlenecks are typical for urban environments with periodic traffic congestion. In this paper, we focus on the investigation of how traffic congestion is related with complicated crossings. First, we select an approach for the classification of the complexity of road partitions and the derivation of complicated crossings based on geodata from OpenStreetMap (OSM). Second, we calculate traffic congestions using Floating Taxi Data (FTD) from Shanghai in 2007. Then, we develop a matching technique to link the congestion and complicated crossings, and subsequently define the concept of traffic bottlenecks represented by polygons. The bottlenecks indicate locations where the transportation infrastructure is complex and traffic congestion appears periodically. Finally, we select suitable cartographic representations of traffic bottlenecks in potential thematic vehicle traffic maps.

Keywords Floating taxi data (FTD) · Volunteered geographic information (VGI) · Complicated crossings · Traffic bottleneck · Traffic congestion · Transportation infrastructure · Traffic maps

---

A. Keler (✉) · J.M. Krisp · L. Ding
Institute of Geography, Applied Geoinformatics, University of Augsburg, Augsburg, Bavaria, Germany
e-mail: andreas.keler@geo.uni-augsburg.de

J.M. Krisp
e-mail: jukka.krisp@geo.uni-augsburg.de

L. Ding
e-mail: linfang.ding@geo.uni-augsburg.de

L. Ding
Chair of Cartography, Technical University of Munich, Munich, Bavaria, Germany

M.P. Peterson (ed.), *Advances in Cartography and GIScience*, Lecture Notes in Geoinformation and Cartography, DOI 10.1007/978-3-319-57336-6_34

1 Introduction

The perception of a complicated crossing often includes associations with difficult situations for drivers. Sometimes, this is combined with the assumption of more frequent traffic accidents and congestion. In densely populated cities, there is a strong association between complicated crossings and vehicle traffic congestion. The question addressed here is whether perceptions of complicated crossings are actually influenced by traffic congestion.

In this paper, we use Floating Taxi Data (FTD) to derive traffic congestion and OpenStreetMap (OSM) to detect complicated crossings. Afterwards, we develop a matching technique to link complicated crossings and traffic congestion for selected times of day. The resulting traffic bottlenecks represent a starting point for examining its cartographic representation.

2 State of the Art

There is a wide spectrum of related research on the analysis of moving objects. In this paper, we focus on Floating Car Data (FCD) and its analysis and applications. We then review methods for classifying traffic congestion based on movement data. Following this, we describe the complexity of the urban transportation infrastructure and clarify the possibilities for how to visually represent vehicle traffic congestion.

2.1 Vehicle Tracking: The Floating Car Data (FCD) Method

FCD is a relatively new technology for collecting data from numerous vehicles equipped with GNSS devices. Based on FCD datasets, it is possible to infer road elements (Li et al. 2015a), examine commuting patterns (Dewulf et al. 2015), model typical periodical traffic flow patterns (Körner 2011), and answer emission-related questions (Gühnemann et al. 2004). Similar to Duan et al. (2009), we intend to derive traffic congestions from FCD and detect those areas that are periodically influenced by traffic congestion.

2.2 Traffic Congestion Detection

According to Li et al. (2015b), traffic flow and traffic congestion with highways is well-studied. Examining traffic for an entire city has proven to be much more difficult. This scale of the network is essential to connect local flows with the global flow in a city.

Yuan et al. (2014) argue that traffic congestion can be observed with mobile sensors, static sensors, loop detectors and video cameras. However, usually these are only sparsely installed. This makes it difficult to infer detailed information on traffic and daily changes in its dynamics.

A computational method for detecting traffic congestion might include the detection of stops and movement trajectory segments (Kang et al. 2004). The principles of inferring this information can be manifold, since clustering (Rinzivillo et al. 2008; Andrienko et al. 2015), regression (Haworth et al. 2014) and classification (Wen et al. 2014) might deliver comparable results for movement trajectories that have not been preprocessed.

More beneficial, especially when aiming to connect entity movement with its context, is the inclusion of additional properties. For Brakatsoulas et al. (2004), additions include relations between trajectories and their environment and among themselves. The former is important for generating semantic trajectories (Yan et al. 2013) and for this reason Brakatsoulas et al. (2004) define five basic relations (stay within, bypass, leave, enter and cross). The latter is important for inferring group dynamics as flocks, convoys and stay points (or sometimes meeting points or locations).

Traffic congestion appears usually on roads. Therefore, Long et al. (2008) make use of the connectivity of road links and detect congestion propagation patterns. Often, it is difficult to distinguish between traffic light stop patterns of taxis (Protschky et al. 2015), stop-and go traffic (Ranacher et al. 2016) and real congestion, where vehicles are still standing. Besides mean travel times for road links for specific times of the day (Sohr et al. 2010), we can detect typical and abnormal traffic (Lan et al. 2014; Long et al. 2008) based on historical and recent movement data. This differentiation is based on statistics of instantaneous and average velocities and, resulting from the previous, travel times and congestion durations (traffic delays).

Other approaches are focused on using clustering techniques. For the case of massive movement data generated by vehicles, Rinzivillo et al. (2008) are using OPTICS (Ankerst et al. 1999) for detecting traffic congestion of vehicles, which are characterized by their higher vehicle densities. The individual vehicle trajectories are previously aggregated into general flows between bigger areas or whole road segments. Research has also been done on clustering near real-time movement data as, for example, by Costa et al. (2014). Liu and Ban (2013) are using hierarchical clustering for selecting low-speed and stop points of FTD in Wuhan.

2.3 Complexity of Urban Transportation Infrastructure

For Barthélemy (2011), spatial transportation and mobility networks are examples where the spatial component is relevant and important in addition to the topology information. Besides the representations of road lanes, OSM also has information on different road types including highways, minor roads and connection roads. The latter might indicate selected transportation infrastructure elements such as on-ramps and acceleration lanes. A missing piece of information in OSM is the capacity of road segments although it can be estimated based on road type and number of lanes. Locations with lower capacities might correlate with potential traffic bottlenecks.

We use the approach of Krisp and Keler (2015), where the complexity of road networks is based more on assumed human perception. The latter is based on simple extractions from the OSM geodatabase: extracted nodes of road polylines are used for point density estimation. Nodes from different road types are weighted more highly since the existence of different road types at one location is assumed as being confusing for a driver.

Additionally, we attempt to observe complicated crossings in connection with structural and dynamic bottlenecks. Structural bottlenecks are defined by Li et al. (2015b) as those bottleneck links that can be found by traditional network analysis. This means a structural bottleneck is described as a road segment, a lane of a road segment or as an intersection. The latter might then be a node or, in some cases, a hub that affects many different connection roads. In contrast to this, the focus of this work is to define dynamic bottlenecks that might describe a changing location.

2.4 *Traffic Data Visualization*

Visualization methods for traffic have been used since the very beginnings of traffic and transportation engineering. The connection to cartographic methods is often weak. Neglecting the spatial component in traffic is an unfortunate fact that results from the particular historical development of traffic engineering.

Traffic data sets are multifaceted and spatio-temporal (Chen et al. 2015). As a result, traffic data visualization is not merely scientific visualization but involves aspects of information visualization and visual analytics (Chen et al. 2015). Pioneers in this field are Gennady and Natalia Andrienko, who first introduced useful and self-explaining geo-visual analytical methods for movement data (Andrienko and Andrienko 2007, 2011, 2013). One base of Andrienko's work includes the aggregation of movement in time and space (Andrienko and Andrienko 2011; Andrienko et al. 2015, 2016).

Other methods for handling massive movement data in the form of time-stamped points are clustering techniques. One frequently used method for this type of data is density-based clustering with DBSCAN and OPTICS (Ankerst et al. 1999).

Similar to Rinzivillo et al. (2008) we want to classify congested regions based on inferred density-connected clusters. The aim of our approach is to find density-connected points of preselected movement positions with lower instantaneous velocities.

3 Test Data for Applying the Cartographic Traffic Bottleneck Visualization Method

In this work, we examine the vehicle traffic dynamics and the transportation infrastructure of Shanghai, China. This involves the use of Floating Car Data (FCD) of more than 7000 taxis in Shanghai to detect areas of vehicle traffic congestion. We make use of time-based extracts from Floating Taxi Data (FTD) of an anonymous taxi fleet in Shanghai, and OpenStreetMap (OSM) road network data within the administrative boundaries of Shanghai.

3.1 Shanghai Floating Taxi Data from 2007

The FTD from Shanghai was acquired between February and March 2007. It is the base of the findings in Keler and Krisp (2016), where periodical variations in travel time were detected between 13 prominent crossings in Shanghai. There are a total of around 10,000 different taxi identifications with an average of 7120 frequently observed vehicles. The data has 10 original attributes of which we only use car ID, longitude, latitude, time and instantaneous velocity. The temporal sampling interval differs and varies between 1 and 30 s. The average is around 12 s for each inspected hour of the data set.

In preprocessing, the data is partitioned based on time. We define *time windows* that will represent the temporal resolution of our results.

3.2 OSM Road Network of Shanghai

OpenStreetMap (OSM) was founded in 2004 by Steve Coast as a means to create an open geodatabase of the world by volunteered mappers. The common term in the literature for this type of data is Volunteered Geographic Information (VGI). For Stanica et al. (2013) OSM road network features are one of the most accurate data within available digitized road networks making vehicle routing possible. Nevertheless, computing reasonable routes is often challenging because of geodata issues such as the missing connectivity of digitized road segments. Since we focus on extracted nodes from road segment polylines, we will not introduce any procedure for testing reasonable connectivity. In this work, we extract any possible road types that are accessible via the vehicle. The different road types include highways, major and minor arterial roads, and connection roads between these. Additionally, we include any type of local roads, together with cycle ways and pedestrian paths.

4 Method for Detecting and Visualizing Vehicle Traffic Bottlenecks

Our framework consists of two parallel processes for calculating (1) complicated crossings from OSM data, and (2) traffic congestion from FTD. These result in a following-up process for detecting traffic bottleneck locations within the area. Figure 1 pictures the workflow of our approach.

The blue colored background in Fig. 1 summarizes the method of Krisp and Keler (2015), whereas the beige background shows the elements of detecting FTD-based traffic congestion. Our results, detailed below, are based on the matching and intersection of the outcomes of the two methods.

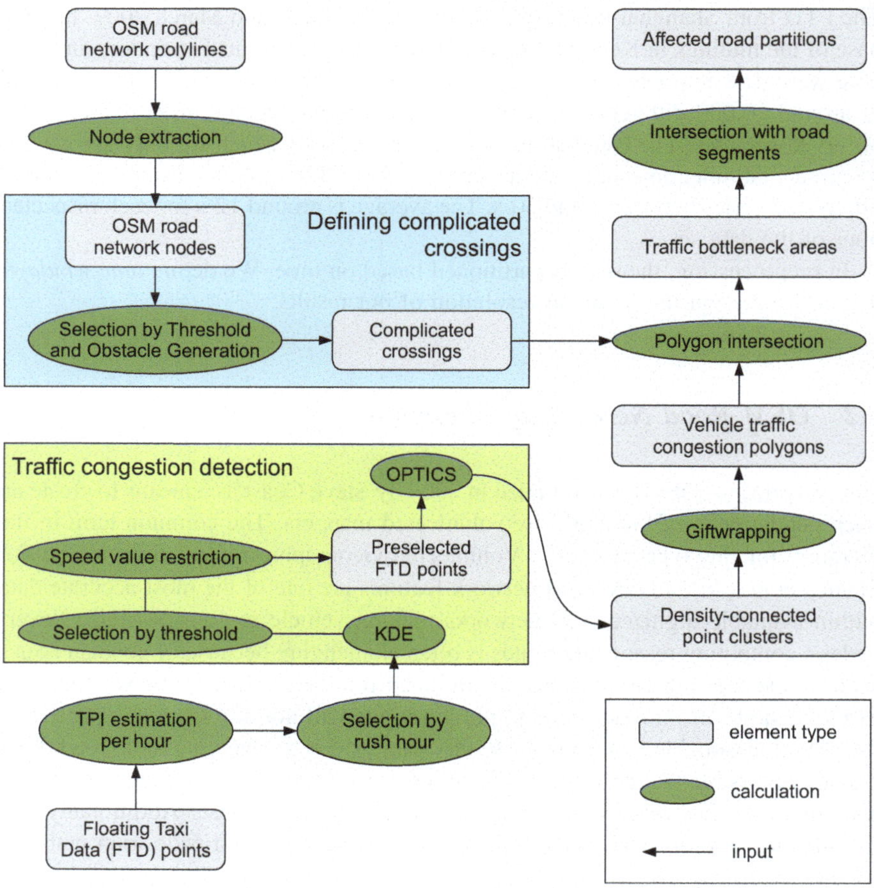

Fig. 1 Workflow of the approach for defining complicated crossings (*blue*), detecting traffic congestion (*yellow*) and defining traffic bottleneck areas

4.1 Detection and Classification of Complicated Crossings

The derivation of complicated crossings within the road network is very difficult since no survey of locals' opinions is available. In this paper, Volunteered Geographic Information (VGI) of the vehicle road network from the OSM project is used to classify complicated crossings in Shanghai.

In Krisp and Keler (2015), complicated crossings in Munich were defined for inexperienced drivers. We test the same method on the road network of Shanghai using the average distance of roundabouts (in this case, 60 m) for selecting a threshold in the creation of obstacle polygons. These obstacle polygons detected complicated crossings and are based on two-dimensional Kernel Density Estimation of street nodes. After inspecting the transportation infrastructure in Shanghai, we keep the search radius of 60 m.

For the road network of Shanghai, which consists of 83,797 polylines, we extract 777,377 nodes. These nodes are used for detecting complicated crossings. From these nodes, we calculate 3,080 complicated crossings in Shanghai.

4.2 Computation of Traffic Congestion and Bottlenecks Based on Floating Taxi Data

Our idea for identifying FTD-based bottlenecks uses rush hour data from preselected times and lower vehicle speeds. It also uses density-based clustering that results in areas of congestion and their influence on selected road segments. This means that we do not apply previous matching of traffic information on road segments. Our premise is that road segments are not spatially accurate enough to represent traffic congestion. Therefore, our approach can be regarded as an alternative method for detecting and representing vehicle traffic congestion.

The first step in computing in location of traffic congestion and bottlenecks is the introduction of a global congestion value for the whole network for one day of FTD. Using Keler et al. (2016), we define a congestion value c, which we classify into 10 classes of equal range:

$$c = k/v; \text{with vehicle density } k\,[\text{vehicles/km}] \text{ and average velocity } v\,[\text{km/h}]$$

By using this index, influenced by the traffic performance index (TPI) as explained in Wen et al. (2014), we can detect rush hour for selected working days within our investigation area. We select data partitions of morning and afternoon rush hours on working days. Afterwards, we segment these partitions into smaller subsets of 10-min time windows.

The subsequent step consists of a Kernel Density Estimation (KDE) of FTD points. We make use of the quartic kernel as in the method for detecting complicated crossings. The following step is to extract kernel densities above a specified threshold.

The selection of the density threshold for detecting traffic congestion should be performed after previous observation. This step comes after the preselection based on the global index, the subdivision into 10-min steps, and its visual inspection within a GIS application. The latter can help the potential analyst to select optimal thresholds.

After its selection, we apply the thresholds for extracting dense FTD point concentrations. Since the results of KDEs are continuous fields, we define polygons with characteristic rounded edges and extract only the records that are inside these areas. One additional restriction for extracting FTD points is the introduction of a threshold for the instantaneous velocity attribute. Here we define 20 km/h as the maximum speed for an ongoing traffic congestion event. This is motivated by a previous study of Robinson (1984) who states that vehicle emissions of hydrocarbons and carbon monoxide are especially high in speeds under 20 km/h.

In a subsequent step, we apply density-based clustering: OPTICS algorithm (Ankerst et al. 1999). Our input parameters for creating density-based clusters are the following: for MinPts we are using 3 points and the search distance Eps is 50 m in Euclidean space. Derived traffic congestion clusters are subsequently converted into convex hulls by using the gift-wrapping algorithm (Jarvis 1973).

4.3 Cartographic Representation of Vehicle Traffic Bottlenecks

The detected traffic congestion polygons and complicated crossings are the inputs for detecting traffic bottlenecks. In our work, we define vehicle traffic bottlenecks as those complicated crossings affected by traffic congestion. The matching results deliver one starting point for further inspection, mainly dependent on the distributions of detected congestion events and complicated crossings. Since there are individual results of bottlenecks for each time window, we conceive a visual representation of its spatio-temporal variations.

For avoiding visual overload, we introduce 2-D representations with polygons of 50% opacity. The second option is the 2.5D-view with extruded bottleneck polygons. The extrusion is based on FTD point density of selected 10-min partitions.

5 Results

A high number of complicated crossings are detected, mainly resulting from multiple intersections of different road segments. A cutout of detected complicated crossings is presented in Fig. 2b. The map needs some explanation from the literature: 30% of Shanghai's overall traffic load is from its elevated road network (Li and Zuo 2004; Zhang 2004; Xu et al. 2012). This makes it difficult to identify

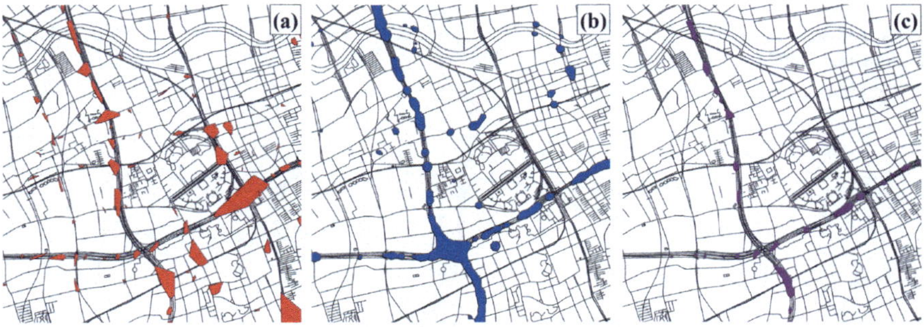

Fig. 2 Congested areas (**a**), complicated crossings (**b**) and detected bottlenecks (**c**) in Shanghai

elevation levels of vehicle trajectories (Xu et al. 2012). For deriving the traffic states, it is difficult to determine if traffic congestion is on or below the elevated highway segment.

As an illustration, we will focus on selected 10-min partitions of the overall FTD set. Detected congested areas of 10 min in the morning rush hours of one selected working day (Monday the 12th of February 2007, from 8:30 to 8:40 AM) are pictured in Fig. 2a.

By matching the spatio-temporal varying congestion polygons with detected complicated crossings, we infer places of potential bottlenecks for the same time windows as the traffic congestion polygons. The result of detected bottlenecks is pictured in Fig. 2c.

Figure 2a includes relatively angular polygons due to the gift-wrapping algorithm. Figure 2b has more rounded polygons that mainly appear at more complex intersections. The matched areas between these two mentioned polygon types are traffic bottleneck polygons on parts of the underlying road network (Fig. 2c). This shows that no previous Map Matching (MM) technique is needed for connecting traffic information with road segments.

We also inspect the variation of the traffic bottlenecks in time and space. Figure 3a shows the detected bottleneck polygons for the three time windows between 8:30 and 9:00 AM (19th of February 2007) within a cartographic view. We can see 50% opacity of the colors that serves for showing the polygon boundaries of intersecting polygons.

The second option for cartographic representation is the extrusion of the polygons within a 2.5D view. By means of representing bottleneck propagation between consecutive time windows, we enrich each polygon by the number of FTD points with instantaneous speed lower than 20 km/h.

The next step is to extrude congestion polygons based on the FTD point density of selected 10 min time periods as pictured in Fig. 3b. This shows that there are high variations in vehicle densities of the detected traffic bottlenecks. The bottlenecks in the eastern part of Shanghai in Fig. 3b show a decrease in vehicle density, whereas in the western part an increase in vehicle density.

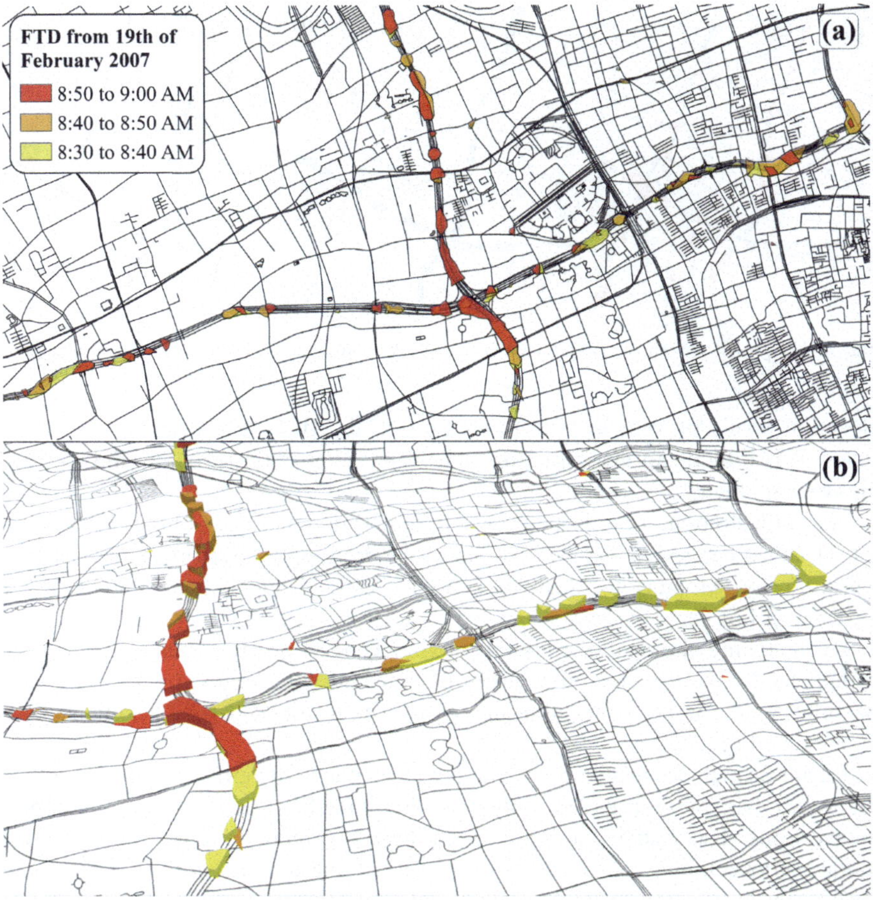

Fig. 3 Cartographic visualization possibilities for three consecutive traffic bottleneck polygons within (**a**) two-dimensional view and (**b**) extruded three-dimensional view

6 Conclusion

In this paper, we propose techniques to detect and visualize traffic bottlenecks. This helps planners in organizing traffic or designing additional elements for urban environments. Additionally, our results might be included in car navigation services, especially the cartographic representation of road networks. As an additional layer for potential car navigation systems, identified bottlenecks might inform users about possible locations of frequent traffic congestions.

The practical use of the presented results might also benefit the understanding of traffic congestion occurrence and propagation. The latter is the best way to evaluate our results, since our method is heavily data dependent and scalable. By scalable, we mean that we are able to set a threshold for distinguishing between slow moving traffic and actual traffic congestion, which might be perceived differently. In general, we can say that the techniques presented here are computationally efficient and easy to perform. This is beneficial for potential analysts since less time and effort is invested into data preprocessing.

Acknowledgements The described taxi Floating Car Data set of Shanghai ('SUVnet-Trace Data,' http://wirelesslab.sjtu.edu.cn/taxi_trace_data.html) was obtained from the Wireless and Sensor networks Lab (WnSN) at Shanghai Jiao Tong University. We would like to thank the Laboratory for Wireless and Sensor Networks at Shanghai Jiao Tong University, especially Prof. Min-You Wu and Jia Peng, for providing access to this data.

References

Andrienko, N., & Andrienko, G. (2007). Designing visual analytics methods for massive collections of movement data. *Cartographica, 42*(2), 117–138.

Andrienko, N., & Andrienko, G. (2011). Spatial generalization and aggregation of massive movement data. *IEEE Transactions on Visualization and Computer Graphics, 17*(2), 205–219.

Andrienko, N., & Andrienko, G. (2013). Visual analytics of movement: An overview of methods, tools, and procedures. *Information Visualization, 12*(1), 3–24.

Andrienko, N., Andrienko, G., & Rinzivillo, S. (2015). Exploiting spatial abstraction in predictive analytics of vehicle traffic. *ISPRS International Journal of Geo-Information, 4*(2), 591–606.

Andrienko, N., Andrienko, G., & Rinzivillo, S. (2016). Leveraging spatial abstraction in traffic analysis and forecasting with visual analytics. *Information Systems, 57,* 172–194.

Ankerst, M., Breunig, M. M., Kriegel, H.-P., & Sander, J. (1999). OPTICS: Ordering points to identify the clustering structure. In *ACM SIGMOD International Conference on Management of data* (pp. 49–60). ACM Press.

Barthélemy, M. (2011). Spatial networks. *Physics Reports, 499*(1–3), 1–101.

Brakatsoulas, S., Pfoser, D., & Tryfona, N. (2004). Modeling, storing and mining moving object databases. *Proceedings of the International Database Engineering and Applications Symposium 2004 (IDEAS '04)* (pp. 68–77).

Chen, W., Guo, F., & Wang, F. Y. (2015). A survey of traffic data visualization. *IEEE Transactions on Intelligent Transportation Systems, 16*(6), 2970–2984.

Costa, G., Manco, G., & Masciari, E. (2014). Dealing with trajectory streams by clustering and mathematical transforms. *Journal of Intelligent Information Systems, 42,* 155–177.

Dewulf, B., Neutens, T., Vanlommel, M., Logghe, S., De Maeyer, P., Witlox, F., et al. (2015). Examining commuting patterns using floating car data and circular statistics: Exploring the use of new methods and visualizations to study travel times. *Journal of Transport Geography, 48,* 41–51.

Duan, Z., Liu, L., & Sun, W. (2009). Traffic congestion analysis of Shanghai road network based on floating car data. *International Conference on Transportation Engineering 2009* (pp. 2731–2736).

Gühnemann, A., Schäfer, R.-P., Thiessenhusen, K.-U., & Wagner, P. (2004). monitoring traffic and emissions by floating car data. *WORKING PAPER* ITS-WP-04-07.

Haworth, J., Shawe-Taylor, J., Cheng, T., & Wang, J. (2014). Local online kernel ridge regression for forecasting of urban travel times. *Transportation Research Part C: Emerging Technologies, 46,* 151–178.

Jarvis, R. A. (1973). On the identification of the convex hull of a finite set of points in the plane. *Information Processing Letters, 2,* 18–21.

Kang, J. H., Welbourne, W., Stewart, B., & Borriello, G. (2004). Extracting places from traces of locations. In *Proceedings of the 2nd ACM International Workshop on Wireless Mobile Applications and Services on WLAN hotspots (WMASH '04)* (pp. 110–118).

Keler, A., & Krisp, J. M. (2016). Visual analysis of floating taxi data based on interconnected and timestamped area selections. In G. Gartner, M. Jobst, & H. Huang (Eds.), *Progress in cartography, EuroCarto 2015* (pp. 115–131). Cham: Springer International Publishing.

Keler, A., Ding, L., & Krisp, J. M. (2016). Visualization of traffic congestion based on floating taxi data. *Kartographische Nachrichten, 66*(1), 7–13.

Körner, M. (2011). Nutzungsmöglichkeiten von Floating Car Data zur Verkehrsflussoptimierung. In J. Strobl, T. Blaschke, G Griesebner (Eds.), *Angewandte Geoinformatik 2011, Beiträge zum 23. AGIT-Symposium Salzburg* (pp. 381–386). Berlin/Offenbach: Wichmann.

Krisp, J. M., & Keler, A. (2015). Car navigation—computing routes that avoid complicated crossings. *International Journal of Geographical Information Science, 29*(11), 1988–2000.

Lan, J., Long, C., Wong, R. C.-W., Chen, Y., Fu, Y., Guo, D., Liu, S., Ge, Y., Zhou, Y., & Li, J. (2014). A new framework for traffic anomaly detection. In *Proceedings of the 2014 SIAM International Conference on DATA MINING* (pp. 875–883).

Li, J., & Zuo, L. (2004). Shanghai urban elevated roads and metropolis traffic. *Urban Roads Bridges & Flood Control, 21*(6), 14–17.

Li, D., Fu, B., Wang, Y., Lu, G., Berezin, Y., Stanley, H. E., et al. (2015a). Percolation transition in dynamical traffic network with evolving critical bottlenecks. *Proceedings of the National Academy of Sciences, 112*(3), 669–672.

Li, H., Kulik, L., & Ramamohanarao, K. (2015b). Robust inferences of travel paths from GPS trajectories. *International Journal of Geographical Information Science, 29*(12), 2194–2222.

Liu, X., & Ban, Y. (2013). Uncovering spatio-temporal cluster patterns using massive floating car data. *ISPRS International Journal of Geo-Information, 2,* 371–384.

Long, J. C., Gao, Z. Y., Ren, H. L., & Lian, A. P. (2008). Urban traffic congestion propagation and bottleneck identification. *Science in China Series F: Information Sciences, 51*(7), 948–964.

Protschky, V., Ruhhammer, C., & Feit, S. (2015). Learning traffic light parameters with floating car data. In *IEEE 18th International Conference on Intelligent Transportation Systems (ITSC 2015)* (pp. 2438–2443).

Ranacher, P., Brunauer, R., Van der Spek, S. C., & Reich, S. (2016). A model to estimate and interpret the energy-efficiency of movement patterns in urban road traffic. *Computers, Environment and Urban Systems, 59,* 152–163.

Rinzivillo, S., Pedreschi, D., Nanni, M., Giannotti, F., Andrienko, N., & Andrienko, G. (2008). Visually driven analysis of movement data by progressive clustering. *Information Visualization, 7*(3–4), 225–239.

Robinson, R. (1984). Problems in the urban environment: traffic congestion and its effects. In *Wollongong studies in geography* (Vol. 14). Department of Geography, University of Wollongong: NSW, Australia.

Sohr, A., Brockfeld, E., & Krieg, S. (2010). Quality of floating car data. In *Conference Proceedings, paper nr 02392, 12th World Conference on Transport Research (WCTR).* Lisbon, Portugal. Retrieved July, 11–15, 2010 from www.digitalpapers.org.

Stanica, R., Fiore, M., & Malandrino, F. (2013). Offloading floating car data. *Proceedings of the 2013 IEEE 14th International Symposium and Workshops on a World of Wireless, Mobile and Multimedia Networks (WoWMoM)* (pp. 1–9).

Wen, H., Sun, J., & Zhang, X. (2014). Study on traffic congestion patterns of large city in China taking Beijing as an example. *Procedia—Social and Behavioral Sciences, 138,* 482–491.

Xu, X., Li, X., Hu, Y., & Peng, Z. (2012). A novel algorithm to identifying vehicle travel path in elevated road area based on GPS trajectory data. *Frontiers of Earth Science, 6*(4), 354–363.

Yan, Z., Chakraborty, D., Parent, C., Spaccapietra, S., & Aberer, K. (2013). Semantic trajectories: Mobility data computation and annotation. In *ACM transactions on intelligent systems and technology (TIST)—Special sections on paraphrasing; intelligent systems for socially aware computing; social computing, behavioral-cultural modeling, and prediction*, 4(3), 1–38.
Yuan, Q., Liu, Z., Li, J., Zhang, J., & Yang, F. (2014). A traffic congestion detection and information dissemination scheme for urban expressways using vehicular networks. *Transportation Research Part C: Emerging Technologies, 47*(2), 114–127.
Zhang, Y. M. (2004). The thinking of relieving the congestion of elevated road system. *Traffic and Transportation, 5,* 37–38.

Psychogeography in the Age of the Quantified Self—Mental Map Modelling with Georeferenced Personal Activity Data

Sebastian Meier and Katrin Glinka

Abstract Personal and subjective perceptions of urban space have been a focus of various research projects in the area of cartography, geography, and related fields such as urban planning. This paper illustrates how personal georeferenced activity data can be used in algorithmic modelling of certain aspects of mental maps and customised spatial visualisations. The technical implementation of the algorithm is accompanied by a preliminary study which evaluates the performance of the algorithm. As a linking element between personal perception, interpretation, and depiction of space and the field of cartography and geography, we include perspectives from artistic practice and cultural theory.

Keywords Mental maps · Location-based services · Activity data · Spatial analysis · Spatial behaviour

1 Introduction

The ways in which we experience and memorise our interactions with the physical world are highly personal and subjective. To a certain extent, mobile and web-based systems already make use of personalised map approaches that account for subjective views and different modes of representation. Nonetheless, most of the tools that we use to navigate and explore geographic space are standardised mapping tools. This paper will briefly highlight the discourses on personal and mental maps with examples from a cartographic perspective as well as informed by cultural theory and artistic practice. Building upon those discourses, we developed algorithmic approaches to aggregate, analyse, and visualise personal activity data. The

S. Meier (✉)
Interaction Design Lab, University of Applied Sciences, Potsdam, Germany
e-mail: meier@fh-potsdam.de

K. Glinka (✉)
Urban Complexity Lab, University of Applied Sciences, Potsdam, Germany
e-mail: glinka@fh-potsdam.de

M.P. Peterson (ed.), *Advances in Cartography and GIScience*, Lecture Notes in Geoinformation and Cartography, DOI 10.1007/978-3-319-57336-6_35

visual output is constructed as explorative visualisations, moreover, it also acts as a discursive artefact in order to foster discussion and reflection on individual spatial knowledge, perception, and behaviour.

This paper thus revolves around the research question of *how we can model* spatial behaviour *based on personal georeferenced activity data, while (a) building on mental map theories about the influence of mode of transport, and (b) reflecting artistic and political aspects of* psychogeography.

2 Personal (Mental) Maps

Several academic disciplines, ranging from sociology to psychology and philosophy, have explored the subjective perception of the physical world and its social and mental manifestations. At the same time, the embodiment and representation of political and social implications of (urban) space have been a recurring topic in literature, visual art, and philosophy. In the following two sections we will first discuss research from cartography and cognitive sciences and then take up artistic and theoretic approaches.

2.1 Academic Perspectives: Cartography and Cognitive Sciences

A large body of research has been focused on the question of how people construct spatial knowledge and how this knowledge is used to make sense of the physical world, for instance when performing navigational tasks. A dominant concept used in this discourse is the mental map. A general hypothesis within these discourses assumes that an aggregation of the entirety of our experiences form and influence our mental representations of space (Kitchin 1994; Montello 2002). Tversky extended the metaphoric concept of the mental map to *mental collages*, as those mental representations are not solely of map-like forms (Tversky 1993).

More generally, the correlation between map use and its impact on the shaping of mental maps has been a strong research focus, unravelling findings like the interdependency of orientation in maps and mental maps (mental rotation) (Tversky 1992; Hintzman et al. 1981). Among these influencing factors, the mode of transport has been identified as one that has significant impact on our perception of space and on how we interact with our environment, an assumption that we will explore further below.

Scientists have tried to capture and visualise mental representations of space, for instance by letting people draw or describe maps (Lynch 1960; Vertesi 2008). Research in neuroscience has produced valuable results in visualising neuronal maps that indicate where spatial information is stored in our brain (Maguire et al.

1997). However, visualising the resulting mental maps themselves has proven difficult. Taking these constraints into account, this paper will not try to visualise a whole mental map or mental collage, but instead focus on one of the previously mentioned influencing factors that form our mental maps: *the mode of transport*.

Within the research area of mental maps and navigational tasks, a specific strand of research is dedicated to so-called landmarks; a construct which was introduced to the science community by Lynch (1960). Landmarks are physical constructs, such as buildings, that are of certain significance for a user, for instance, as visual aids in order to memorise a route (Foo et al. 2005). Researchers have been trying to understand what makes a physical construct a landmark and how people use them when navigating and memorising urban structures.

Most research that focuses on understanding the impact of landmarks includes analysing the given physical structures, their visual features, and how a building stands out from its surroundings (e.g., salience approaches by Caduff and Timpf 2008). This approach does not account for the individual and his or her experiences within this space.

The concept of landmarks is, like many other cartographic approaches, characterised by the objective to generalise spatial information insofar as it seeks to identify objects that might be of significance for the general public. Nonetheless, the concept of landmarks also comes into play in our approach. However, we use landmarks in a highly personalised algorithmic spatial representation as a canvas for data mapping and visualisation, as will be further described in Sect. 3.

2.2 Psychogeography as Political Practice

Within the context of Post-Marxism and the spatial analysis of culture, political theorist Frederic Jameson elaborates on "cognitive mapping" from a much more political stance. He contextualises the concept in reference to the notion of an "(unrepresentable, imaginary) global social totality that was to have been mapped" (Jameson 1988: 356) and links it to the context of a socialist agenda. When applied to the design space of an actual rendering of a cognitive map, his envisioning of a new space could be translated into a conceptual model for spatial representation. Jameson describes a "suppression of distance […] and the relentless saturation of any remaining voids and empty places" (Jameson 1988: 351). The distortion that arises from this suppression of distance can serve as one of the implications for the design of a visual representation of a mental map. At the same time, Jameson points out that "[a]esthetics is something that addresses individual experience rather than something that conceptualizes the real in a more abstract way" (Jameson 1988: 358). Building on this notion, we deliberately distance our visualisation of personal movement data from seemingly objective and cartographic representations and put an emphasis on metamorphic shapes and aesthetic representations. This, as will be shown in Sect. 4, is done to a degree that explicitly disregards geographic accuracy and submerges into unique formations of a psychogeographical space that is the subjective city.

The term psychogeography was most prominently coined by the French Situationist International in the 1950s and is understood as "the study of the [...] specific effects of the geographical environment, whether consciously organized or not, on the emotions and behavior of individuals" (Debord 1955). With his work, Debord joined the ranks of writers and artists that, before him, had revolved around the flâneur as a metaphoric figure that strolls the (post) modernist city. This figure has since been subject to significant scrutiny and even re-emerged in the guise of the cyberflâneur (Shaw 2015) which opens a point of reference for our model in the age of the quantified self.

Self-tracking cultures have introduced disciplinary monitoring on a seemingly voluntary level into formerly nondisciplinary (private) spaces (Whitson 2015). With the visualisations described below, we return to Debord's depiction of personal experiences within urban space by using data gathered in self-tracking apps. Reflecting on these references, our mental maps should re-introduce the subjective view to a practice which often claims to provide objective results. The visualisation deliberately applies distortion and metamorphic shapes to a cartographic and geographic canvas. As a result, our visualisation of personal movement data only fully makes sense in reference to the individual experience of urban space.

2.3 *Duality*

The two previous sections introduced a framework for the concept of mental maps which builds on cartography as well as critical theories. Our approach is informed by both of these discourses and aims at introducing a new perspective on representations of personal spatial behaviour. Thereby, we seek to bridge the gap between the specialised disciplines and practices of cartography and visualisation, and more artistic and experimental approaches. We believe that visualisations can be extended beyond their function as visual output of (quantitative) data in that they can be staged as individual and aesthetic discursive artefacts, as will be illustrated in the next section. These visual representations can help to identify meaning in data as well as critically reflect on the algorithmic mechanics. Experimental techniques open a window for dialogic interpretation that surpasses abstract information visualisation and traditional cartographic representations.

2.4 *Explorative Tools for Personal Spatial Data Analysis*

Recent developments in the context of openly accessible web technologies have led to a democratisation of coding tools and means of production. This development has also brought about a variety of tools for visualising, exploring, and reflecting

Fig. 1 *Visits* showing a location history of 6 months using a map-timeline approach. This map/timeline hybrid segments the data into stays at places, revealing more information about the temporal aspects of the data (Thudt et al. 2013: 1)

upon personal data. The majority of these tools and experimental techniques focuses on information visualisation (see overview by Huang et al. 2015) and interfaces for personal photo collections (e.g., Hyunmo and Shneiderman 2000; Thudt et al. 2013). A small subset of tools explores visualisations as a means of accessing personal spatio-temporal data. We have identified two directions or types of projects within this subset that are important to our approach. Of these two types, we describe an exemplary project each in more detail. The first group of projects focuses on *moments*, similar to many personal photo collection interfaces, for instance, the prototype *Visits* by Thudt et al. (2013, 2016), shown in Fig. 1.

Through the aggregation and clustering of spatio-temporal data, timeline-like visualisations are created that highlight specific time periods. In *Visits,* those time periods are split into smaller periods based on spatial activity. Next to a contextualising map that helps the reader understand the spatial relationships, the linearity of time dictates the appearance of the main visual elements.

This first group of tools, exemplified by *Visits*, emanates from specific moments in space and time. The second group of tools also uses data aggregation but instead of focusing on singular spatio-temporal events, they provide an overview and summaries of the data. This puts them in a direct relationship with life-logging visualisations. A good example for such an application is the *Shifted Maps* prototype (Otten et al. 2015). *Shifted Maps* uses activity data to create spatial clusters and visualise the relationships between them in three different dimensions: the position of a location or location cluster; the time it takes to travel between nodes; as well as how frequently the links between locations are used (Fig. 2).

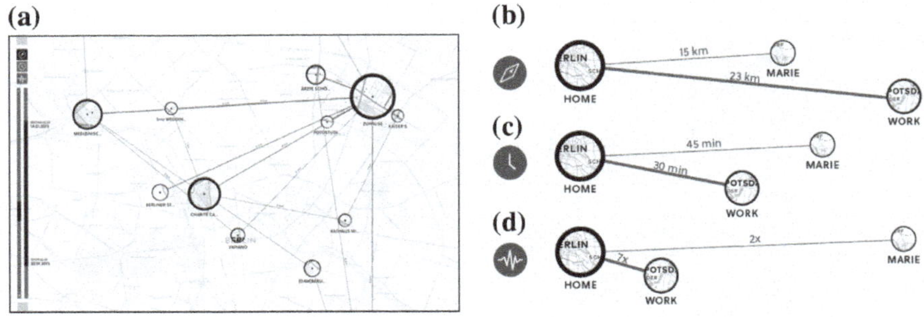

Fig. 2 The *Shifted Maps* visualisation (**a**) and three viewing modes for spatial (**b**), temporal (**c**), and frequency (**d**) networks of places (Otten et al. 2015: 1)

While our approach to some extent bears a similarity to these tools, we are neither interested in summaries nor in a spatial aggregation. Instead, we emphasize the temporal dimension and map the results onto the geographic space, resulting in an overview of spatial behaviour.

The projects described above, as well as our own endeavour strongly build upon principles of time geography. This concept was largely conceptualised in the 1960s and 70s by the social geographer Hägerstrand (1970) and investigates social dynamics in a spatio-temporal context. This new wave of projects takes a quantitative, data-driven approach to time geography. Therefore, one might say, that the trajectories we are analysing in the next chapter are digital counterparts to Hägerstrand's "life paths" (Hägerstrand 1970: 10–11).

3 Algorithmic Approach

Informed by the theories outlined in Sect. 2, we base our approach on the supposition that mental maps are influenced by many external stimuli, one of them being the way we move through a city (Chorus and Timmermans 2010; Mondschein et al. 2010). Our movements change the way we experience the physical world and thus influence the shaping of mental maps. In our approach, we focus on the mode of transport as one such experiential factor, which builds upon work by Mondschein et al. who reported that active or passive navigation influences the quality (richness of detail) of mental maps (Mondschein et al. 2010). Following this supposition, our algorithm tries to calculate the personal level of *potential memorisation* or *spatial knowledge* of a certain area. The calculated value is based on trajectory data and the trajectory's mode of transport: from the highest value for walking, to cycling, to the lowest value for motorised transport. The term (potential) memorisation might at first seem ambiguous. The term memorisation is often used in contexts of intentional learning, in such, describing how information is moved from the working memory to the long-term memory. In our case, we rather describe the acquisition of knowledge of a certain area. Still, the repetitive interactions and the type of interactions with a certain area help us estimate the quality of spatial knowledge. Therefore, we decided to use the term (potential) memorisation.

The trajectories are sourced from *personal activity data*, sometimes referred to as *lifelogging data*. The term, as it is currently being used, gained attention with the rise of smartphones that were logging data on their users, including GPS trajectories, compass and accelerometer data. Using machine-learning, these datasets are now interpreted and grouped into activities. By calculating speed, rhythm, and movement patterns, algorithms detect the mode of transport and can differentiate between walking, running, cycling, riding a train, or travelling on an airplane (Lau 2012; Pennanen and Kyrölä 2013). While previous research on personal activity data was primarily reliant on GPS trajectories (Spek et al. 2009), this new generation of activity data allows more in-depth analysis of individual's activity. Even though the latest tracking devices store a richer set of dimensions in activity data, in this paper we will only focus on GPS trajectories combined with the classification of the mode of transport for those trajectories, in order to get insights into one aspect of a person's mental map, the knowledge or rather memorisation of certain spatial areas. To be precise, for our prototypical implementation and study, we used data gathered through the Moves App (ProtoGeo 2016).

3.1 Data Aggregation, Cleaning and Organisation

The data collected by the Moves application can be downloaded as JSON files. Each file contains the user's trajectories for a specified time period. The Moves application is constantly creating those trajectories based on a GPS signal. Technically, every trajectory consists of a GeoJSON linestring (Butler et al. 2008), a timestamp, and the classification of the mode of transport.

Most activity logging devices and applications have to be cautious with their power consumption. Consequently, the precision of the generated georeferenced trajectories is not ideal. One type of error is *variation in the data*, which means that even though a set of trajectories represent trips on the same road, they will be off by a few meters in any direction. We have factored this in by implementing buffers as described in the next section.

The second error are points that are completely off the track. To remove such points, we ran statistics on the whole trajectory set of the individual user and calculated average distances between points for each mode of transport, which allowed us to remove outliers and replace them with new points between the prior point and the next. Even though the new point will still likely be off track, it gives us a better dataset than with the original outliers.

In addition, the Moves App tries to snap start- and end-points of trajectories to locations gathered from Foursquare (Foursquare 2014). In cases where no mapped start- and end-points existed, we used a buffer of 20 m to generate new locations and mapped them onto the dataset. The 20-m buffer is needed since points for one location will not be exactly the same due to GPS inaccuracy. This procedure created a cleaned dataset of locations as well as trajectories connecting those locations.

3.2 Temporal Data Clustering and Network Analysis

On the basis of the cleaned locations and trajectories, we set out to map the level of assumed spatial knowledge over certain areas by deriving a value for the potential level of memorisation based on the mode of transport. Mondschein et al. (2010) did not develop a formula to calculate these levels, their findings rather indicate an order of memorisation potential for various transport modes. We used their findings and developed a formula that translates visits of a certain part of town by a certain mode of transport into a potential memorisation level that can be used in algorithmic calculations (henceforth referred to as “Potential Memorisation Index”, or *PMI*). This formula serves only as a starting point for our research and has to be refined by more user studies as discussed in Sect. 3.3. Based on the order introduced by Mondschein et al. we used a logarithmic curve that represents an exponential learning curve with a limit. At a certain level of knowledge, each additional spatial interaction will only increase the value of the *PMI* by a very small amount (Fig. 3).

When combining the *PMI* from each transport mode, the highest value is chosen. If a user passes a location x times, for example, with motorised transport, the formula only generates a very low index. But, if a user walked through the area y times, a higher value is generated. The higher value derived from the walking activity is chosen for further calculations. With these *PMIs* at hand, we created a 50-m raster and then intersected the grid cells with the trajectories using a buffer of additional 20 m to overcome the precision issue (Sect. 3.1), which resulted in a heat-map-like visualisation. Since cities are not made up of 50-m grid cells, we chose buildings as our final projection canvas and calculated the *PMI* for each building. The building’s geometries are sourced from OpenStreetMap (OpenStreetMap 2016).

Looking at the resulting visualisation (see Fig. 4), we identified areas with high *PMIs*, mainly resulting from walking activities, which we called “walking islands”. Those islands were located across the city and connected by other modes of motorised transport (e.g., underground). To explore those connections of known areas, we created a network graph on top of the islands, highlighting the connections between islands and their respective strength. Strength was calculated through the number of trips, as well as modes of transportation and time spent on trips between islands (Fig. 5).

$$\text{index} = \log(\mathit{times\_visited} + 1)^{0.1} * \mathit{transport\_mode}$$

Fig. 3 PMI, per transport mode (motorised transport = 1, cycling = 2 and walking = 3)

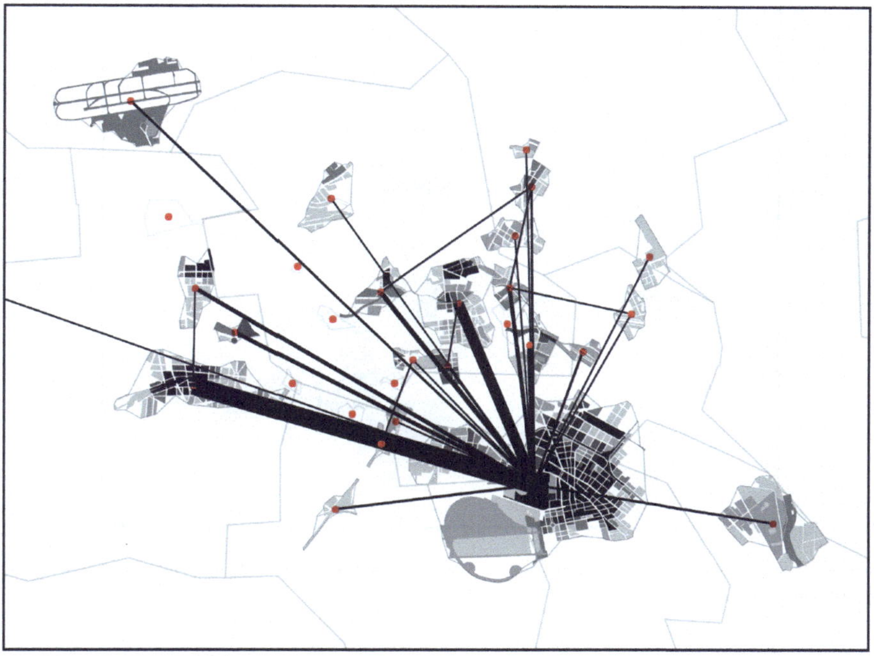

Fig. 4 Trajectories mapped onto buildings. *On top* of the buildings, the network structure between the walking islands is visualised. Colour-scale: *white* is unknown to well-known buildings in *black*. All exemplary visualisations are generated from the first author's personal activity data from the last 12 month reduced to the extent of the city of Berlin

Fig. 5 The same network represented in a situationist fashion with a manual projection and weighted links (*arrows*)

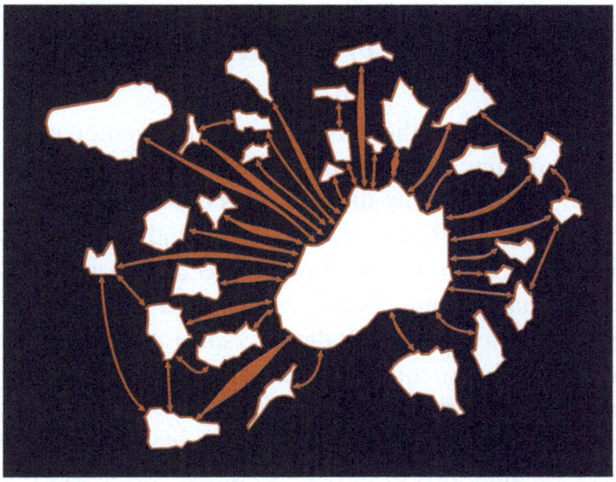

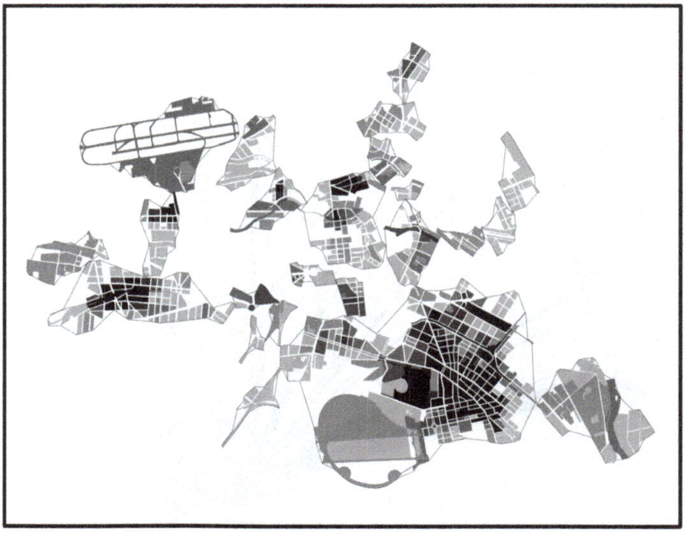

Fig. 6 Using a force directed graph method, the city layout is rearranged and distorted by the suppression of distance

This network graph was then translated into a force-directed graph-like concept, allowing us to reorganise the islands depending on their edges. As a result, a new city layout emerged (see Fig. 6) that represents the movement and the associated personal experience inspired by the visualisations of the Situationist International.

3.3 Evaluation of Clusters and Model

As mentioned in the previous section, several uncertainties become apparent when implementing the concept of Mondschein et al. in an algorithmic approach (apart from the uncertainty of the data as described in Sect. 3.1). In order to improve our modelling approach, we therefore conducted a preliminary user study with our initial algorithm that incorporates feedback on the assumed level of memorisation.

Recruiting participants for this kind of experiment poses a challenge, as they need to provide us with their personal activity data. Nonetheless, we were able to recruit five participants, three male, two female, between the ages of 25 and 45, all currently living in Berlin.

Patterns of movement through urban space change across time, for example, due to switching jobs or moving to a new place. Thus, we segmented the data on the time axis, whereby the dataset for each participant represents only their personal data from the last 12 months. This means that for every one of the five participants, we used one year of data for the metropolitan region of Berlin/Brandenburg. We also accounted for how long each participant has already been living in Berlin, which was shown to have no impact on the evaluation results.

For the test, we selected 22 locations with a wide range of algorithmically-generated potential memorisation values derived from the mix of transport modes. In addition, we added three locations presumably unknown to the user (at least based on the data we received). We created a web interface in which the user would receive a 360° image of each location, using Google Street View. The user could change the angle of the image, but not the location. The users were then asked if they knew the location. If so, they were asked to pin-point the location on a map. If the user did not know the location, the correct position was disclosed on the map. If the user recognised the area from the image, or knew the area indicated on the map, they were asked to rank their knowledge on a scale from 1 to 6. From user feedback, we learned that the Google Street View images were, in some cases, quite old and therefore hard to identify. Thus, we allowed people to also rank their knowledge solely based on the position indicated on the map. This applied in cases when they were not able to identify the location based on the image, but later saw on the map that they indeed knew the location.

Our study investigated the performance of our algorithmic approach by comparing the *PMI* of our algorithm and the reported response from the users. With such a small sample, it is difficult to discernibly identify a correlation or a significant effect, but there is a clear trend in the data that indicates that our algorithm performs well, even in this untrained phase. Figures 7 and 8 show that for a high memorisation response, our algorithm also calculated a high *PMI*. Returning to Mondschein et al. theory, we see in Fig. 9 that areas with low memorisation responses show less walking and cycling and more motorised transport modes.

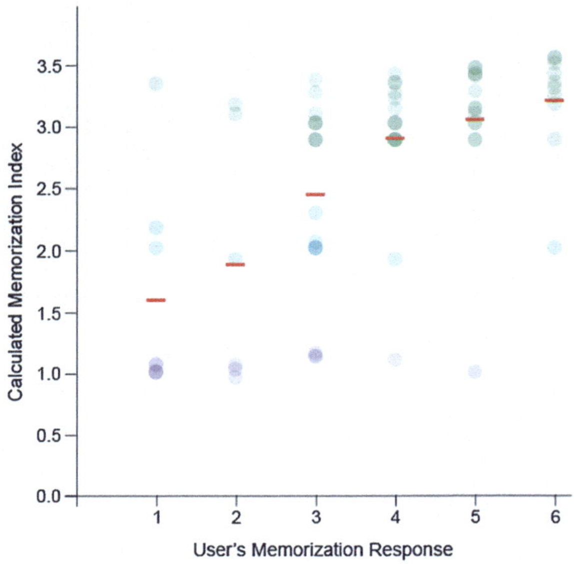

Fig. 7 The x-axis represents the memorisation response from participants. The y-axis represents the algorithmically-calculated potential memorisation index. Figure shows the raw data points with averages (*red*)

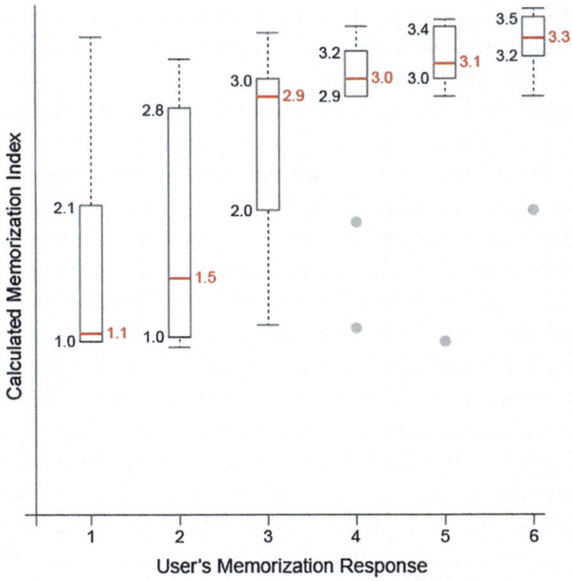

Fig. 8 The x-axis represents the memorisation response from participants. The y-axis represents the algorithmically-calculated potential memorisation index. Figure shows an analogous box-plot with medians (*red*)

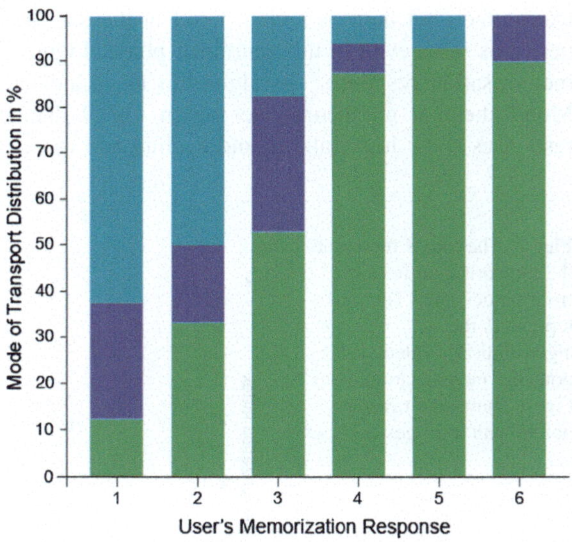

Fig. 9 The x-axis represents the memorisation response from participants. Figure visualises the percentage for modes of transport for each memorisation response value. *Green* walking, *purple* cycling, and *blue* motorised transport

4 Reflective Practices

In addition to the visualisations in Figs. 4, 5, 6, we mapped the *PMI* onto buildings as their height, creating a three-dimensional space. As laid out in Sect. 2, our approach is informed by research on mental maps and psychogeography. Drawing on this, we developed a visualisation artefact that deliberately disregards cartographic rules and representational precision. The 3-dimensional physical model of the transformed space (see Fig. 10) was created by applying rapid prototyping techniques. With this physical model, we hope to encourage reflection on the perception of urban structures and the identification of spatial manifestations. The city that served as the backdrop on which the subjective experience of urban space was made and gathered in the form of movement data is now embodied by the distorted proportions and metamorphic shapes that are a dominant feature of the artefact. The artefact should rather trigger personal reflection than analytic exploration. It can be read in the reflection of Jameson's stance on cognitive mapping as the "suppression of distance […] and the relentless saturation of any remaining voids and empty places" (Jameson 1988: 351), while it also addresses the individual experience by means of aesthetic interpretation (Figs. 10 and 11).

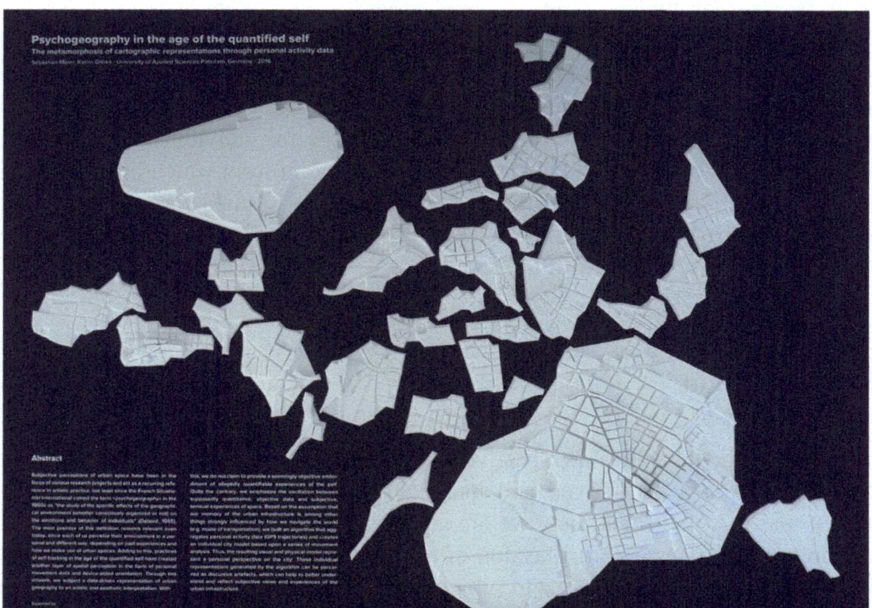

Fig. 10 Images of the 1 m × 1 m in size physical model, generated from the potential memorisation indices

Fig. 11 Images of the 1 m × 1 m in size physical model, generated from the potential memorisation indices

5 Applications and Future Works

In our next step of evaluation, we will include a larger sample group to help us improve our algorithm. Furthermore, we want to focus on long-time potential memorisation approaches (forgetting areas, rememorising areas, changes in the (built) environment), to improve the modelling. In addition, we believe our approach, of choosing the mode of transport with the highest resulting *PMI*, should be reconsidered. As Tversky suggested, our mental maps are derived from all interactions and experiences (Tversky 1993). Therefore, we should be combining all *PMIs* from all different modes of transport.

When we are able to increase the precision of our method, we see numerous possible applications that go beyond personal reflection. Those applications include improved recommendation methods for location-based services or personalised routing services. Furthermore, the comparison of several individual visualisations of personal activity data could also act as a vantage point from which the social and political conditions of accessibility could be traced. By visualising certain aspects of mental maps based on personal activity data, we could be able to highlight "projections and repressions in the form of 'go' and 'no-go' space" (Jenks 1995).

6 Conclusion

Informed by the theoretical perspective of mental maps presented in Sects. 2.1 and 2.2, we provide a data-driven and quantitative perspective of the discourse on mental maps. At the same time, we are aware of the complexities of mental maps and do not try to answer the question of how to visualise mental maps in general. Thus, we have limited our scope to subjective experiences of the urban space on the basis of movement through the city. We presented an algorithm to create a *PMI* based on mode of transport, inspired by the works of Mondschein et al. The algorithm showed promising results in the preliminary user study.

Through the visualisations, especially the physical model, we hope to enable new perspectives on personal spatial data and behavioural patterns in urban space.

Acknowledgements The code for the methods described in this paper are available on GitHub: https://www.github.com/sebastian-meier/MentalMaps.

References

Butler, H., Daly, M., & Doyle, A. (2008). GeoJSON specification. Retrieved November 1, 2014, from http://geojson.org/geojson-spec.html.

Caduff, D., & Timpf, S. (2008). On the assessment of landmark salience for human navigation. *Cognitive Process*, *9*, 249–267.

Chorus, C., & Timmermans, H. (2010). Determinants of stated and revealed mental map quality: An empirical study. *Journal of Urban Design*, *15*(2), 211–226.

Debord, G. (1955). Introduction to a critique of urban geography. Les Lèvres Nues #6. Translated by Ken Knabb. http://www.cddc.vt.edu/sionline/presitu/geography.html.

Foo, P., Warren, W., Duchon, A., & Tarr, M. (2005). Do humans integrate routes into a cognitive map map-versus landmark-based navigation of novel shortcuts. *Journal of Experimental Psychology*, *31*(2), 195–215. American Psychology Association.

Foursquare. (2014). Foursquare for developers. Retrieved November 4, 2014, from https://developer.foursquare.com/.

Hägerstrand, T. (1970). What about people in regional science. *Papers of the Regional Science Association*, *24*(1), 6–21.

Hintzman, D. L., O'Dell, C., Arndt, D. (1981). Orientation in cognitive maps. *Cognitive Psychology*, *13*, 149–206. Academic Press Inc.

Huang, D., Tory, M., Aseniero, B. A., Bartram, L., Bateman, S., Carpendale, S., et al. (2015). Personal visualization and personal visual analytics. *TVCG*, *21*(3), 420–433.

Hyunmo, K., & Shneiderman, B. (2000). Visualization methods for personal photo collections: Browsing and searching in the PhotoFinder. *ICME*, *3*, 1539–1542.

Jameson, F. (1988). Cognitive mapping. In *Marxism and the interpretation of culture* (pp. 347–357). UK: Macmillan Education.

Jenks, C. (1995). Watching your step: The history and practice of the. In C. Jenks (Ed.), *Visual culture* (pp. 142–160). New York.

Kitchin, R. M. (1994). Cognitive Maps: What are they and why study them. *Journal of Environmental Psychology*, *14*, 1–19. Academic Press Ltd.

Lau, S. L. (2012). *Towards a user-centric context aware system: Empowering users through activity recognition using a smartphone as an unobtrusive device*. Kassel University Press.

Lynch, K. (1960). *The image of the city*. Cambridge: The Technology Press & Harvard University Press.

Maguire, E. A., Frackowiak, R. S., & Frith, C. D. (1997). Recalling routes around London: Activation of the right hippocampus in taxi drivers. *Journal of Neuroscience*, *17*, 7103–7110.

Mondschein, A., Blumenberg, E., & Taylor, B. (2010). Accessibility and cognition: The effect of transport mode on spatial knowledge. *Urban Studies*, *47*, 845–866.

Montello, D. R. (2002). Cognitive map-design research in the twentieth century: Theoretical and empirical approaches. *Cartography and Geographic Science*, *29*(3), 283–304.

OpenStreetMap. (2016). OpenStreetMap. Retrieved June 1, 2016, from http://www.openstreetmap.org.

Otten, H., Hildebrandt, L., Nagel, T., et al. (2015). Are there networks in maps: An experimental visualization of personal movement data. IEEE VIS 2015.

Pennanen, J., & Kyrölä, A. (2013). User activity tracking system. Patent: US 20150004998 A1.

ProtoGeo. (2016). Moves—Activity diary for iPhone and Android. Retrieved June 24, 2016, from https://www.moves-app.com/.

Shaw, D. B. (2015). Streets for cyborgs the electronic flâneur and the posthuman city. *Space and Culture, 18*(3), 230–242.

Thudt, A., Baur, D., & Carpendale, S. (2013). Visits: A spatiotemporal visualization of location histories. EUROVIS—Short Papers.

Thudt, A., Baur, D., Huron, S., & Carpendale, S. (2016). Visual mementos: Reflecting memories with personal data. *IEEE Transactions on Visualization and Computer Graphics*, *22*, 369–378.

Tversky, B. (1992). Distortions in cognitive maps. *Geoforum*, *23*(2), 131–138. Pergamon Press Ltd.

Tversky, B. (1993). Cognitive maps, cognitive collages, and spatial mental models. *COSIT*, *716*, 14–24.

Van der Spek, S., Van Schaick, J., De Bois, P., & De Haan, R. (2009). Sensing human activity: GPS tracking. *Sensors*, *9*(4), 3033–3055. Molecular Diversity Preservation International.

Vertesi, J. (2008). Mind the gap: The London underground map and users' representations of urban space. *Social Studies of Science*, *38*, 7–33.

Whitson, J. R. (2015). Foucault's fitbit: Governance and gamification. In S.P. Walz & S. Deterding (Eds.), *The gameful world* (pp. 339–358).

Part VIII
Final Reflections

In Search of the Essence of Cartography

Marek Baranowski, Dariusz Gotlib and Robert Olszewski

Abstract Discussed here is the essence of cartography as well as the specifics and the uniqueness in the cartographic modelling process. During the analysis of the theory of cartography and relations to other fields of knowledge, especially philosophy, eight properties of cartographic modelling have been defined which distinguish the process from other methods of space modelling. In attempting to define essential properties of cartographic modelling which would distinguish the process from other methods of space modelling, the authors introduce the notion that the method of cartographic presentation in the form of a map is not limited to graphic representation only. The language of graphic symbols used in classic methods of cartographic presentation is only one form of the cartographic notation of spatial information.

Keywords Cartographic modelling · Map · Cartography

1 Introduction

The tradition of modelling and visualizing of geographic space is much older than formal cartographic techniques. Totemic paintings on the walls of Lascaux cave (French Aquitania), representing Paleolithic art, indicate significance of spatial relations for cognition and exploration of the world. Besides spoken language and writing, a map belongs to the most important tools for learning about reality and

M. Baranowski (✉)
Institute of Geodesy and Cartography, Warsaw, Poland
e-mail: marekleonard@yahoo.com

D. Gotlib · R. Olszewski
Faculty of Geodesy and Cartography,
Warsaw University of Technology, Warsaw, Poland
e-mail: dariusz.gotlib@gmail.com

R. Olszewski
e-mail: r.olszewski@gik.pw.edu.pl

M.P. Peterson (ed.), *Advances in Cartography and GIScience*, Lecture Notes
in Geoinformation and Cartography, DOI 10.1007/978-3-319-57336-6_36

communication between people (Ostrowski 2008). A map had been used by the human species even before the first civilizations were established, and as a component of culture and civilization has accompanied the development of humanity since primal times up to the current scientific and technical revolution. For centuries, maps have provided images of the state of knowledge of geographic space and reflected the intellectual and technological development of their epoch.

At the dawn of history, sharing information about explored space made it possible for homo sapiens to survive as a species, and also shaped the foundation for cartographic modelling, understood as the abstraction of features of surrounding reality and its visualization. Technical methods of expression and the media used for communication have acted as means of preserving the conceptual model and have naturally evolved through the ages, from cave paintings and clay tablets through analog maps to multi-resolution and multi-representational spatial databases, GIS and mobile maps. This does not, however, change one of the missions of cartography, which is the modelling of geographic space.

The discussion on the role and place of cartography in the information era has intensified during last 30 years. Each year brings some new technological solutions and social arrangement, which keep this discussion vital. In 2016, a Cartographic Summit was organized by the International Cartographic Association and ESRI under a subtitle "Future of Mapping." Thus, the final answer on the role of cartography in the contemporary world is still open. Authors of this paper would like to join the discussion and present their views arising from the Polish cartographic tradition.

2 Significance of Maps and Cartography

Cartography is a science, a technique and an art which enables us to discover and understand space through its cartographic model—to analyze objects, phenomena, relations between them and their variability defined in space as well as in time (Imhof 1982). As Saliszczew (1998, p. 313) argued "the essence of cartographical method consists in the inclusion into the process of studying reality of an intermediary link—a geographical map—as a model of the studied phenomena. Here a map plays a double role: (1) a research tool; and (2) a subject in the form of a model representing real phenomena".

Maps are "holistic representations of spatial reality" (Visvalingam 1989). According to Robinson (1995) "all maps are abstractions of reality" and at the same time "all maps involve transformations of various kinds." This means that in order to elaborate a map, it is necessary to purposefully transform "raw" source data into the form of a useful space model. Therefore, we can agree that a map is a certain generic term and that it can be applied not only to traditional analog atlases, plans, globes or 3D models, but also to spatial databases of appropriate structure. A map is a model-based representation of reality created using symbolism which is comprehensible to users. However, this does not mean that the process of symbolization is limited exclusively to graphic representation.

Visvalingam (1989) recognizes that the role of cartography is not to focus on the visual product, but rather on the process of mapping, which covers intentional gathering of data, its transformation, generalization, symbolization, analysis and interpretation. In this sense, a contemporary geospatial database cannot be treated as a data repository, but rather as a model of reality. The development of digital technologies does not stand only (nor mainly) for modernization of the process of map preparation (still often understood to be a flat graphic representation of a fragment of space), but allows us to extend "the way in which we can communicate, explore and understand spatial information through maps".

As early as in 1984, Guptill and Starr (1984) defined cartography as "an information transfer process that is centered about a spatial database which can be considered, in itself, a multifaceted model of geographic reality. Such a spatial database then serves as the central core of an entire sequence of cartographic processes, receiving various data inputs and dispensing various types of information products."

The process of map (as well as spatial database) creation must be preceded by a process of learning and understanding spatial aspects of reality (Ostrowski 2008). The sense of such an understanding is not only to master the methods and algorithms of spatial data transformations, but mainly to learn geographic reality, comprehend spatial distribution of objects and phenomena, their interactions, as well as processes which shape geographic space and show different intensity depending on the scale of observation. Such an approach complies with the idea of Makowski (2001, 2005), who proposed that a map is a system, a model and an image. In this understanding, the process of cartographic modelling relies on intentional selection and generalization appropriate for the scale and purpose of the elaboration. This agrees with the psychological conditioning of humans that use pictures (figurative information or abstract cognitive information as a representation of perceived reality) in all spatial endeavors. Makowski (2005) claims that "a map is an informative whole, a purposeful time-space concept of systemic (reaching a goal), as well as model (information) and image (data medium, message) character. It is therefore a triad formation. The map is a product of psyche and intellect, therefore, it is not bound to the matter of the medium. The physical (material) representation of the map is a vehicle for imagination (mental image) during data transmission; interpretation of the contents relies on systemic reasoning supported by appropriate knowledge."

In the light of the above considerations, contemporary cartography cannot be limited exclusively to the role of a formal discipline dealing with descriptions of the theory and methods of conveying graphic information about spatial distribution of objects and phenomena. Cartography is the science of modelling and imaging of time-space information structures in the form of maps describing multi-dimensional reality (Makowski 2005). The essence, and a fundamental advantage of cartographic modelling, is the fact that a map, despite significant reductions in its spatial dimensions in relation to reality, makes it possible to present and analyze mutual distribution of significant objects and phenomena regardless of the size of the modeled area (Ostrowski 2008). Thanks to such a reduction "a map allows us to significantly extend our natural range of perception, therefore creating opportunity to study spatial relations on vast areas" (Robinson 1988).

3 Cartography in Philosophical Context

In 2001 Prof. Michael Wood, the former President of the ICA, posed a rhetorical question: Is the future of the 21st century world possible without cartography? Besides giving a negative answer one could also ask further questions: What is cartography now? What role can it play in the development of digital society?

For the authors, cartography is a field of knowledge dealing with modelling of geographic space and its imaging—including, in the particular case also geovisualization (but not only). The subject of interest of cartography is the description of the real world, organizing information related to such description in spatial databases as well as the rules of visualization and sharing of geographic data. Cartographic methodology forms the foundation of transforming spatial data into a classical analogue map, digital spatial database or an online service presenting geographical information.

Du (2011, p. 2) claims that

> … traditionally, Theoretical Cartography has been closely related to various school of Philosophy. It is difficult to exhaust all philosophical approaches in Theoretical Cartography, Epistemology and Semiotics are main two of them. Epistemology is concerned with cognitive and aesthetic aspects of humans to map symbol, color, pattern and layout, while semiotics with the conceptual model of maps and its surroundings. Metaphysics is also adopted to build the theory of Cartography (e.g., Meta-Cartography), where Ontology plays a role.

For the analysis of the links between philosophy and the process of cartographic modelling it is therefore important to refer to both ontology of the nature of being and reality and epistemology (gnoseology), which deals with the process of cognition, knowledge, truth and their relations to reality.

The first theory of knowledge in European philosophy was by Plato, where true knowledge refers to ideal objects. Language does not refer only to concrete entities, but mostly to general categories. Concrete entities are only examples of general terms. According to the theory of knowledge proposed by Plato in "The Republic" (Politea, about 360 BCE), cognition relates to unchanging general categories rather than to elusive forms which are material representations of "ideal objects." Just as the language does not refer only to concrete entities, but mostly to general entities, also cartographic space models are examples of general terms. Plato presents this problem with an allegory of a cave where people take shadows from the bonfire light for real objects; similarly, a man trusting his senses sees only "shadows" and not real entities. In this reasoning one can find a certain analogy to exploring space with a map which is its model.

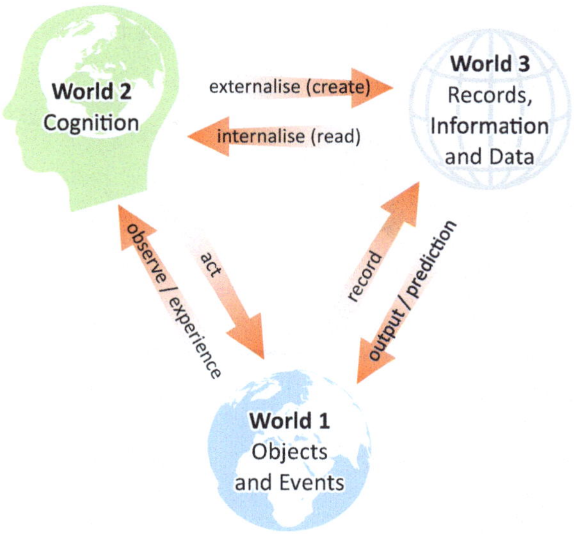

Fig. 1 Developed based on Popper's three worlds

The evolution of contemporary cartography from an epistemological point of view can be considered to be equivalent to Wittgenstein's trajectory (his First and Second Philosophy). On the other hand, cartographic products and the different stages of map creation can be linked to Popper's Three World Theory (Azocar Fernandez and Buchroithner 2014). Popper split the world into three categories (see Fig. 1):

World 1: the world of physical objects and events,
World 2: the world of mental objects and events
World 3: objective knowledge.

The theory of interaction between World 1 and World 2 is an alternative theory to Cartesian dualism, which is based on the theory that the universe is composed of two essential substances: *Res Cogitans* and *Res Extensa*. Popper's World 3 contains the products of thought—scientific theories, stories, myths, works of art and maps. The interaction of World 2 and World 3 is based on the theory that World 3 is partially autonomous. The process of learning causes World 3 to change World 2. Popper's theory of the three worlds establishes a distinction among the world in itself—the subjective world and the objective world (Mejia Soto 2004).

As written by Azocar Fernandez and Buchroithner (2014) Fig. 2 …

> depicts the "Earth" as an object belonging to *World 1* which is knowable through cartography. The subject is represented by the interaction between the mapmaker and the user through the map, and the latter belongs to *World* 2. The mapmaker designs and creates the map and the user reads and interprets it. Then the user generates a map image which belongs to *World* 3. From this mental image, new images are generated by the user. These images are regarded as new objects or new realities (belonging to *World 3*) because they replace the phenomena of *World 1*.

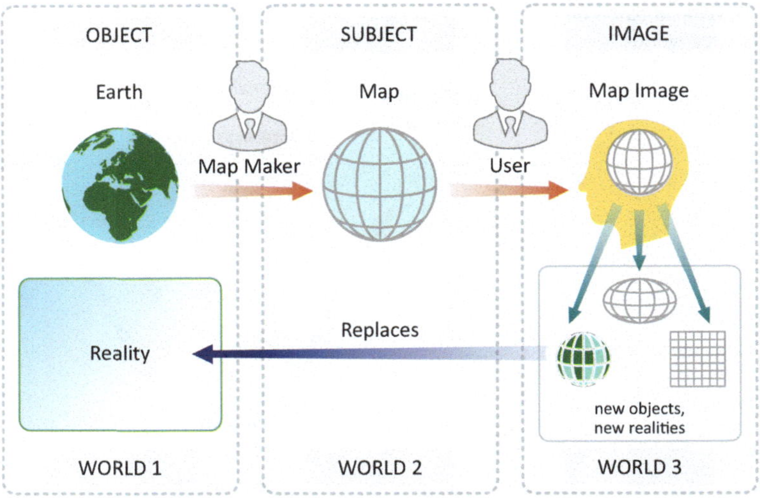

Fig. 2 Interactions between Popper's three worlds based on: Azocar Fernandez and Buchroithner (2014)

Being aware of deep implications of Popper's philosophical theories for cartographic modelling, it is worth asking to what extent modern cartography with its methodology of modelling geographic space makes it possible to get to know "Three Worlds"? To what extent a map, which belongs to *World* 2, affects our cognition of both the real *World 1*, of which it is a model, and of *World 3*, the image of which it creates? The *World 3* is a product of a man in which it fundamentally differs from Plato's world of ideas or Hegel's objective spirit, however, through its autonomy it reflexively affects *World 2*, and even *World 1*. In this process, a map plays an exceptionally important role – of a model of *World 1*, a vehicle for imagination and a base of spatial knowledge. In a thusly defined approach, a map becomes not only a model or "image of the World," but also a tool for its study.

In the considerations on the properties of cartographic modelling the authors focused mainly on the *World 2*, assuming that creation of this "World" is a fundamental task of cartography. It is not only "a World" of the flat image or multi-dimensional geovisualization, it is much more—a medium-independent information structure specific only to cartography.

4 Properties of Cartographic Modelling

According to Ostrowski (2008), in the 1960s, as a result of interdisciplinary research on the border of cartography, information theory, semiotic, modelling theory, IT and philosophy, there appeared three basic theoretical concepts (problem orientations) which shaped further development of cartography as an independent science:

- communication concept, treating a map as an element of information transmission,
- concept of cartographic semiotics, identifying a map as a specific language or system of signs,
- model-cognitive concept, playing a special role at the time of development of geo-informational techniques, creation of spatial databases and data analysis.

The creation of the concept of cartographic modelling in 1960s is the achievement of the cognitive orientation. The breakthrough came with Board's "Maps as Models" (1967), where the author considers maps as image models, representation models and concept models, containing a significant part of a certain generalization of reality. In this role, maps are useful analytic tools which enable us to see reality in new light. The cartographic model is identified with a generalized projection of a fragment of geographic space. The process of modelling is realized through the subjective selection of significant components of geographic space and "intelligent generalization of them" (Imhof 1982).

It should be stressed, that the role of cartographic modelling is to appropriately present locations of objects, but also to present their significance in spatial context and reflect relations between them. Cartographic modelling has a huge impact on the way geographic space is perceived and on how we recognize the role and character of objects and relations between them.

Guided by the above convictions the authors attempted to define properties which are of key importance for modelling processes and analyzed several dozen various methods of structuring knowledge about geographic space and relations within it (Baranowski et al. 2016). It allowed them to select a set of properties which distinguish cartographic modelling from other methods of describing the world, e.g., photographs, paintings or literary descriptions. In the initial phase, the authors analyzed a wide spectrum of properties significant for cartographic modelling. Following the principle of Ockham's razor, "*Entia non sunt multiplicanda praeter necessitate,*" in the later phase it was assumed that to explain the essence of cartographic modelling one should strive for simplicity, selecting explanations which are based on as few assumptions and terms as possible. The authors consider eight properties to be of key importance for cartographic modelling. Only a process realized in compliance with all these properties can be considered as cartographic modelling. It should be also stressed that following Aristotle's concept that *the whole is much more than a simple sum of its elements,* the essence of cartographic modelling is also decided by mutual interactions between the properties of cartographic modelling defined by the authors:

C1. Portrayal of space.
C2. Identification of types of objects and phenomena.
C3. Description of spatial relationships between objects.

C4. Localization in an applied reference system.
C5. Deliberate choice of a certain level of generalization compliant to the map's objective.
C6. Symbolization based on knowledge.
C7. Aware authorship of the message.
C8. Explicitness of communication (conformity between intention and reception).

- Portrayal of space (C1) is understood here as describing a specific arrangement of objects or phenomena in the real or virtual world. In this sense, a cartographic modelling significantly differs from Plato's theory of defining ideal objects (models).
- Identification of types of objects and phenomena (C2) enables unequivocal recognition of an object and its classification into the appropriate group within defined conceptual classification. Important features of the cartographic model are generalization of concepts and categorization of objects and phenomena.
- Description of spatial relations between objects (C3) allows the direct (or indirect) defining of relations between objects and phenomena, e.g., drainage system and relief, or transportation network and settlement.
- Locating within the applied system of reference (C4) allows clear spatial placement of modeled objects and phenomena, where the defined system of reference can be of global or local character. The ability to locate concrete objects rather than unidentified representatives of a certain group is of key importance.
- Intentional consideration of a given level/levels of generalization in terms of function and purpose (C5).
- Generalization of spatially located information is a feature of cartographic modelling, so the purpose of its elaboration is an important factor for geometric accuracy and/or the level of conceptual generalization (information definition) of the resulting model.
- Symbolization based on knowledge (C6) allows us to consciously use a wide range of symbols (graphic, lexical, sound, bit etc.) adequate for the aim and purpose of the resulting model and to the perception ability of the user. In this approach, the essence of cartographic modelling is the structuring of spatial information rather than only its visual rendering.
- Being aware of the authorship of the message (C7) means that elaboration of a cartographic model requires a purposeful selection of source data and its cartographic transformation. The cartographic model is characterized by the subjectivity of its maker.
- The explicitness of communication (conformity between intention and reception) (C8) means that the cartographic model is elaborated for utilitarian purpose, possibility to use it according to its objective and is therefore important for evaluation of the model itself. In this sense, the cartographic model should be explicitly interpreted both by the maker and the potential user.

Referring to the concept of cartographic modelling Gotlib (2011, p. 12) suggested a definition of a map as

> … a model of reality presenting location and selected features of objects and phenomena in relation to the surface of the Earth, another celestial body or other objects, e.g., a building, human body, as well as spatial relations between objects and phenomena. At the same time it was acknowledged that this model can refer to current, past or future state, as well as represent changes in time. The model can be communicated to the user in various forms: graphic, sound, text, tactile. It can be also processed and analyzed only on the basis of a digital recording. Classification of a process into cartographic modelling does not therefore depend on its form, but rather on the character of the message, i.e., the set of properties that set apart this process from other processes.

Cartographic modelling understood this way concentrates mainly on the method of structuring of spatial information rather than only on the visual imaging per se. Symbols can be written in different ways, e.g., ink on paper, magnetic state of a disc or electric impulses, but they represent something else than their physical form (Duch 2005). Symbolic signs used in the process of map elaboration point at certain representations but knowledge is also contained in relations between signs and their relative location.

Maps are a form of symbolization governed by set of conventions, the aim of which is to communicate information about a given location. Representation requires certain substrate, material, structures which can appear in an unlimited number of configurations so that it can describe the complex real situations.

5 The Case of Cartographic and Art Models

During verification of the concept of cartographic modelling defined by the set of its eight obligatory properties, the authors analyzed a variety of space models (e.g., work of art, literary description, architectural plan, satellite image, etc.) and compared them to the cartographic model (Baranowski et al. 2016). Comparison of the cartographic model and a work of art can be considered as significant when taking the cognitive viewpoint. The literature on the subject contains many opinions on the debate on "realism" *versus* "symbolism" of maps. Since a map is not only a model but also a symbolic representation of reality, the authors agree with the statement that "cartography is more concerned with discovering and communicating the quintessence of reality, rather than with achieving a photographic record of the real thing. A cartographic map is thus a "précis of reality" (Visvalingam 1989).

A realist work of art presents reality in a way referring to our visual perception of it. An artistic image of an object or group of objects usually presents a fragment of real or virtual space. To do so, it applies various means of an artistic expression leading to, among others, description of spatial relations between portrayed objects. Before the age of photography, realistic works of art acted as images of reality. In such cases, the fidelity of representation of spatial relations was crucial, although most often they were shown using perspective projection. In an artistic image, objects are presented in a relatively faithful manner, without taking account of differing object types but focusing on individual objects. Therefore, in this case, we can speak about identifiability of objects rather than their types. Location of objects within the applied system of reference is not formally defined, and the scale of object presentation is not regular.

In the monograph "Cartography and Art" (Cartwright et al. 2009) an important question was asked about relations of cartographic modelling and the artistic process. The inherent feature of both cartography and art is a specific transformation of space (into a map form or into a work of art), but the transformation process is realized in different ways. In his work, "Rejecting Illusionism: Transforming Space into Maps and into Art," Fairbairn (2009, p. 24) states that "cartographic process is consequently data-driven (..) cartography has always rejected the illusionism of photography," while creation of a work of art is "emotion-driven." The role of a work of art is not to create easily identifiable representations of reality, but through appropriately chosen means to express "the deepest mystery of being." The process of cartographic modelling involves a collection of transformations such as generalization, modifications, emphasizing, interpretation, symbolization, while at the same time preserving semantic and topological correctness of the cartographic model. What is of key importance for a work of art is a subjective portrayal of space by the artist and evocation of varied emotions among the audience. "Romantic idea that art is the language of emotions has a long and complex tradition and may be more firmly entrenched today than in earlier period" (Gombrich 1978). Therefore, one can claim that what makes a work of art stand out (or, what makes it a good emotional representation of portrayed space) is the situation when "feeling evoked by the artifact resembles the feeling evoked by the original". Summing up the above comparison, we can say that "the major purpose of works of art is to arouse our emotions, whilst the products of cartography are intended to represent and communicate spatial data" (Fairbairn 2009).

6 Conclusions

Here we have posed the question: What is cartography today? How it is distinctive and autonomous? We focused on the fundamental methodological identity of cartography, i.e., on the cartographic modelling of space and its features. As presented, all eight of the analyzed properties of the processes of space representation appear together only in the case of cartographic modelling, that definitely separate it from other space modelling processes.

The impact of the digital technology on cartography is evident, unquestionable and obviously very strong. However contemporary cartography is still a cornerstone of modelling the surrounding geographical reality—in form of a classical analogue map, as well as a spatial database—fundamental to the development of geospatial infrastructure. Progress in technological development has not changed that important and unique role of cartography in modelling of geographical space.

While the development of information systems contributes to perfecting the digital processing of geoinformation, the legacy of cartographic knowledge determines the methodology of space modelling, as well as of the visualization of the collected data. The fundamental advantage of cartographic modelling is its capacity for imaging and analyzing the mutual arrangement of portrayed objects and events, regardless of the size of the modelled area. This is true for the classical analogue map and the model stored in electronic form.

The map as a product of the process of cartographic modelling (with all properties specified here), makes it possible not only to take stock of the surrounding space methodically, but also to analyze it, make predictions regarding the changes or a multi-criteria evaluation, and—finally—understand the space. The ability to abstract features of a space on a given level of conceptual generalization still remains fundamental.

The authors believe that it is possible to unambiguously answer the question whether the role of modern cartography is to only visualize geoinformation or to model geoinformation on different levels of generalization. Cartography is a science, technique and art, feeding the methodology of object and event modelling along with their corresponding space-time relations into the geospatial information infrastructure. The process of cartographic modelling makes it possible to describe and understand the space, abstract the space, as well as to enable the observer to visualize the essence of the surrounding space by means of a functional cartographic model.

Further investigations could focus on finding other properties of cartographic modelling of space resulting in stronger distinction from other modelling. There could be also defined complementary features of cartography (not only cartographic modelling) discriminating the discipline from others. Searching of identity of contemporary cartography. It is especially important at the time of its fast changes that on the one hand leads to extension of its research and practical domain, but also to its diffusion among other disciplines, with the special focus on the geographic information science, that is so tightly related to cartography.

References

Azocar Fernandez, P. I., & Buchroithner. M. (2014). *Paradigms in cartography. An epistemological review of the 20th 21st centuries*. Springer.

Baranowski, M., Gotlib, D., Olszewski, R. (2016). Properties of cartographic modelling under contemporary definitions of a map. In *Polish Cartographical Review, Polskie Towarzystwo Geograficzne*, 48(3), 91–100, doi:10.1515/pcr-2016-0011

Board, Ch. (1967). Maps as models. In *Models in geography*, London.

Cartwright, W., Gartner, G., & Lehn, A. (2009). *Cartography and art*. Springer.

Du, Q. (2011). How can theoretical cartography contribute to GIScience: A philosophy perspective. In *Proceedings of the 25th international cartographic conference,* Paris, France, 03–08 July 2011.

Duch, W. (2005). Dokąd zmierza inteligencja obliczeniowa? http://www.phys.uni.torun.pl/publications/kmk/03-CI-przyszlosc.pdf.

Fairbairn, D. (2009). Rejecting illusionism: Transforming space into maps and into art. In W. Cartwright, G. Gartner & A. Lehn (Eds.), *Cartography and art*. Springer.
Gombrich, E. H. (1978). *The Story of Art*. London, New York: Phaidon Press Ltd
Gotlib, D. (2011). Metodyka prezentacji kartograficznych w mobilnych systemach lokalizacyjnych i nawigacyjnych. „Prace naukowe Politechniki Warszawskiej – Geodezja", z.48.
Guptill, S. C., & Starr, L. E. (1984). The future of cartography in the information age. In L.E. Starr (Ed.), *Computer assisted cartography research and development report*. ICA Commission C.
Imhof, E. (1982). In H.J. Steward (Ed.), *Cartographic relief presentation*. Berlin: de Gruyter.
Makowski, A. (2001). *Na obrzeżach cybernetyki*, w: Mapa w systemach komputerowych, Materiały XXVIII Ogólnopolskiej Konferencji Kartograficznej, Szczecin.
Makowski, A. (2005). *System informacji topograficznej kraju. Teoretyczne i metodyczne opracowanie koncepcyjne*, Oficyna Wydawnicza Politechniki Warszawskiej, Warszawa.
Mejía Soto, E. (2004). *Aproximación al Concepto de Objetividad en Karl Popper*. Retrieved from http://www.gestionpolis.com.
Ostrowski, W. (2008). *Semiotyczne podstawy projektowania map topograficznych na przykładzie zabudowy*, praca habilitacyjna, Uniwersytet Warszawski.
Robinson, A. H. (1995). *Elements of cartography*. New York: Wiley.
Saliszczew, K. A. (1998). *Kartografia ogólna*, Wydawnictwo naukowe PWN, Warszawa.
Visvalingam, M. (1989). Cartography, GIS and maps in perspective. *The Cartographic Journal, The World of Mapping*, *26*: 26–32.

Author Index

A
Albert, Gáspár, 131, 239
Allen, Christopher, 85
Arikawa, Masatoshi, 117

B
Baranowski, Marek, 525
Behr, Franz-Josef, 99
Brito, Patricia Lustosa, 255
Brittell, Megen, 479
Buttenfield, Barbara P., 359

C
Cheng, Yan, 187
Chen, Qinying, 21
Chmelařová, Kateřina, 283
Coetzee, Serena, 99
Cooper, Antony, 99

D
de Jesus, Elaine Gomes Vieira, 255
de Menezes, Paulo M.L., 3
Demirel, Hande, 75
Deng, Xiaolong, 375
de Oliveira Fernandest, Vivian, 255
Ding, Linfang, 493
Dominguès, Catherine, 223
dos Santos, Deivison F., 3
Dozier, Jessica, 85
Dumpor, Dzenan, 269

E
Edwards, Geoffrey, 463

F
Farmer, Carson J.Q., 359
Fernandes, Manoel do C., 3
Fougeyrollas, Patrick, 463

G
Ghandehari, Mehran, 359
Gharebaghi, Amin, 463
Giraud, Timothée, 173
Glinka, Katrin, 507
Gonçalves, Duarte, 325
Gotlib, Dariusz, 525
Grummon, Christine, 479

H
Han, Su Yeon, 85
Haunert, Jan-Henrik, 389
Huang, Lina, 375

I
Ilyés, Virág, 131, 239
Iwaniak, Adam, 99

J
Jacobs, Merin, 325
Jiang, Wenping, 375
Jolivet, Laurence, 47, 223
Jung, Chin-Te, 85

K
Kaczmarek, Iwona, 99
Katumba, Samy, 99
Keler, Andreas, 493
Kis, Dávid, 131, 239
Köbben, Barend, 99
Kraak, Menno-Jan, 207
Krisp, Jukka M., 493
Kveladze, Irma, 207

L
Laeta, Tainá, 3
Lambert, Nicolas, 173
Leblond, Jean, 463

M.P. Peterson (ed.), *Advances in Cartography and GIScience*, Lecture Notes in Geoinformation and Cartography, DOI 10.1007/978-3-319-57336-6

Li, Qianyi, 295
Liu, Tao, 405
Liu, Yanfang, 295
Liu, Zixi, 295
Li, Xiaojiang, 341
Lobben, Amy, 479

M
Meier, Sebastian, 507
Midtbø, Terje, 269
Moellering, Harold, 99
Moore, Antoni B., 159
Morales-Coayla, Patrick, 463
Mostafavi, Mir-Abolfazl, 463
Murphy, Christian E., 417

N
Noreau, Luc, 463

O
Olszewski, Robert, 525
Olteanu-Raimond, Ana-Maria, 47
Omri, Masrudy, 479

P
Pánek, Jiří, 61
Parallel, Charlotte, 159
Pászto, Vít, 61
Peng, Dongliang, 389
Perdue, Nicholas, 479
Pitura, Mateusz, 145
Pődör, Andrea, 35
Poon, Sheung-Hung, 117

R
Rapant, Petr, 99
Ratti, Carlo, 341
Routhier, François, 463

S
Šašinka, Čeněk, 283
Schiewe, Jochen, 449
Schmitz, Peter M.U., 309, 325
Schoof, Govert, 99
Seiferling, Ian, 341
Şeker, Dursun Zafer, 75
Shah, Syed Attique, 75
Sharifi Noorian, Shahin, 417
Sinvula, Kisco, 99
Song, Ying, 295
Spitzberg, Brian H., 85
Stachoň, Zdeněk, 283
Steiniger, Stefan, 99
Szigeti, Csaba, 131, 239

T
Takahashi, Shigeo, 117
Tian, Jing, 187
Tsou, Ming-Hsiang, 85

V
van Elzakker, Corné P.J.M., 207
Várkonyi, Dávid, 131, 239
Vatseva, Rumiana, 99

W
Wang, Zhonghui, 405
Wolff, Alexander, 389
Wu, Hsiang-Yun, 117

X
Xi, Daping, 375
Xu, Guoguang, 295

Y
Yang, Jiue-An, 85
Yang, Weifang, 405
Yan, Haowen, 405
Ye, Ning, 187
Ying, Shen, 375
Yu, Yang, 295

Z
Zentai, László, 35
Zhang, Liming, 405
Zhao, Bo, 21
Zhou, Liang, 405
Zhou, Mengjie, 187
Zhou, Qi, 435

Subject Index

A

Academic SDI, 100, 109–112
Accessibility, 101, 354, 425, 463–466, 468, 470, 473–475, 479, 490, 491, 520
Accessibility systems, 480, 485–488
Accessible rich internet applications, 491
Active range optimization, 118, 123, 128
Activity data, 507, 511, 513, 515, 516, 520
Actor-network theory, 22, 23, 29
Android, 23, 24
Application Programmer Interface (API), 27, 75, 77–79, 83, 84, 88–90, 270, 343, 345, 486
Apple maps, 480
ArcGIS, 3, 9, 39, 63, 92, 132, 152, 175, 313, 376, 437, 440, 454, 484
ArcMap, 145, 150, 152, 153, 156
ArcObjects SDK, 398
Area aggregation, 393
Artificial Reality (AR) glasses, 284, 286, 287, 289–291
Accessible Rich Internet Applications (ARIA) markup, 484, 487
Augmented reality, 155, 283, 284, 287, 288
Automated map generalization, 405–407
AXS map network, 480

B

Basemaps, 269, 270, 272, 274, 276, 278, 279
Big data, 159, 160, 162, 163, 165, 169, 170
Bing maps, 269
Buffering approach, 435–443, 445, 447

C

Cartesian dualism, 529
Cartogram, 309, 311–319, 321
Cartographic modelling, 525–528, 530–535
Cartography package, 173, 174, 176
Choropleth, 177, 179–181, 188, 271, 316, 321, 449–451, 458
Choropleth mapping, 452
Classification, 10, 39, 79, 83, 94, 96, 263, 341, 348, 350, 407, 449, 451, 493, 495, 499, 513, 532, 533
CloudMade, 224
Cognitive mapping, 509, 519
Combinatorial optimization, 419
Compactness, 394, 396, 418, 431
Complicated crossings, 493, 494, 496, 498–501
Computational methods, 481
Connecting ridge points, 377, 380
Contiguity configuration, 359, 361–364, 366, 369, 370
Continuous map generalization, 403
Crowdsourcing, 38, 47, 49, 54, 61, 62, 70, 87, 94, 353
Council for Scientific and Industrial Research (CSIR), 106, 309, 312, 339
Cascading Style Sheet (CSS), 89, 486
Custom-made map, 223–225, 227, 228, 236

D

Data classification, 449–451, 453, 458
Data quality, 21, 22, 27, 31, 47, 48, 50, 51, 58, 59, 78, 102
Design science, 78, 84
Digital elevation models (DEM), 154, 359–363, 367, 369, 371, 372, 375, 376, 378, 383, 385
Disability, 464, 466, 488
Disaster management, 75–80, 83
3D maps, 309, 312, 315, 317, 321
Dynamic labeling, 119
Dynamic segmentation, 468
Dynamic weighting, 398

M.P. Peterson (ed.), *Advances in Cartography and GIScience*, Lecture Notes in Geoinformation and Cartography, DOI 10.1007/978-3-319-57336-6

E

Effectiveness, 84, 133, 188, 190, 198, 204, 212, 218, 239, 241, 271, 391, 417, 424, 437, 449, 450
Efficiency, 131, 132, 190, 198, 204, 208, 212, 218–220, 224, 239, 256, 271, 391, 423, 427
Election mapping, 312
Emergency response, 86, 96
Emotional mapping, 63–65
Epistemology, 528
Experimental cartography, 133, 252

F

Facebook, 76, 77
Fisheye image, 348–351
Flâneur, 510
Floating taxi data, 493, 494, 497
Foursquare, 513

G

Generalized area partitioning tree, 390
Genetic algorithm, 120, 417, 419, 425, 431
Geocache, 23
Geoinformation, 100, 101, 535
Geomorphologic shape, 376
Georeferencing, 3, 5, 7–11, 13, 16–18, 152
GeoTIFF, 136
Geoviewer, 85, 87–96
Geovisual analytics (GVA), 207, 208
Geovisualization, 90, 91, 96, 528, 530
GIS, 35–39, 41, 42, 45, 47, 62, 76, 102, 107, 111, 112, 133, 174, 176, 177, 256, 257, 284, 314, 449, 451, 464, 484, 500, 526
GitHub, 182
Gnoseology, 528
Google, 24, 25, 29, 95, 97, 102, 258, 259, 270, 343, 354
Google maps, 24, 132, 269, 270, 343, 445, 480
Google street view (GSV), 341–343, 345, 346, 352, 517
Google translate, 95
GPS, 22, 23, 25, 38, 87, 224, 284, 290, 513
GPS spoofing, 25
Graphical partitioning process, 458
Graph partitioning, 417, 431, 453, 456, 457
Green field planning, 424, 425, 431

H

Happiness index, 295, 296, 298–301, 303–305
Head-mounted display (HMD), 284, 285
Heads-up display (HUD), 285
Hebrus Valles, 145–147, 149–152, 154–156
HTML5, 89, 196
Human-centered computing, 88
Human-computer interaction (HCI), 174
Hydrographic analysis, 376

I

Inverse Distance Weighted (IDW) interpolation, 39
Institut Géographique National (IGN), 47, 49, 58, 224
Illustrator, 150, 151
Individualized maps, 132, 139, 140, 142
Inertial measurement units (IMU), 284
Information and communication technologies (ICT), 284
Interactive maps, 141, 155, 213, 256–259, 261–267
Interfaces, 163, 208, 255–257, 259, 265–267, 490, 491, 511
iPad, 23
iPhone, 23

J

JavaScript, 63, 88, 89, 91, 141, 486
Java Mission-planning and Analysis for Remote Sensing (JMARS), 145, 150, 152, 153, 156
JQuery, 89

K

Kernel density estimation, 92, 499
Koeler map, 3–8, 10, 15, 16

L

Land cover, 59, 101, 241, 245, 251
Leader lines, 117–119, 121–125, 128
Leaflet, 89, 91
LIDAR, 315, 346, 359, 361–363, 367
Likert scale, 483
Land Information New Zealand (LINZ), 167, 171, 440
Location-based services, 520
Location spoofing, 21, 22, 24, 26–31

M

Map projection, 146, 151
Map-related information, 284
Map spaces, 405, 407–409, 411, 412, 414
Mean absolute error, 368
Mental maps, 62, 507–510, 512, 519, 520
Metadata extraction, 82
Mobile maps, 526
Mobile phone, 22, 25
Model-cognitive concept, 531
Moves App, 513

Multi-criteria routing, 480, 481
MySQL, 63

N
Navigation tools, 257, 261
Neocartography, 70
Network analyst, 481, 483, 484, 488, 490
Neuro-linguistic programming (NLP), 82
National Mapping Agency (NMA), 48, 226
National Oceanic and Atmospheric Administration (NOAA), 92
Noise mapping, 35, 40
National Weather Service (NWS), 92

O
Ontology, 52, 57, 528
Openlayers, 131, 141, 246, 250
Open source, 42, 85, 87, 89, 133, 141, 142, 176, 251, 481
OpenStreetMap (OSM), 39, 42, 47, 48, 94, 131–133, 135, 269, 435, 493, 497, 514
Optical see-through display, 283
Ordnance survey (OS), 224, 437
Organised criminals, 338
Overland flow simulation, 375–377

P
Paragliding, 283, 284, 286–291
Pedestrian network, 463–467, 475, 487
Personalization, 479, 481–483, 486, 488
Photographical survey, 7, 8
Photoshop, 145, 152–154, 156
PHP, 486
Poaching, 325–327, 330, 332, 334, 338, 339
Pokémon Go, 21, 22, 28
Polylines, 136, 496, 497, 499
Positional accuracy, 48, 435–437, 439–441, 445–447
Preservation of archives, 4
Project-based learning, 36, 41, 42
Psychogeography, 508–510, 519
Python, 175

Q
QGIS, 27, 39, 131, 133, 135–137, 141, 142, 175, 437
QGIS server, 131, 141
Quality assessment, 47, 48, 59
Questionnaire, 192, 196, 242, 467

R
Representational State Transfer (REST), 484, 486
Ridge lines, 375–385
Rigid pixel paradigm, 359, 360, 364, 371
Root Mean Square Error (RMSE), 361, 362, 367–372
Road matcher, 440
Routing systems, 481
R software, 173, 174
Rstudio, 182

S
Sahana, 87
Steepest Ascent Method Based on Constrained Direction (SAMBCD), 375, 377–379, 382, 384, 386
Spatial Data Infrastructure (SDI), 99, 101–103, 106–112
Semiotics, 528
Similarity metrics, 408
Situation awareness, 85, 88, 94, 96
Sky view factor (SVF), 348–353
Short Message Services (SMS), 76, 77
Social media, 75–81, 83, 85–88, 94, 97
Socio-economic indicator, 335
Sound map, 159, 163, 164, 166, 170
Space-time cube (STC), 207, 208, 210
Spatial analysis, 256, 509
Spatial behaviour, 508, 510, 512
Spatial data infrastructure, 99, 101, 406
Spatial reasoning, 405, 408
Spatial similarity, 405–409, 411, 413, 414
Spatial-temporal variations, 295, 298
Spatio-temporal data, 190, 209, 474, 511
Shuttle Radar Topography Mission (SRTM), 131, 133, 136, 362, 383
Statistical maps, 173, 225
Steepest ascent method, 375–377, 384, 385
Street greenery, 341–343, 345, 346, 348, 352, 353
Street openness, 353
Sub-pixel variations, 359
Suppression of distance, 509, 519
Surface-adjusted analysis, 360, 366, 371, 372
Scalable Vector Graphics (SVG), 136, 137, 141

T
Tactile, 159, 163, 170
Territory design problem, 417, 431
Theil index, 295, 297, 301, 306
Thermal Emission Imaging System (THEMIS), 152
Theoretical cartography, 528
Think aloud, 258
Three world theory, 529
Time graph (TG), 210

Time wave (TW), 210
Topological correctness, 534
Traffic bottleneck, 493, 494, 496, 498, 500–502
Traffic congestion, 493–495, 498–501, 503
Traffic maps, 493
Transduction, 160, 161, 164
Transductive relations, 162
Transportation infrastructure, 493–495, 497, 499
Traveling salesman problem, 419
Treemap, 187, 188, 190, 192, 193
Tweets, 85, 87–92, 94–96
Twitter, 76, 77, 86–88, 90, 91, 94–96

U
UDig, 210, 213
Urban landscape, 341, 342, 345, 346, 352, 353
Urban planning, 3, 62, 70, 353, 507
Usability, 94, 101, 133, 207–210, 213, 220, 241, 251, 255–258, 260, 262, 266, 270, 490
User-centered design (UCD), 89
User-oriented maps, 240
User study, 142, 187, 189, 190, 194, 196, 197, 516, 520
User test, 133, 223–225, 237, 286, 290, 291
United States Geological Survey, 77, 362, 372
Ushahidi, 87

V
Visual map, 159, 160, 163, 165, 167–169, 227
Visual property, 235, 236
Volunteered geographic information (VGI), 47, 48, 58, 435, 482, 488, 490, 497, 499
Virtual private network (VPN) spoofing, 25

W
Webhosting, 63
Web-mapping, 61, 70, 87, 89, 152, 170, 270, 273, 279
Weighted linear model, 465, 471
Web Feature Service (WFS), 92
Wheelchair user, 464, 465, 467, 470, 473–476
WheelCon, 467, 468, 470, 475
Whole-of-society, 325–329, 331, 332, 338, 339
Wheelchair Outcome Measure (WhOM), 470
WiFi spoofing, 22, 25
Wildlife crime, 325, 328
WMS, 92, 104, 152

The manufacturer's authorised representative in the EU is Springer Nature Customer Service Centre GmbH, Europaplatz 3, 69115 Heidelberg, Germany. If you have any concerns regarding our products, please contact ProductSafety@springernature.com

Printed and bound by CPI Group (UK) Ltd, Croydon, CR0 4YY
30/09/2026
02225567-0020